Estimating Construction Costs

Sixth Edition

Robert L. Peurifoy

Garold D. Oberlender

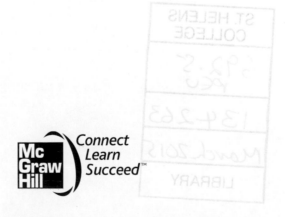

ESTIMATING CONSTRUCTION COSTS, SIXTH EDITION
International Edition 2014

Published by McGraw-Hill Education, 2 Penn Plaza, New York, NY 10121. Copyright © 2014 by McGraw-Hill Education. All rights reserved. Previous editions © 2002, 1989, 1975, 1958, and 1953. No part of this publication may be reproduced or distributed in any form or by any means, or stored in a database or retrieval system, without the prior written consent of publisher, including, but not limited to, in any network or other electronic storage or transmission, or broadcast for distance learning.
Some ancillaries, including electronic and print components, may not be available to customers outside the United States

10 09 08 07 06 05 04 03 02
20 15 14
CTP   SLP

When ordering this title, use ISBN 978-1-259-01082-8 or MHID 1-259-01082-1

Printed in Singapore

www.mhhe.com

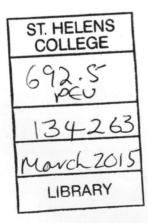

Estimating Construction Costs

ABOUT THE AUTHORS

Robert L. Peurifoy (deceased) was a distinguished author, professor, and consulting engineer in the engineering and construction profession. He received his B.S and M.S. degrees in civil engineering from the University of Texas. He taught civil engineering at the University of Texas and Texas A&I College and construction engineering at Texas A&M University and Oklahoma State University. Mr. Peurifoy served as a highway engineer for the U.S. Bureau of Public Roads and was a contributing editor of *Roads and Streets magazine*. He was the author of two other books: *Construction Planning, Equipment, and Methods* and *Formwork for Concrete Structures,* 4th ed., coauthored with Garold D. Oberlender. He also wrote over 50 magazine articles dealing with construction. He was a long-time member of the American Society of Civil Engineers, which presents an award that bears his name.

Garold D. Oberlender, Ph.D, P.E. is professor emeritus of civil engineering at Oklahoma State University, where he served as coordinator of the Graduate Program in Construction Engineering and Project Management. He received his B.S. and M.S. degrees in civil engineering from Oklahoma State University and his Ph.D in civil engineering from the University of Texas at Arlington. He has more than 40 years of experience in teaching, research, and consulting engineering related to the design and construction of projects. He is author of the McGraw-Hill publication *Project Management for Engineering and Construction*, 2nd ed., and *Formwork for Concrete Structures*, 4th ed., coauthored with Robert L. Peurifoy. Dr. Oberlender is a registered professional engineer in several states, a member of the National Academy of Construction, a fellow in the American Society of Civil Engineers, and a fellow in the National Society of Professional Engineers.

BRIEF CONTENTS

About the Authors v

Preface xv

CHAPTER **1**
Introduction 1

CHAPTER **2**
Bid Documents 22

CHAPTER **3**
Estimating Process 44

CHAPTER **4**
Conceptual Cost Estimating 64

CHAPTER **5**
Cost of Construction Labor and Equipment 82

CHAPTER **6**
Handling and Transporting Material 108

CHAPTER **7**
Earthwork and Excavation 126

CHAPTER **8**
Highways and Pavements 174

CHAPTER **9**
Foundations 208

CHAPTER **10**
Concrete Structures 231

CHAPTER **11**
Steel Structures 279

CHAPTER **12**
Carpentry 295

CHAPTER **13**
Roofing and Flashing 328

CHAPTER **14**
Masonry 341

CHAPTER **15**
Floor Systems and Finishes 361

CHAPTER **16**
Painting 380

CHAPTER **17**
Plumbing 387

CHAPTER **18**
Electric Wiring 404

CHAPTER **19**
Sewerage Systems 414

CHAPTER **20**
Water Distribution Systems 431

CHAPTER **21**
Total Cost of Engineering Projects 446

CHAPTER **22**
Computer Estimating 452

APPENDIX
Example Bid Documents 513

Index 564

CONTENTS

About the Authors   v

Preface   xv

CHAPTER 1
Introduction   1

Purpose of This Book   1

Estimating   2

Importance of the Estimator and the
    Estimating Team   3

Purpose of Estimating   3

Types of Estimates   4

Approximate Estimates   5

Detailed Estimates   7

Organization of Estimates   8

Building Construction Projects   9

Heavy Engineering Construction
    Projects   10

Quantity Takeoff   11

Labor and Equipment Crews   12

Checklist of Operations   13

Bid Documents   13

Addenda and Change Orders   14

Overhead   14

Material Taxes   15

Labor Taxes   15

Worker's Compensation Insurance   16

Labor Burden   16

Bonds   16

Insurance   16

Representative Estimates   17

Instructions to the Readers   17

Production Rates   18

Tables of Production Rates   18

Computer Applications   18

Forms for Preparing Estimates   20

CHAPTER 2
Bid Documents   22

Bid Documents and Contract
    Documents   22

Contract Requirements   23

Arrangement of Contract Documents   23

Building Construction Specifications   24

Heavy/Highway Specifications   24

Bidding Requirements   25

Negotiated Work   32

Addendum   32

Alternates   32

Change Order   32

Warranties   33

General Conditions of the Contract
    Documents   33

Bonds   33

Insurance   35

Heavy/Highway Drawings   36

Building Construction Drawings   37

Symbols and Abbreviations   42

Problems   43

CHAPTER 3
Estimating Process   44

Decision to Bid   44

Estimating Process   44

The Estimating Team   46

Estimate Work Plan   47

Methods and Techniques   49

Preparing Estimates   50

Estimating Procedures   51

Estimate Checklists   52

Documentation of Estimate   53

Estimate Reviews   54

Risk Assessment   57

Risk Analysis   57

Contingency   57

Traditional Methods of Assigning
    Contingency   58

Estimate Feedback for Continuous
    Improvement   62

Problems   63

CHAPTER **4**

Conceptual Cost Estimating   64

Accuracy of Conceptual Estimates   64

Liability of Conceptual Cost
    Estimates   65

Preparation of Conceptual
    Estimates   65

Parametric Estimating   66

Broad-Scope Conceptual
    Estimates   68

Time Adjustments for Conceptual
    Estimates   70

Adjustments for Location   71

Adjustment for Size   71

Combined Adjustments   71

Unit-Cost Adjustments   72

Narrow-Scope Conceptual Cost
    Estimates   74

Factors Affecting Cost
    Records   74

Conceptual Costs for Process
    Industry   74

Problems   81

CHAPTER **5**

**Cost of Construction Labor and
Equipment**   82

Construction Labor   82

Sources of Labor Rates   82

Cost of Labor   83

Social Security Tax   83

Unemployment Compensation Tax   83

Workers' Compensation and Employer's
    Liability Insurance   84

Public Liability and Property Damage
    Insurance   84

Fringe Benefits   84

Production Rates for Labor   87

Construction Equipment   90

Sources of Equipment   90

Renting versus Owning Equipment   90

Equipment Costs   91

Depreciation Costs   92

Methods of Depreciation   92

Investment Costs   96

Ownership Costs   98

Operating Costs   100

Problems   105

CHAPTER **6**

**Handling and Transporting
Material**   108

Introduction   108

Cycle Time and Production Rate
    Calculations   109

Transporting Sand and Aggregate with
    Tractor Loaders   113

Transporting Material with Conveyors   118

Handling Cast-Iron Pipe   119

Handling Lumber   120

Handling and Transporting Bricks   122

Problems   124

CHAPTER **7**
Earthwork and Excavation   126

Job Factors   126
Management Factors   126
Estimating Production Rates of
    Equipment   127
Methods of Excavating and Hauling
    Earth   128
Physical Properties of Earth   129
Excavating by Hand   133
Excavating with Trenching
    Machines   135
Ladder-Type Trenching Machines   138
Excavating with Draglines   139
Handling Material with a Clamshell   142
Excavating with Hydraulic
    Excavators   143
Front Shovels   143
Hauling Excavated Materials   147
Backhoes   149
Dozers   152
Excavating and Hauling Earth with
    Scrapers   154
Graders   159
Shaping and Compacting
    Earthwork   161
Preparing the Subgrade for Highway
    Pavements   165
Drilling and Blasting Rock   166
Cost of Operating a Drill   168
Problems   171

CHAPTER **8**
Highways and Pavements   174

Operations Included   174

Clearing and Grubbing Land   174
Land-Clearing Operations   174
Rates of Clearing Land   176

Disposal of Brush   180
Demolition   181

Concrete Pavements   181
General Information   181
Construction Methods Used   181
Batching and Hauling Concrete   182
Placing Concrete Pavements   183
Concrete Pavement Joints   186
Curing Concrete Pavement   189

Asphalt Pavements   195
Aggregates   195
Asphalts   196
Asphalt Plants   196
Transporting and Laying Asphalt Mixes   199
Compacting Asphalt Concrete Mixes   201
Equipment for Hot-Mix Asphaltic-Concrete
    Pavement   202
Cost of Hot-Mix Asphaltic-Concrete
    Pavement   202
Computer Estimating of Highway
    Projects   205
Problems   206

CHAPTER **9**
Foundations   208

Types of Foundations   208
Footings   209
Sheeting Trenches   209
Pile-Driving Equipment   211
Sheet Piling   214
Wood Piles   217
Driving Wood Piles   217
Prestressed Concrete Piles   219
Cast-in-Place Concrete Piles   220
Steel Piles   221
Jetting Piles into Position   224
Drilled Shaft Foundations   224
Problems   230

CHAPTER **10**
Concrete Structures   231

Cost of Concrete Structures   231
Forms for Concrete Structures   231
Materials for Forms   232
Labor Required to Build Forms   236
Forms for Slabs on Grade   236
Materials for Footings and Foundation
    Walls   236
Quantities of Materials and Labor-Hours
    for Wall Forms   239
Prefabricated Form Panels   244
Commercial Prefabricated
    Forms   245
Forms for Concrete Columns   246
Material Required for Concrete Column
    Forms   248
Quantities of Materials and Labor-Hours
    for Column Forms   250
Economy of Reusing Column
    Forms   252
Column Heads, Capitals, and Drop
    Panels   254
Shores and Scaffolding   254
Material and Labor-Hours for Concrete
    Beams   256
Forms for Flat-Slab Concrete Floors   259
Patented Forms for Floor Slabs   262
Material and Labor-Hours Required for
    Metal-Pan Concrete Floors   263
Corrugated-Steel Forms   264
Cellular-Steel Floor Systems   265
Concrete Stairs   265

Reinforcing Steel   267
Types and Sources of Reinforcing
    Steel   267
Properties of Reinforcing Bars   268
Estimating the Quantity of Reinforcing
    Steel   268
Cost of Reinforcing Steel   269

Labor Placing Reinforcing Steel
    Bars   270
Welded-Wire Fabric   272

Concrete   272
Cost of Concrete   272
Quantities of Materials for
    Concrete   273
Labor and Equipment Placing
    Concrete   274
Lightweight Concrete   276
Perlite Concrete Aggregate   276

Tilt-Up Concrete Walls   277
Problems   278

CHAPTER **11**
Steel Structures   279

Types of Steel Structures   279
Materials Used in Steel Structures   279
Estimating the Weight of Structural
    Steel   280
Connections for Structural Steel   280
Estimating the Cost of Steel
    Structures   280
Items of Cost in a Structural-Steel
    Estimate   281
Cost of Standard Shaped Structural
    Steel   281
Cost of Preparing Shop Drawings   281
Cost of Fabricating Structural
    Steel   283
Cost of Transporting Steel to
    the Job   283
Cost of Fabricated Structural Steel
    Delivered to a Project   284
Erecting Structural Steel   286
Labor Erecting Structural Steel   291
Field Painting Structural Steel   293
Problems   294

CHAPTER **12**
Carpentry 295

Introduction 295
Classification of Lumber 295
Plywood 298
Cost of Lumber 299
Nails and Spikes 299
Bolts and Screws 300
Timber Connectors 303
Fabricating Lumber 304

Rough Carpentry 305
House Framing 305
Sills 306
Floor Girders 306
Floor and Ceiling Joists 307
Studs for Wall Framing 309
Framing for Window and Door
    Openings 311
Rafters 311
Prefabricated Roof Trusses 314
Roof Decking 315
Wood Shingles 315
Subfloors 316

Exterior Finish Carpentry 317
Fascia, Frieze, and Corner
    Boards 317
Soffits 318
Wall Sheathing 318
Aesthetic Exterior Siding 318

Heavy Timber Structures 319
Interior Finish, Millwork, and Wallboards 320
Interior Finish Carpentry 320
Labor-Hours Required to Set and Trim
    Doors and Windows 321
Wood Furring Strips 322
Gypsum Wallboards 322
Wall Paneling 323
Interior Trim Moldings 324

Finished Wood Floors 324
Problems 325

CHAPTER **13**
Roofing and Flashing 328

Roofing Materials 328
Area of a Roof 328
Steepness of Roofs 329
Roofing Felt 329

Roofing Shingles 329
Wood Shingles 329
Asphalt Shingles 330
Slate Roofing 333
Clay Tile Roofing 334

Built-Up Roofing 334
Felt for Built-Up Roofing 335
Pitch and Asphalt 335
Gravel and Slag 335
Laying Built-Up Roofing on Wood
    Decking 335
Laying Built-Up Roofing on Concrete 336
Labor Laying Built-Up Roofing 337

Flashing 338
Metal Flashing 338
Flashing Roofs at Walls 338
Flashing Valleys and Hips 340
Labor Required to Install Flashing 340
Problems 340

CHAPTER **14**
Masonry 341

Masonry Units 341
Estimating the Cost of Masonry 341
Mortar 342

Bricks 343
Sizes and Quantities of Bricks 343
Pattern Bonds 343

Types of Joints for Brick Masonry   345
Estimating Mortar for Bricks   345
Quantity of Mortar for Brick Veneer
  Walls   346
Accessories for Brick Veneer Walls   347
Cleaning Brick Masonry   347
Solid Brick Walls   348
Labor Laying Bricks   349

Concrete Masonry Units   352
Labor Laying Concrete Masonry
  Units   353

Stone Masonry   356
Bonds for Stone Masonry   356
Mortar for Stone Masonry   356
Weights of Stone   356
Cost of Stone   357
Labor Setting Stone Masonry   358
Problems   359

CHAPTER **15**
Floor Systems and Finishes   361

Floor Systems   361
Steel-Joist System   361
Combined Corrugated-Steel Forms and
  Reinforcement for Floor
  System   368

Floor Finishes   372
Concrete-Floor Finishes   372
Terrazzo Floors   376
Vinyl Tile   378
Problems   379

CHAPTER **16**
Painting   380

Materials   380
Covering Capacity of Paints   381

Preparing a Surface for Painting   382
Labor Applying Paint   383
Equipment Required for Painting   383
Cost of Painting   383
Problems   386

CHAPTER **17**
Plumbing   387

Plumbing Requirements   387
Plumbing Code   388
Piping Used for Plumbing   389
Steel Pipe   390
Copper Pipe   390
PVC Plastic Water Pipe   391
Indoor CPVC Plastic Water Pipe   392
Labor Installing Plastic Water Pipe   392
Soil, Waste, and Vent Pipes   393
House Drain Pipe   393
Fittings   394
Valves   394
Traps   394

Roughing in Plumbing   394
Estimating the Cost of Roughing in
  Plumbing   394
Cost of Materials for Rough
  Plumbing   395
Cost of Lead, Oakum, and Solder   395
Plastic Drainage Pipe and Fittings   395
Labor Required to Rough in Plumbing   396

Finish Plumbing   400
Labor Required to Install Fixtures   402
Problems   403

CHAPTER **18**
Electric Wiring   404

Factors That Affect the Cost of Wiring   404
Items Included in the Cost of Wiring   405

Roughing in Electrical Work 405
 Types of Wiring 405
 Rigid Conduit 406
 Flexible Metal Conduit 406
 Armored Cable 406
 Nonmetallic Cable 407
 Electric Wire 407
 Accessories 407
 Cost of Materials 408
 Labor Required to Install Electric
  Wiring 409

Finish Electrical Work 410
 Labor Required to Install Electric
  Fixtures 411
 Problems 412

CHAPTER **19**
Sewerage Systems 414

 Items Included in Sewerage
  Systems 414
 Sewer Pipes 414
 Construction Operations 415
 Constructing a Sewerage
  System 416

Trenchless Technology 423
 Microtunneling 423
 Microtunnel Boring
  Machine 424
 Remote Control System 426
 Active Direction Control 427
 Automated Spoil
  Transportation 427
 Jacking Pipe 427
 Advantages and Disadvantages
  of Microtunneling
  Methods 428
 Microtunneling Process 429

CHAPTER **20**
**Water Distribution
Systems** 431

 Cost of Water Distribution
  Systems 431
 Types of Pipe Material 431
 Valves 436
 Service Lines 436
 Fire Hydrants 437
 Tests of Water Pipes 437
 Sterilization of Water Pipes 437
 Labor Required to Lay Water
  Pipe 437
 Cost of a Water Distribution
  System 438
 Horizontal Directional
  Drilling 440
 Procedure for Horizontal Directional
  Drilling 442
 Production Rates 445

CHAPTER **21**
**Total Cost of Engineering
Projects** 446

 Cost of Land, Right-of-Way,
  and Easements 446
 Legal Expenses 447
 Bond Expense 447
 Permit Expenses 447
 Bringing Off-Site Utilities to
  a Project 448
 Engineering Expense 448
 Cost of Construction 449
 Interest During Construction 449
 Contingency 450
 Example Estimate for Total Cost
  of an Engineering
  Project 450

CHAPTER **22**

Computer Estimating   452

Introduction   452

Importance of the Estimator   453

Use of Computers in Estimating   453

Electronic Media   455

Using Spreadsheets for Estimating   455

Disadvantages of Spreadsheets   457

Commercially Available Estimating
    Software   457

Advantages of Commercial Estimating
    Software   459

Management of Data   460

Typical Steps in Computer
    Estimating   462

Starting an Estimate   463

Biditems   464

Quantity Takeoff   467

Resource Types   468

Resources   470

Labor Resources   471

Precision in Labor Costing   474

Equipment Resources   475

Material   478

Crews   480

Structuring the Estimate   481

Entering the Estimate   483

Copying from Past Estimates   490

Alternates   492

Showing How Costs Were
    Calculated   493

Reviewing the Estimate   494

Checking the Estimate for
    Reasonableness   495

Turning a Cost Estimate into a Bid   496

Bid Pricing   499

Taking Quotes   501

Turning in The Bid   504

Loading Heavybid From the
    Internet   505

Problems   505

APPENDIX

Example Bid Documents   513

Index   564

PREFACE

This book presents the basic principles of estimating the time and cost of construction projects. To prepare an accurate estimate the estimator must perform a careful and thorough analysis of the work to be performed. The analysis includes the type and quantity of work, type and size of equipment to install the work, and jobsite conditions that are unique to the project that can impact the time and cost of construction.

This book emphasizes the thought process that is required of the estimator to analyze job conditions and assess the required labor, equipment, and methods of construction that will be necessary to perform the work. To assemble a complete estimate for bid purposes, the estimator must combine his or her knowledge of construction methods and techniques into an orderly process of calculating and summarizing the cost of a project.

Estimating is not an exact science. Knowledge of construction, common sense, and judgment are required. Experienced estimators agree that the procedures used for estimating vary from company to company, and even among individuals within a company. Although there are variations, fundamental concepts that are universally applicable do exist. The information contained in this book presents the fundamental concepts to assist the reader in understanding the estimating process.

In preparing this sixth edition, the author has retained the fundamental concepts of estimating that have made this book successful for many years. Example problems have been revised with more explanations regarding assumptions used in the calculations. Throughout the book, the example problems are presented in a consistent format, beginning with the quantities of materials and production rates through pricing labor, material, and equipment. All example problems are revised in this edition of the book. Also, homework problems are updated.

This revised edition of the book has reorganized and consolidated chapters to increase the clarity of the subject matter for the reader. The previous chapter on interior finish, millwork, and wallboards is merged into the chapter on carpentry. The previous chapter on floor finishes is merged into the chapter on floor systems. The chapter on glass and glazing is deleted. New material on unit cost estimating is added to the section of approximate estimates in Chapter 1. The new 50 divisions of the Construction Specifications Institute (CSI) are added and discussed in Chapter 2. The material is expanded on expected net risk analysis for assigning contingency in Chapter 3 and a new section on parametric estimating is added in Chapter 4, Conceptual Cost Estimating.

Major revisions to Chapter 5, Cost of Construction Labor and Equipment, include additional examples of calculating labor costs. New sections are added with example problems for calculating the depreciation cost of equipment by several methods, including straight-line, double-declining-balance, and sum-of-the-years digits. New sections are added on investment costs, ownership costs and comparing costs of renting versus owning equipment based on hours used per year. A new section on transporting sand and aggregate with track loaders is added to Chapter 6, Handling and Transporting Materials.

Chapter 7, Earthwork and Excavation, has updated photos and many new sections, including production rates, efficiency factors, dozers, graders, and preparing subgrades with GPS guidance systems. Chapter 8, Highways and Pavement, has revised example problems that are reorganized to show more details in preparing cost estimates for asphalt and concrete pavements. New photos are added to Chapter 9, Foundations, including pile driving rig, drilled shaft drilling rig, and steel casings. Chapter l0, Concrete Structures, has updated photos of formwork and revised example problems. Chapter 11, Steel Structures, has added photos of a truck-mounted crane erecting steel trusses and an iron worker erecting steel beams to better illustrate the process of erecting steel structures.

Chapter 12, Carpentry, is revised to include carpentry, interior finish, millwork, wallboards and Chapter 13, Roofing and Flashing, has a new photo of laborers installing asphalt shingles. Chapter 14, Masonry, retains the information of the previous edition with revised example problems. Chapter 15 is revised to include floor systems and finishes. A new section is added to Chapter 16 on painting that presents content on the covering capacity of paints and the table of approximate labor-hours required to apply paint is updated. The section on types of pipe is reorganized for clarity in Chapter 17, Plumbing. Updated prices of electrical material is presented in Chapter 18, Electrical Wiring, and a revised example problem is presented in Chapter 19, Sewage Systems.

Chapter 20, Water Distribution Systems, is completely rewritten with new sections added on pipe currently used in the construction industry, including polyethylene (PE) pipe, polyvinyl chloride (PVC) plastic pipe, ductile iron pipe (DIP), and reinforced concrete pipe (RIP). The section on horizontal directional drilling (HDD) is revised with a new photo of a horizontal directional drilling machine. The topics in Chapter 21, Total Cost of Engineering Projects, are expanded, including the cost of right-of-ways and easements, and engineering expenses. New sections are added on permit expenses and bringing off-site utilities to a project.

Chapter 22, Computer Estimating, is revised with additional material on the use of computers in preparing estimates for bidding purposes. A new section is added on typical steps in computer estimating to give the reader a fundamental introduction to computer estimating. Calculation routines and alternates are other new sections added to the chapter which address the process of integrating spreadsheets into commercial software and the process of evaluating the cost of alternate construction methods during the estimating process and before submitting a bid, respectively. Another new section on reviewing the estimate provides a system of checking for errors in an estimate. The HCSS/Student software can be obtained at the website: www.mhhe.com/peurifoy_oberlender6e.

The appendix in the book is an example bid document that contains drawings for a simple project that can be used for applying the estimating processes presented in this book. The project is ideal for class presentations of quantity takeoff calculations and for applying the principles of estimating presented in this book. The drawings can also be used for assigning estimating homework problems to students or assignment of semester projects for students to work in teams to prepare an estimate. The material in the appendix can be obtained at the McGraw-Hill website: www.mhhe.com/peurifoy_oberlender6e.

The website also includes a password-protected instructor's manual. For students, technical specifications for a sample project are included. Also available for students is a student edition computer estimating software program, HeavyBid/Student, courtesy of HCSS (Heavy Construction Systems Specialists, Inc.).

I would like to thank HCSS for permission to include their HeavyBid/Student software with this book. In particular, a special thanks to Mike Rydin of HCSS for his assistance in revising material in the computer estimating chapter. Also, the author appreciates the many equipment manufacturers, dealers, and construction contractors who generously provided updated photographs presented throughout this revised edition of the book.

I would like to thank Carisa Ramming for her careful review, helpful comments, and advice in the development of this sixth edition. McGrawHill and I would also like to thank the following reviewers for their many comments and suggestions: Maury Fortney, Walla Walla Community College; Raymond Gaillard, Northeast Mississippi Community College; Amine Ghanem, California State University Northridge; Alex Mills, Oklahoma State University – Tulsa; and Neil Opfer, University of Nevada Las Vegas.

I wish to recognize and pay tribute to the late Robert L. Peurifoy for his pioneering work as an author and teacher of construction engineering in higher education. Throughout my career, Mr. Peurifoy was an inspiration to me as a teacher, mentor, colleague, and friend.

Finally, I greatly appreciate the patience and tolerance of my wife, Jana, for her understanding and support during the writing and editing phases of the sixth edition of this book.

Garold D. Oberlender, Ph.D, P.E.

McGraw-Hill Create™

Craft your teaching resources to match the way you teach! With McGraw-Hill Create, you can easily rearrange chapters, combine material from other content sources, and quickly upload content you have written like your course syllabus or teaching notes. Find the content you need in Create by searching through thousands of leading McGraw-Hill textbooks. Arrange your book to fit your teaching style. Create even allows you to personalize your book's appearance by selecting the cover and adding your name, school, and course information. Order a Create book and you'll receive a complimentary print review copy in 3–5 business days or a complimentary electronic review copy (eComp) via email in minutes. Go to www.mcgrawhillcreate.com today and register to experience how McGraw-Hill Create empowers you to teach *your* students *your* way.

Electronic Textbook Option

This text is offered through CourseSmart for both instructors and students. CourseSmart is an online resource where students can purchase the complete text online at almost half the cost of a traditional text. Purchasing the eTextbook allows students to take advantage of CourseSmart's web tools for learning, which include full text search, notes and highlighting, and email tools for sharing notes between classmates. To learn more about CourseSmart options, contact your sales representative or visit www.CourseSmart.com.

1

Introduction

PURPOSE OF THIS BOOK

The purpose of this book is to enable the reader to gain fundamental knowledge of estimating the cost of projects to be constructed. Experienced estimators agree that the procedures used for estimating vary from company to company, and even among individuals within a company. Although there are variations, fundamental concepts that are universally applicable do exist. The information contained in this book presents the fundamental concepts to assist the reader in understanding the estimating process and procedures developed by others. This book can also serve as a guide to the reader in developing his or her own estimating procedures.

There are so many variations in the costs of materials, labor, and equipment from one location to another, and over time, that no book can dependably give costs that can be applied for bidding purposes. However, the estimator who learns to determine the quantities of materials, labor, and equipment for a given project and who applies proper unit costs to these items should be able to estimate the direct costs accurately.

This book focuses on basic estimating, not workbook estimating. Basic estimating involves analysis of the work to be performed, including the type and quantity of work, type and size of equipment to be used during construction, production rates of labor and equipment to install the work, and other job-site conditions that are unique to the project and that can impact the time and cost of construction. After these items are defined, both the estimated time and cost to build the project can be determined. The typical question in basic estimating is: How much time will our labor and equipment be on the job? Once that question is answered, the cost of labor and equipment can easily be calculated by multiplying the hourly or daily rates of labor and equipment times the time they will be on the job. Based on the quantity and quality of material, the cost of material can easily be obtained from a material supplier. Both the time and cost to do the work are obtained in basic estimating.

Workbook estimating involves performing a material quantity takeoff, obtaining unit costs from a nationally published cost manual, and multiplying the quantity of work times the unit cost of material and labor to determine the cost estimate. Workbook estimating is quick, simple, and easy to perform. However, unit costs can vary widely, depending on the volume of work, weather conditions, competition in pricing, variations in the skill and productivity of workers, and numerous other factors. In workbook estimating, the cost is determined directly from the material quantities, without regard to how long the workers and/or equipment will be on the job. Workbook estimating emphasizes obtaining the cost to do the work, rather than the time to do the work.

Subsequent chapters of the book present the principles and concepts of estimating construction costs. Emphasis is placed on the thought process that is required of the estimator to analyze job conditions and assess the required labor, equipment, and method of construction that will be necessary to perform the work. These are functions that can be performed only by the estimator because experience and good judgment are required to prepare reliable estimates.

To assemble a complete estimate for bid purposes, the estimator must combine his or her knowledge of construction methods and techniques into an orderly process of calculating and summarizing the cost of a project. This process requires the assembly of large amounts of information in an organized manner and numerous calculations must be performed. The computer is an ideal tool to facilitate this process by decreasing the time and increasing the accuracy of cost estimating. The computer can retrieve data and perform calculations in seconds, enabling the estimator to give more attention to alternative construction methods, to assess labor and equipment productivity, to obtain prices from subcontractors and material suppliers, and to focus on bidding strategies. Today, computers are used extensively for cost estimating.

The final chapter of this book is devoted to computer estimating. Although the computer is an effective tool for estimating, the estimator must still control the estimating process. In simple language, the computer should work for the estimator, the estimator should not work for the computer. The estimator must know the software that he or she is using, both its capabilities and its limitations. The results of a computer estimate are only as good as the estimator using the computer. The computer cannot exercise judgment; only the estimator has that capability.

ESTIMATING

Estimating is not an exact science. Knowledge of construction, common sense, and judgment are required. Estimating material costs can be accomplished with a relatively high degree of accuracy. However, accurate estimating of labor and equipment costs is considerably more difficult to accomplish.

Estimating material costs is a relatively simple and easy task. The quantity of materials for a particular job can be accurately calculated from the dimensions on the drawings for that particular job. After the quantity of material is calculated, the estimator can obtain current unit prices from the supplier and then multiply the quantity of material by the current unit price to estimate the cost of materials. Applying a percentage for material waste is the only adjustment of the material cost that may require judgment of the estimator.

Estimating labor and equipment costs is considerably more difficult than estimating material costs. The cost of labor and equipment depends on productivity rates, which can vary substantially from one job to another. The estimator and his or her team should evaluate the job conditions for each job and use their judgment and knowledge of construction operations. The skill of the laborers, job conditions, quality of supervision, and many other factors affect the productivity of labor. The wage rate of laborers can be determined with relative accuracy. Also, the quantity of work required by laborers can be determined from the plans and specifications for a job. However, the estimator must use his or her judgment and knowledge about the job to determine the expected productivity rate of laborers. Likewise, the estimator's judgment and knowledge of equipment and job conditions are required to determine the expected equipment productivity rate.

IMPORTANCE OF THE ESTIMATOR AND THE ESTIMATING TEAM

Whether using computers, or not, the estimator and his or her team play a vital role in preparing estimates. Information must be assembled, organized, and stored. Cost records from previously completed projects and cost quotes from suppliers, vendors, and subcontractors must be gathered. Assessment of job conditions and evaluation of labor and equipment and productivity rates must be performed. The estimator must review and check all parts of an estimate to ensure realistic costs. The estimator must also document the estimate so it can be used for cost control during the construction process.

The computer can assist in these activities, but the estimator must manage and control the estimating process. The estimator has to be able to work under pressure because most estimates are prepared in stringent time frames. The quality and accuracy of an estimate is highly dependent on the knowledge and skill of the estimator.

PURPOSE OF ESTIMATING

The purpose of estimating is to determine the forecast costs required to complete a project in accordance with the contract plans and specifications. For any given project, the estimator can determine with reasonable accuracy the direct costs for materials, labor, and equipment. The bid price can then be determined by adding to the direct cost the costs for overhead (indirect costs required to build the project), contingencies (costs for any potential unforeseen work), and profit (cost for compensation for performing the work). The bid price of a project should be high enough to enable the contractor to complete the project with a reasonable profit, yet low enough to be within the owner's budget.

There are two distinct tasks in estimating: determining the probable real cost and determining the probable real time to build a project. With an increased emphasis on project planning and scheduling, the estimator is often requested to provide production rates, crew sizes, equipment spreads, and the estimated time required to perform individual work items. This information, combined with costs, allows an integration of the estimating and scheduling functions of construction project management.

Because construction estimates are prepared before a project is constructed, the estimate is, at best, a close approximation of the actual costs. The true cost of the project will not be known until the project has been completed and all costs have been recorded. Thus, the estimator does not establish the cost of a project: he or she simply establishes the amount of money the contractor will receive for constructing the project.

TYPES OF ESTIMATES

There are many estimates and re-estimates for a project, based on the stage of project development. Estimates are performed throughout the life of a project, beginning with the first estimate and extending through the various phases of design and into construction, as shown in Fig. 1.1. Initial cost estimates form the basis to which all future estimates are compared. Future estimates are often expected to agree with (i.e., be equal to or less than) the initial estimates. However, too often the final project costs exceed the initial estimates.

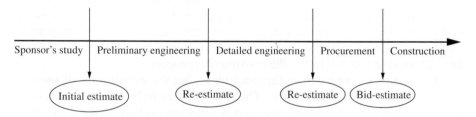

Figure 1.1 | Estimates and re-estimates through phases of project development.

Although each project is unique, generally three parties are involved: the owner, the designer, and the contractor. Each has responsibility for estimating costs during various phases of the project. Early in a project, prior to the design, the prospective owner may wish to know the approximate cost of a project before making a decision to construct it. As the design of the project progresses, the designer must determine the costs of various design alternatives to finalize the design to satisfy the owner's budget and desired use of the project. The contractor must know the cost required to perform all work in accordance with the final contract documents, and the plans and specifications.

Cost estimates can be divided into at least two different types, depending on the purposes for which they are prepared and the amount of information known when the estimates are prepared. There are approximate estimates (sometimes called feasibility, screening, authorization, preliminary, conceptual, order-of-magnitude, equipment-factored, or budget estimates) and detailed estimates (sometimes called final, bid/tender, or definitive estimates).

There is no industry standard that has been established for defining estimates. Individual companies define estimate names and percent variations that they use based on their experience with a particular type of construction and operating procedures within the company. Various organizations have also defined classifications of cost estimates. For example, Table 1.1 is the cost estimation classification by the Association for the Advancement of Cost Engineering (AACE) International. In general, an early estimate is defined as an estimate that has been prepared before completion of detailed engineering. This definition applies to Class 5, Class 4, and early Class 3 estimates of AACE International.

TABLE 1.1 | AACE International cost estimation classifications (18R-97).

Estimate class	Level of project definition	End usage—Typical purpose of estimate	Expected accuracy range
Class 5	0% to 2%	Concept screening	−50% to 100%
Class 4	1% to 5%	Study or feasibility	−30% to +50%
Class 3	10% to 40%	Budget, authorization, or control	−20% to +30%
Class 2	30% to 70%	Control or bid/tender	−15% to +20%
Class 1	50% to 100%	Check estimate or bid/tender	−10% to +15%

APPROXIMATE ESTIMATES

The prospective owner of a project establishes the budget for a project. For example, a government agency will need to know the approximate cost before holding a bond election, or it may need to know the approximate cost to ensure that the cost of a project does not exceed the funds appropriated during a fiscal year. The prospective owner of a private construction project will generally conduct a feasibility study during the developmental phase of a project. As a part of the study, an economic analysis is undertaken to compare the cost of construction with potential earnings that can be obtained upon completion of the project. Privately owned utility companies, and other types of multiple-builder owners, prepare annual construction budgets for all projects proposed during a fiscal year. An approximate estimate is sufficiently accurate for these purposes.

The designer of a project must determine the costs of various design alternatives to obtain an economical design that meets the owner's budget. An architect may reduce a building to square feet of area, or cubic feet of volume, and then multiply the number of units by the estimated cost per unit. An engineer may multiply the number of cubic yards of concrete in a structure by the estimated cost per cubic yard to determine the probable cost of the project. Considerable experience and judgment are required to obtain a dependable approximate estimate for the cost, because the estimator must adjust the unit costs resulting from the quantities of material, workmanship, location, and construction difficulties. Approximate estimates are sufficiently accurate for the evaluation of design alternatives or the presentation of preliminary construction estimates to the owner, but are not sufficiently accurate for bid purposes.

The unit cost method of estimating is commonly used to prepare approximate estimates. For building type projects, they are often referred to as square foot estimates. Other examples of unit cost estimating are cost per mile of electrical transmission line, cost per lineal foot of pipeline, cost per mile for each lane of highway pavement, cost per square yard of site grading, etc.

After the size and configuration of a project has been developed, the quantities of work can be calculated. Examples of quantity of work include square yards of site grading, number of pile foundations, cubic yards of concrete, square feet of floors, lineal feet of pipe, and number of doors in a building. After the quantity of work is defined, the cost can be calculated by simply multiplying the cost per unit of work times the quantity of work for each item. For example, if there are 900 sy of floor covering in a building and the unit cost for installed carpet is $30/sy, then the cost of floor covering is 900 sy $\times$ $30/sy = $27,000. Similar calculations can be performed for other types of work.

The quantity of work is derived from the drawings for a project. Values of unit costs are sometimes obtained from a national pricing manual. Examples of national pricing manuals include: *ENR Contracting Cost Books* from McGraw-Hill Construction, *Building Construction Cost Data* from RSMeans, and *The Building Estimator's Reference Book* from Frank R. Walker Company. The cost data from national publications are averages from several cities. Therefore, it is necessary to adjust the cost information. Adjustments for location, time, and size are presented in Chapter 4 of this book.

Values of unit costs may also be obtained from company records of previously completed projects of a similar type of work. The unit costs are typically an average of the values obtained from recent jobs. It should be recognized that crew composition and production rates are unique to each individual job. Sometimes the experience, skills, and intuition of the estimator must be applied to adapt the previous cost records to the conditions of the job being estimated.

When the time for completion of a project is important, the owner may select a construction contractor before the design has been completed. A contractor may be asked to provide an approximate cost estimate based on limited information known about the project. The contractor determines the approximate costs for various work items, such as the cost per cubic yard for foundations, the cost per pound of structural steel, and the cost per square foot for finished rooms. Based on preliminary quantities of work, the contractor calculates a preliminary cost for construction of the project. The owner may then negotiate a construction contract with the contractor based on the approximate cost estimate. Construction projects of this type require owners who are knowledgeable in project management and contractors who have developed good project record keeping. Screening of design/build contractors is often based partly on the approximate estimate for a project.

DETAILED ESTIMATES

A detailed estimate of the cost of a project is prepared by determining the costs of materials, labor, equipment, subcontract work, overhead, and profit. Contractors prepare detailed estimates from a complete set of contract documents prior to submission of the bid or formal proposal to the owner. The detailed estimate is important to both the owner and the contractor because it represents the bid price—the amount of money the owner must pay for completion of the project and the amount of money the contractor will receive for building the project.

The preparation of the detailed estimate generally follows a systematic procedure that has been developed by the contractor for his or her unique construction operations. The process begins with a thorough review of the complete set of contract documents—the bidding and contract requirements, drawings, and technical specifications. It is also desirable to visit the proposed project site to observe factors that can influence the cost of construction, such as available space for storage of materials, control of traffic, security, and existing underground utilities.

The compilation of costs begins with a well-organized checklist of all work items necessary to construct the project. The estimator prepares a *material quantity takeoff* of all materials from the drawings. This involves tabulating the quantity and unit of measure of all work required during construction. Upon completion of the quantity takeoff, an extension of prices is performed. The quantity of material multiplied by the unit cost of the material yields the material cost. The quantity of work required of equipment is divided by the equipment production rate and then multiplied by the unit cost of equipment to obtain the total cost of equipment. Similarly, dividing the quantity of work required for labor by the labor production rate and then multiplying by the unit cost of labor obtains the cost of labor.

For many projects, a significant amount of work is performed by subcontractors who specialize in a particular type of work. For building type projects, examples include clearing, drywall, painting, and roofing contractors. For heavy/highway type projects, examples of subcontract work include guardrails, striping, signs, and fences. The estimator provides a set of drawings and specifications to potential subcontractors and requests a bid from them for their particular work. Subcontractors and suppliers normally quote the cost of their work by e-mail, fax, or phone just prior to final submission of the bid.

The direct cost of a project includes material, labor, equipment, and subcontractor costs. Upon completion of the estimate of direct costs, the estimator must determine the indirect costs of taxes, bonds, insurance, and overhead required to complete the project. Taxes on labor and materials vary, depending on geographic locations. Bond requirements are defined in the specifications for the project. Examples include bid bonds, material and labor payment bonds, and performance bonds. Insurance requirements are also defined in the contract bid documents. Examples include workmen's compensation, contractor's public liability and property damage, and contractor's builders risk insurance. The base estimate is the total of direct and indirect costs.

A risk analysis of uncertainties is conducted to determine an appropriate contingency to be added to the base estimate. Regardless of the effort and amount of

TABLE 1.2 | Steps for preparing a detailed estimate.

1. *Review the scope of project.* Consider the effect of location, security, traffic, available storage space, underground utilities, method of payment, etc., on costs.
2. *Determine quantities.* Perform a material quantity takeoff for all work items in the project, and record the quantity and unit of measure for each item.
3. *Obtain suppliers' bids.* Receive and tabulate the cost of each supplier on the project.
4. *Price material.* Extend material costs:

$$\text{Material cost} = \text{Quantity} \times \text{Unit price}$$

5. *Price labor.* Based on probable labor production rates and crew sizes, determine labor costs:

$$\text{Labor cost} = (\text{Quantity/Labor production rate}) \times \text{Labor rate}$$

6. *Price equipment.* Based on probable equipment production rates and equipment spreads, determine equipment costs:

$$\text{Equipment cost} = (\text{Quantity/Equipment production rate}) \times \text{Equipment rate}$$

7. *Obtain specialty contractors' bids.* Receive and tabulate the cost for each specialty contractor on the project.
8. *Calculate taxes, bonds, insurance, and overhead.* Tabulate the costs of material and labor taxes, bonds, insurance, and job overhead.
9. *Contingency and markup.* Add costs for potential unforeseen work based on the amount of risk.
10. *Profit.* Add costs for compensation for performing the work in accordance with the bid documents.

detailed estimating of a project, there is almost always some unforeseen work that develops during construction. Caution must be used in assigning contingency to an estimate. A contingency that is too low might reduce the profits in a project, while a contingency that is too high may prevent the bid from being competitive.

Upon calculation of the direct and indirect costs, analysis of risk, and assignment of contingency, a profit is added to the estimate to establish the bid price. The amount of profit can vary considerably, depending on numerous factors, such as the size and complexity of the project, amount of work in progress by the contractor, accuracy and completeness of the bid documents, competition for work, availability of money, and volume of construction activity in the project area. The profit may be as low as 5 percent for large projects or as high as 30 percent for small projects that are high risks or remodels of existing projects. Table 1.2 lists the steps required to compile a detailed estimate for a project.

ORGANIZATION OF ESTIMATES

A comprehensive and well-defined organization of work items is essential to the preparation of an estimate for any project. Each contractor develops her or his own procedures to compile the cost of construction for the type of work the company performs. The contractor's system of estimating and use of forms develop over years of experience in that type of work.

Two basic approaches have evolved to organize work items for estimating. One approach is to identify work by the categories contained in the project's written specifications, such as those of the Construction Specification Institute (CSI) for building construction projects. The other approach uses a *work break-down structure* (WBS) to identify work items by their location on the project.

BUILDING CONSTRUCTION PROJECTS

Building construction contractors usually organize their estimates in a format that closely follows the Construction Specifications Institute (CSI) numbering system, which divides work into major divisions. Each major division is subdivided into smaller items of work. The numbering system is developed for architects writing specifications that are unique to each project. However, contractors typically use CSI numbering as a checklist and guide for quantity takeoff, price extensions, and summary of cost for the final estimate.

For years the CSI used a 16 division numbering system, but recently expanded it to 50 divisions as discussed in Chapter 2 of this book. It doesn't matter which numbering system is used as long as the information that is needed to assist in the bidding process is included and can be found easily.

An illustrative example of cost summary for a building construction project is shown in Table 1.3, which follows the CSI 16 division numbering system. This table

TABLE 1.3 | Example of bid summary for a building construction project.

Item	Description	Material	Labor	Subcontract	Total
01	General requirements	$ 164,350	$ 363,550	$ 48,820	$ 576,720
02	Sitework	150,700	201,230	1,461,860	1,813,790
03	Concrete	970,176	515,240	0	1,485,416
04	Masonry	0	0	2,127,240	2,127,240
05	Metals	2,132,340	593,210	0	2,725,550
06	Wood and plastics	387,530	104,960	49,080	541,570
07	Thermal and moisture	0	0	1,380,720	1,380,720
08	Doors and windows	368,210	321,150	0	689,360
09	Finishes	1,725,870	1,879,220	0	3,605,090
10	Specialties	157,480	111,040	96,250	364,770
11	Equipment	0	0	457,290	457,290
12	Furnishing	0	0	0	0
13	Special construction	0	0	0	0
14	Conveying systems	0	0	1,283,346	1,283,346
15	Mechanical	0	0	5,133,384	5,133,384
16	Electrical	0	0	3,546,610	3,546,610
	Total for Project	$6,056,656	$4,089,600	$15,584,600	$25,730,856
Add-ons for final estimate:					
	Material tax (5%)	$302,883			$26,033,689
	Labor tax (18%)		$736,128		26,769,817
	Contingency (2%)			$ 535,397	27,305,214
	Bonds and insurance			$ 340,910	27,646,124
	Overhead and profit (15%)			$4,146,919	31,793,043
				Bid price =	$31,793,043

TABLE 1.4 | Item 2 Estimate for sitework.

Cost Code	Cost Item	Quantity	Material	Labor	Subcontract	Total
2110	Clearing	Lump sum	$ 0	$ 0	$ 36,940	$ 36,940
2222	Excavation	8,800 cy	0	118,800	94,160	212,960
2250	Compaction	960 cy	0	22,230	7,220	29,450
2294	Handwork	500 cy	0	17,500	0	17,500
2281	Termite control	Lump sum	0	0	34,750	34,750
2372	Drilled piers	1,632 ft	145,800	28,000	145,240	319,040
2411	Foundation drains	14 each	4,900	14,700	0	19,600
2480	Landscape	Lump sum	0	0	87,220	87,220
2515	Paving	4,850 sy	0	0	1,056,330	1,056,330
		Totals =	$150,700 +	$201,230 +	$1,461,860 =	$1,813,790

shows the summary costs of material, labor, and subcontract for Item 02, Sitework. Table 1.4 shows a breakdown of the sitework costs, which include clearing, excavation, compaction, handwork, termite control, drilled piers, foundation drains, landscape, and paving. Preparation of the estimate would involve further breakdown of items in Table 1.4. For example, the cost of excavation work would be broken down into the labor and equipment costs for the crew performing excavation.

HEAVY ENGINEERING CONSTRUCTION PROJECTS

Heavy engineering construction contractors generally organize their estimates in a WBS unique to the project to be constructed. An example of the WBS organization of an estimate for an electric power construction project is illustrated in Tables 1.5 to 1.7. Major areas of the project are defined by groups: switch station, transmission lines, substations, etc., as shown in Table 1.5. Each group is subdivided into divisions of work required to construct the group. For example, Table 1.6 provides a work breakdown for all the division of work required to construct group 2100, transmission line A: steel fabrication, tower foundations,

TABLE 1.5 | Example of electric power construction bid summary using the WBS organization of work group–level report for total project.

No.	Group	Material	Labor and Equipment	Subcontract	Total
1100	Switch station	$1,257,295	$ 323,521	$3,548,343	$ 5,129,159
2100	Transmission line A	3,381,625	1,259,837	0	4,641,462
2300	Transmission line B	1,744,395	0	614,740	2,359,135
3100	Substation at Spring Creek	572,874	116,403	1,860,355	2,549,632
4200	Distribution line A	403,297	54,273	215,040	672,610
4400	Distribution line B	227,599	8,675	102,387	338,661
4500	Distribution line C	398,463	21,498	113,547	533,508
	Total for project	$7,985,548 +	$1,784,207 +	$6,454,412 =	$16,224,167

TABLE 1.6 | Division-level report for transmission line A—Code 2100.

Code	Description	Material		Labor		Equipment		Total
2100	TRANSMISSION LINE A							
2210	Fabrication of steel towers	$ 692,775		$ 0		$ 0		$ 692,775
2370	Tower foundations	83,262		62,126		71,210		216,598
2570	Erection of steel towers	0		144,141		382,998		527,139
2620	Insulators and conductors	2,605,588		183,163		274,744		3,063,495
2650	Shield wire installation	0		78,164		63,291		141,455
	Total for 2100	$3,381,625	+	$467,594	+	$792,243	=	$4,641,462

TABLE 1.7 | Component-level report for tower foundations—Code 2370.

Code	Description	Quantity	Material		Labor		Equipment		Total
2370	TOWER FOUNDATIONS								
2372	Drilling foundations	4,198 lin ft	$ 0		$25,428		$44,897		$ 70,325
2374	Reinforcing steel	37.5 tons	28,951		22,050		15,376		66,377
2376	Foundation concrete	870 cy	53,306		13,831		10,143		77,280
2378	Stub angles	3,142 each	1,005		817		794		2,616
	Total for 2370		$83,262	+	$62,126	+	$71,210	=	$216,598

steel erection, etc. Each division is further broken down into components of work required to construct each division. For example, Table 1.7 provides a work breakdown for all the components of work required to construct division 2370, tower foundations: drilling, reinforcing steel, foundation concrete, and stub angles. The WBS provides a systematic organization of all the information necessary to derive an estimate for the project.

Other types of heavy engineering projects, such as highways, utilities, and petrochemical and industrial plants, are organized in a WBS that is unique to their particular types of work. The total estimate is a compilation of costs in a WBS that matches the project to be constructed.

Regardless of the system of estimating selected, either CSI or WBS, to each work item in the estimate a code number should be assigned that is reserved exclusively for that work item for all estimates within the contractor's organization. This same number should also be used in the accounting, job cost, purchasing, and scheduling functions, to enable one to track the work items during construction.

QUANTITY TAKEOFF

To prepare an estimate the estimator reviews the plans and specifications and performs a quantity takeoff to determine the type and amount of work required to build the project. Before starting the quantity takeoff, the estimator must know how the project is to be constructed and must prepare a well-organized checklist of all items required to construct the project.

The quantity of material in a project can be accurately determined from the drawings. The estimator must review each sheet of the drawings, calculate the quantity of material, and record the amount and unit of measure on the appropriate line item in the estimate. The unit costs of different materials should be obtained from material suppliers and used as the basis of estimating the costs of materials for the project. If the prices quoted for materials do not include delivery, the estimator must include appropriate costs for transporting materials to the project. The cost of taxes on materials should be added to the total cost of all materials at the end of the estimate.

Each estimator must develop a system of quantity takeoff that ensures that a quantity is not omitted or calculated twice. A common error in estimating is completely omitting an item or counting an item twice. A well-organized checklist of work will help reduce the chances of omitting an item. A careful recheck of the quantity calculations will detect those items that might be counted twice. The estimator must also add an appropriate percentage for waste for those items where waste is likely to occur during construction. For example, a 5 percent waste might be added to the volume of mortar that is calculated for bricklaying.

The material quantity takeoff is extremely important for cost estimating because it often establishes the quantity and unit of measure for the costs of labor and the contractor's equipment. For example, the quantity of concrete material for piers might be calculated as 20,000 cubic yards (cy). The labor-hours and the cost of labor required to place the concrete would also be based on 20,000 cy of material. Also the number and the cost of the contractor's equipment that would be required to install the concrete would be based on 20,000 cy of material. Therefore, the estimator must carefully and accurately calculate the quantity and unit of measure of all material in the project.

LABOR AND EQUIPMENT CREWS

Prior to preparing the quantity takeoff, an assessment of how the project will be constructed must be made. The assessment should include an analysis of the job conditions, labor and equipment crews, and appropriate subdivisions of the work. To illustrate, consider a grading and paving project. Typically, the quantity takeoff for grading is cubic yards of material. An evaluation of the job conditions considers the type of soil, including presence of rock, and the required haul distance to select the appropriate equipment to perform the work. If both soil and rock are present, the work may be subdivided into ordinary earth excavation and rock excavation. For ordinary earth excavation with long haul distances, scrapers may be used for excavating, hauling, and distributing the earth. For the rock excavation, the equipment crew may include drilling and blasting, or ripping the rock with a dozer, and a loader with a spread of trucks to haul the loosened rock. Table 1.8 is an example of a labor and equipment crew for a grading operation.

The paving work can be subdivided into three categories: main line paving, short run paving, curbs and gutters, and handwork paving. For each of these subdivisions a different crew and equipment spread would be required. Table 1.9

TABLE 1.8 | Construction crew for a grading operation.

Supervisor and pickup
Small self-loading scraper
Medium size scraper
Small dozer for assisting scraper
Dozer with ripper
Front end loader, 3 cy
Hydraulic excavator
Small tractor backhoe/box blade
Self-propelled vibrator sheepfoot roller
Self-propelled nine-wheel pneumatic roller
Motor grader
Water truck
Single axle flatbed dump truck
Tandem axle dump truck
Common laborer
Skilled laborer
Traffic flag person

TABLE 1.9 | Asphalt crew for urban work.

Labor
  Supervisor
  Paving machine operator
  Screed operator
  Steel wheel roller operator
  Asphalt raker
  Asphalt laborer
  Asphalt distributor operator
  Traffic flag person
Equipment
  Supervisor's pickup
  Service truck and hand tools
  Asphalt paver
  Asphalt roller
  Asphalt distributor truck
  Front-end loader/box blade tractor
  Self-propelled sweeper

TABLE 1.10 | Curb and gutter crew for concrete paving.

Labor
  Supervisor
  Concrete finisher—three total
  Concrete finisher helper
  Form setter—stringline setter
  Form setter—stringline setter helper
  Laborer
Equipment
  Supervisor's pickup
  Flatbed truck and hand tools
  Small tool storage van
  Slipform curb and gutter machine
  Miscellaneous steel forms/stringline
    equipment
  Concrete saw/cure spray machine
  Air compressor
  Small tractor

shows a typical crew and equipment mix for asphalt work in an urban area, and Table 1.10 shows a typical crew and equipment mix for a curb and gutter concrete paving crew.

CHECKLIST OF OPERATIONS

To prepare an estimate, an estimator should use a checklist that includes all the operations necessary to construct the project. Before completing an estimate, one should check this list to be sure that no operations have been omitted. The CSI master format provides a uniform approach for organizing project information for building projects. Tables 1.3 and 1.4 illustrate the cost summary of an estimate following the CSI format.

It is desirable for the operations to appear, as nearly as possible, in the same order in which they will be performed during construction of the project. Other checklists, serving the same purpose, should be prepared for projects involving highways, water systems, sewerage systems, etc.

The checklist can be used to summarize the costs of a project by providing a space for entering the cost of each operation, as illustrated in Tables 1.3 to 1.7. A suitable symbol should be used to show no cost for those operations that are not required. The total cost should include the costs for material, equipment, and labor for the particular operation, as determined in the detailed estimate.

BID DOCUMENTS

The end result of the design process is the production of a set of bid documents for the project to be constructed. These documents contain all the drawings and written specifications required for preparing the estimate and submitting the bid. Written specifications can be divided into two general parts: one part addresses the legal aspects between the owner and contractor, while the other part addresses the technical requirements of the project. The legal part of the written specifications contains at least four items that are important to the estimator: procedures for receipt and opening of bids, qualifications required of bidders, owner's bid forms, and bonds and insurance required for the project. Bid documents are discussed further in Chapter 2.

ADDENDA AND CHANGE ORDERS

An addendum is a change in the contract documents during the bidding process. Sometimes the designer or owner may wish to make changes after the plans and specifications have been issued, but before the contractors have submitted their bids. Also, a contractor may detect discrepancies in the plans and specifications while preparing an estimate. When an error is detected, the contractor should notify the designer to note any discrepancies.

An addendum is issued by the designer to make changes or correct errors in the plans and specifications. An addendum may be a reissue of a drawing or pages in the written specifications. Each addendum is given a number. After it is issued, an addendum becomes a part of the contract document. Thus, it is important for the estimator to ensure that the cost of all addenda are included in the estimate before submitting a final bid.

Any change in the contract documents after the contract is signed is called a change order. For most projects, changes are necessary during the construction process. Examples include additions or modifications requested by the owner, substitution of materials, or adverse weather conditions. A change order may add or delete work in the project, increase or decrease costs, or increase or decrease the time allowed for construction. In some situations, a change order may make a change that does not include any adjustment in time or cost.

OVERHEAD

The overhead costs chargeable to a project involve many items that cannot be classified as permanent materials, construction equipment, or labor. Some firms divide overhead into two categories: job overhead and general overhead.

Job overhead includes costs that can be charged specifically to a project. These costs are the salaries of the project superintendent and other staff personnel and the costs of utilities, supplies, engineering, tests, drawings, rentals, permits, insurance, etc., that can be charged directly to the project.

General overhead is a share of the costs incurred at the general office of the company. These costs include salaries, office rent, permits, insurance, taxes, shops and yards, and other company expenses not chargeable to a specific project.

Some contractors follow the practice of multiplying the direct costs of a project, materials, equipment, and labor by an assumed percentage to determine the probable cost of overhead. Although this method gives quick results, it may not be sufficiently accurate for most estimates.

While it is possible to estimate the cost of job overhead for a given project, it is usually not possible to estimate accurately the cost of general overhead chargeable to a project. Since the cost of general overhead is incurred in operating all the projects constructed by a contractor, it is reasonable to charge a portion of this cost to each project. The actual amount charged may be based on the duration of the project, the amount of the contract, or a combination of the two.

EXAMPLE 1.1

This example illustrates a method of determining the amount of general overhead chargeable to a given project.

$$\text{Average annual value of all construction} \ = \$6,000,000$$

$$\text{Average annual cost of general overhead} = \ \ \ \ 240,000$$

Amount of general overhead chargeable to a project:

$$= \frac{\$240,000 \times 100}{\$6,000,000} = 4\% \text{ of total project cost}$$

MATERIAL TAXES

After the direct costs for materials and labor have been determined, the estimator must include the applicable taxes for each. The tax rate for materials will vary depending on the location. Generally a 3 percent state tax and a 2 to 3 percent city or county tax are assessed on materials. Therefore the tax on materials will range from 3 to 6 percent. It is the responsibility of the estimator to include the appropriate amount of tax in the summary of the estimate.

Some states and cities charge a tax on the value of equipment used on the project. It is necessary for the estimator to obtain information on required taxes for the particular location where the project is to be constructed.

Some owners, such as churches and schools, are tax exempt provided the owner makes the purchase of material. Thus, the owner may wish to purchase all materials and equipment that will be permanently installed in the project by the contractor. For this situation, the contract documents will have a section that describes how the owner will issue purchase orders for payment by the owner.

LABOR TAXES

There are two basic types of taxes on labor. The federal government requires a 7.65 percent tax on all wages up to $76,000 per year. In addition, an unemployment tax of approximately 3 percent may be required. Therefore, the total tax on labor is approximately 11 percent. The estimator must determine the appropriate tax on labor and include that amount in the summary of the estimate. A discussion of labor costs is presented in Chapter 5.

WORKERS COMPENSATION INSURANCE

Contractors pay workers compensation insurance for workers who may become injured while working on the project. This insurance provides medical expenses and payment of lost wages during the period of injury.

Each state has laws and regulations that govern workers compensation insurance. The cost also varies, depending on the type of work that is performed by each worker and by geographic location within the state. The cost for this insurance can range from 10 to 30 percent of the base cost of labor.

LABOR BURDEN

The term *labor burden* refers to the combined cost of labor taxes and insurance that a contractor must pay for workers. Some contractors multiply the base cost of labor by a certain percentage to estimate the total cost of labor. The total labor burden for construction labor generally ranges from 25 to 35 percent.

BONDS

Generally bidders are required to submit bonds as qualifications for submitting a bid for a project. To provide financial and legal protection for the owner, the contractor secures bonds from a surety company on behalf of the owner. Thus, bonding is a three-party arrangement. It is issued by the security company, paid by the contractor, for the protection of the owner.

Three types of bonds are commonly required in construction contracts: bid bond, performance bond, and payment bond. The bid bond ensures the owner that the contractor will sign the contract for the bid amount. The performance bond ensures the owner that the contractor will perform all work in accordance with the contract documents. The payment bond ensures the owner that all material and labor will be paid. If the contractor defaults, the bonding company agrees to fulfill the contract agreement.

The purchase cost of bonds depends on the total bid price and the success of the contractor in completing previous jobs. Chapter 2 provides further information on bonds and the estimating process.

INSURANCE

There are many risks involved in construction. Many types of insurance are available for protection to contractors, employees, subcontractors, and the general public. The two most common types of insurance that are secured by the contractor are *basic builder's risk,* which covers the project that is being constructed, and *public liability and property damage,* which covers actions of the contractor's employees while performing their work at the jobsite.

Basic builder's risk insurance affords a contractor protection against loss resulting from fire and lightning damage during the period of construction.

Public liability and property damage insurance protects the contractor against injuries to the general public or public property due to actions of the employees while performing work during construction.

Chapter 2 provides a detailed discussion of insurance costs that should be included in preparing an estimate.

REPRESENTATIVE ESTIMATES

Numerous examples of estimates are presented in this book to illustrate the steps to follow in determining the probable cost of the project. Nominal amounts are included for overhead and profit in some instances to give examples of complete estimates for bid purposes. In other instances, only the costs of materials, construction equipment, and labor are included. The latter three costs are referred to as *direct costs*. They represent the most difficult costs to estimate, and they are our primary concern.

In preparing the sample estimate, unit prices for materials, equipment, and labor are used primarily to show how an estimate is prepared. Note that these unit costs will vary with the time and location of a project. An estimator must obtain and use unit prices that are correct for the particular project. Estimators do not establish prices; they simply use them.

Remember that estimating is not an exact science. Experience, judgment, and care should enable an estimator to prepare an estimate that will reasonably approximate the ultimate cost of the project.

INSTRUCTIONS TO THE READERS

In the examples in this book, a uniform method is used for calculating and expressing the time units for equipment and labor and the total cost. For equipment, the time is expressed in equipment-hours, and for labor it is expressed in labor-hours. A labor-hour is one person working 1 hour or two people each working $\frac{1}{2}$ hour.

If a job that requires the use of four trucks lasts 16 hours, the time units for the trucks are the product of the number of trucks and the length of the job, expressed in hours. The unit of cost is for 1 truck-hour. The calculations are as follows:

Trucks: 4 trucks × 16 hr = 64 truck-hours @ $45.00/hr = $2,880.00

In a similar manner, the time and cost for the truck drivers are:

Truck drivers: 4 drivers × 16 hr = 64 labor-hours @ $28.00/hr = $1,792.00

The terms "64 truck-hours" and "64 labor-hours" can be shortened to read 64 hr without producing ambiguity.

PRODUCTION RATES

To determine the time required to perform a given quantity of work, it is necessary to estimate the probable rates of production of the equipment or labor. These rates are subject to considerable variation, depending on the difficulty of the work, skill of the laborer, job and management conditions, and the condition of the equipment.

A production rate is the number of units of work produced by a unit of equipment or a person in a specified unit of time. The time is usually 1 hr. The rate may be determined during an interval when production is progressing at the maximum possible speed. It is obvious that such a rate cannot be maintained for a long time. There will always be interruptions and delays that reduce the average production rates to less than the ideal rates. If a machine works at full speed only 45 minutes per hour (min/hr), the average production rate will be 0.75 of the ideal rate. The figure 0.75 is defined as an efficiency factor.

A backhoe with a 1-cy bucket may be capable of handling 3 bucket-loads per minute under ideal conditions. However, on a given job the average volume per bucket may be only 0.8 cy and the backhoe may be actually operating only 45 min/hr. For these operating conditions, the average output can be calculated as follows:

> The ideal output: 3 cy/min $\times$ 60 min/hr = 180 cy/hr
>
> The bucket factor: 0.8
>
> The efficiency factor: 45/60 = 0.75
>
> The combined operating factor: 0.8 $\times$ 0.75 = 0.6
>
> The average output: 0.6 $\times$ 180 cy/hr = 108 cy/hr

The average output should be used in computing the time required to complete a job.

TABLES OF PRODUCTION RATES

In this book, numerous tables give production rates for equipment and laborers. In all tables the rates are adjusted to include an operation factor, usually based on a 45- to 50-min working hour. If this factor is too high for a given job, the rates should be reduced to more appropriate values.

When preparing an estimate for a project, if access is available to production rates obtained from actual jobs constructed under similar conditions, an estimator should use them instead of rates appearing in tables that represent general industry averages.

COMPUTER APPLICATIONS

An estimator must assemble a large amount of information in an organized manner and perform numerous calculations to prepare a cost estimate. The estimator can use the computer to organize, store, and retrieve information and to perform the many calculations necessary to prepare an estimate. It can be an effective tool for decreasing preparation time and increasing the accuracy of cost estimating.

The computer is used for estimating in at least five different applications: quantity takeoff, price extensions and bid summary, historical cost database, labor and equipment productivity database, and supplier database. Each application can be subdivided and should be linked together in an overall, integrated system.

Electronic spreadsheets are widely used for preparing cost estimates. A spreadsheet program can be used to perform the numerous price extension calculations and the bid summary of an estimate. Spreadsheet programs can be easily developed by the estimator using her or his system of estimating for the particular type of construction.

Many companies have developed commercial software for estimating construction costs. The software is developed for specific types of work, such as residential, building, or infrastructure type projects. Most commercial software is supplied with cost and productivity databases that can be used by the estimator.

An electronic digitizer can be used to obtain the quantities of materials from the construction drawings of a project. The estimator can use a digitizer pin to trace the lines on the drawings to obtain information such as the square yards of paving, square feet of brick, linear feet of pipe, or number of windows. Using a digitizer to calculate the quantity takeoff automates the process and provides the information to the estimator in an organized form.

Numerous computer databases can be developed by the estimator to automate and standardize the estimating function. The estimator can develop a historical cost database from the cost records of projects that have been completed by the company. This information can be stored as unit costs in the database and organized in a CSI or WBS system with cost codes for each item. The estimator can retrieve information from the historical cost database for the preparation of estimates for future projects. As new information is obtained from current projects, the estimator can update the historical cost database.

Labor and equipment productivity databases can be developed from records of previously completed projects. For example, the labor-hours per square foot of formwork, the number of cubic yards of earth per equipment-day, etc., can be organized and stored for specific job conditions. The estimator can retrieve the labor or equipment productivity figures from the database for the preparation of a cost estimate for a prospective project. Adjustments to the stored productivity can be made by the estimator to reflect unique job conditions.

The estimator can prepare a database of information for the material suppliers and subcontractors who perform work for the company. The database can be organized by type, size, and location of the supplier. During the preparation of an estimate for a project, the estimator can retrieve supplier and subcontractor information pertinent to the project. For example, the estimator can sort and list all drywall contractors capable of performing $80,000 of work at a particular job location. The name, address, and phone number of the contact person for each potential supplier or subcontractor can be retrieved from the computer.

Different individuals access much of the information used in the operation of a construction company at different times. For example, a project is planned and scheduled based on the time and cost information prepared by the estimator. Likewise, the project budget control system is developed from the cost estimate. The computer can be used to link the information from the estimating function to the planning function and to the budget control function of a contractor's operation. Common cost codes can be used for each operation to integrate, automate, and standardize the operations of a construction firm.

Chapter 22 presents a comprehensive discussion of computer estimating. The software at the McGraw-Hill website for this book has example problems that enable the reader to gain knowledge and experience in computer methods for estimating.

FORMS FOR PREPARING ESTIMATES

Experienced estimators will readily agree that it is very important to use a good form in preparing an estimate. As previously stated, the form should treat each operation to be performed in a construction project. For each operation, there should be a systematic listing of materials, equipment, labor, and any other items, with space for all calculations, number of units, unit costs, and total cost.

Each operation should be assigned a code number, and this number should be reserved exclusively for that operation on this estimate as well as on estimates for other projects within a given construction organization. For example, Table 1.4, item 2250 refers to compaction, whereas item 2372 refers to drilled piers. The accounting department should use the same item numbers in preparing cost records.

Table 1.11 illustrates a form that might be used in preparing a detailed estimate. When a project includes several operations, the direct costs for material, equipment, labor should be estimated separately for each operation, then the indirect costs.

TABLE 1.11 | Form to estimate construction costs.

Item no.	Description	Calculations	Number of units	Unit cost	Material cost	Equipment cost	Labor cost	Total cost
2350-0	Furnish and drive 200 creosote-treated piles. Drive piles to full penetration into normal soil. Piles size: 50-ft length, 14-in. butt, 6-in. tip	200×50 ft = 10,000 lin ft	10,000 lin ft					
-10	**Materials** Piles; add 5 for possible breakage	205×50 ft = 10,250 lin ft	10,250 lin ft	$10.80/lin.ft	$110,700			$110,700
-20	**Equipment** Moving to and from the job	lump sum	lump sum			$7,000		$7,000
	Crane, 12-ton		80 hr	$145/hr		$11,600		$11,600
	Hammer, single-acting, 15,000 foot pound	200 piles/($2\frac{1}{2}$ piles/hr) = 80 hr	80 hr	$15/hr		$1,200		$1,200
	Air compressor equipment		80 hr	$9/hr		$720		$720
	Leads and sundry equipment		80 hr	$5/hr		$400		$400
-30	**Labor**							
-32	(add 16 hr to set up and take down equipment)	80 + 16 = 96 hr						
-32	Foreman	96 hr	96 hr	$34/hr			$3,264	$3,264
-34	Crane operator	96 hr	96 hr	$28/hr			$2,688	$2,688
-36	Laborer (1 total)	96 hr	96 hr	$21/hr			$2,016	$2,016
-38	Workers on hammer (2 total)	96 hr × 2 = 192 hr	192 hr	$23/hr			$4,416	$4,416
-39	Helpers (2 total)	96 hr × 2 = 192 hr	192 hr	$19/hr			$3,648	$3,648
-40	**Subtotal direct costs**				$110,700	$20,920	$16,032	$147,652
-50	**Indirect costs**							
-51	Material taxes							
-511	State sales tax	5% × $110,700						$5,535
-512	County sales tax	1% × $110,700						$1,107
-520	Labor taxes							
-521	FICA (social security tax)	7.65% × $16,032						$1,226
-522	Unemployment tax	3% × $16,032						$481
-530	Insurance							
-531	Workers' compensation insurance	9% × $16,032						$1,443
-532	Contractor's liability insurance	4% × $16,032						$641
-540	Overhead							
-541	Job overhead	8% × $147,652						$11,812
-542	Office overhead	2% × $147,652						$2,953
-60	**Subtotal indirect costs**							$25,198
-70	**Total direct and indirect costs**							$172,850
-80	**Add-ons**							
-811	Contingency	5% × $172,850						$8,643
-812	Profit	10% × $172,850						$17,285
	Subtotal of add-ons							$25,928
-90	Performance bond	1% × ($172,850 + $25,928)						$1,988
-91	**Total cost**, amount of bid	$172,850 + $25,928 + $1,988						$200,766
-92	Cost per lin ft	$200,766/10,000 lin ft						$20.08 per lin ft

Bid Documents

BID DOCUMENTS AND CONTRACT DOCUMENTS

The terms *bid documents* and *contract documents* are often used interchangeably, although they are not exactly the same. The bid document applies to before the contract is signed, whereas the contract document applies to after the contract agreement is signed by the owner and contractor.

The bid documents consist of the invitations to bid, instructions to bidders, bid forms, drawings, specifications, requirements for bonds and insurance, and all addenda. This information is necessary for the contractor to prepare an estimate and submit a bid. After the owner has made the decision to accept the bid, the owner and contractor sign the contract agreement, which forms the contract documents. The contract documents consist of the signed agreement, bonds, insurance, drawings, and specifications. Thus, the bid documents become the contract documents at the time the contract is signed. Any change orders that are approved during construction also become a part of the contract documents.

The purpose of the contract is to provide a legal document for construction and completion of the project. The contract includes the contract agreement, bonds, insurance, plans, specifications, and all change orders that are required to complete the construction work.

Designers are responsible for producing the contract documents, which are the plans and specifications from which the contractor can build the project. The prime designer for building type projects is the architect, whereas the prime designer for heavy/industrial type projects is the engineer. The construction contractor is required to perform all work in accordance with the contract documents.

The contractor is responsible for providing all labor, materials, tools, transportation, and supplies required to complete the work in a timely and workmanlike manner and in accordance with the plans, specifications, and all terms of the contract. This chapter provides a brief description of the contract documents with emphasis on aspects that can impact the estimator and the estimating process.

No useful estimate can be prepared without a thorough knowledge of the drawings and specifications of the contract documents.

CONTRACT REQUIREMENTS

The contracting requirements section includes the contract agreements, requirements of bonds and certificates, general and supplementary conditions, and addenda. Each of these items must be included in the final estimate and subsequent bid that is submitted to the owner.

The agreement is the actual contract that will be awarded to the contractor and signed by the owner and contractor. Contracts are awarded to the contractor from the owner by either competitive bidding or negotiation. Most municipal, state, and federal projects are awarded by competitive bidding. This process involves an advertisement for bids in the public media with an announced bid opening date. Bid opening is performed in the open public. The contractor performs a detailed estimate and submits a bid price for building the project in accordance with the contract documents. The engineer and owner perform the evaluation of bids at a later date, then a final decision is made to award the contract. Most government agencies are required by law to award the contract to the lowest bidder, provided all conditions of the contract have been met. Private owners are not required to award the contract to the lowest bidder, but may select the contractor deemed most desirable for the project.

Negotiated contracts are awarded to a contractor independent of the competitive bidding process. The owner, or agent of the owner, reviews the plans and specifications with various contractors. Each contractor prepares a cost estimate and quotes the price to perform the various work in the project. Adjustments in the scope of work, schedule, and cost of the project are negotiated to satisfy both the owner and contractor. Award of the contract is then made to the contractor the owner feels can provide the best total performance, which may or may not be the lowest initial cost.

ARRANGEMENT OF CONTRACT DOCUMENTS

Contract documents for construction projects consist of two general categories: information related to business/legal matters and information related to technical matters. Illustrative examples of the business/legal portions of the contract documents are invitation to bid, instruction to bidders, information available to bidders, bid forms, bond and insurance requirements, general conditions, and supplementary conditions of the contract.

The technical portion of the bid documents consists of the plans (drawings) and written specifications that describe the material, workmanship, and methods of construction that are required to build the project. It is practically impossible to describe by words alone how a project is to be built. The drawings are the pictorial directory of how the project is to be built. To further clarify the quality of materials and workmanship, the written specifications are used to supplement the drawings. Where conflicts exist between the drawings and written specifications, it is common practice that the written specifications govern. A clause to this effect is usually presented in the contract documents.

BUILDING CONSTRUCTION SPECIFICATIONS

The written specifications for a project are as important as the drawings in defining the total project. The quality of material, performance rating of equipment, level of workmanship, and warranty requirements are all defined in the written specifications. Information about the size, type, and performance of installed equipment (pumps, generators, etc.) or components (valves, switches, etc.) are defined in the written specifications rather than on the drawings.

A unique set of specifications is written for each project. For building construction type projects, the specifications generally follow the Construction Specifications Institute (CSI) MasterFormat numbering system, which consists of 50 major divisions. Each of the 50 divisions is subdivided into smaller items of work. There are six digits for each item in the numbering system. The first two digits of the numbering system identify the division number, such as 00, 01, 02, 03, 04...........48, 49. The next two digits of the numbering system refer to sections within the division, and the last two digits are the subsections within a section. Thus, a complete listing of all numbers in the CSI MasterFormat is extensive. The 50 divisions of CSI MasterFormat are shown in Table 2.1.

Division 00 applies to the business/legal matters and the remaining 49 divisions apply to technical specifications. Division 00 is particularly applicable to the estimating team as they prepare an estimate for bidding purposes. It contains information about bid solicitation, instructions to bidders, information available to bidders, bid submissions, bid forms, and requirements for bonds and insurance.

HEAVY/HIGHWAY SPECIFICATIONS

For heavy/highway projects the technical specifications are standard for each state department of transportation. Thus, all highway construction projects within a state will use the same standard set of technical specifications. Typical arrangement of standard specifications for highway projects include these divisions:

1—General Provisions: general information, definition, and terms

2—Excavation and Embankment: soils, sodding

3—Base Courses: aggregate, lime, fly ash, subbases

4—Surfaces Courses: bituminous and concrete pavements

5—Structures: concrete, steel, and timber bridges and foundations

6—Materials: aggregate, asphalt, concrete, conduits, testing

7—Traffic Control: signals, lighting, signs, striping

8—Incidental Construction: riprap, gabions, guard rails, fences

TABLE 2.1 | CSI MasterFormat 50 Divisions

Division 00 Procurement & Contracting Requirements	Division 26 Electrical
Division 01 General Requirements	Division 27 Communications
Division 02 Existing Conditions	Division 28 Electronic Safety and Security
Division 03 Concrete	Division 29 Reserved
Division 04 Masonry	Division 30 Reserved
Division 05 Metals	Division 31 Earthwork
Division 06 Wood, Plastics, and Composites	Division 32 Exterior Improvements
Division 07 Thermal and Moisture Protection	Division 33 Utilities
Division 08 Openings	Division 34 Transportation
Division 09 Finishes	Division 35 Waterway and Marine Construction
Division 10 Specialties	Division 36 Reserved
Division 11 Equipment	Division 37 Reserved
Division 12 Furnishings	Division 38 Reserved
Division 13 Special Construction	Division 39 Reserved
Division 14 Conveying Equipment	Division 40 Process Integration
Division 15 Reserved	Division 41 Material Processing & Handling Equip.
Division 16 Reserved	Division 42 Process Heating, Cooling, and
Division 17 Reserved	Drying Equipment
Division 18 Reserved	Division 43 Process Gas and Liquid Handling,
Division 19 Reserved	Purification, and Storage Equipment
Division 20 Reserved	Division 44 Pollution Control Equipment
Division 21 Fire Suppression	Division 45 Industry-Specific Manufacturing Equip.
Division 22 Plumbing	Division 46 Reserved
Division 23 Heating, Ventilating, & Air Conditioning	Division 47 Reserved
Division 24 Reserved	Division 48 Electrical Power Generation
Division 25 Integrated Automation	Division 49 Reserved

Although these technical specifications are standard, a unique set of bidding requirements, drawings, and list of pay quantities are provided for each specific project. The bidding requirements pertain to business and legal matters, such as advertisement for bids, prequalification of bidders, proposal forms, bid opening date and time, and award of contract. Of particular interest to the estimator is the date for submission of bid and the list of pay quantities.

Submission of the bid by the bid opening date and time is mandatory, otherwise the bid will be disqualified. Therefore, the estimator must develop a schedule for preparing the estimate that meets the required bid opening date and time.

BIDDING REQUIREMENTS

The information contained in this section of the contract documents is extremely important to the estimating team. Typical information in the bidding requirements includes:

Bid Solicitation

Instructions to Bidders

Information Available to Bidders

Bid Forms and Supplements

Bidding Addenda

Bid Solicitation

The bid solicitation, sometimes called the invitation to bid, contains the date and time that bids must be submitted. The estimator must establish a plan to complete all work to meet the deadline of the bid date, otherwise the bid will be disqualified. The bid solicitation also gives the names and addresses of the owner and design organization, with instructions on how to obtain the bid documents. Table 2.2 lists typical items in the invitation to bidders. Figure 2.1 is an example bid solicitation.

TABLE 2.2 | Typical contents of an invitation to bid

1. Name, location, and type of project
2. Name and address of owner and designer
3. Date, time, and location for receipt of bids
4. Disqualification criteria for late submission of bids
5. Required amount of bid bond, certified check, or cashier's check
6. Procedure for obtaining plans and specifications
7. Clarification of owner's right to reject any and all bids

Instructions to Bidders

The instructions to bidders section describes vital information that is required to submit a bid. For example, bidders may be required to attend a prebid conference. Another example that may appear in the instructions to bidders is a statement that the contractor must make provisions during execution of the contract documents to allow the owner to take advantage of the owner's tax exempt status for materials and equipment purchased for the project. Thus, no taxes would be applied during preparation of the estimate. Early in the estimating process the estimating team must become thoroughly aware of the requirements that are placed on preparing the bid in accordance with the contract documents. Table 2.3 lists typical information in the instructions to bidders.

Figure 2.2 is an abbreviated example of instructions to bidders. This example does not contain all of the items typically covered, but it is provided to illustrate the wording and to show the type of information that must be considered by the estimator in preparing the estimate and submitting the bid.

Information Available to Bidders

The information available to bidders section may include such items as referencing a subsurface exploration report that has been prepared for the project, but is not shown in the contract documents. The information available to bidders may include information about the project from a material testing laboratory. This section may also include reference to standards, such as the American Institute of Architects (AIA) or the Engineers Joint Contracts Documents Committee (EJCDC) general conditions of the contract, that apply to the project but are not included in the contract documents.

SECTION 00001 – **ADVERTISEMENT FOR BIDS:**

The OWNER, will receive bids on:

**MAINTENANCE FACILITY PROJECT
1234 STREET
CITY, STATE, ZIP**

at the office of **OWNER, located in Room 200 at 1234 Street in City, State until 3:00 P.M., MAY 26, 20XX,** and then publicly opened and read aloud at a designated place. Bids received after this time or more than ninety-six (96) hours excluding Saturdays, Sundays, and holidays before the time set for the opening of bids will not be accepted. Bids must be turned in at the above office during the time period set forth. All interested parties are invited to attend.

The contract documents may be examined in the office of ABC DESIGN FIRM, 3456 Street, City, State, Zip Code, and copies may be obtained there upon receipt of a letter of intent to pay **$100.00** for two (2) sets of drawings and specifications in the event that they are not returned in good condition within ten (10) days of the date of bid opening. Additional sets may be obtained by paying the cost of printing the drawings and assembly of specifications.

The OWNER reserves the right to reject any or all bids or to waive any minor informalities or irregularities in the bidding.

Each bidder must deposit, with his bid, security in the amount and form set out in the contract documents. Security shall be subject to the conditions provided in the Instruction to Bidders.

No bidder may withdraw his bid within sixty (60) days after the date of opening thereof. Attention of bidders is particularly directed to the statutory requirements and affidavit concerning nondiscrimination, nonsegregated facilities, noncollusion, and business relationship provisions.

The OWNER reserves the right to reject any and all bids.

OWNER

By: _____
            Chairman's Signature
            Chairman Name

FIGURE 2.1 I Example bid solicitation.

TABLE 2.3 I Typical contents of instructions to bidders.

1. Instructions on submission, receipt, opening, and withdrawal of bids
2. Qualifications of bidders and subcontractors
3. Requirement of bidders to use bid forms in the contract documents
4. Instructions to use both words and figures in submitting cost amounts
5. Statement that all addenda will become a part of the contract
6. Request to state the number of calendar weeks for completing the work
7. Requirement of contractors to notify design of any discrepancies
8. Criteria for extra work that may be specified by the owner or designer
9. Request for unit cost of labor and equipment for extra work
10. Requirement of contractor to abide by local ordinances and regulation

SECTION 00002 – **INSTRUCTIONS TO BIDDERS:**

RECEIPT AND OPENING OF BIDS:
Bids will be received by the Owner at the time and place set forth in the Advertisement for Bids and then at said place be publicly opened and read aloud.

BID SUBMISSION:
Each bid must be submitted in a sealed envelope bearing on the outside the name of the bidder, his address, and the name of the project for which the bid is submitted, and addressed as specified in the Bid Form.

SUBCONTRACT:
Any person, firm, or other party to whom it is proposed to award a subcontract under this contract must be acceptable to the Owner.

WITHDRAWAL OF BIDS:
Bids may be withdrawn on written or telegraphic or facsimile request received from bidders prior to the time fixed for opening. Negligence on the part of the bidder in preparing the bid confers no right for the withdrawal of the bid after it has been opened.

BASIS OF BID:
The bidder must include all unit cost items and all alternatives shown on the Bid Forms; failure to comply may be cause for rejection. No segregated bids or assignments will be considered.

QUALIFICATION OF BIDDER:
The Owner may make such investigation as he deems necessary to determine the ability of the bidder to perform the work, and the bidder shall furnish to the Owner all such information and data for this purpose as the Owner may request.

TIME OF COMPLETION:
Bidder must agree to commence work on or before a date to be specified in a written "Notice to Proceed" of the Owner and to substantially complete the project **within 180 consecutive calendar days thereafter.**

CONDITIONS OF THE WORK:
Each bidder must inform himself fully of the conditions relating to the construction of the project and employment of labor hereon. Failure to do so will not relieve a successful bidder of his obligation to furnish all material and labor necessary to carry out the provisions of his contract.

ADDENDA AND INTERPRETATIONS:
All addenda so issued shall become a part of the contract documents. No interpretation of the meaning of the plans, specifications, or other contract documents will be made to any bidder orally.

LAWS AND REGULATIONS:
The bidder's attention is directed to the fact that all applicable State Laws, Municipal Ordinances, and the rules and regulations of all authorities having jurisdiction over construction of the project shall apply to the contract throughout, and they will be deemed to be included in the contract the same as though herein written out in full.

OBLIGATION OF BIDDER:
At the time of the opening of bids, each bidder will be presumed to have inspected the site and to have read and to be thoroughly familiar with the plans and contract documents (including all addenda).

FIGURE 2.2 | Example instructions to bidders.

Bid Forms

The bid form defines the format that is required for submission of the bid. The format of the bid form impacts the assembly and summary of costs in the final estimate. The owner may request the bid as a single lump sum, unit prices based on predefined pay quantities in the bid documents, or a combination of both lump and unit prices. For construction projects, the work may be priced by several methods: lump sum, unit price, cost plus a fixed fee or cost plus a percentage of construction, cost plus a guaranteed maximum price, or a combination of these pricing methods. The method selected depends on the distribution of risk between the owner and contractors.

Bid form for lump-sum contracts   For projects where a complete set of plans and specifications have been prepared prior to construction and the quantity of work is well defined, the estimate is normally prepared for the purpose of submitting a lump-sum bid on the project. Building type projects are usually bid on a lump-sum basis. When the cost of a project is estimated on this basis, only one final total-cost figure is quoted. Unless there are changes in the plans or specifications, this figure represents the amount that the owner will pay to the contractor for the completed project. Since the contractor provides the lump-sum price before construction starts, he or she is exposed to uncertainties during construction and thereby assumes risk for the project.

It is common practice for projects to have one or more "alternates" attached to the bid documents of lump-sum contracts. The alternate may be to add or deduct a work item from the base lump-sum bid. This allows the owner the option of selecting the number of alternates so that the total bid cost will be within the amount specified by the owner's budget.

A lump-sum estimate must include the cost of all materials, labor, equipment, overhead, taxes, bonds, insurance, and profit. It is desirable to estimate the costs of materials, labor, and equipment separately for each operation; to obtain a subtotal of these costs for the entire project; and then to estimate the cost of overhead, taxes, bonds, insurance, and profit.

For a building type project the bid form consists of a page in the specifications for bidders to record the total bid amount. If numerous alternates are in the bid documents, the base bid is recorded and the bid price for alternates is listed separately. Figure 2.3 is an example bid form for a building type project with one alternate.

Bid forms for unit-price contracts   Most heavy engineering construction projects are bid on a unit-price basis. Such projects include pavements, curbs and gutters, earthwork, various kinds of pipeline, clearing and grubbing land, etc. These projects are bid unit-price because the precise quantities of material may not be known in advance of construction. For example, the actual quantity of material to be excavated, hauled, and compacted in the fill area may vary substantially from the calculated quantity in the bid forms due to unknown settlement of the soil and other factors. The contractor bids the work on a unit-price basis and is paid based on the actual quantity of work. The final cost is determined by multiplying the bid price per unit by the actual quantity of work completed by the contractor. Therefore, for a unit-price contract the cost that the owner will pay to the contractor is not determined until the project has been completed. Thus, the owner assumes the risk for uncertainty in the actual quantity of work.

SECTION 00003 – **BID FORM:**

Place: _____   Date: _____

PROPOSAL OF _____
                                              (hereinafter called "Bidder")

TO: The Owner.

Gentlemen:
The bidder, in compliance with your invitation for Bids for the:

MAINTENANCE FACILITY PROJECT
1234 STREET
CITY, STATE

having examined the plans and specifications with related documents and the site of the proposed work, and being familiar with all of the conditions surrounding the work, including the availability of materials and labor, hereby agree to furnish all labor, materials, equipment, and supplies, and to perform the work required by the project in accordance with the contract documents, within the time set forth in Instructions to Bidders, and at the prices stated below. These prices are to cover all expenses incurred in performing the work required by the contract documents, of which this bid is a part.

Bidder acknowledges receipt of the following addenda: _____.

All Bid amounts shall be shown in both WORDS and FIGURES. In case of discrepancy, the amount shown in words will govern.

BASE BID:
Bidder agrees to perform all of the work described in the plans and specifications as being in the Base Bid for the sum of: _____ Dollars.
                        $ _____

The Bidder agrees to perform all of the work described in the Drawings and Specifications, which has been designated as Alternate Bids. Alternate Bids shall be add amounts to the Base Bid. Changes shall include any modifications of the work or additional work that the Bidder may require to perform by reason of Owner's acceptance of any or all Alternate Bids.

ALTERNATE NO. 1 :
Add 3" thick Type "C" asphalt paving to compacted crushed limestone as shown on Sheet C-1.
ADD: _____ Dollars $_____.

The Bidder hereby agrees to commence work under this contract on or before a date to be specified in a written "Notice to Proceed" by the Owner and to substantially complete the project **within 180 consecutive calendar days thereafter**. The Owner reserves the right to reject any and all bids. Bidder agrees that this bid shall be good and shall not be withdrawn for a period of sixty (60) calendar days after the opening thereof.

In the event a contract is awarded by the Owner to the Successful Bidder, it shall be executed within thirty (30) days. The Bidder shall return with his executed contract the Performance Payment Bond, Statutory Bond, and Warranty Bond as required by the Supplementary Conditions.

(SEAL) if bid is   Respectfully submitted by: _____
corporation   FEI/SS Number: _____

FIGURE 2.3 | Example bid form for lump-sum contract.

The cost per unit, submitted in a bid, includes the furnishing of materials, labor, equipment, supervision, insurance, taxes, profit, and bonds, as required, for completely installing a unit. Typical designated units include square yards (sy), cubic yards (cy), linear feet (lin ft) or feet (ft), tons, acres, etc. A separate estimate should be prepared for each type or size of unit.

The costs of materials, equipment, and labor are determined for each unit. These are called *direct costs*. To these costs must be added a proportionate part of the indirect costs, such as moving in, temporary construction, overhead, insurance, taxes, profit, and bonds, since indirect costs are not bid separately.

Figure 2.4 is an example bid form for a unit-price highway project. The estimator must carefully check the pay quantities to ensure that they are accurate. The list of bid items is crucial to the estimator because the contractor will be paid only for work that is specified in this list of pay quantities.

Item	Description	Quantity	Unit	Unit Price	Total
1	Mobilization	lump sum	job		
2	Water control	lump sum	job		
3	Common excavation	20,000	cy		
4	Rock excavation	12,000	cy		
5	Drilling piers	5,350	lin ft		
6	Concrete foundation	250	cy		
7	Anchor bolts	185	each		
8	Rock anchors	17,500	lb		
9	Slope protection	260	each		
10	Compacted fill	620,000	cy		
11	Random fill	1,790,730	cy		
12	Riprap – 24 in.	3,700	cy		
13	Drainage structure #1	lump sum	job		
14	Drainage structure #2	lump sum	job		
15	Lime-treated subgrade	25,500	sy		
16	Aggregate base course	3,350	cy		
17	Asphalt pavement	40,100	sy		
18	Concrete pavement	150	cy		
19	Bermuda sodding	2,080	sy		
20	Traffic control	lump sum	job		
21	Demobilization	lump sum	job		

FIGURE 2.4 | Bid form for unit-price contract.

During preparation of the estimate, the contractor must place all costs in the pay quantity list. If an item of work required to complete the project is not on the bid list, such as traffic control or laboratory testing, then the estimator must include the cost for this work in one of the other unit-cost bid items. Thus, the bid form is important to the contractor because it impacts the preparation and summarization of the cost estimate. This part of the specifications impacts the requirements of summarizing the estimate in an acceptable format for submission of the bid to the owner.

NEGOTIATED WORK

Sometimes the owner will negotiate the work for a project with a construction firm prior to completion of a set of plans and specifications. This is usually done when the owner wants to start construction at the earliest possible date to benefit from an early completion and use of the project. A representative of the owner works with the contractor to evaluate alternatives to obtain a project configuration that meets the needs of the owner, yet with a cost within the owner's allowable budget.

The final contract agreement usually is a cost plus a fixed fee with a guaranteed maximum amount. If the cost is above or below the guaranteed maximum amount, then the owner and the contractor agree to a splitting of the difference. For example, the contractor might incur 70 percent and the owner 30 percent of the amount that exceeds the guaranteed maximum price, or the contractor may receive 60 percent and the owner 40 percent of the amount that is below the guaranteed maximum price.

ADDENDUM

An addendum is a change in the contract documents during the bidding process, before award of the contract. An addendum is sometimes called a bulletin. Typically addenda are issued to correct errors in the contract documents or clarify an issue, or it may concern addition to the work at the request of the owner. The estimate and resulting bid must include all costs required to complete the work in accordance with the drawings and specifications, together with all of the addenda. Therefore, the estimating team must be certain that the costs of all addenda are included in the estimate.

ALTERNATES

An alternate is an addition or subtraction to a base bid price for substitutions requested by the owner during the bidding process. Each alternate is listed and numbered separately in the bid documents. For example, an Alternate No. 1 may be add the parking lot and Alternate No. 2 may be deduct the sidewalks.

CHANGE ORDER

A change order is issued by the designer, but signed by the owner and contractor, to make a change in the contract documents during construction. Although change orders usually add time and cost to the original contract, some change orders are issued without extensions of time or changes in cost. Upon approval by the owner, change orders become a part of the contract documents.

WARRANTIES

Warranties are guarantees by the contractor that specific components of the project will be free from defects due to materials or workmanship for a specified warranty period. Warranties cover specific items, such as a roofing warranty or an equipment warranty, that are part of the construction project. The type of warranty and warranty period are defined in the written specifications of the contract documents.

GENERAL CONDITIONS OF THE CONTRACT DOCUMENTS

The general conditions is that part of the contract document that establishes the rules for administration of the construction phase of the project. Most contract documents incorporate standard general conditions that have been developed by such organizations as the AIA or the EJCDC. The information contained in the general conditions applies more to the construction phase of a project rather than the bidding phase.

BONDS

Most owners require prospective bidders to submit bonds as qualifications for submitting a bid for a project. The bonding company that issues the bond requires the contractor to show financial stability and previous experience in performing the type of work that is required to build the project. In addition, it is common practice for the owner to require all prospective bidders to be prequalified before submitting a bid. The prequalification process usually involves a review of the prospective contractor's financial and safety records and an evaluation of performance on previous projects.

The contractor secures bonds from a bonding company on behalf of the owner as financial and legal protection for the owner. Three types of bonds are commonly required in construction contracts: bid bond, performance bond, and payment bond. The bid bond ensures the owner that the contractor will sign the contract for the bid amount. The performance bond ensures the owner that the contractor will perform all work in accordance with the contract documents. The payment bond ensures the owner that all material and labor will be paid. If the contractor defaults, the bonding company agrees to fulfill the contract agreement.

Bid Bond

It is common practice to require each bidder on a project to furnish with the bid a bid bond, a cashier's check, or a certified check in the amount equal to 5 to 20 percent of the amount of the bid. In the event that the contract to construct the project is tendered to a bidder and the bidder refuses or fails to sign the contract, the owner may retain the bond or check as liquidated damages.

For some projects, cashier's checks are specified instead of bid bonds. The bidder purchases these checks, which are issued to the owner of the project by a bank. They can be cashed easily, whereas it is necessary for the owner to secure payment on a bid bond through the surety, and the surety may challenge the payment. The use of cashier's checks requires bidders to tie up considerable sums of money for periods that may vary from a few days to several weeks in some instances.

There is no uniform charge for cashier's checks. Some banks charge $0.23 per $100.00 for small checks, with reduced rates for large checks, while others make no charge for checks furnished for regular customers. However, the interest cost for a cashier's check for $100,000 for 2 weeks at 6 percent interest will amount to approximately $240.00.

Performance Bond

All government agencies and many private owners require a contractor to furnish a performance bond to last for the period of construction of a project. The bond is furnished by an acceptable surety to ensure the owner that the work will be performed by the contractor in accordance with the contract documents. In the event a contractor fails to complete a project, it is the responsibility of the surety to secure completion. Although the penalty under a performance bond is specified as 25, 50, or 100 percent of the amount of the contract, the cost of the bond usually is based on the amount of the contract and duration of the project.

The cost for a contractor to purchase a performance bond varies, depending on the capacity of the contractor to perform the work and the financial stability of the contractor. Representative costs of performance bonds per $1,000.00 are:

For buildings and similar projects:

First	$ 500,000 =	$14.40
Next	$2,000,000 =	8.70
Next	$2,500,000 =	6.90
Next	$2,500,000 =	6.90
All over	$7,500,000 =	5.75

For highways and engineering construction:

First	$ 500,000 =	$12.00
Next	$2,000,000 =	7.50
Next	$2,500,000 =	5.75
Next	$2,500,000 =	5.25
All over	$7,500,000 =	4.80

Material and Labor Payment Bond

The contractor secures the material and labor payment bond from a surety company. The bond is issued to ensure the owner that all wages and bills for materials will be paid upon completion of the project. In the event the contractor fails to pay labor and material costs, the surety company assumes that responsibility. Typically, the performance bond and the material and labor payment bond are secured by the contractor as a package.

INSURANCE

The contractor, as required in the contract documents, secures insurance. The contractor supplies a certificate of insurance to verify insurance coverage. Due to the many risks involved in construction, most contractors carry insurance in addition to the requirements in the contract documents. The precise losses that are reimbursable from insurance are based on predetermined losses named in the insurance policy. Many types of insurance for construction projects are available.

Basic Builder's Risk Insurance

General builder's risk, which covers damages to the project due to fire, wind, and hail, is the insurance commonly required on construction projects. Numerous other types of insurance are available that cover losses, including workers on the job, equipment, public liability, and property damage to parties that are not affiliated with the project. Under the terms of this coverage, the insurance is based on the estimated completed value of the project. However, because the actual value varies from zero at the beginning of construction to the full value when the project has been completed, the premium rate usually is set at 60 percent of the completed value. In the event of a loss, the recovery is limited to the actual value at the time of the loss.

Although the cost of protection will vary with the type of structure and its location, these rates are representative costs for basic builder's risk insurance:

Type of construction	Cost of premiums per $100.00 of value for basic builder's risk
Framed construction	$4.539
Joisted masonry	$1.817
Noncombustible	$0.626

Basic Builder's Risk, Extended Coverage

This insurance is a supplement to the basic builder's risk insurance, and it provides protection against wind, smoke, explosion, and vandalism. The premium rate for this insurance is about $1.93 per $100.00 of completed value of the project.

Public Liability and Property Damage Insurance

This insurance protects the contractor against injuries to the general public or public property due to actions of the employees while performing work during construction. The cost for this insurance will depend on the type of work and the safety record of the contractor. The cost can range from 2 to 8 percent of the base cost of labor.

Workers Compensation Insurance

Contractors pay workers compensation insurance for protection of workers against injury or death on a project. This insurance provides medical expenses and payment of lost wages during the period of injury.

Regulations governing workers compensation insurance are established by state laws and vary widely from one state to another. The cost also varies, depending on the type of work that is performed by each worker. For example, different rates exist for bricklayers, carpenters, ironworkers, equipment operators, painters, etc. The cost can range from 10 to 30 percent of the base cost of labor.

Comprehensive General Liability Insurance

As the result of construction operations, it is possible that persons not employed by the contractor may be injured or killed. Also, property not belonging to the contractor may be damaged. Comprehensive general liability insurance should be carried as a protection against loss resulting from such injuries or damage. This insurance provides the protection ordinarily obtainable through public liability and property damage insurance. The coverage should be large enough to provide the necessary protection for the given project. The premium rate varies with the limits of liability specified in the policy.

Contractor's Protective Liability Insurance

This is a contingent insurance that protects a contractor against claims resulting from accidents caused by subcontractors or their employees, for which the contractor may be held liable.

Contractors' Equipment Floater

This insurance provides protection to the contractor against loss or damage to equipment because of fire, lightning, tornado, flood, collapse of bridges, perils of transportation, collision, theft, landslide, overturning, riot, strike, and civil commotion.

The cost of this insurance, which will vary with the location, should be about $1.50 per $100.00 of equipment value per year.

HEAVY/HIGHWAY DRAWINGS

The drawings for highway projects consist of plan and profile sheets, earthwork cross sections, pavement cross sections, details, reinforcing steel schedules, and any supplemental drawings that show the location, dimensions, and details of the work to be performed.

The plan and profile sheets, combined with the earthwork cross sections, provide a pictorial description of the three dimensions of the work to be performed. The top half of the plan and profile sheets show a vertical view looking down on the highway project and the bottom half shows the corresponding horizontal view along the centerline of the highway. The cross section sheets show horizontal views along the highway, perpendicular to the centerline of the highway. The plan and profile sheets are arranged to show the project from the beginning to the final station.

Cross sections of pavements show the width, thickness, and slope of the paving surfaces. Details are views that show specific areas of the project.

BUILDING CONSTRUCTION DRAWINGS

Although the drawings are prepared specifically for each project, the arrangement of drawings follows a general pattern. The first sheet in a set of drawings is the title sheet. This sheet contains the name of the project, owner, designer, and other pertinent information related to the project. Following the title sheet is the index sheet, which provides a summary of all the remaining sheets in the drawings.

The index sheet is analogous to the table of contents of a book. It shows the list of drawings. For building type projects, the remaining sheets generally follow this arrangement: civil, architectural, structural, mechanical, electrical, plumbing, and fire protection. A letter before the sheet number identifies each drawing. For example, the civil drawings are numbered C1, C2, C3, etc.; the architectural drawings are numbered as A1, A2, A3, etc.; and structural drawings are numbered as S1, S2, S3, etc.

The estimator must thoroughly review the drawings for discrepancies. Errors in the drawings lead to errors in the quantity takeoff and ultimately errors in the final cost estimate. A clear understanding of the drawings is a prerequisite to good estimating.

Plans, Elevations, and Sections

The pictorial presentations used on drawings are shown from several viewing angles. Views looking vertically down on the object are called *plan views*. Views looking horizontally at an object are called *elevation views*. Views looking at an object from a point that is not perpendicular to any face of the object are called *perspective views*. A *section view* of an object is a view of the object as seen by passing a cutting plane through the object. A *detail* is an enlargement of a specific area of a project. Connections of structural components are generally shown as *details*.

Figure 2.5 shows exterior elevations of all four sides of a building and Fig. 2.6 shows a section through the building. Figure 2.7 is a floor plan of the structure, with a schedule for doors, windows, and room finishes.

Some items in a project are difficult to show by plan, elevation, or section views. Items such as the routing of conduits for electrical or mechanical work

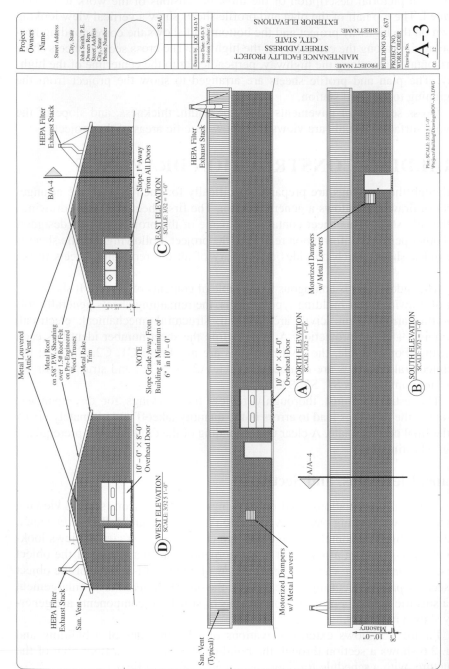

FIGURE 2.5 | Illustration of exterior elevations.

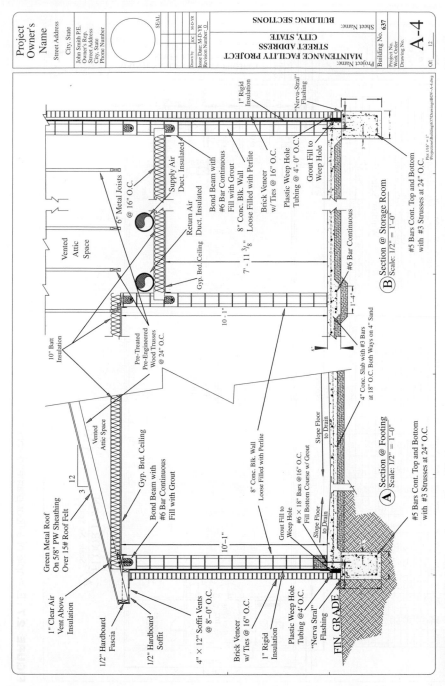

FIGURE 2.6 | Illustration of a section view.

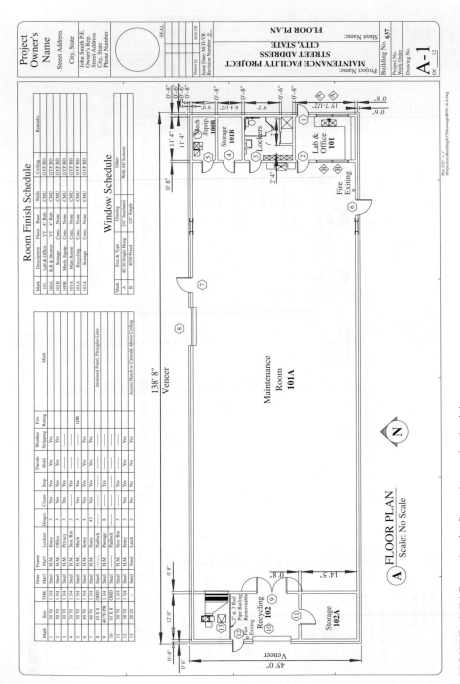

FIGURE 2.7 | Example of a floor plan and schedules.

are examples. It is a common practice to show the routing in three-dimensional pictorial drawings called *isometric* or *pictorial* drawings.

Line Work

Several types of line work that are used in preparing drawings include:

Thick lines—outline edges of objects (plans and elevations)

Thin lines—denote lengths (dimension lines)

Long lines with short dashes—denote the centerline of an object

Phantom lines—show the path of moving parts

Broken lines—denote the object is longer than shown

Short dashed lines—denote edges on the object that cannot be seen from the near side surface of the object

Scales

Usually the type of scale used is identified at the bottom of the page on the drawings. However, several different scales may be used on a single sheet. Therefore, it is important to be certain that the correct scale is used in determining the quantities for estimating. Occasionally, a portion of the drawing is shown without any scale. When this condition exists, there should be a note that states the drawing is not to scale. The estimator must also be cautious because sometimes the drawings are reduced in size, such as a half-size drawing, to permit easier handling during construction.

Two types of scales are used in construction drawings, the architect's scale and the engineer's scale. The architect's scales that are commonly used for plans and elevations include $\frac{1}{32}$ in. = 1 ft and $\frac{1}{16}$ in. = 1 ft. Complicated areas often use the $\frac{1}{4}$ in. = 1 ft, or $\frac{1}{2}$ in. = 1 ft on the architect's scale. Special applications that require high details are $1\frac{1}{2}$ in. = 1 ft, or 3 in. = 1 ft, or may be drawn to a half or even a full scale.

The engineer's scale, graduated in tenths of an inch, is often used for civil, structural, and mechanical drawings. The scales are 10, 20, 30, 40, 50, and 60. The higher numbers of 50 or 60 are used for large areas, such as site-work or plot plans. Details of drawings often use the smaller numbers, such as 10, 20, or 30. The 40 on the engineer's scale is equivalent to the $\frac{1}{4}$ in. on the architect's scale.

Schedules

To simplify the presentation of repetitious items (such as footings, columns, doors, windows, room finishes, etc.) a schedule is frequently used. A schedule is a tabular listing of the repetitive items in the project. For example, a room schedule is a tabular list of all rooms in the structure, showing the type of floor covering, wall covering, and type of ceiling. Likewise, a footing schedule is a listing of all footings in the foundation, showing the diameter, depth, type, size, and number of reinforcing bars. Thus, a schedule is a concise and convenient method to show all common types of items in one location. Table 2.4 is a room finish schedule for a building type project.

TABLE 2.4 | Room finish schedule.

Room	Description	Floor	Base	North Mat.	North Fin.	East Mat.	East Fin.	South Mat.	South Fin.	West Mat.	West Fin.	Ceiling Mat.	Ceiling Fin.
A	Corridor	VCT	4″ VC	Gyp	Pt	Gyp	Pt	Gyp	Pt	Ex	Pt	AC	8′-0″
B	Office	VCT	4″ VC	Gyp	Pt	Gyp	Pt	Gyp	Pt	Ex	Pt	AC	8′-0″
C	Office	VCT	4″ VC	Gyp	Pt	Gyp	Pt	Gyp	Pt	Ex	Pt	AC	8′-0″
D	Office	VCT	4″ VC	Gyp	Pt	Gyp	Pt	Gyp	Pt	Ex	Pt	AC	8′-0″
E	Storage	VCT	4″ VC	Gyp	Pt	Gyp	Pt	Gyp	Pt	Ex	Pt	Conc	varies
F	Storage	VCT	4″ VC	Gyp	Pt	Gyp	Pt	Gyp	Pt	Ex	Pt	Ex	Ex
G	Corridor	VCT	4″ VC	Gyp	Pt	Gyp	Pt	Gyp	Pt	Ex	Pt	Ex	Ex

Abbreviations: VCT—3/32″ vinyl clay tile to match existing   Pt—paint, latex
VC—vinyl cove base to match existing   Ex—existing items
Gyp—5/8″ fire rated gypsum wall board   Conc—concrete
AC—2′ × 4′ lay-in acoustical ceiling tile

SYMBOLS AND ABBREVIATIONS

Symbols are used on the drawings to identify the types of materials and work required during construction. Examples are valves, pumps, type of welds, electrical outlets, etc. Abbreviations of organizations that produce technical information and standards for materials and construction procedures are frequently referenced in the specifications of contract documents. A list of common abbreviations in bid documents includes:

AASHTO—American Association of State Highway and Transportation Officials

ACI—American Concrete Institute

AISC—American Institute of Steel Construction

AITC—American Institute of Timber Construction

ANSI—American National Standards Institute

ASA—American Standards Association

ASHRAE—American Society of Heating, Refrigerating, and Air Conditioning Engineers

ASTM—American Society of Testing and Materials

AWI—Architectural Woodwork Institute

AWS—American Welding Society

AWWA—American Water Works Association

BIA—Brick Institute of America

CRSI—Concrete Reinforcing Steel Institute

CSI—Construction Specification Institute

EPA—Environmental Protection Agency

ISO—International Standards Organization

NCMA—National Concrete Masonry Association

NDS—National Design Specification

NEC—National Electrical Code

NEMA—National Electrical Manufacturers Association

NFPA—National Fire Protection Association

OSHA—Occupational Safety and Health Administration

PCA—Portland Cement Association

PCI—Prestressed Concrete Institute

SJI—Steel Joist Institute

UL—Underwriter's Laboratory

PROBLEMS

2.1 Review the sections of *Instructions to Bidders* and *Bid Forms* in Division 00 of the Bid Documents in the appendix of this book and prepare a list of items the estimating team must address as they prepare a bid for the project.

2.2 As discussed in this chapter, the bid document becomes a legal document when the owner and contractor sign the contract agreement. Prepare a list of items that are in the bid documents that are no longer applicable when the contract is signed.

2.3 Obtain copies and compare the general conditions of the contract from three sources: the American Institute of Architects (AIA), the Associated General Contractors (AGC), and the Engineers Joint Contract Documents Committee (EJCDC).

2.4 Obtain the specifications from three sources: your local city government, your state department of transportation (DOT), and the American Association of State Highway and Transportation Officials (AASHTO). Summarize the list of major divisions of these three sources.

2.5 Discuss the problems that may arise during contract negotiations between the owner and contractor when the contractor's bid summary is by materials (such as concrete, metals, etc.), rather than by physical facilities (such as Building A, Building B, Parking Lot, etc.).

2.6 List problems that may occur for the owner during the bidding phase of a unit price contract when the estimated pay quantities are lower than the actual quantities.

2.7 List problems that may occur for the contractor during construction when the actual quantity of work is lower than the pay quantities in the bid documents.

2.8 Sometimes "Lump-sum Contracts" are called "Hard-dollar Contracts" or "Fixed-cost Contracts," which infers the bid amount will be equal to the final cost of the project. Prepare a list of methods a contractor may use to obtain costs above the bid amount.

3

Estimating Process

DECISION TO BID

Prior to starting the estimating process, upper management must assess the desirability to bid the job. A contractor has to consider the investment of time and money to prepare an estimate and compare that investment to the probability of winning the bid, the risks that may occur during construction, and the potential of making a profit after the project is completed.

Numerous factors are considered in determining which projects to bid. Contractors must determine whether or not the skills and availability of their labor and equipment will match the requirements of the job to be bid. For a job to be profitable, the contractor should have experience in building projects similar to the one being considered for bidding.

They must also assess the dollar amount of work currently under construction by their company to ensure the amount of available bonding credit for bidding additional work. Thus, the contractor must consider the amount of work in progress and the available bonding capacity, compared to the size of the project that is being considered for bidding.

Other factors that play a role in determining which project to bid include geographic location, complexity of the project, owner's and designer's reputation, available working capital, and how much the contractor wants to expand their construction operations.

After the decision has been made to bid the project, the estimating process can be started. Management assigns the project to an estimator who will handle the estimating process.

ESTIMATING PROCESS

Estimating is a process, just like any endeavor that requires an end product. Information must be assembled, evaluated, documented, and managed in an organized manner. For any process to work effectively, key information must be

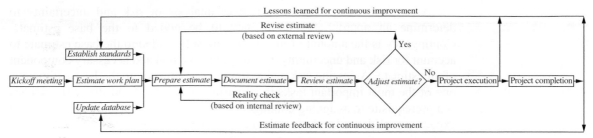

FIGURE 3.1 | Estimating work process.

defined and accumulated at critical times. Standardization is a key word in preparing reliable estimates. The estimating team must establish a standard procedure for preparing estimates to ensure organization and consistency, and to reduce the potential for errors.

People involved in preparing the estimate must know their role, responsibility, and authority. Effective communications among members of the estimating team are essential to selecting the estimate methodology commensurate with the desired level of accuracy, collecting project data and confirming historical cost information, organizing the estimate into the desired format, reviewing and checking the estimate, and documenting the estimate after it is complete.

The estimating work process is illustrated in Fig. 3.1. The first step is the kickoff meeting, which provides background information about the project to the estimating team, expectations of the team, and any pertinent information that may be needed to prepare the estimate. After orientation of the estimating team, a work plan for preparing the estimate is established before starting the estimating process. Preparing the estimate involves calculating the quantities of materials, pricing labor and material, soliciting subcontractor bids, summarizing direct costs, and adding indirect costs to determine the base estimate. After the estimate is prepared, there must be a clear documentation and thorough review, followed by necessary adjustments.

Effective communication is necessary during the estimating process. A support document should be developed and available for presentation, review, and future use of the estimate. A thorough documentation of the estimate forms a baseline for project control, so decisions during project execution can be made with a better awareness of the budget, thereby improving the overall outcome of the project.

Well-executed estimate reviews will increase the credibility and accuracy of the estimate. They also help the team and project management to know the level of scope definition and the basis of the estimate. The review of estimates is an important part of the estimating process because it helps the customer to understand the contents and level of accuracy of the estimate, enabling the customer to make better business decisions.

Adjustment to the estimate involves analysis of risk and uncertainty to determine an appropriate contingency to be added to the base estimate. Contingency is the amount of money that must be added to the base estimate to account for risk and uncertainty. Contingency is a real and necessary component of an estimate. Assessing risk and assigning contingency to the base estimate is one of the most important tasks in preparing estimates. Typically, risk analysis is a prerequisite to assigning contingency. Based on the acceptable risks and the expected confidence level, a contingency is established for a given estimate. The lead estimator for a project must assess the uniqueness of each project and select the technique of risk analysis that is deemed most appropriate. After the base estimate and contingency have been determined, the markup for profit can be added to obtain the bid price for the project.

No estimating process is complete without the feedback loops shown in Fig. 3.1. The estimating process must be a continuous cycle. During project construction, a cost control system should be established to record actual costs for comparison to the estimated costs. Actual cost information from completed projects must be captured in a feedback system that can be integrated into the cost database for use in preparing future estimates. Lessons learned during project execution must also be documented and incorporated into estimating standards and procedures. The lessons learned during construction must be communicated back to the estimating team, to enable them to establish better standards for preparing future project estimates.

THE ESTIMATING TEAM

Upper management and the lead estimator, who has overall responsibility for preparing an estimate, should perform a thorough review of the bid documents to make a preliminary evaluation of possible construction methods and the project schedule, and to identify potential difficulties and any special requirements. The review should be made before assigning estimating responsibilities to members of the estimating team. This review provides information regarding the scope and methods to be used for constructing the project and the experience required of the estimating team.

An assessment must be made to determine the workloads and capabilities of the in-house estimating staff. If sufficient in-house estimating staff is not available, additional assistance can be obtained from field personnel with experience on similar projects, or outside consultants for special requirements that cannot be handled in-house.

The lead estimator should break down the work and make a preliminary work plan for preparing the estimate. Development of an estimate work plan for detailed work assignments is discussed later in this chapter.

It may be necessary to make early arrangements to obtain additional copies of the bid documents. Additional sets of plans will be needed by material suppliers, vendors, and subcontractors to provide cost estimates for specialty work.

One of the key responsibilities of the lead estimator is to ensure the team has all of the necessary resources to perform their work.

Kickoff Meeting of the Estimating Team

The kickoff meeting is the first meeting of the estimating team. The purpose of the kickoff meeting is to ensure alignment between the customer and the estimating team before starting an estimate. The customer is the party requesting the estimate, which may be upper management of the estimating team or an outside organization. There must be a clear understanding of the customer's expectations and the estimating team's ability to meet those expectations to mitigate estimate inaccuracies that can result from misunderstandings and miscommunications. A clear understanding also enables establishment of the estimate work plan and staffing requirements.

Preparation of estimates requires extensive time and money. Early in the process there must be a review of the notification to bidders to determine the bid requirements. The bid documents must be analyzed to determine the feasibility of bidding the project as well as to review bid forms, general and special conditions, and the drawings. There should also be a review of the bidding documents to determine the time and effort required to prepare the estimate, the availability of qualified personnel to prepare the estimate, and the cost to prepare the estimate. Other factors that should be addressed include bonding capacity, the company's capability to finance the cost of construction, and available working capital.

There must also be an assessment of external requirements to determine if the risks are worth the potential reward. There are times when it is best to not invest time and costs in preparing the estimate to bid a job. Factors to consider include competition of other bidders; experience with the architect, engineer, construction manager, and/or owner; availability of material suppliers and subcontractors; time allowed for construction; liquidated damages; and the owner's ability to pay. The quality and completeness of the bid documents are key factors in the ability to produce a reliable estimate. The estimate kickoff meeting is an effective method of addressing these issues. Table 3.1 is an illustrative checklist of issues to be discussed at the kickoff meeting of the estimating team.

ESTIMATE WORK PLAN

Effective management of the estimating effort requires planning, scheduling, and control. Prior to starting the estimate, a work plan for preparing the estimate should be developed. The estimate work plan identifies the work that is needed to prepare the estimate, including who is going to do it, when it is to be done, and the budget for preparing the estimate. The plan also includes the tools and techniques that are appropriate for the level of scope definition and the expected accuracy of the estimate.

TABLE 3.1 | Checklist of issues for the estimate kickoff meeting.

1. What are the driving principles and expectations for the project?
2. What is the level of quality and completeness of the plans and specifications?
3. What level of accuracy and detail is expected in the estimate?
4. Who are potential competitors?
5. What are the experiences with the designer and/or owner?
6. Does the project have unique or unusual characteristics?
7. What is the estimate due date and the anticipated project start and completion dates?
8. What level of confidentiality is required of the team?
9. Who are the customer's contacts with the team?
10. What is the availability of material suppliers and subcontractors?
11. What is the availability of labor at the construction site?
12. Are there other information sources that can aid the estimating?
13. Have similar projects/estimates been developed previously?
14. What owner-furnished items are to be excluded from the estimate?
15. What owner-furnished costs are to be included in the estimate?
16. Are there specific guidelines to be used in preparing the estimate?
17. Are there special permitting requirements that may affect the cost and schedule?
18. Are there any special regulations that might influence the final total installed cost?
19. Are there other issues that could affect the cost or schedule of the project?
20. What are the company's current commitments of personnel and equipment?
21. Are there any special licensing requirements?
22. What is the time allowed for construction?
23. Are there liquidated damages?
24. What is the owner's ability to pay?
25. What are the impacts of community relations?

The leader of the estimating team is responsible for developing an estimate work plan for the project. The estimate work plan is a document to guide the team in preparing accurate estimates and improving the estimating process. The estimate work plan is unique for each project, based on specific project parameters and requirements. Figure 3.2 illustrates the type of information that should be included in an estimate work plan. The work plan should contain sufficient detail to enable all members of the estimating team to understand what is expected of them. After the work plan is finalized it serves as a document to coordinate the estimating work and as a basis to control and maintain the estimating process.

In almost all situations, professional cost estimators with technical backgrounds can produce higher quality estimates. Like any technical specialty, estimating requires specific skills, training, and experience. Involvement of the estimating team early in the project is essential in the business development process.

Individuals with many different job titles, responsibilities, and functions prepare cost estimates for projects in the engineering and construction industry. Depending on the size and needs of each company, those preparing cost estimates may be working alone or as part of a group. They may be centralized in one location or in multiple locations. In some situations, they may be integrated with different departments of the company or they may work in one department.

There are advantages and disadvantages to centralizing or decentralizing the estimating staff. *Where* an estimate is prepared is not as important as *who* is

Estimate Work Plan

Project Name: _____

Project Number: _____

Customer's Name:_____

Type of Estimate Required
Desired Level of Accuracy
Level of Effort Required
Deliverables of Estimate

Estimating Services to be Provided
Deliverables of Estimate by In-house Resources
Deliverables of Estimate by Outside Resources

Budget for Preparing Estimate
Anticipated Work-hours for Estimating Staff
Dollars Budgeted for Nonsalary Estimating Work

Required Staffing for Preparing Estimate
Principal Estimator (leader of estimating team)
In-house and Outside Resources
Availability of Personnel for Staffing

Schedule for Preparing Estimate
Anticipated Start Date
Requirements of Review Date
Customer Due Date

Estimating Methodology
Tools
Technique
Method
Procedures

Estimate Control
Level of Scope Definition
Checklists
Review Process

Presentation
Format for Presenting Estimate
Audience of Presentation

FIGURE 3.2 I Typical information to be addressed in the estimate work plan.

preparing the estimate and the *process* used in preparing the estimate. It is important to implement and maintain effective control over the estimating process. Procedures must be in place for

■ Disseminating knowledge and sharing expertise among the estimating staff.

■ Assigning and sharing the workload among estimators to improve efficiency.

■ Reviewing, checking, and approving work for quality control.

METHODS AND TECHNIQUES

Selection of the method for preparing an estimate depends on the level of scope definition, time allowed to prepare the estimate, desired level of accuracy, and intended use of the estimate. For example, the level of scope definition may be a simple plot plan that shows only the location and types of building and the owner wants only a quick cost estimate to evaluate the economic feasibility of pursuing the project. For this situation, a conceptual cost estimate using the square-foot method of estimating would be appropriate. Methods of conceptual estimating are discussed in Chapter 4.

The level of scope definition may be a complete set of final design drawings and specifications that define and show everything that must be in the completed building, including foundations; structural, mechanical, electrical, and architectural finishes; etc. The final drawings and specifications are provided to a list of contractors for bidding the job to select a contractor to build the project. For this situation, a detailed estimate would be appropriate. This book illustrates methods of detailed estimating for various types of specific construction work, such as earthwork, concrete structures, roofing, plumbing, painting, etc.

PREPARING ESTIMATES

While preparing an estimate, there must be two-way communications between the estimating team and management. The estimating team must keep management informed of the work being performed and management must respond to questions that may arise from the estimating team. The estimating process can assist management in identifying areas of uncertainty and additional information that may be needed, or assumptions that must be made in lieu of definitive information about the project.

As the estimate is being prepared, it is important to perform periodic "reality checks" to make sure the costs developed are within reason. Based on estimator experience and familiarity with the project, this can include

- Simple "intuitive" checks for reasonableness.
- Comparisons with similar projects.
- Comparisons with industry data (dollars/square foot, cost/megawatt, indirect/direct costs, etc.).
- Check ratios such as lighting costs/fixture, fire protection costs/sprinkler, etc.

Timely exchange of information is critical to ensure current price data, databases, and feedback. Preparing estimates requires expertise from multiple disciplines. An effective organization includes experienced field personnel, estimators, and managers who are knowledgeable in estimating. An effective team must be organized to prepare, review, check, and approve the work. This same team must also capture lessons learned to improve the estimating process and improve efficiency.

In preparing early estimates, the skill level of the estimator and his or her experience with the type of facility to be estimated is extremely important. The quality of any estimate is governed by these major considerations:

- Quality and amount of information available for preparing the estimate.
- Time allocated to prepare the estimate.
- Proficiency of the estimator and the estimating team.
- Tools and techniques used in preparing the estimate.

The plans and specifications of the project are determined by the design organization, and the project sponsor determines the completion date for an estimate. Therefore, these two elements may be beyond the control of the estimator. However, the estimator does have control over the selection of the tools and methodology to be used in preparing the estimate. The approach to selecting the method of estimating should be commensurate with the expected level of accuracy of the estimate and the constraints of time.

ESTIMATING PROCEDURES

To prepare an estimate for the entire project, a procedure must be established to assemble all of the costs. The preparation of a comprehensive detailed estimate involves five overlapping activities that must be closely coordinated and integrated. These activities are development of construction methods, preparation of the construction schedule, material quantity takeoffs, the estimate of costs, and assessment of risks for contingency.

To develop an accurate estimate, a site investigation must be conducted. The primary objective of a site visit is the investigation of the physical characteristics of the project site. The investigation team collects information and evaluates any constraints or limitations observed during the investigation, including capacity of roads and bridges for transporting material and equipment to the job, available storage space, and any other factors that can impact the cost of building the project.

The estimate must be based on the planned method of construction. The selection of construction methods must be established early in the estimating process because the construction method will affect all other aspects of the estimate. A decision must be made to plan the construction operations to be followed, to allow the estimating team to separate operations to be performed by contractor's personnel and the work to be subcontracted to specialty contractors. These decisions will depend on availability of contractor-owned equipment and the skills and experience of the contractor's personnel. For some projects, it may be necessary to prepare a preliminary estimate for each of the potential construction methods, to determine the relative cost of each scheme. These studies should be in sufficient detail to identify the most economical methods.

The schedule is an integral part of the estimate. It should be prepared following the decision on construction method and before estimating the costs. In preparing the schedule, the estimating team must consider the requirements of the project, including milestones established by the bid documents. The sequence of work may be dictated by the needs of the owner, weather, safety, lead-time for materials or special equipment, and many other factors. Each of these factors may have a significant impact on the costs. For example, placing concrete during the winter season would cost more than placing concrete in the summer season.

The time required or the duration of each activity will be consistent with the productivity rates selected for the estimate and resource limitations of the contractor, including labor, equipment, material, and money. The total time for each phase of the project and for the project as a whole will affect the direct and indirect costs. The construction schedule can be used to determine resource requirements. Crews may be scheduled for each of the work items to determine average peak workload requirements as well as total work-hours (WHs) required. Based on payment provisions in the bid documents, the schedule can be used to develop a cash flow projection form in which the cost of contractor financing can be established. In each case, the schedule must integrate with and confirm determinations made in the detailed cost estimate.

The material quantity takeoff is the systematic breakdown of the project into units of work for the purpose of evaluating the required cost and time to build the project. Prior to starting the quantity takeoff, the plans and specifications must be thoroughly reviewed for requirements that will affect the way in which quantities will be taken off and assembled. Consideration must be given to the breakdown required in the bid documents, alternates required, unit prices required, method of measurement for payment, and bills of materials. The material quantity takeoff should be handled in a consistent order for all estimates, generally following the sequence of construction. Each estimator must develop a system of quantity takeoff to ensure that a quantity is not omitted or calculated twice, which is a common error in estimating. A well-organized checklist of work will help reduce the chances of making such an error.

After the quantity takeoff is completed, the direct costs can be calculated. Direct costs consist of material, labor, equipment, and subcontracts. Material cost is calculated by multiplying the quantity of material by the unit cost of the material. The unit cost of material can be obtained from material suppliers. The cost of labor is calculated by dividing the quantity of work by the production rate of laborers, multiplied by the labor rate. Labor rates include base wages plus payroll taxes and insurance. The cost of equipment is calculated by dividing the quantity of work by the production rate of equipment, multiplied by the equipment rate. Equipment rates include both ownership cost and operation expenses. Each individual subcontractor submits the cost of subcontractors' work. Contractors develop and continually update a cost database for their particular type of work. The estimating work process shown in Fig. 3.1 illustrates the feedback loops during construction to record actual costs and productivity rates into the database.

The indirect costs are estimated after the direct costs have been completed. Many of the indirect costs depend on the results of the direct costs. Indirect costs consist of mobilization, field office expenses, insurance, bonds, taxes, final cleanup, and other expenses required to complete the project.

After the direct and indirect costs are completed and summarized, a risk analysis of uncertainties is made to determine an appropriate contingency for the estimate. Contingency is discussed later in this chapter.

ESTIMATE CHECKLISTS

Checklists are valuable tools to reduce the potential of overlooking a cost item. Checklists act as reminders to the estimator by

■ Listing information required to prepare the early estimate.
■ Listing miscellaneous other costs that may be required in the estimate.

■ Listing work items that may be required to build the project in accordance with the plans and specifications.

A listing of information required to prepare an estimate in the process industry may include type of unit, feed capacity, and project location. For a computer-generated estimate in the process industry, the required information includes soil and site data, building requirements, plot plan dimensions, and other specific engineering requirements. For projects in the building sector, a listing of information to prepare an estimate may include type of building, functional use of building, number of occupants, building location, and project location.

Typical examples of miscellaneous cost items may include spare parts, permits, training, or other items. Typical scope items that may be required, but are not identified in the definition provided for the estimate, may include certain utility and auxiliary systems. Checklists are also useful during initial client-customer meetings where they serve as agenda items for discussion. Checklists also assist in preparing an "estimate work plan" by identifying important points to emphasize in the write-up for the execution of the estimate. Table 3.2 is an illustrative example of an early estimate checklist for a project in the process industry. Table 3.3 is an example checklist for a detailed estimate for concrete work.

TABLE 3.2 | Checklist for an early estimate in the process industry.

1. Process unit description (delayed coker, hydrogen plant, etc.)
2. Process licenser
3. Feed capacity
4. Production capacity
5. Product yield
6. Utility levels at process unit location
7. Feedstock specifications
8. Integration of multiple units
9. Process pressure and temperature operating levels
10. Provision for future expansion of capacity
11. Provision for processing multiple or different feedstocks
12. Single-train versus multiple-train concept
13. Project location
14. Miscellaneous costs (spare parts, training, chemicals, etc.)
15. Other items, such as unusually high or low recycle rates

TABLE 3.3 | Checklist for detailed estimate for concrete work.

1. Are anchor bolts included?
2. Is reinforcing steel included?
3. Is insulation of the slab included?
4. Is vapor barrier under the slab included?
5. Is cost of pumping concrete included?
6. Are shop drawings for tilt-up panels included?
7. Is caulking of the tilt-up panels included?
8. Is hardware and rigging included for tilt-up panels?
9. Is the lifting crane included?
10. Are form liners included?
11. Is grouting under the wall panels included?
12. Is waste included in the concrete?
13. Are architectural finishes included?
14. Are concrete curing compounds included?
15. Is the portable compressor included?

DOCUMENTATION OF ESTIMATE

After the estimate is completed, a document should be prepared that defines the basis of the estimate. Estimate documentation is essential for presentation, review, and future use of the estimate. The documentation for an estimate improves communications between the estimating team and management, establishes a mechanism for estimate reviews, and forms a basis for early project cost control. The estimating team should develop a standard format for presentation of the estimate that is easily understood by management.

Inaccurate cost estimates are often the result of omissions in the estimate, miscommunications of project information, or nonaligned assumptions. Documenting the estimate will minimize these inaccuracies by

- Improving communications among all project participants.
- Establishing a mechanism for review of the estimate.
- Forming a solid basis for project controls.

As the estimate is being developed, the act of preparing documentation facilitates communications between the estimating team and management. Estimate documentation improves the outcome of the estimate through

- Sharing information.
- Identifying items that require clarification.
- Helping the estimator obtain and organize information needed for the estimate.
- Avoiding confusion over what is covered, and not covered, by the estimate.
- Providing useful information for future estimates.
- Highlighting weak areas of the estimate.
- Increasing the credibility of the estimate.

A portion of the documentation may be developed by sources other than the estimator. For example, procurement personnel may obtain material quotes, labor information may be obtained from field personnel, etc. However, the estimator has overall responsibility for collecting and organizing this information. Reviewing and clarifying the information with the originator improves the estimate accuracy.

A standard default format or outline should be developed to organize and prepare documentation for the cost estimates. A different standard can be developed for different types of estimates. The process of developing, utilizing, and storing the documentation for future use should be built into the cost estimate work process. The items that should be documented are shown in Table 3.4.

ESTIMATE REVIEWS

Once the estimate is complete, a detailed review should be made of the entire estimate package, including the backup materials, assumptions, unit prices, productivity rates, etc. The estimate should also be checked against the project schedule requirements to ensure they are compatible with such elements as overtime rates assumed during outages, price escalation, etc.

The number of reviews will vary depending on the size of project, type of estimate, length of time allowed for preparing the estimate, and other factors. For any estimate there should be at least two reviews: an internal review during development of the estimate and a final review at or near completion of the estimate.

TABLE 3.4 | Recommended documentation of estimates.

1. Standard format for presenting cost categories (codes)—summary and backup levels
2. Basis of estimate—clear understanding of what constitutes the estimate
3. Level of accuracy—expected for the estimate
4. Basis for contingency—risk analysis, if applicable
5. Boundaries of the estimate—limitations of the estimate
6. Scope of work—the quality and completeness of the plans and specifications
7. Labor rates—breakdown and basis of labor rate
8. Assumed quantities—conceptualized etc.
9. Applied escalation—dates and basis of escalation
10. Work schedule—shifts, overtime, etc. to match the milestones (not contradictory)
11. Other backup information—quotes, supporting data, assumptions
12. Checklists used—a list of completed checklists
13. Description of cost categories—codes used in preparing the estimate
14. Excluded costs—list of items excluded from the cost estimate

About halfway through the development of the estimate, a "reality check" should be scheduled. The purpose of the midpoint check is to avoid spending unnecessary time and money in pursuing an estimate that may be unrealistic or based on assumptions that are no longer valid.

The internal midpoint estimate review is brief. Typically the lead estimator and project manager are involved. This review is intended as a reality check of the data being developed to assess whether to proceed with the estimate. This is a "go–no go" point when the results of the review will guide the estimator and the team in these two steps:

1. Either report to upper management that the costs have gotten outside of the boundaries established as a target for the project, or
2. Give the team the "go ahead" to proceed with the remaining estimate process to complete the estimate.

The final estimate review is a more structured process. The depth of the review depends on the type or class of estimate that is being prepared. The meeting is intended to validate assumptions used in preparing the estimate, such as construction sequence, key supplier selection, and owner's cost.

The final estimate review may be a lengthy meeting. For a final estimate review, the attendees should include the lead estimator, experienced field personnel, and a representative of upper management. To be effective, the final estimate review meeting should be conducted with a written agenda. The meeting should be documented with written minutes that are distributed to all attendees. The estimator must come to the review meeting prepared with this information for comparisons:

- Historical data used in preparing the estimate.
- Actual total installed costs of similar projects.
- Percentage of total installed cost on key cost accounts.

Comparisons of the estimate with this information provide useful indicators for the estimate review. The estimator needs to assess each estimate to determine the appropriate checks that should be included in an estimate review meeting.

In some situations it may be desirable to use outside assistance for estimate reviews. For example, it may be helpful to obtain a review of the estimate by an experienced peer group to validate assumptions, key estimate accounts, construction sequence, potential omissions, etc. In other situations, it may be advantageous to engage a third party to perform an independent review. This will provide a check to compare the estimate with past similar estimates from the perspective of a different team.

Estimate reviews should focus on the big picture and follow Pareto's law, separating the significant few from the trivial many. Generally, an estimate is prepared bottom up, whereas the review is conducted top down. Table 3.5 is an illustrative example of review items for an estimate.

TABLE 3.5 | Review items for an estimate.

1. Size, type, and location of facility
2. Key assumptions used
3. Major undecided alternatives
4. Estimate methodology
5. Historical data used
6. Estimate exclusions
7. Estimator's experience
8. Checklists used to prepare estimate
9. Analysis of risks and uncertainties
10. Amount of applied contingency

The estimating team should develop a standard cost estimate presentation format that includes the level of detail and summary of engineering design, engineered equipment, bulk materials, construction directs and indirects, owner's costs, escalation, taxes, and contingency. Computer methods, including spreadsheets or estimating programs, provide consistent formats for preparing and presenting estimates. Uniform formats provide these benefits:

- Reduce errors in preparing estimates.
- Enhance the ability to compare estimates of similar projects.
- Promote a better understanding of the contents of an estimate.
- Provide an organized system for collecting future cost data.

Presentation of the estimate is important. The estimating team must develop a format that is easily understood by management. Using a standard format for presentations promotes better communication among all participants in the project and a better understanding of what is included in the estimate. This understanding is necessary so good decisions can be made based on the estimate.

RISK ASSESSMENT

Assessing risk and assigning contingency to the base estimate is one of the most important tasks in preparing early estimates. Risk assessment is not the sole responsibility of the estimators. Key members of the project management team must provide input on critical issues that should be addressed by the estimators in assessing risk. Risk assessment requires a participatory approach with involvement of all project stakeholders including the business unit, engineering, construction, and the estimating team.

The owner is responsible for overall project funding and for defining the purpose and intended use of the project. The design organization is responsible for producing the contract documents, the plans and specifications, to construct the project. The estimating team is responsible for preparing an estimate of the probable final cost to construct the project, including direct and indirect costs, and assessing risk and assigning contingency.

RISK ANALYSIS

Typically, risk analysis is a prerequisite to assigning contingency. Based on the acceptable risks and the expected confidence level, a contingency is established for a given estimate. Risk analysis and the resultant amount of contingency help management to determine the level of economic risk involved in pursuing a project. The purpose of risk analysis is to improve the accuracy of the estimate and to instill management's confidence in the estimate.

Numerous publications have been written to define risk analysis techniques. Generally, a formal risk analysis involves either a Monte Carlo simulation or a statistical range analysis. There are also numerous software packages for risk analysis. The lead estimator for a project must assess the uniqueness of each project and select the technique of risk analysis that is deemed most appropriate. For very early estimates, the level of scope definition and the amount of estimate detail may be inadequate for performing a meaningful cost simulation.

CONTINGENCY

Contingency is a real and necessary component of an estimate. Engineering and construction are risk endeavors with many uncertainties, particularly in the early stages of project development. Contingency is assigned based on uncertainty. Contingency may be assigned for many uncertainties, such as pricing, escalation, schedule, omissions, and errors. The practice of including contingency for possible scope expansion is highly dependent on the attitude and culture toward changes, particularly within the business unit.

In simple terms, contingency is the amount of money that should be added to the base estimate to better predict the total installed cost of the project. Contingency can be interpreted as the amount of money that must be added to the base estimate to account for work that is difficult or impossible to identify at the time a base estimate is being prepared. In some owner or contractor organizations, contingency is intended to cover known unknowns. That is, the estimator knows there are additional costs, but the precise amount is unknown. However, sometimes an allowance is assigned for known unknowns and a contingency is assigned for unknown unknowns.

AACE International document 18R-97 defines contingency as "An amount of money or time (or other resources) added to the base estimate to: (a) achieve a specific confidence level; or (b) allow for changes that experience shows will likely be required."

TRADITIONAL METHODS OF ASSIGNING CONTINGENCY

The most effective and meaningful way to perform risk analysis and assign contingency is to involve the project management team. Estimators have insights and can assess imperfections in an estimate to derive an appropriate contingency. However, the interaction and group dynamics of the project management team provide an excellent vehicle to assess the overall project risk. The integration of the project management team's knowledge and the estimator's ability to assign contingency provides management with an appreciation of and confidence in the final estimate. The end result is an estimate that represents the judgment of the project management team, not just the estimator's perspective.

Figure 3.3 illustrates the risk assessment process. The estimator must select the method deemed most appropriate for each project, based on information provided by the project management team and based on the intended use of the estimate by the business unit. The estimator must communicate the method selected, risk, accuracy, and contingency for the estimate.

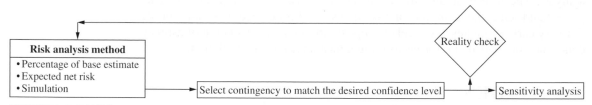

FIGURE 3.3 | Risk assessment process.

Percentage of Base Estimate

For some situations, contingency may be assigned based on personal past experience. A percentage is applied to the base estimate to derive the total contingency. Although this is a simple method, the success depends on extensive experience of the estimator and historical cost information from similar projects. It is less accurate than other more structured methods.

Some organizations apply standard percentages for contingencies based on the class of estimate. Company policy rather than a numerical analysis governs this method. In some situations, contingency is applied as a percentage of major cost items rather than applying a percentage of the total base estimate. Typically, the amount of the percentage is based on the level of scope definition or on the stage of project development.

In some instances, a percentage can be applied to major cost items that make up the estimate, rather than applying a percentage to the total base estimate. The method typically relies on the personal experience and judgment of the estimator, but the percentage can also be from established standard percentages based on historical data. This method has the advantage of considering risk and uncertainty at a lower level than applying contingency on the total base estimate.

The personal experience and judgment of the estimators and engineers should not be overlooked in the process of assigning contingency. Even the most advanced computers are not a substitute for the knowledge and experience of the human mind. Estimators with many years of experience with a particular type of work can often be quite accurate in assigning contingency. They assign contingency based on how they "feel" about the level of uncertainty and risk associated with a project, their knowledge of the cost data used in preparing the estimate, and the thoroughness of the effort in preparing the base estimate.

Expected Net Risk

The estimator may determine contingency based on expected maximum risk and likelihood. After the evaluation of normal contingency of each estimate element, an individual element may also be evaluated for any specific unknowns or potential problems that might occur. The first step involves determining the maximum possible risk for each element, recognizing that it is unlikely that all the risk will occur for all elements.

The next step involves assessing the percentage probability that this risk will occur. The expected net risk then becomes a product of the maximum risk times the probability. The sum of all the expected net risks provides the total maximum risk contingency required. Table 3.6 illustrates an expected net risk analysis.

TABLE 3.6 | Expected net risk analysis.

Estimate item	Base estimate	Maximum cost	Maximum risk	Percentage probability	Expected net risk
1.	$40,000	$50,000	$10,000	20%	$2,000
2.	8,000	12,000	4,000	40%	1,600
3.	100,000	150,000	50,000	30%	15,000
4.	250,000	320,000	70,000	50%	35,000
5.	72,000	95,000	23,000	70%	16,100
6.	237,000	320,000	83,000	60%	49,800
7.	12,000	28,000	16,000	10%	1,600
8.	94,000	135,000	41,000	30%	12,300
9.	730,000	870,000	140,000	40%	56,000
10.	43,000	72,000	29,000	80%	23,200
11.	572,000	640,000	68,000	50%	34,000
12.	85,000	97,000	12,000	20%	2,400
	$2,243,000	$2,789,000	$546,000		$249,000

The risk analysis data in Table 3.6 shows a base estimate of $2,243,000 with a maximum anticipated cost of $2,789,000, a difference of $546,000. However, it is unlikely that all of the risk will occur for each item in the estimate. Therefore, the expected net risk for each bid item is calculated by multiplying the maximum risk by the percentage probability of each item. The total expected net risk for the project is $249,000, which is calculated as the sum of the net risk for each item in the estimate. The $249,000 represents a contingency markup of 11.1% on the base estimate of $2,243,000, calculated as ($249,000/$2,243,000 = 11.1%). For this project the final estimate is calculated as the base estimate plus the total expected net risk, $2,243,000 + $249,000 = $2,492,000.

Simulation

A formal risk analysis for determining contingency is usually based on simulation. A simulation of probabilistic assessment of critical risk elements can be performed to match the desired confidence level. Monte Carlo simulation software packages are useful tools for performing simulation. However, knowledge of statistical modeling and probability theory is required to use these tools properly.

Range estimating is a powerful tool that embraces Monte Carlo simulation to establish contingency. Critical elements are identified that have a significant impact on the base estimate. For many estimates, there are less than 20 critical elements. The range of each critical element is defined and probability analysis is used to form the basis of simulation. Using this method, noncritical elements can be combined into one or a few meaningful elements.

Range estimating is probably the most widely used and accepted method of formal risk analysis. In range estimating the first step requires identification of the critical items in the estimate. The critical items are those cost items that can affect the total cost estimate by a set percentage, for example ± 4 percent. Thus, a relatively small item with an extremely high degree of uncertainty may be critical, whereas a major equipment item for which a firm vendor quote has been obtained would not be considered critical. Typically, no more than 20 critical items are used in the analysis. If more than 20 critical items are identified, the set percentage can be increased to reduce the number of critical items.

Once the critical items are identified, a range and a target are applied to each item. For example, the range may include a minimum value so there is only a 1 percent chance that the cost of the item would fall below that minimum. Similarly, an upper value may be established so there will be only a 1 percent chance of going over that value. The target value represents the anticipated cost for that item. The target value does not need to be the average of the minimum and maximum values. Usually the target value is slightly higher than the average.

After the critical items have been identified and ranged, a Monte Carlo simulation is performed. The Monte Carlo analysis simulates the construction of the project numerous times, as many as 1,000 to 10,000, based on the ranges given to the critical items and the estimated values of the noncritical items.

The results of the simulation are rank ordered, then presented in a cumulative probability graph, commonly called an S curve. The cumulative probability graph typically shows the probability of underrun on the horizontal x axis and either the total project cost or contingency amount on the vertical y axis. The decision maker can then decide the amount of contingency to add based on the amount of risk.

Caution also must be exercised because it is possible to seriously underestimate the cost of a project when using range estimating. There is a risk of understating the true risk of a project due to statistical interdependencies among the critical items in the analysis. Whenever two or more cost items are positively correlated, meaning they increase together or decrease together, the Monte Carlo simulation may cause one to be high and the other low so that they cancel each other out. Thus, the true risk would be understated. Also, underestimating the ranges on the critical items can have a profound impact on the results, also leading to an understatement of the true risks inherent in the design and construction of the project.

When used properly, formal risk analysis using Monte Carlo simulation range estimating can be an extremely valuable tool because it requires a detailed analysis of the components of the estimate, a process that can identify mistakes and poor assumptions. However, precautions must be taken when using simulation methods for early estimates. For many early estimates there is not enough detailed information or an adequate number of cost items for a valid simulation.

Assessing Estimate Sensitivity

The contingency applied to an estimate includes the combined impact of all risk elements. The accuracy of an estimate can be improved by assessing high cost impact factors, increasing the level of scope definition, or a combination of both. A sensitivity analysis can be performed to illustrate how a specific risk element can impact the total estimate.

The sensitivity analysis evaluates the impact of only one risk element at a time. It is frequently used in conjunction with an economic analysis. During the process of determining contingency, the risk elements that can have the maximum impact on the total installed cost are prime candidates for sensitivity analysis. Tables 3.7 and 3.8 show a sample sensitivity analysis for a $3 million base estimate.

TABLE 3.7 I Base estimate summary.

Equipment cost	$1,200,000
Material cost	$600,000
Labor cost	$1,000,000 ◀── (= $50/WH × 20,000 WH)
Subcontractor cost	$200,000
Total base estimate	$3,000,000

TABLE 3.8 | Sensitivity analysis.

Risk element	Percentage change from estimate	New base estimate
Labor rate ($50/WH)	0	$3,000,000
Labor rate	+10	3,100,000
Labor rate	−5	2,950,000
Total work-hours (20,000 WH)	0	$3,000,000
Total work-hours	+15	3,150,000
Total work-hours	−7	2,930,000
Equipment ($1,200 M)	0	$3,000,000
Equipment	+5	3,060,000
Equipment	−5	2,940,000

For any estimate, it is necessary to apply a contingency to the base estimate. The method used for assigning contingency will vary depending on analysis of risk and other factors that can impact the cost of a project. This section has presented traditional methods used for assigning contingency.

ESTIMATE FEEDBACK FOR CONTINUOUS IMPROVEMENT

It is unfortunate, but the perception of many people is that the estimator's involvement with a project is over when the estimate is finished. In reality, it is an advantage to the management of a project, and in the best interest of the customer, for the estimator to remain connected to the project during execution.

The estimator can be an important asset to project management during the execution phase of a project. Involvement of the estimator during project execution enables the estimator to stay in touch with the project and provide an early warning of any potential cost overruns. Including the estimator in the distribution list of monthly project reports can provide input to the project management team to enable them to make good decisions related to costs.

During project execution, the estimator can also be a valuable resource for recasting the cost estimate into work/bid packages and for analyzing actual bids with the recast estimate. The estimator can also assist in management of changes during project execution, by assessing the impact of changes on cost.

No estimating process is complete without the continuous feedback loops shown in Fig. 3.1. Feedback from project execution provides lessons learned to the estimator, which enables the estimating team to modify estimating standards and practices. Feedback from project completion also enables the estimating team to update the database to improve the accuracy of future estimates. Terminating the estimator's involvement when the estimate is finished prevents continuous improvement of the estimating process.

To provide meaningful feedback, the estimator must explore how the cost will be tracked during project execution. An estimate should be prepared with cost breakdowns in a format that enables easy future cost tracking. A standard code of accounts enables an organization to simplify the estimating process, update the database, and facilitate cost control. This benefits both the estimating team and the project management team.

A final project cost report is an extremely valuable document to capture lessons learned for improving estimates. It provides a real feedback to compare with the original cost estimate. Pitfalls for future estimates can be eliminated or minimized. Both the original estimate and the final project cost reports should be maintained at a central location. A cost reference with reports sorted by project location, type, size, etc., can be used to update the cost database for future estimates.

The best source of data for estimators to develop and enhance the estimating tools and techniques is their own organization. There is an abundance of project data that is available from completed projects and definitive estimates. The key to success is the establishment of a mechanism to capture and retrieve this information in a format that can be useful in developing statistical relationships, such as percentage of breakdowns of total installed cost by cost category, total installed cost to equipment cost ratios, construction indirect costs to direct labor cost ratios, etc. When a project is completed, the actual total installed costs can be added to the database. Estimate feedback is an integral part of the estimating process. It is not an add-on feature. A process for providing feedback loops is necessary to improve the accuracy of early estimates.

PROBLEMS

3.1 Prepare a list and briefly describe the important factors that should be included in estimating procedures.

3.2 Describe the purpose of a kickoff meeting for an estimating team. What are typical issues that should be discussed?

3.3 Review the material in Chapter 7 on Earthwork and develop a checklist of items that should be included in a detailed estimate for a grading operation of a highway project.

3.4 Discuss the difference between internal and external reviews of an estimate and describe a situation when an external review might be especially valuable.

3.5 What value can be obtained from documenting an estimate, other than minimizing inaccuracies?

3.6 Review technical journal articles that pertain to risk analysis and risk assessment. Based on your review, summarize methods of assigning contingency to the time and cost of construction estimates.

3.7 Interview three estimators (one working for an owner, one for a designer, and one for a contractor) and identify factors that each estimator believes is important for preparing an accurate estimate.

4

Conceptual Cost Estimating

At the inception of a project by the owner, prior to any design, only limited information is known about a project. However, the owner must know the approximate cost to evaluate the economic feasibility of proceeding with the project. Thus, there is a need to determine the approximate cost of a project during its conceptual phase.

As discussed in Chapter 1, cost estimates can be divided into at least two different types, depending on the purposes for which they are prepared and the amount of information known when the estimate is prepared: approximate estimates (sometimes called preliminary, conceptual, or budget estimates) and detailed estimates (sometimes called final or definitive estimates). Each of these estimates may be subdivided.

Later chapters of this book discuss methods and procedures of estimating the construction costs of engineering projects for which detailed information is known. For example, the cost of excavation is determined for a known quantity and type of soil, size of excavator, and job conditions. The cost of concrete forms is determined based on specific sizes, shapes, and numbers of concrete columns, beams, walls, etc.

This chapter presents methods and procedures for estimating project costs during the preliminary or conceptual phase. Because there is little definition of a project at this stage, the accuracy of the estimate will be less than that of a detailed estimate. However, the conceptual cost estimate is important because the owner will examine this estimate before continuing with development of the project.

ACCURACY OF CONCEPTUAL ESTIMATES

The accuracy of any estimate will depend on the amount of information known about a project. A conceptual cost estimate should be identified by the information from which the estimate was compiled. For example, a conceptual estimate that is prepared from the project scope (sometimes called the *project charter* or *project mission*), when there is little or no design, may be identified as *level I,* in terms of accuracy. *Level I accuracy* can be defined as accurate within +40 and −10 percent. A *level II* conceptual cost estimate can be defined as the estimate that is prepared upon completion of the preliminary design. Such estimates might be accurate within +25 and −5 percent. *Level III* conceptual estimates might be identified as estimates that are compiled upon completion of the final design. A level III conceptual cost estimate might be accurate within +10 and −3 percent.

It is important to classify the conceptual estimate by predefined levels of accuracy similar to those just described. Such classifications provide a measure of reliability to those who must use the information.

LIABILITY OF CONCEPTUAL COST ESTIMATES

Generally the designers are not obligated under standard-form contracts to guarantee the construction cost of a project. However, as a part of their design responsibility, designers prepare an estimate of the probable construction cost for the project for which they have prepared the design.

Major decisions are often made by the owner from information contained in the conceptual cost estimate. This places a responsibility on the estimator. Although the initial estimate may be prepared from little-known information, it is the duty of the estimator to re-estimate the project as additional information about the project becomes available. This is particularly important for a cost-plus construction contract where the designer has the responsibility to monitor costs during construction.

PREPARATION OF CONCEPTUAL ESTIMATES

There are many variations in the contract arrangements among the three principal parties in a project (owner, designer, and contractor). The conceptual estimate is generally prepared by the owner during the owner's feasibility study or by the designer during the design phase. It may be prepared by the contractor for negotiated work between an owner and a contractor.

Multibuilder owners are involved with the construction of projects on a continual basis. Examples are electrical utilities, oil and gas firms, retail stores with nationwide locations, etc. To initiate a project, these firms generally conduct an owner's study. The owner's study consists of a technical feasibility study and an economic feasibility study of the proposed project. A conceptual cost estimate is prepared by the owner as part of the economic feasibility analysis.

Designers prepare conceptual cost estimates throughout the design process. The estimate is used in the selection of design alternatives and to keep the owner informed of forecast costs. It is the responsibility of the designer to develop a design that will produce a completed project within the amount of money authorized in the owner's budget.

Some projects are of an emergency nature and must be completed in the least time possible. For such projects the owner will negotiate a contract with a contractor. The contractor may be asked to prepare a conceptual cost estimate for the project for which there are no, or limited, plans and specifications.

The preparation of conceptual cost estimates requires knowledge and experience with the work required to complete the project. Cost information from previous projects of similar type and size is essential. The estimator must combine all known information with his or her personal experience and use considerable judgment to prepare a reliable conceptual estimate.

PARAMETRIC ESTIMATING

Parametric estimating is commonly used to prepare conceptual cost estimates of building construction projects in the early stages of project development. Parametric estimating relates the total cost of a project to a few physical measurements, or *parameters*. For example, the gross square floor area of a building is a typical overall parameter for a building project. Sometimes parametric estimating is referred to as *square foot* or *cubic foot* estimating, or simply as *unit cost* estimating.

Unit costs for parametric estimating may be derived from internal company records of past similar projects, or from national pricing manuals that are available from several sources, such as *Engineering News-Record* (ENR) and *R. S. Means Cost Data*. The cost data used in parametric estimating should be current with respect to time, and the cost data should be from projects that are similar in type and size as the project that is being estimated. It may be necessary to adjust the unit costs based on time and size as presented in subsequent sections of this chapter.

Figure 4.1 illustrates data for parametric estimating of a building, based on an office building with 10,000 sf of floor space, 2 stories, and 8-ft walls. Some of the parameter *unit costs* are expressed in the gross enclosed square feet of floor area (such as electrical work), while some unit costs are expressed in square feet of the building components (such as square feet of roof), and some unit costs are expressed in the unit of measure of an individual component (such as linear feet of partition wall).

In parametric estimating the unit costs of all components are converted to equivalent costs per square foot of the building. For example, Figure 4.1 shows the unit cost of partition walls as $11.25 per linear foot of wall based on one linear foot of partition wall for each 6 square feet of floor area in the building. The unit cost of the component of partition walls is converted to square feet of building as: (10,000 sf / 6 sf/lf) × ($11.25/lf / 10,000 sf) = $1.88/sf. An alternate calculation is ($11.25/lf of wall) / (6 sf of building / lf of wall) = $1.88/ft. Thus, the $11.25/lf of partition wall is equivalent to $1.88/sf of floor area of the building.

System/Component	Specification	Unit	Unit cost	Cost per sf	% of total
Office building, 10,000 sf, 2-story, 8-ft walls					
Foundation					
Piers and footings	Poured concrete footings	sf ground	4.36	2.18	
Slab on grade	Concrete, grading, and fill	sf slab	6.95	3.48	7.5
Excavation	Excavation and backfill	sf ground	2.75	1.38	
Exterior Closure					
Exterior walls	Concrete block (70% of floor area)	sf wall	12.27	8.59	
Elevated floor	Concrete, precast double tees	sf floor	9.96	4.98	
Roof	Concrete, precast double tees	sf roof	14.55	7.28	25.7
Exterior doors	Wood/glass (1,250 sf ground/door)	each	2,145.00	0.86	
Exterior windows	Aluminum/glass (300 sf flr/window)	each	724.00	2.41	
Interior work					
Partition walls	Studs/gyp board (6 sf flr/lf partition)	lf wall	11.25	1.88	
Interior doors	Hollow core (75 sf floor/door)	each	260.00	3.46	
Wall finishes	Paint interior walls	sf wall	1.25	3.33	16.4
Floor finishes	Carpet	sy floor	37.98	4.11	
Ceiling finishes	Suspended acoustic ceiling	sf ceiling	2.58	2.58	
Mechanical					
Plumbing	Fixtures/feeds (160 sf flr/fixture)	each	1,450.00	9.06	
Heating/cooling	HVAC system	sf floor	14.15	14.15	26.8
Fire protection	Sprinkler system	sf floor	1.98	1.98	
Electrical					
Service/distribution	400 amp, panel board, leads	sf floor	5.47	5.47	
Lighting & wiring	Fixtures, feeds, switches, wiring	sf floor	15.82	15.82	23.6
Special electrical	Alarm system	sf floor	0.83	0.83	
Subtotal of system/components:				93.83	100.0
Add 20% for overhead and profit:				18.77	
Total building cost = $112.60/sf					

FIGURE 4.1 | Unit costs for parametric estimate of a building.

Another example is interior doors, which are measured as "each" with a unit cost of $260/door. Figure 4.1 shows 1 door for 75 square feet of building. The unit cost of interior doors at $260 each is converted to cost per square feet of building as: (10,000 sf / 75 sf/door) × ($260/door / 10,000 sf) = $3.46/sf. An alternate calculation is ($260/door / 75 doors/sf) = $3.46/sf. Thus the unit cost of $260/door is equivalent to $3.46/sf of floor area of the building.

The interior wall finishes is based on painting 2 sides of the linear feet of 8-ft high interior partition walls. The unit cost for this building component is calculated as: (1 lf of wall / 6 sf of building) $\times$ (8-ft $\times$ 2 sides $\times$ $1.25/sf) = $3.33/sf. Thus, the $1.25/sf of wall painting is equivalent to $3.33/sf of floor area of the building.

The parametric estimate is prepared by multiplying the square feet of the proposed building by the cost per square foot of each component. The sum of all components is the estimated cost. This method of estimating is commonly used by the owners during the economic feasibility phase of a project and by architects and engineers during the design phase to evaluate design configurations and options.

With good historical cost records on comparable structures, the parametric estimating method can give reasonable accuracy for conceptual cost estimates. With this method an experienced estimator with access to well-documented records can quickly prepare an estimate and budget that will help in making decisions during the early phases of a project.

BROAD-SCOPE CONCEPTUAL ESTIMATES

A broad-scope conceptual cost estimate of a proposed project is prepared prior to the design of the project. It is prepared from cost information on previously completed projects similar to the proposed project. The number of units, or size, of the project is the only known information, such as the number of square feet of building area, the number of cars in a parking garage, the number of miles of 345-kV transmission lines, the number of barrels of crude oil processed per day, etc.

The best source of information for preparation of conceptual cost estimates is the cost records from previous projects. Although the range of costs will vary among projects, the estimator can develop unit costs to forecast the cost of future projects.

The unit cost should be developed from a weighting of the data that emphasizes the average value, yet it should account for the extreme maximum and minimum values. Equation [4.1] can be used for weighting cost data from previous projects:

$$UC = \frac{A + 4B + C}{6} \qquad \text{[4.1]}$$

where UC = forecast unit cost
          A = minimum unit cost of previous projects
          B = average unit cost of previous projects
          C = maximum unit cost of previous projects

Example 4.1 illustrates the weighting of the cost data from a previous project to determine the forecast unit cost of a proposed project.

EXAMPLE 4.1

Cost information from 10 previously constructed parking garage projects is shown in the table.

Previous Project	Total cost	Number of cars
1	$933,120	150
2	580,609	80
3	1,050,192	120
4	698,345	90
5	518,780	60
6	1,314,412	220
7	583,436	70
8	1,422,828	180
9	1,512,592	175
10	684,700	95

Use the weighted unit cost method to estimate the conceptual cost for a proposed parking garage that is to contain 170 cars.

Previous Project	Unit cost per car
1	$6,220.80
2	7,257.61
3	8,751.60
4	7,759.39
5	8,646.33
6	5,974.60
7	8,334.80
8	7,904.60
9	8,643.38
10	7,207.37
	Total = $76,700.48

Minimum cost per car, A = $5,974.60

Average cost per car, B = $76,700.48/10 = $7,670.05

Maximum cost per car, C = $8,751.60

From Eq. 4.1, $\text{UC} = \dfrac{\$5,974.60 + 4\,(\$7,670.05) + \$8,751.60}{6}$

$\qquad\qquad\quad = \$7,567.73/\text{car}$

The conceptual cost estimate for 170 cars = 170 @ $7,567.73

$\qquad\qquad\qquad\qquad\qquad\qquad\quad = \$1,286,514$

The technique of conceptual cost estimating illustrated in Example 4.1 can be applied to other types of projects, for example, for apartment units, motel rooms, miles of electrical transmission line, barrels of crude oil processed per day, square yards of pavement, etc.

It is necessary for the estimator to adjust the cost information from previously completed projects for use in the preparation of a conceptual cost estimate for a proposed project. There should be an adjustment for time, location, and size.

TIME ADJUSTMENTS FOR CONCEPTUAL ESTIMATES

The use of cost information from a previous project to forecast the cost of a proposed project will not be reliable unless an adjustment is made proportional to the difference in time between the two projects. The adjustment should represent the relative inflation or deflation of costs with respect to time due to factors such as labor rates, material costs, interest rates, etc.

Various organizations publish indices that show the economic trends of the construction industry with respect to time. The *Engineering News-Record* (ENR) quarterly publishes indices of construction costs. An index can be used to adjust previous cost information for use in the preparation of a conceptual cost estimate.

The estimator can use the change in value of an index between any two years to calculate an equivalent compound interest rate. This equivalent interest rate can be used to adjust past cost records in order to forecast future project costs. Example 4.2 illustrates the use of indices for time adjustments.

EXAMPLE 4.2

The cost indices for building construction projects show these economic trends:

Time	Index
4 yrs ago	3684
3 yrs ago	3795
2 yrs ago	3873
1 yr ago	4148
Current year	4362

There is no method to predict future cost trends with absolute certainty. For this example a compound interest is calculated based on the change in the cost index over the 4-yr period:

$$\frac{4362}{3684} = (1 + i)^4$$

$$1.18404 = (1 + i)^4$$

$$i = 4.31\%$$

Suppose that cost information from a $7,843,500 project completed last year is to be used to prepare a conceptual cost estimate for a project proposed for construction 2 years from now. Thus, there are 3 years between the previous project and the proposed project. Using the 4.31% compound interest, the cost of the proposed project can be adjusted for time as follows:

$$\text{Cost} = \$7,843,500 \, (1 + 0.0431)^3$$

$$= \$8,902,003$$

Another approach is to calculate a separate interest for each of the 4 years of cost indices. An average value of the 4 interest rates could then be calculated and used as the interest rate for time adjustments of future projects. Another approach is to use one of the many methods of curve fitting, such as the least-squares method of curve fitting.

ADJUSTMENTS FOR LOCATION

The use of cost information from a previous project to forecast the cost of a proposed project will not be reliable unless an adjustment is made that represents the difference in cost between the locations of the two projects. The adjustment should represent the relative difference in costs of materials, equipment, and labor with respect to the two locations.

Various organizations publish indices that show the relative differences in construction costs with respect to geographic location. The ENR is an example of a publisher of location cost indices. Example 4.3 illustrates the use of indices for location adjustments.

EXAMPLE 4.3

The location indices for construction costs show this information:

Location	Index
City A	1,025
City B	1,170
City C	1,260
City D	1,105
City E	1,240

Suppose that cost information from a $387,200 project completed in city A is to be used to prepare a conceptual cost estimate for construction of a proposed project in city D. The cost of the proposed project should be adjusted for location as

$$\text{Cost} = \frac{1,105}{1,025} \times \$387,200 = \$417,420$$

ADJUSTMENT FOR SIZE

The use of cost information from a previous project to forecast the cost of a future project will not be reliable unless an adjustment is made that represents the difference in size of the two projects. In general, the cost of a project is directly proportional to its size. The adjustment is generally a simple ratio of the size of the proposed project to the size of the previous project from which the cost data are obtained.

COMBINED ADJUSTMENTS

The conceptual cost estimate for a proposed project is prepared from cost records of a project completed at a different time and at a location with a different size. The estimator must adjust the previous cost information for the combination of time, location, and size.

Example 4.4 illustrates the use of combined adjustments for preparation of a conceptual cost estimate.

EXAMPLE 4.4

Use the time and location indices presented in Examples 4.2 and 4.3 to prepare the conceptual cost estimate for a building with 54,500 sf of floor area. The building is to be constructed 3 years from now in City B. A similar type of building that cost $4,757,500 and contained 43,250 sf was completed 2 years ago in City E. Estimate the probable cost of the proposed building.

Proposed cost

$$= \text{Previous cost} \times \text{Time adjustment} \times \text{Location adjustment} \times \text{Size adjustment}$$

$$= \$4,757,500 \times (1 + 0.0431)^5 \times \frac{1,170}{1,240} \times \frac{54,500}{43,250}$$

$$= \$4,757,500 \times 1.23489 \times 0.94355 \times 1.26012$$

$$= \$6,985,281$$

UNIT-COST ADJUSTMENTS

Although the total cost of a project will increase with size, the cost per unit may decrease. For example, the cost of an 1,800-sf house may be $53.50 per square foot whereas the cost of a 2,200-sf house of comparable construction may be only $48.75 per square foot. Certain items, such as the kitchen appliances, garage, etc., are independent of the size of the project.

Size adjustments for a project are unique to the type of project. The estimator must obtain cost records from previous projects and develop appropriate adjustments for his or her particular project. Example 4.5 illustrates a method of size adjustments for preparation of a conceptual cost estimate.

EXAMPLE 4.5

Cost records from previous projects show this information:

Project	Total cost	Size, no. of units
1	$2,250	100
2	1,485	60
3	2,467	120
4	2,730	150
5	3,401	190

The cost per unit can be calculated as shown:

Project	Cost per unit
1	$22.50
2	24.75
3	20.56
4	18.20
5	17.90

A plot of the cost records can be prepared (see Fig. 4.2).

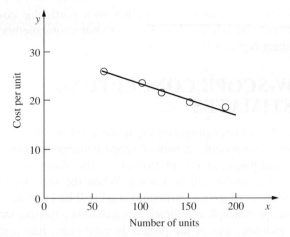

FIGURE 4.2 I Comparison of size and cost per unit.

For a first-order relationship, the general equation for a straight line is

$$Y = b + mX$$

where b = intercept of the line and m = slope of the line. Substituting in values for b and m, we get

$$Y = 24.75 + \left(\frac{17.90 - 24.75}{190 - 60} \right) X$$

$$= 24.75 - 0.05269\,X \qquad \text{where } 60 < X < 190$$

The equation for unit cost with respect to the number of units can be written as:

$$\text{Forecast unit cost} = 24.75 - 0.05269(S - 60)$$

where S = the number of units in the proposed project.

This equation represents the relationship between the unit cost and size of the five previously completed projects. This equation can be used to calculate the cost per unit for future projects whose sizes may range from 60 to 190 units. For example, the unit cost for a 170-unit project would be

$$\text{Unit cost} = 24.75 - 0.05269(S - 60)$$

$$= 24.75 - 0.05269(170 - 60)$$

$$= \$18.95$$

As illustrated in Example 4.5, the adjustment of unit costs based on the size of a project is unique and can be obtained only from previous cost records.

The cost data for some types of projects could be nonlinear, rather than linear as previously illustrated. For example, a second-order equation may better fit the data for some types of projects. The technique presented in Example 4.5 can also be applied for nonlinear data.

The estimator must evaluate his or her own particular cost records and develop a unit cost-size relationship. There are numerous methods of curve fitting, such as linear regression or least squares.

NARROW-SCOPE CONCEPTUAL COST ESTIMATES

As the design of a project progresses and more information becomes known about the various components, a narrow-scope conceptual cost estimate can be prepared. For example, upon completion of the foundation design, the number of cubic yards of concrete will be known. When the structural-steel design is complete, the number of tons of structural steel will be known.

A narrow-scope conceptual estimate is prepared in a manner similar to that for a broad-scope estimate, except the project is subdivided into parameters. For a building construction project the parameters might be square yards of asphalt parking, cubic yards of concrete foundations, tons of structural steel, square feet of finished floor, number of doors, etc. For a steel-pole electric transmission line, the parameters might be acres of clearing land, cubic yards of concrete foundations, tons of steel pole, linear feet of conductor wire, number of insulator strings, etc.

The cost of a proposed project is prepared from historical cost records of previous projects, with an appropriate adjustment for time, size, and location, as already discussed.

FACTORS AFFECTING COST RECORDS

The estimator must be cautious when using historical cost records from completed projects. A proposed project may have features significantly different from those of the completed project from which the cost records are obtained. For example, the cost per square foot for a building with a high ratio of perimeter to floor area will be significantly higher than for a building with a low ratio of perimeter to floor area. Other factors that could affect costs are span lengths, height between floors, quality of furnishings, quality of work, etc.

The estimator must compare the features of the proposed project with those of previous projects and make appropriate adjustments.

CONCEPTUAL COSTS FOR PROCESS INDUSTRY

Selection of the methods for preparing early estimates depends on the level of scope definition, time allowed to prepare the estimate, desired level of accuracy, and the intended use of the estimate. For projects in the process industry, the commonly used methods include:

■ Cost capacity curve
■ Capacity ratios, raised to an exponent

- Plant cost per unit of production
- Equipment factored estimates
- Computer-generated estimates

Cost Capacity Curves

A cost capacity curve is simply a graph that plots cost on the vertical axis and capacity on the horizontal axis. Estimates generated by this method are sometimes called curve estimates. These curves are developed for a variety of individual process units, systems, and services. The minimum information needed to prepare an estimate by cost capacity curves is the type of unit and capacity. For example, the type of unit may be a crude unit or alkaline unit in a refinery. Examples of capacity are barrels of crude oil per day or cubic feet of gas per hour. Additional information that can enhance the quality of a curve estimate includes adjustments for design pressure, project location, project schedule, etc.

Cost capacity curves are normally prepared by a conceptual estimating specialist who develops, maintains, and updates those costs vs. capacity curves on a regular basis. These curves are developed and updated utilizing return cost data from completed jobs. This information is normalized to a location, such as U.S. Gulf Coast, and for a particular time frame expressed as a baseline, such as December of a particular year.

The estimated cost is determined by locating the capacity on the horizontal x axis, then following a straight line up to the point of intersection with the curve. The estimated cost is then read from the vertical y axis by a straight line from the x axis point of intersection with the curve to the y axis. The total installed cost derived from the curve may be adjusted for escalation to present day or future and may be further adjusted to reflect other geographic locations. Example 4.6 illustrates the procedure for preparing a cost capacity curve estimate.

Cost capacity curve estimates are used extensively by owners and engineering-construction companies to prepare early estimates. Each company develops the curves for their particular type of work. For example, a set of curves can be developed to estimate the cost of linear feet of piping, tanks, vessels, and all types of construction. These companies also develop curves for estimating the work-hours of labor.

Early cost estimates can be prepared with minimum effort and in a short amount of time, which is helpful to owners in preparing budgets for capital expenditures and to engineering-construction contractors for negotiating contracts. Since the accuracy of curve estimates is not very high, a reasonable amount of contingency should be applied.

EXAMPLE 4.6

Cost capacity curves for process units in a chemical plant are shown in the figure. What is the estimated cost for a project that has a Process Unit C with a capacity of 3,000 barrels per day?

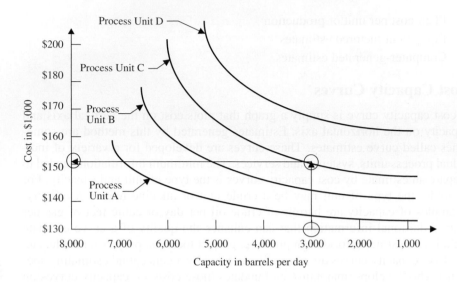

Locating the 3,000 barrels per day capacity along the abscissa, draw a vertical line upward until it intersects the Process Unit C cost curve. Then, draw a horizontal line to the left to read the estimated cost as $151,000.

Capacity Ratios Raised to an Exponent

Capacity ratios raised to an exponent is another estimating technique for conceptual estimating in the process industry. This method takes into account the effect of economy of scale on the total installed cost. For example, if the cost of Process Unit B with capacity B is known, then the estimated cost of Process Unit A is calculated by multiplying the cost of Process Unit B times the ratio of the process unit capacities raised to an exponent (X), as shown in the equation:

$$\text{Cost of Process Unit A} = (\text{Cost of Process Unit B})$$

$$\times \left[\frac{\text{Capacity of Process Unit A}}{\text{Capacity of Process Unit B}} \right]^X \qquad \textbf{[4.2]}$$

Essentially, this method is a mathematical solution to the cost capacity curves, which is a graphical technique. The exponent represented by X is mathematically derived from historical records from completed projects. It represents the nonlinear relationship between cost and size, based on economies of scale.

Historical data can be captured from completed projects and a least-squares fit of the data, or other methods of curve fitting, can be used to determine an appropriate value of X for similar types of projects. Thus, the exponent distinguishes the curve of one process unit from another. Typically the range of the exponent X is between 0.55 and 0.88, depending on the type of process unit. When utilizing this equation to develop a cost estimate, if the exponent for the particular process unit is unknown, an exponent of 0.6 is used, which represents a standard or typical exponent for process plants.

EXAMPLE 4.7

The cost of a 320 cubic feet per hour (cf/hr) process unit is $1,280,000. From historical cost records, the capacity ratio exponent of a process unit is 0.72. Estimate the cost of a similar process unit with a capacity of 450 cf/hr.

$$\text{Cost of Process Unit A} = (\text{Cost of Process Unit B})$$
$$\times \left[\frac{\text{Capacity of Process Unit A}}{\text{Capacity of Process Unit B}} \right]^{x} \quad \text{[4.2]}$$

$$\text{Cost of Process Unit A} = (\$1,280,000) \times \left[\frac{450 \text{ cf/hr}}{320 \text{ cf/hr}} \right]^{0.72}$$

$$= (\$1,280,000) \times (1,2782183)$$

$$= \$1,636,119$$

Plant Cost per Unit of Production

This conceptual estimating method is used to estimate the total plant cost based on the average plant costs per unit of production based on previously completed projects. This method is a very simple and approximate estimating technique where the only information available is the product description and the plant capacity. For example, cost records may show that the average cost per unit for co-generation facilities is $1,000 per kilowatt ($1,000/kW) of production. Thus, for a future 300-megawatt (300 MW) co-generation facility, the estimated cost would be calculated by multiplying the $1,000/kW times the 300 MW of power to drive a total estimated cost of $300,000,000.

This estimating technique assumes that the relationship between plant cost and production capacity is linear and, therefore, would apply best within a fairly narrow range. Ideally, average plant costs per unit of production capacity are best developed over various capacity ranges so that the estimator could select the relationship that is applicable for his or her estimate.

This method of preparing early estimates is similar to the square foot estimating method used for projects in the building sector. For building projects, the total estimated cost of a particular building project is determined by multiplying the average cost per square foot of previous projects by the total square feet in the proposed building.

Equipment-Factored Estimates

For the process industry, equipment-factored estimates are derived by applying various factoring techniques to estimated equipment costs. The factors used are developed and updated utilizing return cost data from completed projects.

This information is normalized to a location, usually the U.S. Gulf Coast, and a base timeline such as December of a particular year. The estimated total installed cost of a normalized unit is defined to include these costs:

- Direct equipment costs
- Direct bulk material costs
- Subcontract costs
- Construction labor costs
- Construction indirect costs
- Home office services costs

One example of the factoring technique is the "equipment cost" to "total installed cost" factor. This factoring technique is relatively simple for projects where equipment costs have been estimated. As the name implies, "total installed cost" factors are developed by dividing the equipment costs of a particular process unit into the total installed cost of that unit. The estimated cost of the project is determined by multiplying the "equipment costs" by the "total installed cost" factor, or multiplier. The factors for process plants generally range between 2.5 and 6.0, depending on the nature of the process unit. Conditions that affect the equipment to "total installed cost" factors are

- Equipment sizes
- Pressure
- Metallurgy
- Degree of prefabrication
- Site conditions
- Equipment costs
- Special conditions (large structures, pits, buildings, etc.)
- Explanation of engineering costs included

Another equipment-factored estimating technique develops equipment costs manually or by utilizing commercially available computer software systems. Bulk material costs are factored from the estimated equipment costs, using historical cost data for the same or similar type units. Field labor work-hours are estimated for each individual equipment item and bulk material installation work-hours are ratioed from the bulk material costs by individual bulk material category. The resultant field labor work-hours are adjusted for productivity and labor costs by applying local labor rates to the estimated construction work-hours. Construction indirect costs are developed for the major categories by percentages of direct labor costs. Home office costs are estimated as a percentage of the total installed cost. The equipment factored estimating techniques described can be utilized when there is sufficient technical definition available, consisting of

- Process flow diagram
- Equipment list
- Equipment specifications

- Project location
- General site conditions (assumed if not specified)
- Construction labor information
- Project schedule

EXAMPLE 4.8

The estimating department of a process industry company has developed the equipment factors from cost records of previously completed projects as shown.

Equipment	Factor
Condensers	2.4
Control instruments	4.1
Compressors	2.5
Fans	2.7
Furnaces	2.0
Generators	1.7
Heat exchangers	4.5
Motors	1.8
Pumps	5.3
Reactors	4.0
Tower vessels	3.5
Tanks	2.7

Engineering design has progressed such that the size and specifications of major equipment have been determined to the level of detail that price quotes can be obtained from the manufacturers of the equipment. Use the listed equipment factors to estimate the cost of the project based on the manufacturer's quotes given next.

Item	Equipment quote from manufacturer		Factor		Plant cost
Condensers	$15,000	×	2.4	=	$36,000
Control instruments	$22,000	×	4.1	=	$90,200
Compressors	$85,000	×	2.5	=	$212,500
Fans	$15,000	×	2.7	=	$40,500
Furnaces	$140,000	×	2.0	=	$280,000
Generators	$25,000	×	1.7	=	$42,500
Heat exchangers	$95,000	×	4.5	=	$427,500
Motors	$55,000	×	1.8	=	$99,000
Pumps	$18,000	×	5.3	=	$95,400
Reactors	$120,000	×	4.0	=	$480,000
Tower vessels	$325,000	×	3.5	=	$1,137,500
Tanks	$140,000	×	2.7	=	$378,000
Total	$1,055,000				$3,319,100

For this project, the delivered equipment cost is $1,055,000, but the estimated cost of the equipment installed in the plant is $3,319,100.

Computer-Generated Estimates

Numerous commercial computer software systems exist for estimating capital costs for different types of industries, including the process industry, building construction industry, and the heavy/highway infrastructure industries. These systems can be simple or very sophisticated. Most of the software packages can operate on a personal computer and are furnished with a cost database, which is updated on an annual basis. The more flexible systems enable the purchaser to customize the database.

These computer software packages are available to assist the estimator in generating conceptual estimates. They can also be used for detailed material quantities as well as equipment and material costs, construction work-hours and costs, indirect field costs, and engineering work-hours and costs. The detailed quantity and cost output enables early project control, which is essential in the preliminary phases of a project, before any detail engineering has started. The accuracy of an estimate can be improved because some systems allow vendor costs, takeoff quantities, project specifications, site conditions, etc., to be introduced into the program. To maximize benefits of these software programs, the use of system defaults should be minimized and replaced with definitions as follows:

- Specifications, standards, basic practices, and procurement philosophy
- Engineering policies
- Preliminary plot plans (if available) and information relating to pipe-rack, structures, buildings, automation, and control philosophy, etc.
- Adequate scope definition
- Site and soil conditions
- Local labor conditions relating to cost, productivity, and indirect costs
- Subcontract philosophy

To become proficient in using computer software programs, frequent usage is required and the user should compare the computer-generated results with other estimating techniques to determine the limitations and shortcomings of the programs. Once the shortcomings are known, corrective action to eliminate or minimize the shortcomings can be taken. To achieve maximum benefits from computer software, the estimator should

- Index or benchmark the unit costs and installation work-hours in the computer software's databases to match the company's cost databases.
- Establish system defaults that correspond to the company's engineering and design standards.
- Create a program that allows conversion of the output of the software programs to the company's account codes and format.

Confidence in the output of estimating software systems will improve by adopting these recommendations. This will result in more consistency and reliability of computer-generated estimates.

There are other noncommercial computer software systems that are used in preparing early estimates, in particular spreadsheet programs. Many owner companies and contractors have developed in-house spreadsheet programs for estimating costs based on their company operations and experience dealing with their particular type of work.

Chapter 22 presents a detailed discussion of computer estimating.

PROBLEMS

4.1   Calculate the weighted unit cost per square foot for the project data shown, and determine the cost of a 30,000-sf project.

Project	Total cost	Size, sf
1	$3,036,400	26,400
2	3,129,700	29,800
3	2,580,300	21,500
4	2,287,500	18,300
5	2,743,200	23,450
6	3,065,300	32,350
7	4,503,600	41,700

4.2   Determine the relationship between unit cost and size for the project data shown in Problem 4.1, to estimate the cost of a 25,000-sf project.

4.3   Use the time and location indices presented in this chapter to estimate the cost of a building that contains 48,000 sf of floor area. The building is to be constructed 3 years from now in City C. The cost of a similar type of building that contained 32,000 sf was completed 2 years ago in City B for a cost of $3,680,000.

4.4   From the perspective of the contractor, give examples of problems that may arise if an early estimate is significantly lower than the final actual cost of a project.

4.5   Early estimates are important to the designer. From the perspective of the designer, give examples of problems that may arise if an early estimate is significantly lower than the final cost of a project.

4.6   Early estimates are extremely important to the owner. From the perspective of the owner, give examples of problems that may arise if an early estimate is significantly lower than the final actual cost of a project.

4.7   The cost of a 650 cf/hr process unit is $9,245,000. From historical cost records the capacity ratio exponent of the process unit is 0.6. Use the capacity ratios raised to an exponent method to estimate the cost of a similar process unit with a capacity of 750 cf/hr.

4.8   Use the cost capacity curves in Example 4.6 of this book to estimate the cost of a Process Unit B with a capacity of 5,000 barrels per day.

5

Cost of Construction Labor and Equipment

CONSTRUCTION LABOR

Construction laborers influence every part of a project. They operate equipment, fabricate and install materials, and make decisions that have a major effect on the project. Most individuals involved in construction will readily agree that people are the most important resource on a project. The cost to hire a laborer includes the straight-time wage plus any overtime pay, workers' compensation insurance, social security, unemployment compensation tax, public liability and property damage, and any fringe benefits.

SOURCES OF LABOR RATES

Wage rates vary considerably with the locations of projects and with the various types of crafts. The hourly rate of construction laborers is determined by one of three means: union wage, open-shop wage, or prevailing wage. Construction workers who are members of a labor union are paid a wage rate established by a labor contract between their local union and the construction contractor's management. Union wage rates usually include fringe benefits, which are paid directly to the union. Construction workers who are not members of a union are paid an open-shop wage agreed to by each individual employee and the employer. For construction employees who work on projects funded with state or federal money, their wage rate is established by the prevailing wage at the project location. The federal government and many states have a government-established prevailing wage for each construction craft. The prevailing wage rate is determined for each craft by a wage survey for each geographic location.

TABLE 5.1 | Representative base wage rates in the United States for 2012.

Craft	Rate, $/hr
Bricklayer	29.33
Carpenter	30.18
Cement mason	28.09
Cement finisher	26.24
Crane operator	30.34
Electrician	32.89
Equipment operator	30.84
Glazier	28.70
Insulation worker	30.68
Ironworker	27.36
Laborer	21.78
Painter	26.60
Pipefitter	29.17
Plumber	30.51
Roofer	26.74
Sheet metal workers	31.88
Truck driver	25.44

COST OF LABOR

The rates listed in Table 5.1 are representative base rates. In addition to paying the base rate, an employer must pay or contribute amounts for such items as social security tax, unemployment tax, workers' compensation insurance, public liability and property damage insurance, and any fringe benefits. Fringe benefits include such items as apprenticeship plans, pension plans, and health and welfare insurance. Base rates normally apply to work done during the 40-hr workweek, 8 hr/day and 5 days/week. For work in excess of 8 hr/day or 40 hr/week, the base rate is generally increased to $1\frac{1}{2}$ or 2 times the base rate. The base rates in Table 5.1 are used to determine the costs in the examples in this book. The effect of fringe benefits, taxes, and insurance is not included in the rates.

SOCIAL SECURITY TAX

The federal government requires an employer to pay a tax for the purpose of providing retirement benefits to persons who become eligible. Currently, the employer must pay 7.65 percent of the gross earnings of an employee, up to $110,100 per year. The employee contributes an equal amount through the employer. This rate is subject to change by Congress.

UNEMPLOYMENT COMPENSATION TAX

This tax, which is collected by the states, is for the purpose of providing funds with which to compensate workers during periods of unemployment. The base cost of this tax is usually 3 percent of the wages paid to the employees, all of which is paid by the contractor. This rate may be reduced by establishing a high degree of employment stability, with few layoffs, during a specified period.

WORKERS' COMPENSATION AND EMPLOYER'S LIABILITY INSURANCE

Most states require contractors to carry workers' compensation and employer's liability insurance as a protection to the workers on a project. In the event of an injury to or death of an employee working on the project, the insurance carrier will provide financial assistance to the injured person or to his or her family. Although the extent of financial benefits varies within several states, in general they cover reasonable medical expenses plus the payment of reduced wages during the period of injury. Each state that requires this coverage has jurisdiction, through a designated agency, over the insurance to the extent of specifying the minimum amounts to be carried, the extent of the benefits, and the premium rates paid by the employer.

The base or manual rates for workers' compensation insurance vary considerably among states, and within a state they vary according to the classification of work performed by an employee. A higher premium rate is charged for work that subjects workers to a greater risk of injury. A contractor who establishes a low record of accidents on jobs for a specified period will be granted a credit, which will reduce the cost of the insurance. A contractor who establishes a high record of accidents over a period will be required to pay a rate higher than the base rate, thus increasing the cost of the insurance.

The premium rate for this insurance is specified to be a designated amount for each $100.00 of wages paid under each classification of work. The rate normally varies from about $10.00 per $100.00 of wages paid for low-risk crafts to approximately $30.00 per $100.00 of wages paid for high-risk crafts. To determine the cost of this insurance for a given project, it is necessary to estimate the amount of wages that will be paid under each classification of work and then to apply the appropriate rate to each wage classification. Since the base rates are subject to changes, an estimator should verify them before preparing an estimate.

PUBLIC LIABILITY AND PROPERTY DAMAGE INSURANCE

This insurance protects the contractor against injuries to the general public or damage to public property due to actions of the employee while performing work during construction. The cost for this insurance is specified as a rate for each $100.00 of base wages. The rate can vary from $2.00 to $5.00 per $100.00 of base wages, depending on the craft and the safety record of the contractor. Due to the large variations in the premium rates for this insurance, it is necessary for the estimator to obtain the rate from an insurance company before estimating the cost of this insurance.

FRINGE BENEFITS

As a part of the agreement of employment, the contractor often agrees to pay benefits for the employee. Examples are health insurance, pension plan, training programs, paid holidays, and vacations. The cost of these fringe benefits will depend on the number of different coverages and the amount of coverage. Generally the costs range from 10 to 20 percent of base wage.

Example 5.1 illustrates the costs that a contractor must incur to hire an employee. For estimating and bidding purposes, the estimator normally determines the total cost of all labor, using the base wages. The final cost of labor is then determined by multiplying the base-wage costs by a percentage to account for taxes and insurance.

EXAMPLE 5.1

An earthmoving equipment operator works straight time, 8 hr/day and 5 days/week, with a base wage of $30.00/hr. The social security tax is 7.65 percent and unemployment tax is 2 percent of actual wages. The rate for workers' compensation insurance is $7.50 per $100 of base wage. Public liability and property damage insurance rate is $2.25 per $100 of base wages. Fringe benefits are $2.50/hr. Calculate the average hourly cost to hire the equipment operator.

Cost to hire the equipment operator will be:

Base wage of operator	= $30.00
Social Security tax, 7.65% × $30.00/hr	= 2.29
Unemployment tax, 2% × $30.00/hr	= 0.60
Workers' compensation, $7.50/$100.00 × $30.00	= 2.25
Public liability/property damage, $2.25/$100.00 × $30.00	= 0.67
Fringe benefits, $2.50/hr	= 2.50
Average hourly cost	= $38.31/hr

Daily cost, 8 hr/day @ $38.31/hr = $306.48/day
Weekly cost, 40 hr/week @ $38.31/hr = $1,532.40/week
Monthly cost, 40 hr/week × (52 weeks/12 months) @ $38.31/hr = $6,640.40/month
Yearly cost, 40 hr/week × 52 weeks/year × $38.31/hr = $79,684.80/year

Production unit cost of earthwork:

Suppose the earthmoving equipment costs $85/hr and can move 400 cy of earth per day. The cost of the earthmoving operation will be:

Daily cost of equipment and operator:

Equipment cost, 8 hr/day @ $85.00/hr	= $680.00/day
Operator cost, 8 hr/day @ $38.31/hr	= 306.48/day
Total	= $986.48/day

Production unit cost = $986.48/day/400 cy/day
                     = $2.46/cy

The normal work week for an employee is usually considered as 40 hours per week. The 40 hours consists of working 8 hours per day for 5 days, Monday through Friday. The rate of pay for a worker during the normal work week is called straight-time pay. For all hours over 40 hours per week the worker is generally paid an overtime rate, which is established by an agreement between the worker and his/her employer, or it is established by an agreement between the worker's union and management of the company. For example, the overtime rate my be one and one half times the straight-time pay rate for all hours over 8 hr/day during the Monday through Friday weekdays, double time for all hours on Saturday, and double time and one half for all work on Sunday. Example 5.2 illustrates the cost of labor for overtime work. For overtime work taxes are paid on actual wage and insurance is paid on base wage as illustrated in Example 5.2.

EXAMPLE 5.2

A concrete worker works 10 hr/day, 6 days/week. A base wage of $26.00/hr is paid for all straight-time work, 8 hr/day, 5 days/week. The overtime rate is time and a half for all hours over 8 hr/day on Monday through Friday and double time for all hours worked on Saturdays. The social security tax is 7.65 percent and unemployment tax is 2 percent of actual wages. The rate for workers' compensation insurance is $5.50 per $100 of base wage. Public liability and property damage insurance rate is $2.25 per $100 of base wages. Fringe benefits are $2.50/hr. Calculate the average hourly cost of the crew and the labor cost per sf of concrete slab.

Pay hours and actual hours:

Actual hours = 10 hr/day × 6 days/week = 60 hr

Pay hours = weekly straight-time + weekly overtime
+ Saturday overtime
= (5 days × 8 hr/day @ 1.0) + (5 days × 2 hr/day @ 1.5)
+ (10 hr × 1 day @ 2.0)
= 40 hr + 15 hr + 20 hr
= 75 hrs

Base wage and average hourly pay:

Base wage = $26.00/hr

Average hourly pay = (pay hours/actual hours) × base wage
= (75 hr/60 hr) × $26.00/hr
= $32.50/hr

Cost to hire the concrete worker will be:

Taxes are paid on actual wage and insurance is paid on base wage.

Average hourly pay, 75/60 × $26.00/hr	= $32.50
Social Security tax, 7.65% × $32.50/hr	= 2.49
Unemployment tax, 2% × $32.50/hr	= 0.65
Workers' compensation, $5.50/$100.00 × $26.00	= 1.43
Public liability/property damage, $2.25/$100.00 × $26.00	= 0.59
Fringe benefits, $2.50/hr	= 2.50
Average hourly cost	= $40.16/hr

Daily cost, 10 hr/day @ $40.16/hr = $401.60/day

Production unit cost:

Suppose there are 5 concrete workers in a crew and the crew can place and finish 2,400 sf of flat slab concrete per day. The cost of labor per sf of concrete in place can be calculated as:

Crew daily cost, 5 concrete workers $\times$ $406.60/day = $2,008.00/day

Crew production rate = 2,400 sf/day

Cost per unit installed = (Crew unit cost)/(Crew production rate)

$$= \$2,008.00/day/2,400 \ sf/day$$

$$= \$0.84/sf$$

PRODUCTION RATES FOR LABOR

A *production rate* is defined as the number of units of work produced by a person in a specified time, usually an hour or a day. Production rates may also specify the time in labor-hours or labor-days required to produce a specified number of units of work, such as 12 labor-hours to lay 1,000 bricks. This book uses hours as the unit of time. Production rates should be realistic to the extent of including an allowance for the fact that a person usually will not work 60 min/hr.

The time that a laborer will consume in performing a unit of work will vary between laborers and between projects and with climatic conditions, job supervision, complexities of the operation, and other factors. It requires more time to fabricate and erect lumber forms for concrete stairs than for concrete foundation walls. An estimator must analyze each operation to determine the probable time required for the operation.

Information on the rates at which work has been performed on similar projects is very helpful. Such data can be obtained by keeping accurate records of the production of labor on projects as construction progresses. For the information to be most valuable to an estimator, an accurate record showing the number of units of work completed, the number of laborers employed, by classification, the time required to complete the work, and a description of job conditions, climatic conditions, and any other conditions or factors that might affect the production of labor should be submitted with each production report. The reports should be for relatively short periods, such as a day or week, so that the conditions described will accurately represent the true conditions for the given period. Reports covering a complete project, lasting for several months, will give average production rates but will fail to indicate varying rates resulting from changes in working conditions. It is not sufficiently accurate for an estimator to know that a bricklayer laid an average of 800 bricks per day on a project. The estimator should know the rate at which each type of brick was laid under different working conditions, considering the climatic and any other factors that might have affected production rates. All experienced construction workers know that the production of labor is usually low during the early stages of construction. As the organization becomes more efficient, the production rates will improve; then as the construction enters the final stages, there will usually be a reduction in the production rates. This is important to an estimator. For a small job it is possible that labor will never reach its most efficient rate of production because there will not be sufficient time. If a job is of such a type that laborers must frequently be transferred from one operation to another or if there are frequent interruptions, then the production rates will be lower than when the laborers remain on one operation for long periods without interruptions.

In this book numerous tables give the rates at which laborers should perform various operations. These rates include an adjustment for nonproductive time, by assuming that a person will actually work about 45 to 50 min/hr. Conditions on some projects may justify a further adjustment in the rates. The frequent use of a range in rates instead of a single rate will enable the estimator to select the rate that she or he believes is most appropriate for the project.

EXAMPLE 5.3

This example illustrates a method of determining the probable rate of placing reinforcing steel for a given project.

Steel bars are to be used to reinforce a concrete slab 57 ft wide and 70 ft long. The reinforcing steel will be #4 bars, $\frac{1}{2}$-in. diameter, with no bends, maximum length limited to 20 ft, and spaced 12 in. apart both ways. All laps will be 18 in. Precast concrete blocks, spaced not over 6 ft apart each way, will be used to support the reinforcing. The bars will be tied at each intersection by bar ties. The steel will be stored in orderly stock piles, according to length, about 80-ft average distance from the center of the slab. The slab will be constructed on the ground.

Quantity of work:

The length of the bars parallel to the 57-ft side will be

Length of the side	=	57 ft
Length of laps: 2 × 18 in.	=	3 ft
Total length of bars per row =		60 ft
Use 3 bars 20 ft long	=	60 ft
Total number of bars required: 3 × 70 =		210
Total length of the bars: 210 × 20 ft	=	4,200 ft

The length of the bars parallel to the 70-ft side will be

Length of side	=	70 ft
Length of laps: 3 × 18 in.	=	4 ft 6 in.
Total length of bars per row	=	74 ft 6 in.
Use 4 bars 18 ft $7\frac{1}{2}$ in. long	=	74 ft 6 in.
Total number of bars required: 4 × 57 =		228
Total length of the bars: 228 × 18 ft $7\frac{1}{2}$ in. =		4,246 ft

The weight of the reinforcing will be

20-ft bars: 4,200 ft @ 0.668 lb/ft	=	2,806 lb
18-ft $7\frac{1}{2}$-in. bars: 4,246 ft @ 0.668 lb/ft	=	2,836 lb
	Total weight =	5,642 lb

The number of intersections of bars will be 57 × 70 = 3,990

The time required to place the reinforcing, using 2 steel setters, should be about

Cycle times:

Carrying bars to slab site:

Time to transport bars, 160 ft @ 100 ft/min	= 1.6 min
Add time to pick up and put down reinforcing bars	= 1.0 min
Time for round trip	= 2.6 min

Assume that 2 persons can carry 6 bars, weighing approximately 80 lb each trip

Number of trips required: 438 bars ÷ 6 bars per trip	= 73 trips
Total time to carry reinforcing: 73 trips × 2.6 min/trip ÷ 60 min/hr = 3.17 hr

Placing the bars on blocks and spacing them:

Assume that 2 persons working together can place 2 bars/min, or 120 bars/hr

Time to place, 438 bars ÷ 120 bars/hr	= 3.67 hr

Tying the bars at intersections:

Assume that a person can make 5 ties per min

2 people will make 2 × 5 × 60 = 600 ties per hr

Time to tie reinforcing, 3,990 ties ÷ 600 per hr	= 6.65 hr

Total time:

The total working time will be

Carrying the reinforcing	= 3.17 hr
Placing the reinforcing	= 3.67 hr
Tying the reinforcing	= 6.65 hr
Total working time	= 13.49 hr

On a project a worker will seldom work more than 45 to 50 min/hr, because of necessary delays. Based on a 45-min hour, the total clock time to handle and place the reinforcing will be

$$13.49 \text{ hr} \times \frac{60}{45} = 18.0 \text{ hr}$$

Total labor-hours for the job: 2 laborers × 18 hr	=	36 hr
No. of tons placed: 5,642 ÷ 2,000	=	2.821 tons
Labor-hours per ton placed: 36 ÷ 2.821	=	12.76 hr/ton
Production rate = 5,642 lb/36 hr	=	157 lb/hr, or 1,253 lb/day

If the reinforcing steel is bent to furnish negative reinforcing over the beams or if it has hooks on the ends and if, in addition, it must be hoisted to the second floor of a building, then it will require extra time to hoist and place it. An estimator should adjust the production rate accordingly instead of using flat rates for all projects.

CONSTRUCTION EQUIPMENT

All projects involve the use of construction equipment to some extent. Equipment may be small power operated tools to large earthmoving equipment and lifting cranes. The equipment can be rented, leased, or purchased. The choice usually depends on the amount of time the equipment will be used in the contractor's operations. Any estimate must include the cost of equipment used on the project, whether it is purchased, rented, or leased.

SOURCES OF EQUIPMENT

If extensive use of the equipment is required by the contractor, the equipment is often purchased, which represents a capital investment by the contractor. The contractor must recover sufficient money to pay the ownership and operating costs of the equipment during its useful life, and at the same time make a profit on using the equipment.

If the equipment is used a short amount of time, it is typically rented from a company that specializes in rental equipment. Generators, compressors, small lifting equipment, and skid loaders are typical examples of rental equipment. Generally the equipment is rented on a daily (8 hr), weekly (40 hr), or monthly (176 hr) basis. Depending on the rental contract the repair costs may be the responsibility of the rental company or the contractor. The contractor is usually responsible for fuel, oil, and lubrication costs. The cost of rental equipment can easily be obtained from rental companies.

Leasing of equipment is similar to rental, except the lease agreement is for long-term use of the equipment. The leasing company owns the equipment and the contractor pays the leasing company to use the equipment. Usually the lease cost does not include fuel, oil, and lubricants. Depending on the lease agreement the contractor may have three options at the end of a lease: (1) buy the machine at fair market value, (2) renew the lease, or (3) return the equipment to the leasing company. Sometimes the lease is a lease-purchase contract, whereby the contractor has the option to purchase the equipment at a later date, with the provision that all or part of the money paid for the lease shall apply toward the purchase of the equipment.

RENTING VERSUS OWNING EQUIPMENT

One of the major factors affecting the cost of equipment (whether rented or owned) is the number of hours the equipment is used. For example, if the total cost of a unit of equipment is $150,000 and it is used 2,000 hours per year over its useful life of 5 years, the hourly cost is $15/hr. However, if the equipment is only used 1,200 hours per year over its 5-year life, the cost is $25/hr. The difference between $15/hr and $25/hr is significant.

Similarly, rental companies typically rent equipment on a monthly (176 hr), weekly (40 hr), or daily (8 hr) basis. For example, a unit of equipment may be rented for $3,000 per month, $1,000 per week, or $400 per day. The hourly costs for these three rental arrangements are $17/hr, $25/hr, and $50/hr respectively.

EXAMPLE 5.4

A contractor owns a unit of equipment that costs $20/hr, based on a usage of 1,800 hr/yr. For this equipment the annual cost of ownership can be calculated as $20/hr × 1,800 hr = $36,000/yr. Suppose this equipment can be rented for $4,000/month, $1,500/week, or $500/day. Evaluate the number of hours of usage to justify rental versus ownership.

The hourly costs of rental:

Monthly rental = $4,000/176 hr = $22.73/hr

Weekly rental = $1,500/40 hr = $37.50/hr

Daily rental = $500/8 hr = $62.50/hr

Hours to justify rental:

Monthly rental = $36,000/$22.73/hr = 1,583 hr

Weekly rental = $36,000/$37.50/hr = 960 hr

Daily rental = $36,000/$62.50/hr = 576 hr

If the equipment is used less than 1,583 hr per year but more than 960 hr per year, then it is more economical to rent the equipment on a monthly basis than to own the equipment. If the equipment is used between 576 hr per year and 960 hr per year, then it is more economical to rent the equipment on a weekly basis. If the equipment is used less than 576 hr per year, then it is more economical to rent the equipment on a daily basis.

EQUIPMENT COSTS

When equipment is purchased, it is necessary to determine the cost of owning and operating each unit. Ownership costs are costs that accrue whether or not the equipment is used. The ownership costs include depreciation of the equipment over its useful life, interest on money required to purchase the equipment, taxes, insurance, and storage of equipment when it is not in use. Operating costs include maintenance and repairs, fuel, oil, lubricants, and tires. Some contractors include major maintenance and repairs in ownership costs while other contractors include them in operating costs. In either case, major maintenance and repairs must be included in the cost of using the equipment. When equipment is purchased the costs include:

Ownership costs:
   Depreciation, includes purchase price, salvage, useful life
   Investment, includes interest on money, insurance, storage, etc.
Operating costs:
   Maintenance and repair
   Fuel, oil, lubrication
   Tires, tracks for crawler units, etc.

When equipment is rented or leased, the ownership costs (depreciation and investment) are included in the rental or lease rate. The rental company owns the equipment and the contractor pays for use of the equipment. The operating costs depend on the rental or lease agreement. Maintenance and repairs may be paid entirely by the rental company or the contractor, or there may be a sharing of the costs as defined in the rental agreement. Fuel, oil, and lubrication are typically the responsibility of the contractor. For rubber-tired equipment, the rental company often will measure tread wear and charge the contractor for tire wear. The rental company usually is responsible for repairs of track-type equipment. When rented or leased the costs include:

Rental (or lease) rate of equipment

Maintenance and repairs (as defined in rental contract)

Fuel, oil, and lubrication

Repairs of tires, tracks (as defined in rental contract)

In this book the general practice will be to charge on an hourly basis for equipment that is owned. Discussions and examples which follow will illustrate methods of estimating the hourly cost of owning and operating equipment. These examples are intended to show the estimator how one may determine the probable hourly cost for any type of construction equipment. The costs which are determined in these examples apply for the given conditions only, but by following the same procedure and using appropriate prices for the particular equipment, the estimator can determine hourly costs that are suitable for use on any project.

DEPRECIATION COSTS

Depreciation is the loss in value of equipment resulting from use and age. With time, the value of a unit of equipment decreases due to wear and tear, deterioration, obsolescence, or reduced need. Thus, at the end of its useful life the value of the equipment is substantially less than its purchase amount. This loss in value is depreciation. For example, if a unit of equipment is purchased for an amount (P) and sold at a future date for a salvage value (S), then the total depreciation is $(P - S)$.

METHODS OF DEPRECIATION

The construction company may select the depreciation method that best suits its operations and financial situation. Although any reasonable method may be used for determining the cost of depreciation, the following three are most commonly used:

1. Straight-line method
2. Declining-balance method
3. Sum-of-the-year's-digits method

The depreciation cost of construction equipment is commonly determined for two purposes. One purpose is to determine an appropriate cost that must be applied to estimates of projects on which the equipment is to be used. The other purpose is to determine the depreciated cost for tax purposes, by using methods that are approved by the U.S. Internal Revenue Service (IRS). Therefore, it is common to utilize two different depreciation methods for a particular unit of equipment, reporting one value to the estimator for use in estimate projects and another value to the IRS to obtain the most favorable tax benefits. The straight-line method is commonly used for allocating equipment costs for projects, whereas the double-declining-balance or sum-of-the-year's-digits method are commonly used for tax purposes.

Straight-Line Depreciation Method

When the cost of depreciation is determined by this method, it is assumed that a unit of equipment will decrease in value from its original total cost at a uniform rate. The depreciation rate may be expressed as a cost per unit of time, or it may be expressed as a cost per unit of work produced. The depreciation cost per unit of time is obtained by dividing the original cost, less the estimated salvage value to be realized at the time it will be disposed of, by the estimated useful life, expressed in the desired units of time, which may be years, months, weeks, days, or hours. The equation for calculating depreciation using the straight-line depreciation method is:

$$D = \frac{(P - S)}{n} \qquad [5.1]$$

where D = Depreciation
 P = Purchase price
 S = Salvage value
 n = Useful life

EXAMPLE 5.5

A unit of equipment is purchased for $100,000, which is expected to be used 2,000 hr/yr. The anticipated salvage value is $20,000 at the end of its 4-year useful life. Calculate the hourly depreciation costs using the straight-line depreciation method.

 The total depreciation will be:
$$(P - S) = \$100,000 - \$20,000$$
$$= \$80,000$$

 The depreciation per year will be:

$$\text{From Eq. [5.1], } D = \frac{(P - S)}{n}$$
$$= \frac{(\$100,000 - \$20,000)}{4 \text{ years}}$$
$$= \$20,000/\text{year annual depreciation}$$

The hourly depreciation per year will be:
$$D = \$20{,}000/\text{year}/2{,}000\ \text{hr/year}$$
$$= \$10.00/\text{hr}$$

Table 5.2 provides the annual cost of depreciation for this equipment over its 4 yr useful life.

TABLE 5.2 | Annual depreciation using straight-line depreciation.

End of year	Depreciation for the year	Book value
0	—	$100,000
1	$20,000	$80,000
2	$20,000	$60,000
3	$20,000	$40,000
4	$20,000	$20,000

Another method of estimating the straight-line cost of depreciation is to divide the original cost, less the estimated salvage value, by the probable number of units of work which it will produce during its useful life, This method is satisfactory for equipment whose life is determined by the rate at which it is used instead of by time. Examples of such equipment include the pump and discharge pipe on a hydraulic dredge, rock crushers, rock-drilling equipment, rubber tires, and conveyor belts.

Declining-Balance Depreciation Method

Using this method of determining the cost of depreciation, the estimated life of the equipment in years will give the average percentage of depreciation per year. This percentage is doubled for the 200 percent declining-balance method. The value of the depreciation during any given year is determined by multiplying the resulting percentage by the value of the equipment at the beginning of that year. While the estimated salvage value is not considered in determining depreciation, the depreciated value is not permitted to drop below a reasonable salvage value.

When the cumulative sum of all costs of depreciation is deducted from the original total cost, the remaining value is designated as the book value. Thus, if a unit of equipment whose original cost is $100,000 has been depreciated a total of $60,000, the book value will be $40,000.

EXAMPLE 5.6

A unit of equipment is purchased for $120,000 with an expected salvage value of $15,000 after 5 years. Calculate the book value over the expected life of the equipment using the double-declining-balance method of depreciation.

Total cost, $P = \$120{,}000$

Estimated salvage, $S = \$15{,}000$

Useful life, $n = 5$ years

Average rate of percentage depreciation per year, $1/5 = 0.20$, or 20%

For double-declining-balance method, $2 \times 20\% = 40\%$

Depreciation first year, 40% × $120,000 = $48,000

Depreciation second year, 40% × ($120,000 − $48,000) = $28,800

Table 5.3 provides the annual cost of depreciation for this equipment over its 5 year useful life.

TABLE 5.3 | Annual depreciation using double-declining-balance method.

End of yr	Percentage of depreciation	Depreciation for the year	Book value
0	—	$0	$120,000
1	40%	40% × $120,000 = $48,000	$72,000
2	40%	40% × $72,000 = $28,800	$43,200
3	40%	40% × $43,200 = $17,280	$25,920
4	40%	40% × $25,920 = $10,368	$15,552
5	40%	40 × $15,552 = $6,221	$9,331
5§	—	$15,552 − $15,000 = $552	$15,000

§ The book value of the equipment may not be depreciated below its reasonable minimum salvage value of $15,000; therefore the depreciation for 5th year is the lower value of $552.

Sum-of-the-Year's-Digits Depreciation Method

Using this method of determining the cost of depreciation, all the digits representing each year of the estimated life of the equipment are totaled. For an estimated life of 6 years, the sum of the digits will be $1 + 2 + 3 + 4 + 5 = 15$. The estimated salvage value is deducted from the total cost of the equipment. During the first year the cost of depreciation will be 5/15 of the cost less salvage value. During the second year, the cost of depreciation will be 4/15 of the cost less salvage value. This process is continued for each year through the fifth year. The equation for calculating the sum of the year's digits is:

$$\text{S.O.Y.D.} = \frac{n(n + 1)}{2} \qquad [5.2]$$

where S.O.Y.D = sum of the year's digits

n = total number of years of depreciation

EXAMPLE 5.7

A unit of equipment is purchased for $180,000 with an expected salvage value of $30,000 after 5 years. Calculate the book value over the expected life of the equipment using the sum-of-the-year's-digits method of depreciation.

Total cost, P = $180,000

Estimated salvage, S = $30,000

Useful life, n = 5 years

S.O.Y.D. = $n(n + 1) / 2 = 5(5 + 1) / 2 = 15$
Total depreciation, $(P - S) = \$180,000 - \$30,000 = \$150,000$
Depreciation for first year, $5/15 \times \$150,000 = \$50,000$
Depreciation for second year, $4/15 \times \$150,000 = \$40,000$
Depreciation for third year, $3/15 \times \$150,000 = \$30,000$

Table 5.4 provides the annual cost of depreciation for this equipment over its 5 yr useful life.

TABLE 5.4 | Annual depreciation using sum-of-the-year's-digits method.

End of yr	Depreciation ratio	Depreciation for the year	Book value
0	—	$0	$180,000
1	5/15	5/15 × $150,000 = $50,000	$130,000
2	4/15	4/15 × $150,000 = $40,000	$90,000
3	3/15	3/15 × $150,000 = $30,000	$60,000
4	2/15	2/15 × $150,000 = $20,000	$40,000
5	1/15	1/15 × $150,000 = $10,000	$30,000

INVESTMENT COSTS

It costs money to own equipment, regardless of the extent it is used. These costs are classified as investment costs, which include interest on the money invested in the equipment, taxes assessed against the equipment, insurance, and storage. Even though the owner may pay cash for equipment, they should still charge interest on the investment, because the money spent for the equipment could be invested in some other asset which would produce interest for the owner if it were not invested in equipment.

Figure 5.1 shows the distribution of book values of equipment with a purchase price (P) and salvage value (S) over a useful life of (n) years using straight-line depreciation.

The equation for calculating the average annual value of the distribution of costs shown in Figure 5.1 is:

$$\text{AAV} = \frac{P(n + 1) + S(n - 1)}{2n} \qquad [5.3]$$

where AAV = Average annual value
P = Purchase price
S = Salvage
n = Useful life

Equation [5.3] represents the average book value of the straight-line depreciation. One method of determining annual investment costs is to apply a percentage to the average annual value of the equipment. The annual investment

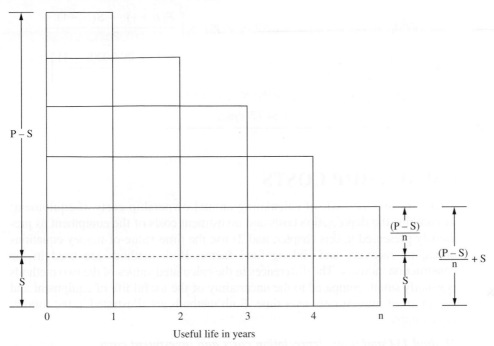

FIGURE **5.1** | Value of equipment by year with salvage value.

cost is calculated by multiplying the average annual value by a percentage, as shown in Equation [5.4].

$$\text{Annual investment cost} = (\text{Percentage}) \times \left[\frac{P(n+1) + S(n-1)}{2n}\right] \qquad [5.4]$$

The percentage value in Eq. [5.4] represents two components of ownership costs: 1) a percentage that represents the annual interest on money borrowed to purchase the equipment, and 2) a percentage that represents the annual cost of insurance, taxes, and storage of the equipment. Most contractors use the current rate of interest for borrowing money and add an additional amount for risk. The average cost for insurance, taxes, and storage during each year can be converted into an equivalent interest rate based on the value of the equipment at any time.

Interest on borrowed money = 6%
Insurance, taxes, and storage = 3%
Total percentage = 9%

EXAMPLE 5.8

Use the percentage of average annual value to determine the annual investment costs of a unit of equipment that is purchased for $75,000 with an expected salvage value of $20,000 after its useful life of 5 years. Assume an interest percentage of 9%. The annual investment cost can be calculated from Equation 5.4.

$$\text{Annual investment cost} = (\text{Percentage}) \times \left[\frac{P(n+1) + S(n-1)}{2n} \right]$$

$$= (9\%) \times \left[\frac{75{,}000(5+1) + 20{,}000(5-1)}{2(5)} \right]$$

$$= (9\%) \times (\$53{,}000/\text{yr})$$

$$= \$4{,}770/\text{year}$$

OWNERSHIP COSTS

There are two methods of calculating annual ownership costs of equipment:
1) combine the depreciation costs and investment costs of the equipment as previously presented in this chapter, and 2) use the time-value-of-money equations of capital recovery and sinking fund factor. Both methods are used in the construction industry. The differences in the calculated values of the two methods is usually small, compared to the uncertainty of the useful life of equipment and variations of interest rates over time. Both methods are illustrated in the following examples.

Method 1) Combining depreciation costs and investment costs

Using this method, the annual depreciation is calculated using straight-line depreciation. Then, the investment cost is calculated by applying a percentage to the average annual value of the equipment. The percentage represents the interest rate applicable to economic conditions over the useful life of the equipment plus a percentage to account for taxes and insurance on the equipment.

$$\textbf{(Eq. [5.1])} \text{ Depreciation costs} = \frac{(P - S)}{n}$$

$$\textbf{(Eq. [5.4])} \text{ Investment costs} = (\text{Percentage}) \times \left[\frac{P(n+1) + S(n-1)}{2n} \right]$$

Ownership costs = Depreciation costs + Investment costs

$$= \frac{(P - S)}{n} + (\text{Percentage}) \times \left[\frac{P(n+1) + S(n-1)}{2n} \right] \qquad \textbf{[5.5]}$$

EXAMPLE 5.9

Use Method 1), the combined depreciation and investment costs, to calculate the ownership cost of a unit of equipment. The equipment has a purchase price (*P*) of $100,000 with a salvage value (*S*) of $20,000 and a useful life (*n*) of 4 years. Use a 9% interest percentage.

From Eq. [5.5],

Ownership costs = Depreciation costs + Investment costs

$$= \frac{(P - S)}{n} + (\text{Percentage}) \times \left[\frac{P(n + 1) + S(n - 1)}{2n} \right]$$

$$= \frac{\$100{,}000 - \$20{,}000}{4} + (9\%) \times \left[\frac{\$100{,}000(4 + 1) + \$20{,}000(4 - 1)}{2(4)} \right]$$

$$= \$20{,}000/\text{year} + \$6{,}300/\text{year}$$

$$= \$26{,}300/\text{year}$$

Method 2) Using time-value-of-money equations

Many books are available that show the development and illustration of time-value-of-money equations. These equations can be used to calculate the annual ownership cost of equipment. Using this method, the purchase price and salvage value are converted to an equivalent annual cost. The capital recovery factor converts the present worth (purchase price) to an equivalent annual cost over the useful life of the equipment. The sinking fund factor converts the salvage value over the useful life. Combining the two values gives the annual ownership costs.

Using the capital recovery factor to convert purchase price to annual costs,

$$\text{Equivalent annual cost of purchase price} = P\left[\frac{i(1 + i)^n}{(1 + i)^n - 1} \right]$$

Using the sinking fund factor to convert salvage value to annual costs,

$$\text{Equivalent annual cost of purchase price} = S\left[\frac{i}{(1 + i)^n - 1} \right]$$

The annual ownership costs can be calculated from these equations by subtracting the sinking fund from the capital recovery as:

$$\text{Ownership costs} = P\left[\frac{i(1 + i)^n}{(1 + i)^n - 1} \right] - S\left[\frac{i}{(1 + i)^n - 1} \right] \qquad \textbf{[5.6]}$$

EXAMPLE 5.10

Use Method 2), time-value-of-money equations, to calculate the ownership costs of a unit of equipment. The equipment has a purchase price (P) of $100,000 with a salvage value (S) of $20,000 and a useful life (n) of 4 years. Use a 9% annual interest rate.

From Eq. [5.6],

$$\text{Ownership costs} = \$100{,}000\left[\frac{0.09(1 + 0.09)^4}{(1 + 0.09)^4 - 1} \right] - \$20{,}000\left[\frac{0.09}{(1 + 0.09)^4 - 1} \right]$$

$$= \$100{,}000\,(0.308668662) - \$20{,}000\,(0.218668662)$$

$$= \$26{,}493/\text{year}$$

Example 5.9 used Method 1) combined depreciation and investment costs, to calculate annual ownership costs as $26,300/year; whereas Example 5.10 used Method 2) time-value-of-money to calculate annual ownership costs as $26,493/year. Both examples were based on the same purchase price, salvage value, useful life, and interest percentage. If the equipment is used 1,800 hours per year, the hourly ownership cost can be calculated as:

For Method 1), Hourly ownership costs = $26,300/1,800 hr = $14.61/hr

For Method 2), Hourly ownership costs = $26,493/1,800 hr = $14.71/hr

The difference in the hourly ownership costs of the two methods is $14,71/hr − $14.61/hr = $0.10/hr, which is small considering the uncertainty of salvage values and variability of interest percentages over the useful life of the equipment. Thus, either method may be considered satisfactory for calculating the hourly ownership cost of equipment for estimating purposes.

OPERATING COSTS

Operating costs accrue only when the unit of equipment is being used, whereas ownership costs accrue whether or not the equipment is used. Operating costs include maintenance and repairs, fuel, oil, and lubricants. Maintenance and repairs may be defined as major or minor. Major maintenance and repairs are often included in ownership costs, whereas minor maintenance and repairs are included in operating costs.

Construction equipment that is driven by internal combustion engines requires fuel and lubricating oil, which should be considered as an operating cost. Whereas the amounts consumed and the unit cost of each will vary with the type and size of equipment, the conditions under which it is operated, and the location, it is possible to estimate the cost with reasonable accuracy for a given condition.

An estimator should be reasonably familiar with the conditions under which a unit of equipment will be operated. Whereas a tractor may be equipped with a 200-hp engine, this tractor will not demand the full power of the engine at all times, possibly only when it is used to load a scraper or to negotiate a steep hill. Also, equipment is seldom, if ever, used 60 min/hr. Thus, the fuel consumed should be based on the actual operating conditions. Perhaps the average demand on an engine might be 50 percent of its maximum power for an average of 45 min/hr.

Maintenance and Repair Costs

The costs for maintenance and repairs include the expenditures for replacement parts and the labor required to keep the equipment in good working condition. These costs vary considerably with the type of equipment, operating conditions, frequency of oil and lubricant services, and skill of the operator. For example, if an excavator is used to excavate soft earth, the replacement of parts will be considerably less often than when the same excavator is used to excavate rock. If a bearing is greased and adjusted at frequent intervals, its life will be much longer than if it is neglected.

Historical cost records of maintaining and servicing equipment are the most reliable guide in estimating maintenance and repair cost. If this information is not available, the manufacturers of construction equipment provide information showing recommended costs for maintenance and repairs for the equipment they manufacture. The annual cost of maintenance and repairs is often expressed as a percentage of the purchase price or as a percentage of the straight-line depreciation costs $(P - S)/n$. For example, the average cost of maintenance and repairs for an excavator may vary from 80 to 120 percent of the depreciation cost, with 100 percent as a fair average value. The annual cost for certain types of rock-crushing equipment may be much higher, whereas the cost for a compressor will be lower.

Fuel Consumption

The amount of fuel consumed by construction equipment depends on the type of fuel (gasoline or diesel), size of engine (maximum rated horsepower), percentage of an hour the equipment is operating (time factor), and percentage of horsepower utilized (engine factor).

Fuel consumption is best determined by record keeping at the jobsite that measures the actual fuel utilized for a particular unit of equipment. When company records are not available, manufacturer's data can be obtained from equipment dealers to estimate fuel consumption for a particular type and size of equipment.

When operating under standard conditions, barometric pressure is 29.9 in. of mercury and a temperature of 68 deg F, the amount of fuel consumed can be estimated by the following equations:

For gasoline engines:
Fuel consumption = (time factor) $\times$ (engine factor) $\times$ hp $\times$ 0.06 gal/hp-hr **[5.7]**

For diesel engines:
Fuel consumption = (time factor) $\times$ (engine factor) $\times$ hp $\times$ 0.04 gal/hp-hr **[5.8]**

EXAMPLE 5.11

A hydraulic excavator with a diesel engine is rated at 160 hp. During a cycle of 20 seconds, the engine is operated at full power for 5 seconds while filling the bucket in hard clay. During the balance of the cycle, the engine will be operated at not more than 50 percent of its rated power. Also, the excavator may not operate more than 45 min/hr on average. Estimate the fuel consumption.

Time factor = 45/60
= 0.75

Engine factor = (5/20 $\times$ 100%) + (15/20 $\times$ 50%)
= 0.25 + 0.375
= 0.625

Operating factor = (Time factor) × (Engine factor)

$$= 0.75 \times 0.625$$

$$= 0.47$$

Maximum rated power = 160 hp

From Eq. [5.8] for diesel engine,

Fuel consumption = (Operating factor) × hp × 0.04 gal/hp-hr

$$= 0.47 \times 160 \text{ hp} \times 0.04 \text{ gal/hp-hr}$$

$$= 3.0 \text{ gal/hr}$$

Lubricating Oil Consumed

The quantity of lubricating oil consumed by an engine will vary with the size of the engine, the capacity of the crankcase, the condition of the pistons, and the number of hours between oil changes. It is common practice to change oil every 100 to 200 hr, unless extreme dust makes more frequent changes desirable. The quantity of oil consumed by an engine during a change cycle includes the amount added at the time of change plus the makeup oil added between changes. Equation [5.9] can be used to estimate the quantity of oil consumed:

$$Q = \frac{\text{hp} \times 0.6 \times 0.006 \text{ lb/(hp·hr)}}{7.4 \text{ lb/gal}} + \frac{c}{t} \qquad [5.9]$$

where Q = quantity consumed, in gallons per hour

hp = rated horsepower of engine

c = capacity of crankcase, in gallons

t = hours between oil changes

Equation [5.9] is based on an operating factor of 0.60, or 60 percent. It assumes that the quantity of oil consumed between oil changes will be 0.006 gal per rated horsepower-hour. Using this equation for a 100-hp engine with a crankcase capacity of 4 gal requiring a change every 100 hr, we find the quantity consumed per hour is

$$Q = \frac{100 \text{ hp} \times 0.6 \times 0.006 \text{ gal/hp·hr}}{7.4 \text{ lb/gal}} + \frac{4 \text{ gal}}{100 \text{ hr}} = 0.089 \text{ gal/hr}$$

Lubricants other than crankcase oil are required for motor-driven equipment. Although the costs of such lubricants will vary, an average cost equal to 50 percent of the cost of the crankcase oil is satisfactory.

Cost of Rubber Tires

Many types of construction equipment use rubber tires, whose life usually will not be the same as the equipment on which they are used. For example, a unit of equipment may have an expected useful life of 6 years, but the tires on the equipment may last only 2 years. Therefore, a new set of tires must be placed on the equipment every 2 years, which would require three sets of tires during the 6 years the equipment will be used. Thus, the cost of depreciation and repairs for tires should be estimated separately from the equipment.

For a unit of equipment, a set of tires, whose cost is $3,800, may have an estimated life of 2,500 hr, with the repairs during the life of the tires costing 15 percent of the initial cost of the tires. The cost is determined as

Depreciation: $3,800/2,500 hr   = $1.52/hr
Repairs of tires: 0.15 × $1.52/hr =   0.23/hr
Total cost = $1.75/hr

EXAMPLE 5.12

Determine the probable cost per hour of owning and operating an earthmoving unit of equipment with a 170-hp diesel engine. Use Method 1 (combined depreciation and investment cost) to calculate ownership costs. The following conditions apply.

Factory delivered price = $250,000

Sales tax = 5% of delivered price

Unloading and assembling = $2,500

Salvage value = $70,000

Useful life = 5 yr

Hours used per year = 1,800 hr

Investment costs = 7% of average annual value

Maintenance and repairs = 50% of straight-line depreciation

Diesel engine = 170 hp

Operating factor = 0.60

Fuel costs = $4.00/gal

Crankcase capacity = 8 gal

Hours between oil changes = 120 hr

Oil costs = $12/gal

Tires = $45,000

Tire life = 4,000 hr

Tire repairs = 10% of tire depreciation

Ownership costs:

Purchase price from factory = $250,000
Sales tax, 5% × $250,000   = + 12,500
Unloading and assembling   = + 2,500
Total delivered cost = $265,000
Less tires = − 45,000
Net purchase price = $220,000

From Eq. [5.1], Annual depreciation, excluding tires:

$$D = \frac{P - S}{n}$$

$$= \frac{\$220,000 - \$70,000}{5 \text{ yr}}$$

$$= \$30,000/\text{yr}$$

From Eq. [5.3], Average annual value,

$$\text{Average annual value, AAV} = \frac{P(n + 1) + S(n - 1)}{2n}$$

$$= \frac{\$220,000(5 + 1) + \$70,000(5 - 1)}{2(5 \text{ yr})}$$

$$= \$160,000/\text{yr}$$

Annual ownership costs,

Depreciation, ($220,000 − $70,000)/5 yr   = $30,000/yr

Investment costs, 7% AAV = 0.07 × $160,000 =   11,200/yr

                            Annual ownership costs = $41,200/yr

Hourly ownership cost = $41,200/yr/1,800 hr/yr

                = $22.89/hr

Operating costs:

From Eq. [5.8], Consumption of diesel fuel = Operating factor × hp × 0.04 gal/(hp-hr)

                         = 0.60 × 170 hp × 0.04 gal/(hp-hr)

                         = 4.08 gal/hr

From Eq. [5.9], Oil consumption, $Q = \dfrac{hp \times 0.6 \times 0.006 \text{ lb}/(hp - hr)}{7.4 \text{ lb/gal}} + \dfrac{c}{t}$

$$= \frac{170 \text{ hp} \times 0.6 \times 0.006 \text{ lb}/(hp - hr)}{7.4 \text{ lb/gal}} + \frac{8 \text{ gal}}{120 \text{ hr}}$$

$$= 0.08 + 0.07$$

$$= 0.15 \text{ gal/hr}$$

Summary of equipment operating costs:

   Annual maintenance and repairs = 50% of depreciation costs

                         = 0.50 × $30,000/yr

                         = 15,000/yr

Hourly maintenance and repairs, $15,000/yr/1800 hr/yr   = $8.33/hr

Fuel, 4.08 gal/hr × $4.00/gal   =   16.32/hr

Lubricating oil, 0.15 gal/hr × $12.00/gal   =   1.80/hr

Tire depreciation, $45,000/4,000 hr   =   11.25/hr

Tire repairs, 10% of tire depreciation, = 0.10 × $11.25/hr =   1.13/hr

                    Total hourly operating costs = $38.83/hr

Hourly ownership plus operating costs:

Total ownership and operating costs = $22.89/hr + $38.83/hr

= $61.72/hr

Assuming an operator for this equipment costs $35.00/hr

Total equipment and operator cost = $61.72/hr + $35.00/hr

= $96.72/hr

Assuming the equipment can move 85 cy/hr of material per hour

The cost per cy moved = $96.72/hr / 85 cy/hr

= $1.14/cy

The hourly cost of owning and operating construction equipment as illustrated in Example 5.12 will vary with the conditions under which equipment is operated. The estimator should analyze each job to determine the probable conditions that will affect the cost and should appropriately adjust costs. As interest rates change, the interest rate applied to equipment costs should be appropriately adjusted and new hourly costs should be applied to charging the equipment to projects and to preparing cost estimates that will involve use of the equipment.

If a crawler tractor is used on rock surfaces, the life of the undercarriage will be significantly reduced. Undercarriage costs can represent a major portion of operating costs of track-type equipment. Undercarriage costs can be determined similarly to tire costs, based on the cost of undercarriage replacement and its useful life. The costs should be based on local costs of parts and labor.

If trucks are operated over smooth and level haul roads that are well maintained, the cost of repairs will be significantly lower than when the same trucks are operated over rough and poorly maintained haul roads. Therefore, the estimator must know the jobsite conditions before appropriate costs can be assigned to construction equipment.

PROBLEMS

5.1 A crew of 5 carpenters and 2 laborers will be used to build formwork for a concrete structure. Work is scheduled for 10 hr/day on Monday through Friday and 8 hr on Saturday. Overtime at a rate of one and one-half will be paid for all hours over 8 hr/day on Monday through Friday and double-time for all Saturday work. The base wage, taxes, and insurance rates are given below. Calculate the hourly, weekly, and monthly cost for the crew.

Item	Carpenters	Laborers
Base wage	$30.18/hr	$20.78/h
Workers' compensation	$5.35/$100	$4.75/$100 of base wage
Public liability/property damage	$2.50/$100	$2.10/$100 of base wage
Social Security (FICA)	7.65%	7.65% of actual wage
Unemployment tax	2%	2% of actual wage
Fringe benefits	$2.40/hr	$2.15/hr

5.2 What is the average annual investment cost for a unit of equipment that costs $176,540 with an estimated useful life of 5 yr and no salvage? Use 7% interest on borrowed money and 2% for taxes, insurance, and storage. What is the average annual investment cost of the equipment if it is assumed to have a salvage value of $54,000 at the end of the 5 yr?

5.3 The purchase price of a unit of equipment is $486,500. The estimated useful life is 5 yr with no salvage value. Determine the book value at the end of 2 yr, using the depreciation methods of (a) straight-line, (b) declining-balance, and (c) sum-of-the-year's-digits.

5.4 If the equipment in Problem 5.3 is assumed to have a salvage value of $45,000 at the end of 5 yr, find the book value at the end of 4 yr using the depreciation methods of straight-line, double-declining-balance, and sum-of-the year's-digits.

5.5 A scraper with a 275-hp diesel engine will be used to excavate and haul earth for a highway project. An evaluation of the jobsite conditions indicates the scraper will operate 40 min/hr. For this project it is anticipated that the total cycle time will be 20 min for a round trip. Previous job records show the scraper operated at full power for the 1.5 min required to fill the bowl of the scraper and at 80% of the rated hp for the balance of the cycle time. Calculate the gallons per hr for fuel consumption of the scraper.

5.6 A crawler tractor is operated by a 180-hp diesel engine. Calculate the probable gallons of fuel consumed per hr for each of the given conditions:

a) When operating at an average of 60 percent of its capacity for 50 min per hr.

b) When operating at 100 percent of its capacity for 15 min per hr, 60 percent capacity for 30 min per hr, and 20 percent of its capacity for 15 min per hr.

5.7 A 210-hp diesel truck is used for hauling gravel from the rock quarry to a stockpile at a paving operation. The crankcase capacity is 6 gallons and the expected time between oil changes is 150 hr. Calculate the gallons per hr of lubricating oil for the truck.

5.8 Use Method 1 (combined depreciation and investment costs for ownership costs) to determine the probable cost per hour of owning and operating a rubber tire unit of equipment with a 250-hp diesel engine. The following conditions apply.

> Factory delivered price = $450,000
> Sales tax = 5% of delivered price
> Unloading and assembling = $2,500
> Salvage value = $120,000
> Useful life = 5 yr
> Hours used per year = 1,500 hr
> Interest on investment = 8%
> Maintenance and repairs = 70% of straight-line depreciation
> Diesel engine = 250 hp
> Operating factor = 0.60
> Fuel costs = $4.00/gal
> Crankcase capacity = 8 gal
> Hours between oil changes = 120 hr
> Oil costs = $16/gal

Tires = \$55,000
Tire life = 3,500 hr
Tire repairs = 10% of tire depreciation

5.9 Determine the probable cost per hour for owning and operating the equipment in Problem 5.8 when the unit will be operated 1,600, 1,800, and 2,000 hr per year. All other conditions will be the same as given in Problem 5.8.

5.10 Use Method 2 (time-value-of-money equations for ownership costs) to determine the probable cost per hour of owning and operating a track-type unit of equipment. The following conditions apply.

Total delivered price = \$220,000
Salvage value = \$75,000
Useful life = 6 yr
Hours used per year = 1,800 hr
Annual interest rate = 9%
Maintenance and repairs = 20% of purchase price
Diesel engine = 150 hp
Operating factor = 0.70
Fuel costs = \$4.00/gal
Crankcase capacity = 6 gal
Hours between oil changes = 150 hr
Oil costs = \$14/gal

CHAPTER 6

Handling and Transporting Material

INTRODUCTION

Although material suppliers generally deliver to the jobsite the construction materials that have been purchased by the contractor, sometimes the materials must be obtained by the contractor at the storage yard of the material supplier. Also, the contractor must move material from stockpiles on the jobsite to the location where the material will be permanently installed.

Some projects require the use of aggregates, sand and gravel, or crushed stone, which are produced from natural deposits or quarries and hauled to the project in trucks. A contractor using his or her laborers and equipment may do the handling and hauling, or it may be accomplished through a subcontractor. Regardless of the method used, it will involve a cost that must be included in the estimate for a project.

When estimating the time required by a truck for a round-trip, the estimator should divide the round-trip time into four elements:

1. Load
2. Haul, loaded
3. Unload
4. Return, empty

These four elements define the cycle time for transporting material. The time required for each element should be estimated. If elements 2 and 4 require the same time, they can be combined. Since the time required for hauling and returning will depend on the distance and effective speed, it is necessary to determine the probable speed at which a vehicle can travel along the given haul road for the conditions that will exist. Speeds are dependent on the vehicle, traffic congestion, condition of the road, and other factors. An appropriate operating factor should be used in determining production rates. For example, if a truck will operate only 45 min/hr, this time should be used in determining the number of round-trips the truck will make in 1 hr.

CYCLE TIME AND PRODUCTION RATE CALCULATIONS

Cycle times and production rates are used throughout this book to estimate the time, cost, and cost per unit of work. Production rates are crucial to estimating the time and cost of construction projects. The time that labor and equipment will be on the job can be calculated by dividing the total quantity of work by the production rate. After the time to perform the work is calculated, the cost of labor and equipment can be determined by multiplying the total time by the hourly cost rate of labor and equipment.

Examples 6.1 through 6.3 illustrate the basic calculations of cycle times and production rates to show the procedures and concepts for estimating the time and cost of handling and transporting materials. These examples are presented in a progressive sequence: 1) using laborers to load a single truck, 2) replacing laborers with an equipment loader to load a single truck, and 3) using the equipment loader to load multiple trucks. It is the analysis process that is important when reviewing these examples. For example, in Example 6.1 it is not practical to load 180 tons of sand using laborers. As shown in this example, the cost and time are high. Example 6.2 replaces the laborers with an equipment loader, which produces a more productive operation at a lower cost. Example 6.3 shows that adding additional trucks to the loading operation will improve the production rate and provide a lower cost per unit for performing the work.

EXAMPLE 6.1

A project requires transporting 180 tons of sand a distance of 8 miles with a 12-cy dump truck. The sand has a density of 95 lb/cy. Two laborers and a driver, at a rate of 1.5 cy/hr each, will load the truck. Assume a haul speed of 30 mph, return speed of 35 mph, and 3 min to dump the load. The cost of the truck is $42.00/hr, the driver is $30.00/hr, and the laborers cost $20.00/hr each. Assume a 45 min/hr as productive for moving the sand. What is the total time, total cost, and cost per unit for transporting the material?

Quantity of work:

$$\text{Quantity of sand} = \frac{180 \text{ T} \times 2{,}000 \text{ lb/T}}{95 \text{ lb/cf} \times 27 \text{ cf/cy}} = 140.4 \text{ cy}$$

Cycle time:

    Load = 12 cy/(3 × 1.5 cy/hr) = 2.667 hr
    Haul = 8 mi/30 mph   = 0.267 hr
    Dump = 3 min/(60 min/hr) = 0.050 hr
    Return = 8 mi/35 mph   = 0.229 hr
              Total cycle time = 3.213 hr/trip

Production rate:

Number of trips per hour = 1.0 trip/3.213 hr

= 0.311 trip/hr

Rate of transporting = 12 cy/trip × 0.311 trips/hr

= 3.73 cy/hr

Assume a 45 min productive hour,

Production rate = 3.73 cy/hr × 45/60 = 2.8 cy/hr

Time:

Using 1 truck and 2 laborers = 140.4 cy/2.8 cy/hr

= 50.1 hr

Cost:

Truck = 50.1 hr × 1 truck @ $42.00/hr   = $2,104.20
Driver = 50.1 hr × 1 driver @ $30.00/hr   =   1,503.00
Laborers = 50.1 hr × 2 laborers @ $20.00/hr =   2,004.00
Total cost = $5,611.20

Unit cost:

Cost per cy = $5,611.20/140.4 cy = $40.00/cy

Cost per ton = $5,611.20/180 tons = $31.17/ton

The time, cost, and cost per unit of work are high using only laborers to load the sand. These high costs are a result of the truck sitting idle while the laborers are loading the truck. Example 6.2 illustrates the economy of bringing equipment to the job to load the sand.

EXAMPLE 6.2

Note that in Example 6.1 the load time is significantly greater than the travel time, $2.667 > 0.546$, which indicates an imbalance between the times of loading and hauling. Assume a small tractor loader can be rented at a cost of $75.00/hr with a load production rate of 95 cy/hr. The loader operator cost is $35.00/hr. Assuming the cost of transporting the loader to and from the job is $400.00, what is the time, cost, and cost per cubic yard for transporting the 180 tons of sand?

Quantity of work:

$$\text{Quantity of sand} = \frac{180 \text{ T} \times 2,000 \text{ lb/T}}{95 \text{ lb/cf} \times 27 \text{ cf/cy}} = 140.4 \text{ cy}$$

Cycle time:

Load = 12 cy/95 cy/hr   = 0.126 hr
Haul = 8 mi/30 mph   = 0.267 hr
Dump = 3 min/(60 min/hr) = 0.050 hr
Return = 8 mi/35 mph   = 0.229 hr
Total cycle time = 0.672 hr/trip

Production rate:

Number of trips per hour = 1.0 trip/0.672 hr

= 1.49 trip/hr

Rate of transporting = 12 cy/trip × 1.49 trips/hr

= 17.9 cy/hr

Assume 45 min/hr as productive,

Production rate = 17.9 cy/hr × 45/60 = 13.4 cy/hr

Time:

Using tractor loader and one truck = 140.4 cy/13.4 cy/hr

= 10.5 hr

Cost:

Loader = 10.5 hr @ $75.00/hr	=	$787.50
Operator = 10.5 hr @ $35.00/hr	=	367.50
Truck = 10.5 hr @ $42.00/hr	=	441.00
Driver = 10.5 hr @ $30.00/hr	=	315.00
Transporting loader to job	=	400.00
Total cost	=	$2,311.00

Unit cost:

Cost per cy = $2,311.00/140.4 cy = $16.46/cy

Cost per ton = $2,311.00/180 tons = $12.84/ton

The time, cost, and cost per unit using the loader are significantly lower than using the laborers in Example 6.1. Thus, it is more economical to rent the tractor loader and pay the cost to bring it to the job, compared to loading by laborers. The number of cubic yards of sand required to justify the loader can be calculated as:

Hourly cost loading by laborers = Truck + Driver + 2 Laborers

= $42.00/hr + $30.00/hr + 2($20.00/hr)

= $112.00/hr

Production rate using laborers = 2.8 cy/hr

Hourly cost using loader = Truck + Driver + Loader + Operator

= $42.00/hr + $30.00/hr + $75.00/hr + $35.00/hr

= $182.00/hr

Production rate using the loader = 13.4 cy/hr

($112.00/hr/2.8 cy/hr) X = ($182.00/hr/13.4 cy/hr) X + $400.00

($40.00/cy) X = ($13.58/cy) X + $400.00

X = 15.1 cy

Thus, if there is less than 15.1 cy of sand it is more economical to use the two laborers and one truck to transport the sand, whereas if there is more than 15.1 cy, it is more economical to use the rented equipment to load the truck.

In this example the loader remains idle for a significant amount of time while the single truck is transporting the sand. Thus, it is desirable to add additional trucks to the operation to balance the cycle time of the loader and trucks, to reduce idle time of equipment on the job. Example 6.3 provides an analysis of using multiple trucks with the loader.

EXAMPLE 6.3

Note that in Example 6.2 the travel time is significantly greater than the load time, $0.546 > 0.126$, which indicates there is an inadequate number of trucks to keep the loader busy. Determine the economical number of trucks to balance the loader. The required number of trucks to balance the loader can be calculated by dividing the cycle time by the load time as:

$$\text{Required number of trucks} = (\text{cycle time})/(\text{load time})$$
$$= 0.672/0.126$$
$$= 5.33$$

Consider using 5 trucks:

If 5 trucks are used, there will be fewer trucks than needed. The production rate will be governed by the production rate of the trucks.

Production rate:

For 1 truck = (12 cy/trip)/(0.672 hr/trip) × 45/60 = 13.4 cy/hr

For 5 trucks = 13.4 cy/hr × 5 trucks = 67.0 cy/hr

Time = 140.4 cy/67.0 cy/hr = 2.1 hr

Cost:

Loader = 2.1 hr @ $75.00/hr	=	$157.50
Operator = 2.1 hr @ $35.00/hr	=	73.50
Trucks = 2.1 hr × 5 trucks @ $42.00/hr	=	441.00
Driver = 2.1 hr × 5 drivers @ $30.00/hr	=	315.00
Transporting loader to the job	=	400.00
	Total cost =	$1,387.00

Cost per cubic yard = $1,387.00/140.4 cy = $9.88/cy

Consider using 6 trucks:

If 6 trucks are used, there will be more trucks than needed. The production rate will be governed by the loader production rate.

Production rate of loader = 95 cy/hr × 45/60 = 71.3 cy/hr

Time = 140.4 cy/71.3 cy/hr = 2.0 hr

Cost:

Loader = 2.0 hr @ $75.00/hr	=	$150.00
Operator = 2.0 hr @ $35.00/hr	=	70.00
Trucks = 2.0 hr × 6 trucks @ $42.00/hr	=	504.00
Driver = 2.0 hr × 6 drivers @ $30.00/hr	=	360.00
Transporting loader to the job	=	400.00
	Total cost =	$1,484.00

Cost per cubic yard = $1,484.00/140.4 cy = $10.57/cy

For this project it is more economical to use 5 trucks.

In Examples 6.1 to 6.3 the calculations were shown to illustrate the procedure of calculating cycle times and production rates for estimating time and cost. For example, using 6 trucks, the time was 2.0 hr and the cost was $1,484.00. Generally, equipment is assigned to a job for a full day, or possibly a half day, rather than only 2.0 hr. If such is the case, the appropriate duration should be used. Also, final estimated costs are usually rounded to a full dollar amount. The exact time and cost calculations were presented in their entirety to illustrate the analysis process.

TRANSPORTING SAND AND AGGREGATE WITH TRACTOR LOADERS

Sand and gravel are mined by companies and stockpiled for use on construction projects. Sand is generally excavated from riverbeds by draglines or clamshells, loaded into trucks, and transported to a central location for later distribution to prospective buyers. Similarly, gravel is mined from a rock quarry, crushed in a rock-crushing machine, screened, and transported by trucks to a central gravel yard. Several types of equipment, such as clamshells, front-end loaders, or portable conveyors can handle sand and gravel.

Tractor loaders are used extensively in construction to handle and transport sand, gravel, blasted rock, and all types of soil. A loader can pick up the material, travel with it, and deposit the load at the desired location. Loaders are used to excavate earth, load trucks, and load aggregate bins at asphalt and concrete plants.

Two types of tractor loaders are available, crawler-type and wheel-type. Crawler loaders are effective for excavating material because of their high traction capabilities. The speed range of track loaders is 2 to 5 mph. Wheel loaders are effective for moving material because of their high maneuverability. The speed range of wheel loaders is 4 to 30 mph.

Loaders are rated by size of bucket and horsepower (hp) of the engine. For most construction operations the bucket sizes range from 2 to 8 cy with engine

FIGURE 6.1 | Wheel loader depositing its load into a truck.

Courtesy: Cornell Construction Company

sizes from 100 to 500 hp. Larger size loaders are available for use in mining operations. The production rate of a loader depends on four factors:

1. Volume of material in the bucket
2. Time required to travel from loading to dumping location
3. Time to return from dumping to the loading location
4. Fixed time of the loader

The volume of material in the bucket depends on the size of the bucket and the bucket fill factor. The bucket fill factor is a function of the type of material that is handled. The volume in the bucket is calculated by multiplying the bucket size times the bucket fill factor. Table 6.1 gives bucket fill factors for wheel and track loaders.

TABLE 6.1 | Bucket fill factors for wheel and track loaders.

Type of material	Wheel loader bucket fill factor (%)	Track loader bucket fill factor (%)
Blasted rock		
well blasted	80–95	80–95
poorly blasted	60–75	60–75
Aggregate		
less than $\frac{1}{2}$-in.	90–100	95–110
greater than $\frac{1}{2}$-in.	85–95	90–110
Other material		
rock/soil mix	100–120	100–120
soil	80–100	80–100

For most loader operations the low speed is about 4 mph, intermediate speed is 10 mph, and high speed is 30 mph. Because of the short distances travelled by loaders it is sometimes desirable to express the speed of a loader in feet per minute as well as miles per hour. A speed of 1 mph is equal to 5,280 ft in 60 min, which equals 88 ft/min.

The fixed time of a loader is the time required to load the bucket, shift gears, turn, and dump the load. The fixed time of track loaders is about 20 sec, which equates to 3 bucket loads per minute. For wheel loaders the fixed time is about 30 sec for bucket sizes from 2 to 4 cy and about 35 sec for bucket sizes from 5 to 8 cy.

EXAMPLE 6.4

An 8 cy wheel loader is used to transport $1\frac{1}{2}$-in. maximum size aggregate from a stockpile to a concrete mix plant that is located 450 ft from the stockpile. The aggregate has a dry weight of 95 lb/cf. The loader will haul aggregate at 75% of its first gear speed of 4 mph and return at 75% of its second gear speed of 9 mph. The fixed time of the loader is 35 sec. Assume a 50-min productive hour and calculate the loader production rate in tons per hour.

Volume of material per trip:

From Table 6.1, for $1\frac{1}{2}$-in. aggregate the average bucket fill factor is 90%.

Volume hauled in the loader bucket = 90% × 8 cy = 7.2 cy

Travel speeds of loader:

Haul speed = 75% × 4 mph × 88 ft/min per mph = 264 ft/min

Return speed = 75% × 9 mph × 88 ft/min per mph = 594 ft/min

Cycle time:

Haul time = 450 ft/264 ft/min $\quad$ = 1.70 min

Return time = 450 ft/594 ft/min $\quad$ = 0.76 min

Fixed time = 35 sec/60 sec/min = 0.58 min

$\qquad\qquad$ Total cycle time = 3.04 min/trip

Production rate:

For 50-min hr, production rate of loader = (7.2 cy/3.04 min) × (50 min/hr)

$\qquad\qquad\qquad\qquad\qquad$ = 118.4 cy/hr

Production rate converting cy/hr to tons/hr:

$$\text{Loader production rate} = \frac{118.4 \text{ cy}}{\text{hr}} \times \frac{27 \text{ cf}}{\text{cy}} \times \frac{95 \text{ lb}}{\text{cf}} \times \frac{\text{ton}}{2{,}000 \text{ lb}}$$

$\qquad\qquad$ = 152 tons/hr

EXAMPLE 6.5

A project requires transporting 14,000 cy of well blasted rock from a stockpile to a jobsite. The jobsite is 11 miles from the stockpile. A wheel loader with a rated bucket capacity of 4 cy will load the rock into 22 cy trucks. The fixed time for the wheel loader is 30 sec. The trucks can maintain an average haul speed of 40 mph and return speed of 45 mph. The truck time at the dump will average 6 min including delays. The cost to move the loader to and from the job will be $1,200. The crew costs are:

> Wheel loader = $75.00/hr
>
> Loader operator = $31.00/hr
>
> Trucks, each = $40.00/hr
>
> Drivers, each = $28.00/hr

Assume a 45-min productive hour and use the most economical number of trucks to estimate the time and cost of transporting the blasted rock.

Quantity of work:

> Gravel to be transported = 14,000 cy

Production rate of loading and hauling operation:

> From Table 6.1, for well blasted rock the average bucket fill factor = 87.5%.
>
> Volume in the loader bucket = 87.5% × 4 cy
>
> $$= 3.5 \text{ cy}$$

Production rate of loader = (3.5 cy/30 sec) × (60 sec/min) × (60 min/hr)

$$= 420 \text{ cy/hr}$$

Cycle time of loading and hauling operation

> Time to load a truck: 22 cy/420 cy/hr = 0.052 hr
>
> Truck haul time: 11 mi/40 mph   = 0.275
>
> Truck dump time: 6 min/60 min/hr   = 0.100
>
> Truck return time: 11 mi/45 mph   = 0.244
>
> Total = 0.671 hr/trip

Number of trucks required = cycle time/load time

$$= 0.671/0.052$$

$$= 12.9$$

Time and cost using 12 trucks:

> If 12 trucks are required, there will be fewer trucks than required. Therefore, the time to complete the job will be determined by the rate at which the trucks haul the gravel because the truck production rate is lower than the production rate of the wheel loader.

For a 45-min productive hour, trucks = (12 trucks × 22 cy)/(0.671 hr × 45/60)

$$= 295 \text{ cy/hr}$$

For a 45-min productive hour, loader = 420 cy/hr × 45/60

$$= 315 \text{ cy/hr}$$

The production rate is governed by trucks at 295 cy/hr.

Time = 14,000 cy/295 cy/hr = 47.5 hr

Cost:

    Wheel loader: 47.5 hr @ $75.00/hr   = $3,562.50
    Loader operator: 47.5 hr @ $31.00/hr   = 1,472.50
    Trucks: 47.5 hr × 12 trucks @ $40.00/hr = 22,800.00
    Drivers: 47.5 hr × 12 drivers @ $28.00/hr = 15,960.00
    Moving loader to and from the job   = 1,200.00
                            Total cost = $44,995.00

    Cost per cubic yard = $44,995.00/14,000 cy
                        = $3.21/cy

Time and cost using 13 trucks:

If 13 trucks are used, there will be more trucks than required. Therefore, the time to complete the job will be determined by the production rate of the wheel loader.

Loader production rate with a 45-min productive hour = 420 cy/hr × 45/60
                                                     = 315 cy/hr

Time = 14,000 cy/315 cy/hr
     = 44.4 hr

Cost:

    Wheel loader: 44.4 hr @ $75.00/hr   = $3,330.00
    Loader operator: 44.4 hr @ $31.00/hr   = 1,376.40
    Trucks: 44.4 hr × 13 trucks @ $40.00/hr = 23,088.00
    Drivers: 44.4 hr × 13 drivers @ $28.00/hr = 16,161.60
    Moving loader to and from the job   = 1,200.00
                            Total cost = $45,156.00

    Cost per cubic yard = $45,156.00/14,000 cy
                        = $3.23/cy

Thus, it is more economical to use 12 trucks. Other factors may influence the choice of selecting 12 versus 13 trucks. For example, there may be only 12 trucks available for transporting the rock. Also, many contractors prefer to have more trucks than needed in case one of the trucks become unavailable, such as due to a breakdown in the truck or a driver is absent from work. In this example, if only 12 trucks are used on the project and 1 truck becomes unavailable, the production rate will decrease from 295 cy/hr to 270 cy/hr, a decrease in production rate of 25 cy/hr. However, if 13 trucks are used and 1 truck becomes unavailable, the production rate would only decrease from 315 cy/hr to 295 cy/hr, a difference of 20 cy/hr. Thus, there would be a lower decrease in production rate for the construction operation.

TRANSPORTING MATERIAL WITH CONVEYORS

Portable belt conveyors are used frequently for handling and transporting material such as sand, gravel, crushed aggregate, earth, and concrete. Because of the continuous flow of material at relatively high speeds, belt conveyors have high capacities for handling material.

The amount of material that can be handled by a conveyor depends on the width and speed of the belt and the angle of repose for the material. Portable belt conveyors are available in lengths of 30 to 50 ft, with belt widths of 18 to 30 in. The maximum speed of conveyor belts ranges from 250 to 450 ft/min. A 300-ft/min belt speed is representative for many jobsites. Table 6.2 gives the areas of cross section of materials with various angles of repose.

TABLE 6.2 | Areas of cross sections of materials for loaded conveyor belts, sf.

Width of belt, in.	Angle of repose		
	10°	20°	30°
18	0.134	0.274	0.214
24	0.257	0.331	0.410
30	0.421	0.541	0.668

EXAMPLE 6.6

A portable belt conveyor is used to load sand from a stockpile into trucks. The conveyor has a 24-in. wide belt that has a travel speed of 300 ft/min. The conveyor will load 12 cy dump trucks that will haul the sand 4 miles at an average travel speed of 30 mi/hr. Assume a dump time of 2 min and an angle of repose of sand of 20°. Determine the number of trucks required to balance the production rate of the belt conveyor.

Production rate of conveyor:

Production rate of conveyer = 300 ft/min $\times$ 0.331 sf

= 99.3 cf/min

Hourly production rate = [(99.3 cf/min)/(27 cf/cy)] $\times$ 60 min/hr

= 220.7 cy/hr

Cycle time of construction operation:

Time to load a truck: 12 cy/220.7 cy/hr = 0.05 hr

Time to haul: 4 mi/30 mph   = 0.13 hr

Time to dump: 2 min/60 min/hr   = 0.03 hr

Time to return: 4 mi/30 mph   = 0.13 hr

Total cycle time = 0.34 hr

Production rate of construction operation:

> Quantity hauled per hour = 12 cy/0.34 hr
> $$= 35.3 \text{ cy/hr}$$

Required number of trucks:

> Number of trucks required = 220.7 cy/hr/35.3 cy/hr
> $$= 6.2$$

Therefore, 6 or 7 trucks are required to balance the production rate of the belt conveyor.

HANDLING CAST-IRON PIPE

A project involves loading 12-in. diameter and 18-ft long cast-iron pipe onto trucks at the supplier and hauling the pipe to the jobsite, where the pipe will be laid on the ground along an open trench. The hauled distance will be 20 miles from the supplier to the jobsite. The trucks will average 45 mph loaded and 50 mph empty.

Trucks will be loaded at the supplier by a truck-mounted rental crane at a rate of 4-min per pipe, or 15 pipes/hr. Trucks will be unloaded at the jobsite by a crawler tractor with a side boom, which will travel beside the trucks and distribute the pipe along the trench at a rate of 5-min per pipe, or 12 pipes/hr. The capacity of each truck is 10 pipes.

The crew at the loading site will consist of 1 crane operator, 2 laborers with the crane, and 2 laborers on the truck. The crew at the unloading site will consist of 1 tractor operator and 2 laborers helping to unload pipe from the truck. Assume a 50-min productive hour and determine the number of trucks required and the direct cost per linear foot for handling the pipe.

Cycle time for 1 truck:

> Loading a truck = 10 pipes/15 pipes/hr $\quad$ = 0.67 hr
> Hauling to jobsite = 20 mi/45 mph $\qquad$ = 0.44 hr
> Unloading a truck = 10 pipes/12 pipes/hr = 0.83 hr
> Returning to supplier = 20 mi/50 mph $\quad$ = 0.40 hr
> $\qquad\qquad\qquad$ Total cycle time = 2.34 hr/trip

Determine number of trucks:

> Number of trips per hour = 1.0 trip/2.34 hr
> $$= 0.43 \text{ trips/hr}$$

> Number of pipes hauled per hour per truck = (0.43 trips/hr) $\times$ (10 pipes/trip)
> $$= 4.3 \text{ pipes/hr}$$

> Number of pipes unloaded per hour = 10 pipes/0.83 hr
> $$= 12.04 \text{ pipes/hr}$$

Required number of trucks = 12.04 pipes/hr/4.3 pipes/hr

= 2.8 trucks

An alternate solution = (total cycle time)/(load time)

= 2.34/0.83

= 2.8 trucks

Assume 3 trucks will be used.

Production rate:

If 3 trucks are used there will be more trucks than needed. Therefore, the production rate will be governed by the largest value in the cycle time, which is unloading a truck at a rate of 12 pipes/hr. Thus, the production rate for a 50-min productive hour can be calculated as:

For a 50-min productive hour, production rate = 12 pipes/hr × (50/60)

= 10 pipes/hr

Costs:

Hourly costs of labor and equipment:

Crane, rental rate = 1 @ $65.00/hr =	$65.00
Crane operator = 1 @ $30.00/hr =	30.00
Tractor with boom = 1 @ $45.00/hr =	45.00
Tractor operator = 1 @ $30.00/hr =	30.00
Trucks = 3 @ $35.00/hr =	105.00
Truck driver = 3 @ $25.00/hr =	75.00
Laborers = 6 @ $22.00/hr =	132.00
Total crew costs =	$482.00/hr

Cost per lineal foot = $482.00/hr/(10 pipes/hr × 18 ft/pipe)

= $2.68/ft

Any cost of moving the equipment to the job and back to the storage yard should be prorated to the total length of pipe handled, and added to the unit cost of pipe just determined, to obtain the total cost per unit length.

HANDLING LUMBER

Lumber is usually loaded onto flatbed trucks by a forklift or laborers at the lumberyard and hauled to the job. At the jobsite, the lumber is unloaded and stacked according to size.

The unit of measure of lumber is board feet (bf). Board feet is a volumetric measure. One board foot is the amount of lumber contained in a 1-in. thick board that is 12-in. wide and 1 ft long. For example, a 2 × 8 piece of lumber that is 14 ft long is equivalent to (2 × 8/12) × 14 = 18.67 bf. A 2 × 12 board that is 20 ft long is equivalent to (2 × 12/12) × 20 = 40 bf.

A laborer should be able to handle lumber at a rate of 2,000 to 4,000 board feet per hour (bf/hr). A reasonable average rate should be about 3,000 bf/hr.

Trucks of the type generally used will haul 2 to 6 tons, corresponding to 1,000 to 3,000 bf per load. The average speed of a truck will vary with the distance, type of road, traffic congestion, and weather.

EXAMPLE 6.7

Estimate the cost of transporting 50,000 bf of lumber from a lumberyard to a job-site. Trucks that can carry 2,000 bf per load will transport the lumber. The jobsite is 7 miles from the lumberyard.

An examination of the haul road indicates an average speed of 35 mph for trucks. Assume a worker will handle 3,000 bf/hr of lumber and a 45-min/hr productivity. Assume 2 workers (the driver and one laborer) will load and unload a truck. Use a truck cost of $45.00/hr, driver cost of $30.00/hr, and laborer cost of $25.00/hr to calculate the time and cost of the job.

Quantity of work:

Lumber to be hauled = 50,000 bf

Cycle time:

Assume both the truck driver and a laborer will load a truck.

Rate of loading a truck, 2 workers × 3,000 bf/hr = 6,000 bf/hr

Loading = 2,000 bf/6,000 bf/hr	= 0.333 hr
Haul time = 7 mi/35 mph	= 0.200 hr
Return time = 7 mi/35 mph	= 0.200 hr
Unloading = 2,000 bf/6,000 bf/hr	= 0.333 hr
Total cycle time	= 1.066 hr/trip

Production rate:

Number of trips per hour = 1.0/(1.066 hr/trip) = 0.938 trips/hr
Rate of hauling lumber = 2,000 bf/trip × 0.938 trips/hr
= 1,876 bf/hr

Assume 45 min/hr as productive,
Production rate = 1,876 bf/hr × 45/60
= 1,407 bf/hr

Time:

Using 1 truck and 1 loader = 50,000 bf/1,407 bf/hr
= 35.5 hr

Cost:

Truck = 35.5 hr @ $45.00/hr	= $1,597.50
Driver = 35.5 hr @ $30.00/hr	= 1,065.00
Laborer = 35.5 hr @ $25.00/hr	= 887.50
Total cost	= $3,550.00

Unit cost:

Cost per bf = $3,550.00/50,000 bf = $0.07/bf

HANDLING AND TRANSPORTING BRICKS

Bricks are generally loaded at the brick supplier by forklifts or small cranes that are mounted on the beds of flatbed trucks and hauled to a jobsite. The capacity of the trucks is normally 2,000 to 3,000 bricks per load. Popular sizes of building bricks weigh about 4 lb each.

Upon arrival at the jobsite, the bricks are unloaded by a small crane that is mounted on the truck, or by a forklift, onto small four-wheel tractors that transport the bricks around the perimeter of the structure where the bricks are to be installed.

Laborers, using brick tongs, can carry 6 to 10 bricks per load to the brick mason for laying the brick. A worker should be able to pick up a load, walk to the location of the brick mason and deposit the load, and return for another load in $\frac{1}{2}$ to 1 min per trip. If it is assumed that the average time for a trip is $\frac{3}{4}$ min, and the laborer carries 8 bricks per trip, in 1 hr he or she will handle 640 bricks. The actual number of bricks that a laborer can handle will depend on the jobsite conditions at a particular project. Table 6.3 provides rates for various job conditions.

TABLE 6.3 | Rates of handling bricks.

Bricks carried per trip	Trip time, min	Bricks hauled per hour	Hours per 1,000 bricks
6	0.50	720	1.39
8	0.50	960	1.04
10	0.50	1,200	0.83
6	0.75	480	2.08
8	0.75	640	1.56
10	0.75	800	1.25
6	1.00	360	2.78
8	1.00	480	2.08
10	1.00	600	1.67

EXAMPLE 6.8

A total of 150,000 bricks have been delivered to a central location on the jobsite. Estimate the cost of transporting the bricks from the central location to where they will be installed by brick masons. The bricks will be distributed in stacks around the jobsite by a skid-loader at a rate of 1,800 bricks per hour.

Laborers will be located around the perimeter of the structure to carry the bricks to the brick masons. Assume each laborer will carry 10 bricks per load and will average $\frac{3}{4}$ min per trip. Determine the cost based on using the economical number of laborers.

Quantity of work:

Material to be transported = 150,000 bricks

Production rate:

Skid-loader production rate = 1,800 bricks/hr

Laborer production rate = (10 brick/trip)/(3/4 min/trip)

= 13.3 bricks/min

Labor hourly production rate = 13.3 bricks/min × 60 min/hr

= 800 bricks/hr

Laborers required:

Loader production rate = 1,800 bricks/hr

Laborer production rate = 800 bricks/hr

Number of required laborers = 1,800/800 = 2.25

Cost using 2 laborers:

If 2 laborers are used, there will be fewer laborers than required to achieve the production rate of the loader. Therefore, the production rate will be governed by the laborers (2 × 800 bricks/hr = 1,600 bricks/hr), rather than the loader production rate of 1,800 bricks/hr.

Time = 150,000 bricks/1,600 bricks/hr = 93.8 hr

Cost using 2 laborers:

Skid-loader = 93.8 hr @ $34.00/hr	=	$3,189.20
Operator = 93.8 hr @ $28.00/hr	=	2,626.40
Laborers = 93.8 hr × 2 laborers @ $21.00/hr	=	3,939.60
	Total cost =	$9,755.20

Cost per 1,000 bricks = $9,755.20/150 = $65.03 per 1,000 bricks

Cost using 3 laborers:

If 3 laborers are used, there will be more laborers than required. Therefore, the production rate will be limited by the loader production rate of 1,800 bricks/hr, rather than the laborers production rate of 3 × 800 bricks/min = 2,400 bricks/min.

Production using 3 laborers = 1,800 bricks/hr

Time = 150,00 bricks/1,800 bricks/hr = 83.3 hr

Cost using 3 laborers:

Skid-loader = 83.3 hr @ $34.00/hr	=	$2,832.20
Operator = 83.3 hr @ $28.00/hr	=	2,332.40
Laborers = 83.3 hr x 3 laborers @ $21.00/hr	=	5,247.90
	Total cost =	$10,412.50

Cost per 1,000 brick = $10,412.50/150 = $69.42/1,000 bricks

Thus, the most economical cost is $9,755.20 using 2 laborers and the skid-loader.

PROBLEMS

6.1   The owner of a sand and rock quarry is considering the purchase of trucks to haul sand and gravel to customers. Two sizes of trucks are being considered, namely, 12- and 18-cy diesel engine dump trucks. The haul distance will vary from 25 to 40 miles, with an average distance of about 30 miles.

   It is estimated the 12-cy trucks can travel at an average speed of 50 mi/hr loaded and 55 mi/hr empty, while the 18-cy trucks can travel at average speeds of 45 mi/hr loaded and 50 mi/hr empty. The trucks will be loaded from stockpiles using a loader that has a production rate of 95 cy/hr. The average truck time at the dump will be 3 min for the 12-cy trucks and 4 min for the 18-cy trucks. Assume the trucks and loader will operate on a 45-min effective hour.

   The cost of a 12-cy truck is $50.00/hr and the 15-cy truck is $65.00/hr. The cost of a truck driver is $28.00/hr. Determine which size truck is more economical.

6.2   Two sizes of loaders are being considered, a 1-cy loader with a 75 cy/hr production rate and a 2-cy loader with a 135 cy/hr production rate. Diesel trucks with 15-cy capacity will be used to haul the material. In addition to the time required by the loader to load a truck, there will be an average delay of 5 min waiting for another truck to move into position for loading. Assume a 50-min productive hour for both the loader and the trucks.

   The cost of the 1-cy loader is $75.00/hr, 2-cy loader is $140.00/hr, and loader operator is $31.00/hr, The cost of a 15-cy truck with a driver is $80.00/hr. Determine which size loader is more economical based on the cost to load material into trucks.

6.3   The operator of a rock quarry is invited to bid on furnishing 42,000 cy of crushed aggregate for a job. The aggregate will be delivered to the job, which is 27 miles from the quarry. The aggregate will be loaded into trucks by a loader that can load at a rate of 75 cy/hr. The trucks will haul 18 cy/load at an average speed of 50 mi/hr loaded and 55 mi/hr empty. The estimated time to dump a load is 5 min. Assume a 45 min effective hour for the loading and hauling operation.

   The cost information for the job is:
   Royalty paid for aggregate = $1.35/cy
   Overhead = $0.75/cy
   Profit = $1.20/cy
   Loader = $85.00/hr
   Operator = $32.00/hr
   Trucks, each = $55.00/hr
   Truck driver, each = $28.00/hr
   Foremen = $35.00/hr

What should be the bid price per cubic yard?

6.4   A highway paving project requires moving 25,000 tons of coarse aggregate from a rock quarry to a stockpile at the jobsite to feed the concrete batch plant. The aggregate has a weight of 3,100 lb/cy. A 5 cy wheel loader with a 35 sec cycle time will load 23-ton trucks at the quarry and haul the aggregate a distance of 14 miles to the jobsite. The average haul speed will be 40 mph and return speed of 45 mph. Seven trucks will be used and each truck with a driver costs $65.00/hr and the loader with an operator costs $105.00/hr. Assume a 45-min productive hour and estimate the total time, total cost, and cost per ton to move the aggregate.

6.5 A portable conveyor is used to load material from a stockpile into trucks. The conveyor has an 18-in.-wide belt that has a travel speed of 250 ft/min. The conveyor will load 15-cy dump trucks that will haul the material 23 mi at an average travel speed of 50 mi/hr. Assume a dump time of 4 min and an angle of repose of 30 degrees for the material. Determine the number of trucks required to balance the production rate of the conveyor.

6.6 A project involves transporting iron pipe to a jobsite that is 43 miles from the supplier. The pipe will be hauled in trucks that can carry 16 pipes per load. Each pipe is 20 ft in length. The average haul speed of a truck will be 50 mi/hr and return speed of 55 mi/hr. A loader at the supplier will load trucks at a rate of 5 min per pipe and a crawler tractor with a side boom at the jobsite can unload the pipe at a rate of 8 min per pipe.

   The cost of the loader, with operator, at the supplier is $85.00/hr and the cost of the tractor, with operator, at the jobsite is $75.00/hr. The cost of a truck, with driver, is $55.00/hr. Determine the economical number of trucks and the direct cost per foot of pipe for transporting the pipe.

6.7 Estimate the total direct cost and cost per 1,000 bf of lumber for transporting 70,000 bf of lumber from the lumberyard to a jobsite, which is 18 miles from the yard. Each truck can haul 3,000 bf per load and can average 40 mi/hr loaded and 45 mi/hr empty. Assume $55.00/hr for a truck with a driver.

   Two laborers will load lumber onto the truck at the yard and another 2 laborers will unload the truck and stack the lumber at the job. Each laborer can handle 1,200 bf/hr. Assume laborers cost $21.00/hr. Assume a 50-min effective hour and determine the cost per 1,000 bf of lumber based on using the economical number of trucks.

6.8 A project requires transporting 150,000 bricks by trucks from a brickyard to a jobsite 53 miles from the yard. A truck, which can haul 3,000 bricks per load, will average 50 mi/hr loaded and 55 mi/hr empty. A forklift at the yard can load bricks at a rate of 2,500 bricks/hr. A skid loader at the jobsite will unload and distribute the brick at a rate of 1,800 bricks/hr. Assume a 45-min effective hour for the loading and hauling operation.

   The cost of the forklift, with operator, is $60.00/hr and the skid loader, with operator, is $45.00/hr. A truck with a driver costs $50.00/hr. Determine the total cost and cost per 1,000 bricks based on using the economical number of trucks.

Earthwork and Excavation

Most projects involve excavation to some extent. The extent of excavation varies from a few cubic yards for footings and trenches for pipes to millions of cubic yards for large earth-filled dams. It is usually a relatively simple operation to determine the quantity of material to be excavated. It is much more difficult to estimate the rate at which earthwork will be handled by excavating equipment. The many factors that can affect the production rate of excavating equipment can be divided into two groups: job and management.

JOB FACTORS

Job factors involve the type or classification of soil, extent of water present, weather conditions, freedom of workers and equipment to operate on the job, size of the job, length of haul for disposal, etc. It is difficult for the contractor to change job conditions. The contractor must analyze each job to determine the conditions that can affect construction operations that will impact the time and cost of performing the work. The analysis of job factors requires a thorough review of the plans and specifications of the bid documents, an evaluation of the soil investigation report, and a visit to the jobsite where the project is to be constructed. For earthwork projects, the bid documents usually contain a soil report that provides geotechnical information about the soil and subsurface conditions. The estimator can also use other sources, such as topographic maps, geologic maps, agriculture maps, or aerial photographs. Information related to job factors must be collected and considered before preparing a cost estimate.

MANAGEMENT FACTORS

Management factors involve organizing for the job, maintaining good morale among workers, selecting and using suitable equipment and construction methods, exercising care in servicing equipment, establishing good field supervisory personnel, and others. These factors are under the control of the contractor. The contractor has control of his or her operations and can often overcome adverse job conditions by good management practices. For example, the haul road can be properly maintained to increase the efficiency and productivity of scrapers or trucks that must operate over the haul roads.

ESTIMATING PRODUCTION RATES OF EQUIPMENT

One source of production rates of earthmoving equipment is data from equipment manufacturers. Most manufacturers publish tables and charts of production rates for their equipment. Those production rates are usually based on ideal working conditions—using new equipment that is operating continuously, without interruptions, using highly skilled operators, and not encountering adverse weather conditions. Equipment manufacturers typically give cycle times based on the size of their equipment. The general equation for calculating ideal production rate is:

$$\text{Production rate} = (\text{Cycles per unit of time}) \times (\text{Volume per cycle}) \quad \textbf{[7.1]}$$

For example, a manufacturer's specification may give a cycle time of 20 sec. for a wheel-type loader with a 2-cy bucket size. The cycle time for the equipment is the amount of time to load the bucket, swing the load for deposit in a truck, and return the bucket to start the next cycle of loading the bucket. The 20 sec of time is equivalent to 3 cycles per minute, or 180 cycles per hour. For this unit of equipment the production rate can be calculated from Eq. [7.1] as:

$$\text{Production rate} = (\text{Cycles per unit of time}) \times (\text{Volume per cycle})$$
$$= (180 \text{ cycles/hr}) \times (2 \text{ cy/cycle})$$
$$= 360 \text{ cy/hr}$$

Another source of production rates of equipment is company job records from completed projects. Companies usually keep job records of production rates of equipment used on their projects. It should be recognized that production rates from previous jobs may vary substantially from one job to another. For example, the operator from the previous job may not be the same operator for a job that is estimated for a future project. Also, the condition of the equipment and the working conditions on the future project may be different than job records of previous projects.

Whether using production rates from manufacturers or production rates from previous projects of companies, efficiency factors should be used to adjust production rates for estimating purposes. There are two methods of adjusting ideal production rates to probable production rates for preparing estimates. One method is to use the effective working minutes per hour. For example, if a unit of equipment is expected to operate effectively for 45 min of each working hour, the calculated efficiency factor = 45/60 = 0.75. The ideal production rate can be multiplied by the 0.75 efficiency factor to estimate the probable production rate of the equipment. Table 7.1 shows efficiency factors based on the number of minutes worked per hour.

TABLE 7.1 | Efficiency factors based on effective hours.

Minutes worked per hour	Efficiency factor
60	1.00
55	0.92
50	0.83
45	0.75
40	0.67
35	0.58
30	0.50

Another method of calculating the probable production rate of equipment is to multiply the ideal production rate by an efficiency factor that is based on job conditions and management conditions. See Table 7.2. The values in this table are from TM 5-331B of the U.S. Department of the Army.

TABLE 7.2 | Efficiency factors based on job and management conditions.

Job conditions[1]	Management conditions[2]			
	Excellent	Good	Fair	Poor
Excellent	0.84	0.81	0.76	0.70
Good	0.78	0.75	0.71	0.65
Fair	0.72	0.69	0.65	0.60
Poor	0.63	0.61	0.57	0.52

[1] Job conditions exclude type of material involved, but include: a) surface and weather conditions, b) topography and work dimensions, and c) specification requirements for work methods or sequence.

[2] Management conditions include: a) skill, training, and motivation of workers, b) selection, operation, and maintenance of equipment, and c) planning, job layout, supervision, and coordination of work.

METHODS OF EXCAVATING AND HAULING EARTH

Methods of excavating vary from hand digging and shoveling for small jobs to that done by backhoes, front shovels, draglines, clamshells, scrapers, bulldozers, loaders, trenching machines, boring machines, and dredges. Some material, such as rock, is so hard that it is necessary to place explosives in drilled holes to loosen it prior to excavating.

The selection of excavating equipment depends on the type of soil and job-site conditions. For example, a dragline may be used for excavating waterway channels where extensive water is present, whereas a front shovel may be selected to excavate rock material that has been loosened in an excavation pit. Scrapers are often used for excavating and hauling earth for highway projects. Scrapers are capable of loading, hauling, and distributing the soil in the compacted fill area. Thus, they can operate as individual units for both the loading and hauling operations.

When earth or rock is excavated by equipment and loaded into trucks to haul to the disposal area, the size of the hauling unit should be balanced with the output of the excavating equipment. As illustrated in Chapter 6, the number of trucks required to balance the production rates of the loader and trucks is calculated by dividing the total cycle time of the earthwork operation by the cycle time of the loader.

PHYSICAL PROPERTIES OF EARTH

To estimate the cost of excavating and hauling earth, it is necessary to know the physical properties of earth because the volume changes during construction operations. For an earthwork operation, the soil is excavated from its natural state, placed in a hauling unit, and transported to the disposal area, where it is distributed and compacted. The earth material can be measured as excavated, hauled, or compacted. For example, 1.0 cy of soil that is excavated from the ground may occupy 1.25 cy after it is loosened and placed in the hauling unit. After the soil is compacted in place it may occupy 0.9 cy.

Earth or soil that is to be excavated is called *cut* or *bank measure*. In its undisturbed condition, prior to excavating, earth normally weighs from 95 to 105 lb/cf, with 100 lb/cf as a median value. The variation in unit weight depends on the type of soil and moisture content. The soil report for a project provides the classification, unit weight, and moisture content of samples from borings that are taken during the soil investigation. Before loosening, solid rock weighs from 130 to 160 lb/cf. For example, shale weighs about 130 lb/cf, whereas limestone weighs about 160 lb/cf.

Earth that is placed in a hauling unit for transportation to the fill area is called *loose measure*. After being loosened during excavation and placed in a hauling unit, earth and rock will occupy a larger volume, with a corresponding reduction in weight per unit of volume. This increase in volume is described as *swell* and is expressed as a percentage gain compared with the original volume. In the haul condition, earth generally will weigh from 80 to 95 lb/cf. The amount of swell depends on the type of soil and the amount of loosening during excavation.

Earth that is to be compacted is called *fill* or *compacted measure*. When the earth is placed in a fill area and compacted with compaction equipment, it occupies a smaller volume than in its natural state in the cut or bank measure. This decrease in volume is described as *shrinkage* and is expressed as a percentage of the original volume, or bank measure. In the compacted fill area, the unit weight of soil can vary from 110 to 120 lb/cf. The amount of shrinkage depends on the type and moisture content of the soil and on the type and number of passes of the compaction equipment. Shrinkage can vary from 5 to 15 percent, depending on these factors.

The correlation between unit weights, volumes, swell, and shrinkage can be obtained from Eqs. [7.2] to [7.5]. Eqs. [7.2] and [7.3] apply to volume measures and Eqs. [7.4] and [7.5] apply to weight measures.

By volume:

$$L = \left(1 + \frac{S_w}{100} \right) B \qquad [7.2]$$

$$C = \left(1 - \frac{S_h}{100} \right) B \qquad [7.3]$$

where S_w = percentage of swell
S_h = percentage of shrinkage
B = volume of undisturbed soil
L = volume of loose soil
C = volume of compacted soil

By weight:

$$L = \frac{B}{1 + S_w/100} \qquad [7.4]$$

$$C = \frac{B}{1 - S_h/100} \qquad [7.5]$$

where S_w = percentage of swell
S_h = percentage of shrinkage
B = unit weight of undisturbed soil
L = unit weight of loose soil
C = unit weight of compacted soil

Distinguishing between the bank, loose, and compacted measure is extremely important to the estimator and the bidding process because the contractor may be paid by any one of the three units of measure. The contract documents define the measure of payment to the contractor for earthwork. For example, the bid form in the contract documents may stipulate that the contractor will be paid based on the amount of soil excavated (bank measure), the amount of soil hauled (loose measure), or the amount of soil compacted (compacted measure).

The estimator can prepare the estimate in any unit of measure, but the final bid amount must correspond to the unit of measure specified in the contract bid documents. Equations [7.2] through [7.5] can be used to convert between these three units of measure. Table 7.3 provides the range of percentage of swell for various soils. Table 7.4 provides representative unit weights for various types of earth and rock.

TABLE 7.3 | Range of swell factors of earth and rock.

Material	Swell, %
Sand or gravel	10–15
Loam	15–20
Common earth	20–30
Hard clay	25–40
Solid rock	50–80

TABLE 7.4 | Approximate unit weights of earth and rock.

Type of material	Unit weight, lb/cf*		
	Loose	Bank	Compacted
Sand, dry	75	95	100
Sand, wet	90	100	105
Clay, dry	75	105	105
Clay, wet	83	110	120
Earth, dry	85	105	115
Earth, wet	95	115	120
Earth and gravel	101	120	125
Gravel, dry	95	105	120
Gravel, wet	110	125	130
Rock, well blasted	105	155	120

* Exact values vary with grain size distribution, moisture content, compaction, and other factors.

EXAMPLE 7.1

Clay soil whose bank unit weight is 105 lb/cf is excavated and hauled by 22-cy (loose measure) trucks that have an empty weight of 67,500 lb. An evaluation of the physical properties of the soil indicates a swell factor of 30 percent. Based on four passes of the compaction equipment it is anticipated the shrinkage factor will be 15 percent. For each 22-cy truckload of soil that is hauled, calculate the equivalent bank measure and compacted measure volume. Also, what is the total vertical weight of the hauling unit and payload of soil that is hauled on each truck-trip?

Equivalent bank and compacted measure volumes, using Eqs. [7.2] and [7.3]:

$$L = \left(1 + \frac{S_w}{100}\right) B$$

$$22 \text{ cy} = \left(1 + \frac{30}{100}\right) B$$

$$B = \frac{22 \text{ cy}}{1 + 0.30}$$

$$= 16.9 \text{ cy bank volume}$$

$$C = \left(1 - \frac{S_h}{100}\right) B$$

$$= (1 - 0.15)\, 16.9 \text{ cy}$$

$$= 14.4 \text{ cy compacted volume}$$

Total vertical weight of a truck and its soil payload, using Eq. [7.4]:

$$L = \frac{B}{1 + S_w/100}$$

$$= \frac{105 \text{ lb/cf}}{1 + 30/100}$$

$$= 80.8 \text{ lb/cy loose weight of soil}$$

Soil payload $= 80.8 \text{ lb/cf} (27 \text{ cf/cy}) \times (22 \text{ cy})$

$$= 47,995 \text{ lb of soil}$$

Total weight $=$ empty truck $+$ payload

$$= 67,500 \text{ lb} + 47,995 \text{ lb}$$

$$= 115,495 \text{ lb}$$

Thus, for each truck-trip the amount of soil that is handled is equivalent to 16.9 cy bank, 22 cy loose, or 14.4 cy compacted. Each truckload weighs 115,495 lb.

EXAMPLE 7.2

A compacted fill is being built by a 2-drum sheep's-foot roller that is pulled at 1.5 mph. Each drum is 5 ft wide. Four passes are required to obtain the 6-in. depth of compacted fill to achieve the required compaction density. An analysis of the soil indicates a swell factor of 25% and a shrinkage factor of 10%.

A spread of 20 cy, loose measure, trucks will haul the fill dirt to the compacted area. The cycle time of a single truck is 20 min. Determine the number of required trucks to balance the production rate of the compaction operation.

Payload of a single truck:

Combining Eqs. [7.2] and [7.3] to convert loose volume to compacted volume:

$$C = \left[\frac{(1 - S_h/100)}{(1 + S_w/100)} \right] L$$

$$= \left[\frac{(1 - 10/100)}{(1 + 25/100)} \right] (20 \text{ cy})$$

$$= 14.4 \text{ cy}$$

Production rate of a single truck:

Truck production rate $=$ Payload/Cycle time

$$= (14.4 \text{ cy})/(20 \text{ min}/60 \text{ min./hr})$$

$$= 43.2 \text{ cy/hr}$$

Production rate of compactor:

$$\text{Rate} = \frac{(\text{compactor speed}) \times (\text{compaction depth}) \times (\text{drum width}) \times (\text{no. drums})}{(\text{number of passes})}$$

$$= \frac{(1.5 \text{ mph} \times 5{,}280 \text{ ft/mi}) \times (6\text{-in.}/12 \text{ in./ft}) \times (5\text{-ft/drum}) \times (2 \text{ drums})}{(4 \text{ passes})}$$

$$= 9{,}900 \text{ cf/hr}$$

Converting cf/hr to cy/hr = 9,900 cf/hr/27 cf//cy

$$= 366.7 \text{ cy/hr}$$

Number of trucks required to balance production rate of compactors:

No. of trucks = (Compactor production rate)/(Truck production rate)

$$= 366.7 \text{ cy/hr}/43.2 \text{ cy/hr}$$

$$= 8.5 \text{ trucks}$$

Therefore 9 trucks are needed.

EXCAVATING BY HAND

Numerous types and sizes of excavating equipment are available. The selection of excavating equipment for a particular project depends on many factors: the type and quantity of material to be excavated, the depth of excavation, the amount of groundwater in the construction area, the required haul distance, and the space available for operation of the equipment. For some jobs, space and working room are not available for excavating equipment to operate, which requires excavation by laborers.

Generally, it is desirable to use excavating equipment instead of excavation by laborers; however, at some jobsites the space is not sufficient for equipment to operate. For example, excavation for a motor-pump foundation for a unit in a refinery may be located in a confined space that prevents access by an excavator. An excavator may be able to access an excavation area, but may not be able to operate because of an overhead pipe rack that blocks the clearance necessary for the equipment to operate. Excavation may be required in an area where there are numerous underground electric or telephone cables. For each of these situations, excavating by hand methods may be necessary.

The rate at which a laborer can excavate varies with the type of material, extent of digging required, height to which the material must be lifted, and climatic conditions. If loosening is necessary, a pick is most commonly used. Lifting is generally done with a round-pointed, long-handled shovel; 150 to 200 shovels of earth are required to excavate 1 cy in its natural state. Representative rates of handling earth by hand are given in Table 7.5.

If the excavated earth is to be hauled distances up to 100 ft, wheelbarrows are frequently used. A wheelbarrow will hold about 3 cf, loose volume. A worker should be able to haul the earth 100 ft and return in about $2\frac{1}{2}$ min if the haul path is reasonably firm and smooth. Filling the wheelbarrow with loose earth will require about $2\frac{1}{2}$ min. Thus, it will require about 5 min to load and haul 3 cf of earth up to 100 ft. This corresponds to about 1 hr/cy, bank measure.

TABLE 7.5 | Rates of handling earth by hand.

Operation	cy/hr
Sand or loam	
Shoveling and loading trucks	0.9–1.1
Excavating from trenches	0.8–1.0
Excavating from pits	0.6–0.8
Backfilling by hand	2.2–2.5
Spreading loose sand	4.4–5.0
Ordinary common earth	
Shoveling and loading trucks	0.6–0.8
Excavating from trenches	0.4–0.6
Excavating from pits	0.3–0.5
Backfilling by hand	2.0–2.3
Spreading loose earth	3.7–4.3
Heavy soil and hard clay	
Shoveling and loading trucks	0.4–0.6
Excavating from trenches	0.3–0.4
Excavating from pits	0.2–0.3
Backfilling by hand	1.5–1.7
Spreading loose clay	2.8–3.3

EXAMPLE 7.3

Excavation of a trench in a confined area of a refinery requires hand excavation and backfilling by three laborers because the space is too small for excavating equipment. The cost of a laborer is $21.00/hr. Assume a 40-min productive hour and use the information in Table 7.5 to estimate the time and cost of excavating and backfilling the trench, which is 3 ft wide, 4 ft deep, and 15 ft long in common earth.

Quantity of work:

Volume of earth = (3 ft × 4 ft × 15 ft)/(27 cf/cy)

= 6.7 cy

Time:

Using average values from Table 7.5:

Excavating earth from trench, 6.7 cy/0.5 hr/cy = 13.4 hr

Backfilling earth into trench, 6.7 cy/2.15 hr/cy = 3.1 hr

Total labor hours = 16.5 hr

For a 40-min productive hour, the total hours = 60/40 × 16.5 hr

= 24.8 hr

Time using 3 laborers = 24.8 hr/3 laborers

= 8.3 hr

Cost:

Labor cost = 3 laborers × 8.3 hr @ $21.00/hr = $522.90

Cost per cubic yard = $522.90/6.7 cy = $78.04/cy

Cost per linear foot = $522.90/15 ft = $34.86/linear ft

EXCAVATING WITH TRENCHING MACHINES

Even though it may be economical to excavate short sections of shallow trenches with hand labor, a trenching machine is more economical for larger jobs. Once the machine is transported to the job and put into operation, the cost of excavating is considerably less than the cost by hand. For a given job, the savings in excavating costs resulting from the use of the machine as compared with hand excavating must be sufficient to offset the cost of transporting the machine to the job and back to storage after the job is completed. Otherwise, hand labor is more economical.

Trenching machines can be purchased or rented. Several types are available. For shallow trenches such as those required for grade beams, underground electric cable, telephone lines, or television cables, a trenching machine with a dozer blade attached to the front, as illustrated in Fig. 7.1, is frequently used. The dozer blade is used for backfilling the trench. This type of equipment is highly maneuverable because of its rubber tires and small size.

Equipment of this type can be used to dig narrow trenches 8 to 12 in. wide to depths of 8 ft, as shown in Table 7.6. The digging speed will depend on the type of soil, width and depth of trench, and horsepower of the trencher. Equation [7.6] can be used to approximate the digging speed for this type of equipment. The soil factor C in Eq. [7.6] can be approximated as shown in Table 7.7.

FIGURE 7.1 | Self-transporting trenching machine.

Courtesy: The Charles Machine Works, manufacturer of Ditch Witch equipment.

$$S = \frac{C \times hp}{D \times W}$$ [7.6]

where S = digging speed, ft/min
  C = soil factor
  D = depth of trench, in.
  W = width of trench, in.
  hp = flywheel horsepower of engine

TABLE 7.6 | Representative sizes of trenches for small trenching machine.

Depth of trench, in.	Maximum width of trench, in.
48	24
60	20
72	16
84	14
96	12 for soft soil
96	10 for firm soil
96	8 for hard soil

TABLE 7.7 | Values of soil factor C for small trenching machine.

Type of soil	C
Sandstone	20
Hard clay	40
Firm clay	60
Soft clay	90

EXAMPLE 7.4

Estimate the total cost and cost per linear foot for excavating and backfilling 1,500 linear ft of trench in hard clay for an underground electrical cable. The trench will be 9 in. wide and the average depth will be 48 in. A rubber-tire trenching machine with 39 hp will be used. The trencher will deposit the excavated soil on each side of the open trench.

After the electrical cable is installed the open trench will be backfilled by a blade that is attached to the front of the trenching machine. The maximum forward speed of the trencher is 4.0 mph. Assume a backfill speed of 20% of the maximum forward speed of the trencher.

The cable will be installed in open terrain with no major obstructions to retard the progress of the trenching operation. Assume a 45-min productive hour for normal delays and estimate the total cost and cost per linear foot of trench.

Probable digging speed:

From Table 7.7, $C = 40$ for hard clay, and using Equation [7.6] for speed:

$$S = \frac{C \times hp}{D \times W}$$

$$= \frac{40 \times 39}{48 \times 9}$$

$$= 3.6 \text{ ft/min}$$

Time required for trenching:

Time to trench = (1,500 ft/3.6 ft/min)/(60 min/hr)

= 6.9 hr

For a 45-min effective hour, time = 6.9 hr × 60/45

= 9.2 hr

Time required for backfilling:

Excavated material is deposited on both sides of trench,

therefore 2 passes of backfill are required.

Assume backfill speed is 20% of maximum forward speed of 4.0 mi/hr.

Speed of backfilling = 20% × 4.0 mi/hr × 5,280 ft/mi/60 min/hr

= 70.4 ft/min

For a 45-min effective hour, speed = 70.4 ft/min × 45/60

= 52.8 ft/min

Time to backfill trench = (1,500 ft × 2 passes)/(52.8 ft/min)

= 56.8 min, or 0.9 hr

Total time on the job:

Trenching time = 9.2 hr

Backfilling = 0.9 hr

Total = 10.1 hr

Cost:

Trenching machine: 10.1 hr @ $48.00/hr = $484.80

Machine operator: 10.1 hr @ $29.00/hr = 292.90

Pickup with trailer: 10.1 hr @ $9.50/hr = 95.95

Laborer: 10.1 hr @ $19.00/hr = 191.90

Foreman: 10.1 hr @ $31.00/hr = 313.10

Moving equipment to and from job: = 250.00

Total cost = $1,628.65

Cost per linear foot = $1,628.65/1,500 ft

= $1.09/ft

Note the estimated time for this project is 2.1 hours more than a normal 8-hr day. The foreman may decide to work overtime in order to complete the work on the same day it is started. That decision would require working overtime for 2.1 hours at a likely rate of 1.5 for the labor, which would increase the cost. Also, some companies work four 10-hr days as a regular work schedule. For that situation the work would be nearly completed in a regular working day.

LADDER-TYPE TRENCHING MACHINES

For deep trenches, such as those required for sewer pipes and other utilities, the ladder-type machine is used. Inclined or vertical booms are mounted at the rear of the machine. Cutter teeth and buckets are attached to endless chains that travel along the boom. As the machine advances, the earth is excavated and cast along the trench. The depth of cut is adjusted by raising or lowering the boom. By adding side cutters, the width of the trench can be increased. This type of machine can excavate trenches from 16 to 36 in. wide and depths up 12 ft. Table 7.8 provides data on ladder-type trenching machines.

TABLE 7.8 | Data on ladder-type trenching machines.

Depth of trench, ft	Width of trench, in.	Digging speed, ft/hr
4–6	16, 20, 22	100–300
	24, 26, 28	75–200
	30, 32, 36	40–125
6–8	16, 20, 22	40–125
	24, 26, 28	30–60
	30, 32, 36	25–50
8–12	18, 24, 30	30–75
	32, 24, 36	14–40

EXAMPLE 7.5

Estimate the total cost and cost per linear foot for excavation of a trench 30-in. wide with an average depth of 7 ft. The trench is 6,400 ft long in common earth. A ladder-type trenching machine will be used. Assume good management conditions and fair job conditions and use the efficiency factor in Table 7.2 to determine the effective production rate of the trenching operation. A foreman and 2 laborers will assist the trenching operation.

Quantity of work:

Linear feet of trench = 6,400 ft

Cubic feet of excavation = 30 in./(12 in./ft) × 7 ft × 6,400 ft = 112,000 cf

Cubic yards of excavation = 112,000 cf/(27 cf/cy) = 4,148 cy

Production rate:

From Table 7.8, digging speed is between 25 and 50 ft/hr

    Assume an average speed of 37.5 ft/hr

From Table 7.2, for good management conditions and fair job conditions

    Efficiency factor = 0.69

Production rate = 0.69 × 37.5 ft/hr = 25.9 ft/hr

Time:

Time for digging trench = 6,400 lf/25.9 ft/hr

= 247.1 hr

Cost:

Trenching machine = 247.1 hr @ $85.00/hr   = $21,003.50

Operator = 247.1 hr @ $31.00/hr   =   7,660.10

Laborers = 247.1 hr × 2 laborers @ $18.00/hr =   8,895.60

Foreman = 247.1 hr @ $34.00/hr   =   8,401.40

Total = $45,960.60

Cost per cubic yard = $45,960.60/4,148 cy = $11.08/cy

Cost per linear foot = $45,960.60/6,400 ft = $7.18/lf

EXCAVATING WITH DRAGLINES

The dragline (Fig. 7.2) is used for excavating earth for drainage channels and building levees where water is present. It can operate on wet ground and can dig earth out of pits containing water because it does not have to go into the pit or hole to excavate. The dragline is designed to operate below the level of the machine. It operates adjacent to the pit, while excavating material from the pit by casting its bucket, which is very advantageous when earth is removed from a ditch, canal, or pit containing water. The dragline cannot excavate rock as well as a hydraulic excavator.

Frequently, it is possible to use a dragline with a long boom to dispose of the material in one operation if the material can be deposited along the canal or near the pit. This eliminates the need for hauling units, thus reducing the cost of

FIGURE 7.2 | Dragline depositing its load in a truck.

Courtesy: Wright Materials, Inc.

handling the material. Draglines are excellent for excavating trenches when the sides are permitted to establish their angles of repose without shoring.

The size of the dragline is indicated by the size of the bucket, expressed in cubic yards. Most draglines can handle more than one size bucket, depending on the length of boom and class and weight of material excavated. If the material is difficult to excavate, the use of a smaller bucket will reduce the digging resistance, which may permit an increase in production rate.

The dragline will produce its greatest output if the job is planned to permit excavation at the optimum depth of cut where possible. Table 7.9 gives ideal outputs, in bank cubic yards, for short-boom draglines when excavating at optimum depth of cut with an angle of swing of 90°, based on a 60-min hour. The upper figure is the optimum depth in feet and the lower number is the ideal output in cubic yards, with no delays. Table 7.10 gives adjustment factors that can be used to determine outputs for other depths and angles of swing. The production rates determined from Tables 7.9 and 7.10 must be corrected for an efficiency less than 60 min/hr.

TABLE 7.9 | Ideal output of short-boom draglines, cy/hr, bank measure.

Class of material	Size of bucket, cy								
	$\frac{3}{8}$	$\frac{1}{2}$	$\frac{3}{4}$	1	$1\frac{1}{4}$	$1\frac{1}{2}$	$1\frac{3}{4}$	2	$2\frac{1}{2}$
Moist loam or light	5.0	5.5	6.0	6.6	7.0	7.4	7.7	8.0	8.5
sandy clay*	70	95	130	160	195	220	245	265	305
Sand and gravel*	5.0	5.5	6.0	6.6	7.0	7.4	7.7	8.0	8.5
	65	90	125	155	185	210	235	255	295
Good common earth*	6.0	6.7	7.4	8.0	8.5	9.0	9.5	9.9	10.5
	55	75	105	135	165	190	210	230	265
Hard, tough clay*	7.3	8.0	8.7	9.3	10.0	10.7	11.3	11.8	12.3
	35	55	90	110	135	160	180	195	230
Wet, sticky clay*	7.3	8.0	8.7	9.3	10.0	10.7	11.3	11.8	12.3
	20	30	55	75	95	110	130	145	175

Source: Power Crane and Shovel Association.

*The upper number indicates the optimum depth in feet. The lower number indicates the ideal output.

TABLE 7.10 | Effect of depth of cut and angle of swing on output of draglines.

Percentage of optimum depth	Angle of swing							
	30°	45°	60°	75°	90°	120°	150°	180°
20	1.06	0.99	0.94	0.90	0.87	0.81	0.75	0.70
40	1.17	1.08	1.02	0.97	0.93	0.85	0.78	0.72
60	1.24	1.13	1.06	1.01	0.97	0.88	0.80	0.74
80	1.29	1.17	1.09	1.04	0.99	0.90	0.82	0.76
100	1.32	1.19	1.11	1.05	1.00	0.91	0.83	0.77
120	1.29	1.17	1.09	1.03	0.98	0.90	0.82	0.76
140	1.25	1.14	1.06	1.00	0.96	0.88	0.81	0.75
160	1.20	1.10	1.02	0.97	0.93	0.85	0.79	0.73
180	1.15	1.05	0.98	0.94	0.90	0.82	0.76	0.71
200	1.10	1.00	0.94	0.90	0.87	0.79	0.73	0.69

Source: Power Crane and Shovel Association.

EXAMPLE 7.6

Estimate the total cost and cost per cubic yard for excavating a drainage ditch that is 12 ft wide at the bottom, 90 ft wide at the top, and 15 ft deep. The ditch is 3,200 ft long and the excavated material is wet sticky clay. A 2 cy dragline will be used with an average angle of swing of 120°. It will cost $2,500 to move the dragline to and from the job and it will require 8 hrs to rig up and 8 hrs to rig down the dragline at the jobsite. Two laborers will assist the excavation operation, which will be supervised by a foreman. Assume the dragline will operate an average of 40 min per hour.

Quantity of material:

Volume of earth $= \frac{1}{2}$(12 ft + 90 ft) × 15 ft × 3,200 ft = 2,448,000 cf

Converting to cubic yards = 2,448,000 cf/27 cf/cy = 90,667 cy

Production rate:

From Table 7.9, the ideal output = 145 cy/hr

From Table 7.9, the optimum depth of cut = 11.8 ft

Calculating percent of optimum = 15 ft / 11.8 ft = 1.27, or 127%

From Table 7.10, for 127% optimum cut and 120° angle of swing, the depth-swing adjustment factor will be 0.89

60-min hour production rate = 0.89 × 145 cy/hr = 129 cy/hr

40-min hour production rate = 40/60 × 129 cy/hr = 86 cy/hr

Time:

Time required to excavate the wet clay = 90,667 cy/86 cy/hr = 1,054 hr

Cost:

Cost during dragline operation

Dragline = 1,054 hr @ $75.00/hr	=	$79,050.00
Operator = 1,054 hr @ $29.00/hr	=	30,566.00
Laborers = 1,054 hr × 2 laborers @ $18.00/hr =		37,944.00
Foreman = 1,054 hr @ $32.00/hr	=	33,728.00

Cost to rig up and rig down dragline

Operator = 16 hr @ $29.00/hr	=	464.00
Laborers = 16 hr × 2 laborers @18.00/hr	=	576.00
Foreman = 16 hr @ $32.00/hr	=	512.00
Cost to move dragline to and from the job	=	2,500.00
	Total direct cost =	$185,340.00

Cost per cubic yard = $185,340.00/90,667 cy = $2.04/cy

HANDLING MATERIAL WITH A CLAMSHELL

Clamshells are used primarily for handling loose materials such as loam, sand, gravel, and crushed stone. They are also used for removing material from pier foundations, sewer manholes, or sheet-lined trenches. Clamshells are especially suited for lifting materials vertically from one location to another, such as depositing material in hoppers and overhead bins. The vertical movement capability can be relatively large when clamshells are used with long crane booms. A clamshell is not effective in loosening solid earth, such as compacted earth, clay, and other solid materials.

The bucket is lowered into the material to be handled with the jaws open. Its weight will cause the bucket to sink into the material as the jaws are closed. Then it is lifted vertically, swung to the emptying position, over a truck or to spoil, and the jaws are opened to permit the load to flow out. See Figure 7.3. The clamshell operator controls all these operations.

The size of the clamshell is indicated by the size of the bucket, expressed in cubic yards. Manufacturers supply buckets either with removable teeth or without teeth. Teeth are used in digging the harder types of materials, but are not required when a bucket is used for handling purposes.

The output of the clamshell is affected by the looseness of the materials being handled, type of material, height of lift, angle of swing, method of disposing of the materials, and skill of the operator. Because of the variable factors that affect the operations of a clamshell, it is difficult to give dependable production rates. For example, if the material must be discharged into a hopper, the

FIGURE 7.3 | Clamshell bucket depositing its load.
Courtesy: The Manitowoc Company, Inc.

TABLE 7.11 I Approximate output of clamshells, cy/hr.

Size of bucket, cy	Angle of swing, deg.	Type of material		
		Light loam	Sand gravel	Crushed stone
$\frac{3}{4}$	45	63	56	49
	90	53	48	42
	180	41	37	32
1	45	81	73	63
	90	68	61	53
	180	54	48	42
2	45	134	120	104
	90	113	102	88
	180	87	78	68

time required to spot the bucket over the hopper and to discharge the load will be much greater than when the material is discharged freely onto a large stockpile. The values given in Table 7.11 are approximate outputs of clamshells of different sizes for various angles of swing. To accurately estimate the production rate of a clamshell, a cycle time should be determined based on job conditions and operating specifications for the equipment. Manufacturers provide operating specifications, such as speed of the hoist line and swing speed.

EXCAVATING WITH HYDRAULIC EXCAVATORS

There are two basic types of hydraulic excavators, depending on their type of digging action. Hydraulic excavators that have their digging action in an upward direction are called *front shovels,* or simply *shovels.* Hydraulic excavators that have their digging action in a downward direction are called by several names, such as *hoe, backhoe,* or *trackhoe.*

FRONT SHOVELS

Front shovels maneuver on tracks, similar to bulldozers, because they are used mostly in pit excavations of rock quarries (Fig. 7.4). When operating on rock, a track has a longer life than rubber tires. Shovels are used mostly for pit excavation where the bucket load is obtained from the vertical face of the excavation pit above and in front of the excavator. The excavated material is loaded into trucks and hauled to another location.

Shovels are excavating machines that can handle all classes of earth without prior loosening; but in excavating solid rock it is necessary to loosen the rock first, usually by drilling holes and discharging explosives in them. The excavated material is loaded into trucks or tractor-pulled wagons, which haul it to its final destination. For the shovel to maintain its maximum output, sufficient hauling units must be provided.

FIGURE 7.4 | Front shovel depositing its load in a truck.

Source: Photo from McGraw-Hill DAL Library.

In estimating the output of a shovel, it is necessary to know the class of earth to be excavated, the height of cut, the ease with which hauling equipment can approach the shovel, the angle of swing from digging to emptying the bucket, and the size of the bucket.

The size of a shovel is designated by the size of the bucket, expressed in cubic yards, loose measure. A bucket can be rated as *struck capacity* or *heaped capacity.* The struck capacity is the volume in the bucket when it is filled even with, but not above, the sides. The heaped capacity is the volume that a bucket will hold when the earth is piled above the sides. The heaped capacity will depend on the depth of earth above the sides and the base area of the bucket. Depending on the type of material, the heaped slope will vary 1:1 or 2:1 above the sides of the bucket. Equipment manufacturers publish *fill factors,* which are percentages that can be multiplied by the heaped capacity to obtain the average payload of the bucket. Table 7.12 provides fill factors for shovels. The actual capacity of a bucket should be determined by measuring the volume of earth in several representative loads, then using the average of these values.

There are four elements in the production cycle of a shovel: load bucket, swing with load, dump load, and return swing. Adding the time of these elements provides the cycle time of the shovel. The production rate can then be determined by dividing the bucket capacity by the cycle time. The best method for determining cycle times is to develop historical data for a particular type of machine and job conditions. Table 7.13 provides the range of element times for shovels with bucket sizes ranging from 3 to 5 cy. A 30- to 45-min productive hour should be applied to these values.

TABLE 7.12 | Fill factors for front shovel buckets.

Material	Fill factor of bucket, % of heaped capacity
Bank clay; earth	100–110
Rock-earth mixture	105–115
Rock—poorly blasted	85–100
Rock—well blasted	100–110
Shale: sandstone	85–100

Source: Caterpillar, Inc.

TABLE 7.13 | Time for elements in a shovel production cycle.

Element of cycle	Range of time, sec
1. Load bucket	7–9
2. Swing loaded	4–6
3. Dump load	2–4
4. Swing empty	4–5

Note: These values are for shovels with 3- to 5-cy buckets.

The production of a shovel should be expressed in cubic yards per hour, based on a bank measure volume. The capacity of a bucket is based on its heaped volume, loose measure. To obtain the bank measure volume of a bucket, the average loose volume should be divided by $(1 + S_w/100)$, reference Eq. [7.2]. For example, if a 2-cy bucket (loose measure) is used to excavate material whose swell is 25 percent, then the average loose volume can be calculated as $(2.0 \text{ cy})/(1 + 0.25) = 1.6$ cy, bank measure. If the shovel can make 2.5 cycles/min, the output will be 2.5 cycles/min $\times$ 1.6 cy/cycle = 4.0 cy/min, bank measure, which is equivalent to 240 cy/hr.

When hard earth is dug, the output will be less than for soft earth. If the face against which the shovel is digging is too shallow, it will not be possible for the shovel to fill the bucket in a single cut, which will reduce the output. If the face is too deep, the bucket will be filled before it reaches the top of the face, which will necessitate emptying the bucket and returning to a partial face cut operation. As the angle of swing for the dipper from the digging to dumping is increased, the time required for a cycle will be increased, which will reduce the output of the shovel.

The optimum height of cut for a shovel is that depth at which the bucket comes to the surface of the ground with a full load without overcrowding or undercrowding the bucket. The optimum depth varies with the class of soil and the size of the bucket. The optimum height of cut ranges from 30 to 50 percent of the maximum digging height. The lower percentage applies to material that can be loaded easily, such as loam, sand, or gravel. A 40 percent would be more applicable to common earth, whereas a 50 percent would likely apply to sticky clay or blasted rock. Table 7.14 gives factors that correct outputs for percentage optimum heights of cut and angles of swing.

TABLE 7.14 | Conversion factors for depth of cut and angle of swing for a power shovel.

Percentage of optimum depth	Angle of swing						
	45°	60°	75°	90°	120°	150°	180°
40	0.93	0.89	0.85	0.80	0.72	0.65	0.59
60	1.10	1.03	0.96	0.91	0.81	0.73	0.66
80	1.22	1.12	1.04	0.98	0.86	0.77	0.69
100	1.26	1.16	1.07	1.00	0.88	0.79	0.71
120	1.20	1.11	1.03	0.97	0.86	0.77	0.70
140	1.12	1.04	0.97	0.91	0.81	0.73	0.66
160	1.03	0.96	0.90	0.85	0.75	0.67	0.62

Source: Power Crane and Shovel Association.

EXAMPLE 7.7

Estimate the probable output of a 4-cy shovel for excavating common earth. For this shovel, the maximum digging height is 34 ft. The average face of cut is 15 ft. Assume a 120° angle of swing for the shovel to load the haul units. Assume an effective 35-min hour for the excavation operation.

Cycle time:

From Table 7.13, using median values

Load bucket	=	8 sec
Swing with load	=	5 sec
Dump bucket	=	3 sec
Return swing	=	4 sec
Total time	=	20 sec/cycle

Cycles per min = (cycle/20 sec) × (60 sec/min) = 3 cycles/min

Production rate:

Bucket fill factor for common earth from Table 7.12 = 105%

Average bucket payload, 4.0 cy × 105% = 4.2 cy

Ideal production rate = (4.2 cy/cycle) × (3 cycles/min) = 12.6 cy/min

Optimum height of cut = 40% × 34 ft = 13.6 ft

Average height of cut = 15 ft

Percent optimum height of cut = 15.0/13.6 = 1.10, or 110%

Adjustment for 110% optimum depth of cut and 120% angle of swing

From Table 7.14, adjustment factor = 0.87

Probable output = 12.6 cy/min × 0.87 × 35 min/hr = 384 cy/hr, loose measure

Assume a 25% swell factor for common earth

Probable production rate = (384 cy/hr)/(1 + 0.25) = 307 cy/hr bank measure

This production rate can be achieved only if a sufficient number of trucks is available to haul the excavated earth away from the shovel. If no truck is available to be loaded, the shovel will remain idle, which will reduce the production rate of the shovel.

To estimate the total cost of excavating a given job, it is necessary to determine the cost of transporting the shovel to and from the job, and the labor cost of setting up the shovel for operation and the cost of removing the shovel at the end of the job. These costs are in addition to the equipment and labor costs of excavating the material during construction. A foreman usually supervises the excavating and hauling. A portion of his or her salary should be charged to excavation.

HAULING EXCAVATED MATERIALS

Trucks are used to haul the material excavated by shovels. See Figure 7.5. The capacity of the hauling unit can be expressed in tons or cubic yards. The latter capacity can be expressed as struck or heaped. The *struck capacity* is the volume that a unit will hold when it is filled even with, but not above, the sides. This volume depends on the length, width, and depth of the unit. The *heaped capacity* is the volume that a unit will hold when the earth is piled above the sides. Although the struck capacity of a given unit is fixed, the heaped capacity will depend on the depth of the earth above the sides and on the area of the bed.

FIGURE   7.5 | Truck for hauling rock and soil.

Courtesy: Sherwood Construction Company.

The Society of Automotive Engineers (SAE) specifies the heaped capacity based on a 1:1 slope of the earth above the sides. The actual capacity of a unit should be determined by measuring the volume of earth in several representative loads, then using the average of these values. Units are available with capacities varying from 12 to 30 cy.

Since shovel production rates are normally calculated in bank measure, it is desirable to express the loose volume of hauling trucks in bank measure. Equation [7.1] can be used for converting the loose measure to bank measure. For example, if a 15-cy truck is hauling ordinary earth that has a swell of 25 percent, the bank measure will be 15 cy/(1 + 0.25) = 12 cy.

The size of the hauling units should be balanced against the bucket capacity of the shovel. For best results, considering output and economy, the capacity of a hauling unit should be four to six times the bucket capacity.

The volume that a truck can haul in a given time depends on the volume per load and the number of trips it can make in that time. The number of trips depends on the distance, speed, time at loading, time at the dump, and time required for servicing. Higher travel speeds are possible on good open highways than on streets with heavy traffic. For example, top speeds in excess of 65 mph may be possible on some paved highways, whereas speeds on crowded city streets may be not more than 10 to 15 mph. Delays can occur at loading, dumping, and servicing trucks. Therefore, the actual operating time may be only 45 or 50 min/hr. An appropriate multiplier should be applied in determining the number of trips per unit of time.

EXAMPLE 7.8

Assume the shovel and job conditions given in Example 7.7 apply to a project that requires excavation and hauling of 58,640 cy, bank measure, of common earth. The production rate of the shovel was calculated as 307 cy/hr, bank measure. The earth will be hauled 4 mi by 20-cy, loose measure, trucks at an average speed of 30 mph. The expected time at the dump is 4 min. The truck time waiting at the shovel to move into loading position will average 3 min. Assume a 45-min hour for trucks and calculate the total cost and cost per cubic yard based on using sufficient trucks to balance the production rate of the shovel.

Quantity of work:

Volume of earth = 58,640 cy, bank measure

Volume of truck = 20 cy/(1 + 0.25) = 16 cy, bank measure

Cycle time of trucks:

Loading truck: 16 cy/307 cy/hr	= 0.05 hr
Traveling: 8 mi/30 mph	= 0.26 hr
Dump time: 4 min/60 min/hr	= 0.07 hr
Waiting to load: 3 min/60 min/hr	= 0.05 hr
Total cycle time	= 0.43 hr/trip

Production rate of trucks:

Number of trips per hour: 1 trip/0.43 hr = 2.3

Number of trips per 45-min hour: 2.3 hr × (45/60) = 1.7 trips/hr

Volume hauled per hour: 1.7 trips/hr × 16 cy/trip = 27.2 cy/hr

Time:

Number of trucks required: 307 cy/hr/27.2 cy/hr = 11.3 trucks required

Use 12 trucks, therefore there will be more trucks than required.

The shovel at 307 cy/hr will govern the excavation production rate.

Time = 58,640 cy/307 cy/hr = 191 hr

Cost:

Equipment excavating and hauling		
Shovel = 191 hr @ $112.50/hr	=	$21,487.50
Trucks = 191 hr × 12 trucks @ $50.00/hr	=	114,600.00
Labor excavating and hauling		
Shovel operator = 191 hr @ $31.00/hr	=	5,921.00
Truck drivers = 191hr × 12 trucks @ $25.00/hr	=	57,300.00
Laborers = 191 hr × 2 laborers @ $18.00/hr	=	6,876.00
Foreman = 191 hr @ $32.00/hr	=	6,112.00
Labor setting up and dismantling shovel		
Shovel operator = 16 hr @ $31.00/hr	=	496.00
Laborers = 16 hr × 2 laborers @ $18.00/hr	=	576.00
Foreman = 16 hr @ $32.00/hr	=	512.00
Transporting shovel to and from job	=	2,850.00
Total cost =		$216,730.50

Cost per cubic yard = $216,730.50/58,640 cy = $3.70/cy

BACKHOES

Backhoes are used to excavate below the natural surface, such as trenches, basements, and general excavation that requires precise control of depths (Fig. 7.6). A backhoe may be a wheel type or track type. Track-type backhoes are commonly called *trackhoes*.

Backhoes have become one of the most widely used types of excavating equipment. Because of their rigidity, they are superior to draglines for loading into dump trucks. Due to their direct pull of the bucket, they can exert greater tooth pressure than shovels. For some job conditions, they are superior to trenching machines, especially for digging utility trenches whose banks are permitted to establish natural slopes and for which trench shoring will not be used. Backhoes can remove the earth as it caves in to establish natural slopes, whereas trenching machines cannot do this easily.

FIGURE 7.6 | Backhoe depositing its load in a truck.

Courtesy: Sherwood Construction Company.

The production rate depends on the bucket payload, average cycle time, and job efficiency. If an estimator can predict the excavator cycle time and the bucket payload, the probable production rate for excavation can be determined. The cycle time will depend on the particular job conditions, such as the difficulty in loosening the soil, the angle of swing, the size of the truck that the backhoe must load, and the skill of the operator. The estimator must use judgment and knowledge of actual job conditions to predict the cycle time for a particular job. Table 7.15 provides representative cycle time for backhoes under average conditions.

The *average bucket payload* is equal to the heaped bucket capacity payload multiplied by the bucket fill factor. The bucket fill factor will depend on the type

TABLE 7.15 | Representative cycle times for backhoes.

Bucket size, cy	Load bucket, sec	Swing loaded, sec	Dump bucket, sec	Swing empty, sec	Total cycle, sec
<1	5	4	2	3	14
1–1½	6	4	2	3	15
2–2½	6	4	3	4	17
3	7	5	4	4	20
3½	7	6	4	5	22
4	7	6	4	5	22
5	7	7	4	6	24

Source: Caterpillar, Inc.

of soil to be excavated. Table 7.16 provides representative values of the bucket fill factor for backhoes.

TABLE 7.16 | Fill factors for backhoe buckets.

Material	Fill factor of bucket, % of heaped capacity
Moist loam or sandy clay	100–110
Sand and gravel	95–100
Hard, tough clay	80–90
Rock, well blasted	60–75
Rock, poorly blasted	40–50

Source: Caterpillar, Inc.

EXAMPLE 7.9

Estimate the probable output of a 2-cy backhoe for excavation of hard, tough clay. Assume a swell of 30 percent and a 45-min hour for the backhoe.

Cycle time:

From Table 7.15,

$$\begin{aligned} \text{Load bucket} &= 6 \text{ sec} \\ \text{Swing loaded} &= 4 \text{ sec} \\ \text{Dump bucket} &= 3 \text{ sec} \\ \text{Return swing} &= \underline{4 \text{ sec}} \\ \text{Total time} &= 17 \text{ sec/cycle} \end{aligned}$$

Cycles per min = (cycle/17 sec) × (60 sec/min) = 3.53 cycles/min

Production rate:

Bucket fill factor for hard, tough clay from Table 7.16 = 85%

Average bucket payload, 2.0 cy × 85% = 1.7 cy

Ideal production rate = (3.53 cycles/min) × (1.7 cy/cycle) = 6.0 cy/min

For 45-min hour = 6.0 cy/min × 45 min/hr = 270 cy/hr loose measure

For bank measure, (270 cy/hr)/(1 + 0.30) = 208 cy/hr bank measure

Table 7.17 provides representative ranges of production rates for excavating with backhoes. The information contained in this table is based on loose measure and a 60-min hour, or 100 percent efficiency. The estimator should apply a job efficiency factor to the factors shown in the table based on personal judgment or knowledge of actual job conditions. The upper limit corresponds to the fastest practical cycle time: easy digging earth, low angles of swing, and no obstructions. The lower limit corresponds to the toughest digging, deep depths, large angles of swing, loading into small trucks, and obstructions in the work area.

TABLE 7.17 | Probable output of backhoes.

Bucket payload, cy loose measure	Output of backhoe, cy/hr loose measure
$\frac{1}{2}$	75–135
$\frac{1}{4}$	90–202
1	120–270
$1\frac{1}{4}$	150–300
$1\frac{1}{2}$	154–360
$1\frac{3}{4}$	180–420
2	205–420
$2\frac{1}{4}$	231–472

Source: Caterpillar, Inc.

DOZERS

Dozers are effective for excavating at shallow depths and pushing soil and rock material over short distances. A shank attached on the back of a dozer is used for loosening materials. A blade on the font of a dozer is used for excavating, pushing, and spreading material. The economical operating distance is usually less than 300 feet. Dozers are also used in push loading scrapers.

Sizes of dozers are rated by weight and horsepower, with weights varying from 15,000 to 250,000 pounds and engines from 100 to 800 hp. Dozers are available with direct drive and power-shift transmissions, which allows shifting while transmitting full engine power. The power-shift transmission provides superior performance in applications involving variable load conditions. Transmissions have 3 to 6 gears that provide operating speeds in forward and reverse directions. Reverse gears have higher speeds than forward gears.

Dozers are classified as either wheel-type or crawler-type. Figure 7.7 shows a crawler dozer with a shank on the back for ripping or loosening hard material and a blade on the front for pushing material. A hydraulic cylinder actuates the shank up and down in the vertical direction for ripping up pavements and loosening rock material. Various sizes and shapes of blades are available. The capacity of a dozer blade can be obtained from the manufacturer's blade rating, or from job records of previously completed projects. The capacity of a dozer blade is rated in loose cubic yards. For most applications the capacity of blades range from 1.5 to 15 cy. However, large size dozers have blade capacities up to 50 cy.

The cycle time of a dozer is the time to push a load of material plus the time to return and maneuver into position to push the next load. The production rate of a dozer is the quantity of material moved divided by the cycle time.

$$\text{Dozer production rate} = \frac{(\text{Quantity moved})}{(\text{Cycle time})} \qquad [7.7]$$

The quantity moved depends on the size of blade and type of soil. The cycle time is a function of forward speed, reverse speed, and the fixed time to maneuver the

FIGURE 7.7 | Crawler tractor dozer with shank on back and blade on front.

Courtesy: Sherwood Construction Company.

dozer into a position to push a load. For dozing operations it is sometimes helpful to convert dozer speed from mph to ft/min. Since there are 5,280 ft per mi and 60 min per hour, the conversion factor for converting 1.0 mi/hr to ft/min is 5,280/60 = 88. Thus, 1.0 mi/hr is equivalent to 88 ft/min.

EXAMPLE 7.10

Estimate the production rate of a crawler dozer that will be used for stripping and leveling earth for preparation of a construction site. The average push distance is 120 ft. The dozer will be operated at 95% of the maximum forward speed in 1st gear for pushing the earth and return at 80% of the maximum reverse speed in 2nd gear. Assume a 50-min productive hour. Information from the dozer manufacturer provides the following data:

Gear	Maximum forward speed	Maximum reverse speed
1	2.2 mph	2.9 mph
2	3.9 mph	5.0 mph
3	6.7 mph	8.6 mph

Rated blade capacity = 5.5 cy, loose measure

Fixed time to maneuver dozer = 0.05 min

Converting the push speed to ft/min = 95% × 2.2 mph × 88 ft/min = 184 ft/min

Converting return speed to ft/min = 80% × 5.0 mph × 88 ft/min = 352 ft/min

Cycle time:

$$\text{Push time} = 120 \text{ ft}/(184 \text{ ft/min}) = 0.65 \text{ min}$$
$$\text{Return time} = 120 \text{ ft}/(352 \text{ ft/min}) = 0.34 \text{ min}$$
$$\underline{\text{Maneuver time} = \qquad\qquad = 0.05 \text{ min}}$$
$$\text{Total} = 1.04 \text{ min}$$

$$\text{Cycle time for a 50-min productive hour} = 60/50 \times 1.04 \text{ min} = 1.25 \text{ min}$$

Production rate:

From Eq. [7.7], dozer production rate $= \dfrac{(\text{Quantity moved})}{(\text{Cycle time})}$

$$= \frac{5.5 \text{ cy}}{(1.25 \text{ min}/60 \text{ min/hr})}$$

$$= 264 \text{ cy/hr, loose measure}$$

EXCAVATING AND HAULING EARTH WITH SCRAPERS

Scrapers are used to excavate and haul earth for highways, airports, dams, canals, and levees. Since these units perform both excavating and hauling operations, they are independent of the operations of other equipment. If one of several units breaks down, the rest of the units can continue to operate, whereas if a shovel or backhoe breaks down, the entire project must stop until the shovel is repaired.

Scrapers are versatile because they load, haul, and dump loose material. They cannot load as productively as a shovel or backhoe and they cannot haul as productively as trucks. However, for off-highway projects with haul distances of less than a mile, the ability of scrapers to both load and haul gives them an advantage. Also, the ability of these machines to deposit their loads in layers of uniform thickness facilitates compaction operations.

Scrapers are classified by their method of loading: push-loaded, push-pull, and self-loading. The push-loaded scraper uses a crawler-tractor to push the scraper to assist the loading (Fig. 7.8). A push-pull scraper allows two scrapers to assist one another during loading by hooking them together. The trailing scraper pushes the lead scraper as it loads. Then, the lead scraper pulls the trailing scraper to assist it in loading. This feature allows two scrapers to work without assistance from a push tractor. A self-loading scraper is sometimes an elevating scraper, which is completely self-contained loading. A chain elevator serves as the loading mechanism. Most scrapers are powered by a diesel engine on a single drive axle. Models are available with tandem powered axles with twin engines, one at the front and one at the back of the scraper.

FIGURE 7.8 | Two-wheel tractor scraper loading with assistance from a dozer.

Courtesy: Sherwood Construction Company.

The volume of a scraper may be specified as either the struck or heaped capacity of the bowl, expressed in cubic yards. The struck capacity is the volume that a scraper would hold if the top of the material were struck off level at the top of the bowl. For the heaped capacity, a 1:1 slope is commonly used for the side slope above the sides of the bowl. For example, a scraper might be designated as 14 cy struck, 20 cy heaped. A wide range of size and capacities of scrapers is available, from 14 cy (struck) to 44 cy (heaped).

Production Rate for Scrapers

The production rate for a scraper will equal the number of trips per hour multiplied by the net volume per trip. The number of trips per hour can be obtained from the cycle time; the time required to load, haul, dump, and return. An appropriate efficiency factor, such as a 45- or 50-min hour, should be used in determining the number of trips per hour, to allow for nonproductive time. Frequently it is desirable to express the speed of a scraper in feet per minute as well as miles per hour. A speed of 1 mph is equal to 5,280 ft in 60 min, which equals 88 ft/min.

Manufacturers provide data on the typical fixed time of scrapers. The fixed time is the amount of time to load, maneuver, and spread or dump a load. The time to load a scraper ranges from 0.4 to 1.1 min, depending on the type of soil and skill of the operator. A load time of about 0.8 min is reasonable for most scraper operations. The time to maneuver and spread or dump a load is about 0.7 min. Thus, a reasonable total fixed time to load and dump a scraper is about 1.5 min. The estimator should also evaluate the job conditions at each project and include a reasonable amount of time for acceleration, deceleration, and turns that are necessary for a scraper to operate.

Equipment manufacturers also publish production rate charts that show the production rate with respect to haul distance for a particular size of scraper. These charts are valuable to the estimator when multiple haul distances are involved in a scraper operation, such as multiple cut and fill balance points in mass diagrams of highway projects.

Estimate the probable production rate for a scraper whose struck capacity is 14 cy, loose measure. The material will be common earth with a swell of 25 percent. An evaluation of the haul road shows an average haul distance of 860 ft and return haul of 940 ft. Based on the condition of the haul road it is anticipated that haul speed is 9 mph loaded and 11 mph empty. From the manufacturer's information the load time is 0.8 min and the dump time is 0.7 min. Assume 0.5 min for turning, acceleration, etc. Assume a 45-min productive hour.

Cycle time:

Haul speed: 9 mph × (5,280 ft/mi) × (1.0 hr/60 min) = 792 ft/min

Return speed: 11 mph × (5,280 ft/mi) × (1.0 hr/60 min) = 968 ft/min

Load time	= 0.80 min
Haul: (860 ft)/(792 ft/min)	= 1.08 min
Dump time	= 0.70 min
Return: 940 ft/968 ft/min.	= 0.97 min
Turns, acceleration, etc.	= <u>0.50 min</u>
Total cycle time	= 4.05 min/trip

Production rate:

Number of trips per hour: (1 trip/4.05 min) × (45 min/hr) = 11.1 trip/hr

Volume per hour: 11.1 trips/hr × 14 cy/trip = 155.4 cy/hr loose measure

For a 25% swell, (155.4 cy/hr)/(1 + 0.25) = 124.3 cy/hr bank measure

Cost of Excavating and Hauling Earth with Scrapers

The cost of handling earth with scrapers can be determined by assuming that the equipment will be operated 1 hr for quantity and cost purposes. The cost per cubic yard can be calculated by dividing the hourly cost by the hourly production rate of the scraper.

If a bulldozer is used to assist several scraper units while they are being loaded, the cost of the bulldozer should be distributed equally among the scraper units that are assisted. The number of scrapers that a bulldozer can assist will be equal to the total cycle time divided by the time required by the bulldozer to assist in loading. For example, consider the scraper unit in Example 7.11.

Total cycle time = 4.05 min

Load time = 0.8 min

Number of scrapers served by a bulldozer = 4.05/0.8 = 5.06 scrapers

Thus, it appears that a bulldozer can assist 5 scrapers. Assuming that a bulldozer can assist 5 scrapers, then one-fifth (0.20) of the cost of a bulldozer should be charged to each scraper. The hourly cost will be

Scraper: 1 hr @ $95.50/hr	=	$95.50/hr
Scraper operator: 1 hr @ $30.00/hr	=	$30.00/hr
Bulldozer: (0.20) @ $65.20/hr	=	$13.04/hr
Bulldozer operator: (0.20) @ $28.00/hr	=	$5.60/hr
	Total cost =	$144.14/hr

From Example 7.11, production rate = 124.3 cy/hr, bank measure

Cost per cubic yard: ($144.14/hr)/(124.3 cy/hr) = $1.16/cy, bank measure

If job records are available from previously completed projects, the production rate of scrapers can be estimated by load counts. A load count is simply the number of loads a scraper can haul in a day of operation. Using this method, historical data are obtained from actual jobs that provide the number of loads per day for a particular size scraper and haul distance. For example, job records may show that an 18-cy scraper can haul 42 loads per day for a haul distance between 1,200 and 1,500 ft. Thus, the production rate can be calculated as

Output: 42 loads/day × 18 cy per load = 756 cy/day

Assume the daily cost of a scraper and operator = $935/day

Cost per cubic yard = ($935/day)/(756 cy/day) = $1.24/cy

The same procedure can be used on an hourly basis. For example, the number of loads per hour can be multiplied by the capacity of the scraper. Then, the cost per cubic yard can be estimated by dividing the hourly cost of a scraper by the hourly production rate to obtain the cost per cubic yard for the scraper operation.

Haul Distances for Scrapers

For short haul distances, the production rate of a scraper depends primarily on the fixed time of the scraper—the time to load, accelerate, decelerate, turn, and dump. As the length of the haul distance increases, the production rate of a scraper will depend primarily on the travel time of the scraper, or the time to haul and return. Therefore, the haul distance must be known to estimate the production rate and the cost of excavation by a scraper for a particular job.

The earthwork required for a highway project is defined by the mass diagram on the plan and profile sheets. Figure 7.9 is a portion of a profile and mass diagram. The amount of material that must be excavated, and the distance it must be hauled, can be determined from the mass diagram. This information enables one to estimate the production rate and cost of excavation.

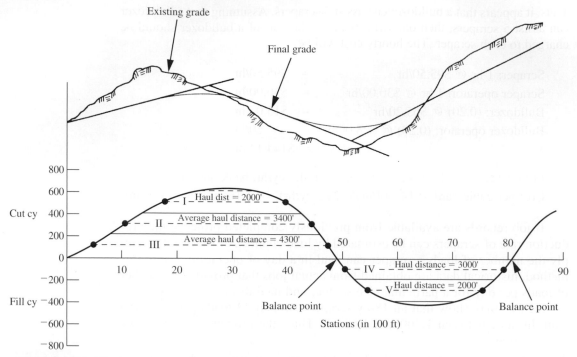

FIGURE 7.9 | Profile and mass diagram for highway project.

EXAMPLE 7.12

Determine the total cost and the cost per cubic yard for the earthwork of the first balance point of the mass diagram in Fig. 7.9. A self-loading scraper with a struck capacity of 15 cy will be used to excavate and haul the soil. The scraper has a load time of 0.8 min and dump time of 0.7 min. Assume an 18 mph haul speed and a 22 mph return speed. The material is common earth, with a swell of 25 percent.

The mass diagram can be arbitrarily divided into three 200-cy sections. For each section the average haul distance can be determined from the diagram. The haul distance, cycle time, number of trips per 45-min hour, quantity hauled per trip, production rate, and time required for each section will be:

Quantity hauled per trip:

Capacity of scraper = 15 cy loose measure

However, from Eq (7.2),

$$L = \left(1 + \frac{S_w}{100} \right) B$$

$$15 \text{ cy} = \left(1 + \frac{25}{100} \right) B$$

$$B = 12 \text{ cy bank measure}$$

Cycle time for a scraper:

Section	Quantity, cy	Distance, ft	Cycle time, min					Quantity hauled per trip, cy	Production rate, cy/hr	Total time, hr
			Load	Haul	Dump	Return	Total			
I	200	2,000	0.8	1.26	0.7	1.03	3.79	12	142.8	1.4
II	200	3,400	0.8	2.14	0.7	1.76	5.40	12	99.6	2.0
III	200	4,300	0.8	2.71	0.7	2.22	6.43	12	84.0	2.4
	600 cy								Total time = 5.8 hr	

Cost of hauling earth:

Scraper: 5.8 hr @ $95.50/hr = $553.90

Operator: 5.8 hr @ $30.00/hr = $174.00

Total cost = $727.90

Cost per cubic yard: $727.90/600 cy = $1.21/cy

GRADERS

Graders are used in many applications of earthwork operations. They are used for spreading, mixing, leveling, and smoothing earth to bring it to the final shape and grade elevation as required in the specifications. See Figure 7.10. Graders can also be used for shallow stripping of soil, cleaning out ditches, dressing the side slopes of embankments, and maintaining haul roads for scraper operations.

A blade (sometimes called moldboard) is attached to the underneath side of the grader for moving earth. Blades are available in several types; including flat, curved, or serrated. The blade can be moved sideways (left and right), moved in the upward and downward position, and rotated about the long axis of the blade to control the pitch of the blade. When the pitch of the blade is rotated forward the grader can be used in a roller action to spread material, or the blade can be rotated backward to increase the cutting action of the blade.

The range of working speeds of a grader is generally from 2 to 6 mi/hr for grading operations on bank slopes, ditches, finish grading, and grading for road maintenance. The range for mixing material and spreading earth is typically 6 to 9 mi/hr. The maximum speed of a grader is about 25 mi/hr, which is the speed for moving the grader from one location to another location.

The production of a grader is usually calculated on a linear basis (such as linear feet or linear mile for highway work), or on a surface area basis (such as sf, sy, or acre for large surface areas). The estimated time for grader production can be calculated by dividing the distance travelled by the speed of the grader. An efficiency factor of 60% is typically used for most grader operations. The time for a grader operation can be calculated from Equation [7.8] and the production rate of a grader can be calculated from Equation [7.9].

FIGURE 7.10 | Motor grader shaping a fill area.

Courtesy: Sherwood Construction Company.

$$\text{Time} = \frac{D \times P}{S \times E}$$ [7.8]

where D = distance

P = number of passes

S = average speed

E = efficiency factor

The production rate for grading surface areas can be calculated by multiplying the speed of the grader by the effective width of the grader blade.

$$\text{Production rate} = \frac{S \times W \times E}{P}$$ [7.9]

where S = average speed

W = effective width of blade

E = efficiency factor

P = number of passes

EXAMPLE 7.13

A grader is used to maintain 4.3 miles of haul road. Based on operator skill and job conditions it is anticipated that 2 passes in first gear at 2.5 mi/hr, 3 passes in second gear at 4.0 mi/hr, and 1 pass in fourth gear at 9.0 mi/hr will be necessary. Assume a 60% efficiency factor and estimate the time and the cost per mile for the grading operation. The cost of the grader, with operator, is $85.00/hr.

Time for grading:

From Eq. [7.8], Time $= \dfrac{D \times P}{S \times E}$

Total time $= \left[\dfrac{4.3 \text{ mi} \times 2 \text{ passes}}{2.5 \text{ mi/hr} \times 60\%} \right] + \left[\dfrac{4.3 \text{ mi} \times 3 \text{ passes}}{4.0 \text{ mi/hr} \times 60\%} \right] + \left[\dfrac{4.3 \text{ mi} \times 1 \text{ pass}}{9.0 \text{ mi/hr} \times 60\%} \right]$

$\qquad = 5.7 \text{ hr} + 5.4 \text{ hr} + 0.8 \text{ hr}$

$\qquad = 11.9 \text{ hr}$

Unit cost:

Cost per mile $= \dfrac{\$85.00/\text{hr} \times 11.9 \text{ hr}}{4.3 \text{ mi}}$

$\qquad = \$235.23/\text{mi}$

EXAMPLE 7.14

A grader is used to prepare the surface area for construction of a new shopping center. The grader will have a 12-ft blade, with an effective width of 10.0 ft. An evaluation of the job indicates the average operating speed of the grader will be 2.5 mi/hr with an efficiency of 60%. Assume 3 passes of the grader will be required for shallow cuts and fills to smooth the ground. Estimate the grader production rate in acres/hr.

From Eq. [7.9], Production rate $= \dfrac{S \times W \times E}{P}$

$\qquad = \dfrac{(2.5 \text{ mi/hr} \times 5{,}280 \text{ ft/mi}) \times 10 \text{ ft} \times 60\%}{3 \text{ passes}}$

$\qquad = 26{,}400 \text{ sf/hr}$

Converting sf per hr to acres per hr $= 26{,}400 \text{ sf/hr}/43{,}560 \text{ sf/acre}$

$\qquad = 0.61 \text{ acres/hr}$

SHAPING AND COMPACTING EARTHWORK

When earth is placed in a fill, it is necessary to spread it in uniformly thick layers and compact it to the density specified in the contract documents. Unless sufficient moisture is present, water should be added to produce the optimum moisture content, which will permit more effective compaction. Spreading can be accomplished with graders or bulldozers, or both, while compaction can be accomplished with self-propelled rollers, tractor-pulled sheep's-foot rollers (Fig. 7.11), tractor-pulled grid rollers (Fig. 7.12), smooth-wheel rollers, pneumatic rollers, vibrating rollers, or other types of equipment. For some projects, the best results are obtained by using more than one type of equipment. Regardless of the type of equipment selected, there should be enough units to shape, wet, and compact the earth at the rate at which it will be delivered.

FIGURE 7.11 | Tractor-pulled sheep's foot roller for compacting clay soils.

Courtesy: Sherwood Construction Company.

FIGURE 7.12 | Tractor-pulled grid roller for compacting rock material.

Courtesy: Sherwood Construction Company.

EXAMPLE 7.15

The compaction rate for a large earthen dam is 1,400 cy/hr. Field tests show a 3% moisture content and a 2,890 lb/cy dry weight for the earth that is hauled to the job-site. However, the specification for moisture density requires a 12% moisture content for the 2,890 lb/cy dry weight.

To increase the moisture content, water will be hauled to the job by 1,000 gal water wagons that have a round-trip time of 50 min. Determine the number of water wagons required to increase the moisture content from 3% to 12% for the 1,400 cy/hr rate of the hauling and compaction operation.

Required increase in water content = 12% − 3%
= 9%

Required water rate = [(9% × 2,890 lb/cy) × 62.4 lb/cf]/7.48gal/cf
= 2,169.8 gal/hr

Available water rate for one water wagon = (1,000 gal/50 min) × 60 min/hr
= 1,200 gal/hr

Number of wager wagons required = 2,169.8 gal/hr/1,200 gal/hr
= 1.8

Therefore 2 water wagons are required to provide the increase in moisture content to service the 1,400 cy/hr compaction rate of the fill operation.

EXAMPLE 7.16

Estimate the total cost and cost per cubic yard for shaping, sprinkling, and compacting earth in a fill area. A spread of scrapers will haul the earth to the fill site at a combined rate of 1,450 cy/hr, bank measure. Assume a 15% shrinkage factor.

It will require 3 passes by a grader to spread and shape the earth. The earth will be compacted in 6-in. thick layers. It is anticipated that 8 passes will be required by sheep's foot rollers to meet the required 3,150 lb/cy density of the earth.

The moisture content will average 8 percent when the earth is placed in the fill, but it must be increased to 12 percent during the compaction operation. The water will be transported to the fill area by trucks with a water sprinkler bar. Estimate the cost based on adequate number of graders, compactors, and water trucks to achieve the production rate of scrapers that will be hauling the earth to the compaction site.

Hauling soil to the compaction area:

Scrapers production rate in bank measure, $B = 1,450$ cy/hr
From Eq. [7.3], the compacted measure, $C = (1 − 15/100) B$
$$= 0.85 (1,450 \text{ cy/hr})$$
$$= 1,232.5 \text{ cy/hr}$$

Surface area placed per hour for 6-in. depth, or 0.5 ft, will be:
Production rate of scrapers = 1,232.5 cy/hr/(0.5 ft/3 ft/yd)
= 7,395 sy/hr

Converting to sf production rate = 7,395 sy/hr × 9 sy/sf

= 66,555 sf/hr

Spreading and shaping earth:

A grader will be used with a 12-ft blade that should cover ground with an effective width of about 8 ft for each pass with an average speed of 2.0 mi/hr.

Grader production rate = [2 mi/hr × 5,280 ft/mi × 8 ft]/3 passes

= 28,160 sf/hr

Number of graders required = scraper production rate/grader production rate

= 66,555 sf/hr/28,160 sf/hr

= 2.36, therefore use 3 graders

Compaction of earth:

The sheep's foot roller is 5-ft wide and will operate at an average speed of 3.5 mph, allowing for lost time. The production rate will be:

Compactor production rate = [3.5 mi/hr × 5,280 ft/mi × 5 ft]/8 passes

= 11,550 sf/hr

Number of compactors required = 66,555 sf/hr/11,550 sf/hr

= 5.8, therefore use 6 compactors

Water trucks:

The required compaction density is 3,150 lb/cy. The water content must be increased from 8 to 12 percent, or 4 percent. The weight of water required per hour will be 4 percent of the weight of earth placed. Therefore, the required production rate of sprinkler trucks will be:

Required water truck production rate = 4% × 1,232.5 cy/hr × 3,150 lb/cy

= 155,295 lb/hr

Water weighs 62.4 lb/cf and there are 7.48 gal/cf, therefore the conversion of lb of water to gal of water = 62.4 lb/cf/7.48 gal/cf

= 8.34 lb/gal

Converting lb/hr to gal/hr = 155,295 lb/hr/8.34 lb/gal

= 18,620 gal/hr

The water trucks have a 3,000 gal capacity and can make one round trip in 35 min from the water source to the jobsite to sprinkle the water. The production rate of a single water truck will be:

Water production rate = [3,000 gal/35 min/trip] × 60 min/hr

= 5,143 gal/hr

Number of water trucks required = 18,620 gal/hr/5,143 gal/hr

= 3.6, therefore use 4 water trucks

Crew costs:

Motor graders: 3 @ $76.30/hr	=	$228.90/hr	
Grader operators: 3 @ $32.00/hr	=	96.00/hr	
Compactors: 6 @ $78.50/hr	=	471.00/hr	
Compactor operators: 6 @ $31.00/hr	=	186.00/hr	
Water trucks: 4 @ $45.00/hr	=	180.00/hr	
Water truck drivers: 4 @ $28.50/hr	=	114.00/hr	
Foreman: 1 @ $34.00/hr	=	34.00/hr	
Pickup truck: 1 @ $7.50/hr	=	7.50/hr	
Total cost per hour	=	$1,317.40/hr	

Cost per sy for shaping and compaction = $1,317.40/hr/7,395 sy/hr

= 0.18/sy

Cost per cy for shaping and compaction = $1,317.40/hr/1,232.5 cy/hr

= $1.07/cy

PREPARING THE SUBGRADE FOR HIGHWAY PAVEMENTS

For construction of an asphalt or concrete highway pavement the subgrade is rough-graded, compacted, and shaped to the approximate elevation. Then, the subgrade must be shaped to the final elevation within the vertical and horizontal tolerances that are specified in the design of the project. The preparation of placing the pavement will include fine-grading the subgrade to exact shape and elevation, and possible wetting and compacting after the fine-grading is complete.

For fine-grading operations motor graders can be equipped with Global Positioning System (GPS), or laser augmentation, that provide final grade position and elevation information. Figure 7.13 shows a grader with an attached arm for GPS. The GPS system communicates to a base station that provides position information by radio signals to a receiver on the grader. An on-board computer compares this information to the design information to compute cut and fill grade. This information is displayed inside the grader on a control box screen in plan, profile, or cross-section views, or by text. The cut and fill data is used to drive hydraulic valves for automatic control of the grader blade.

For large-scale fine-grading projects, such as large multiple-lane highways or airport runways, fine-grading can also be accomplished by a self-propelled trimmer machine that is guided by string-line, GPS, or laser augmentation. The trimmer machine removes the soil to the required depth by a rotating cutter drum and moves the soil with a conveyor belt that deposits the soil in a windrow outside the paving area. The soil deposited in the windrow may be relocated to areas

FIGURE 7.13 | Motor grader with GPS system for three-dimensional grade control.
Courtesy: Cornell Construction Company, Inc.

that require additional soil, or completely removed. Pneumatic and steel wheel rollers are used to compact the sub-grade after the fine-grading is completed, with water added if necessary.

DRILLING AND BLASTING ROCK

Before rock can be excavated, it must be loosened and broken into pieces small enough to be handled by the excavating equipment. The most common method of loosening is to drill holes into which explosives are placed and detonated.

Holes can be drilled by one or more of several types of drills (see Figs. 7.14 and 7.15), such as jackhammers, wagon drills, drifters, churn drills, rotary drills, etc. The selection of equipment is based on the size of the job, type of rock, depth and size of holes required, production rate required, and topography at the site.

Jackhammers can be used for holes up to about $2\frac{1}{2}$ in. diameter and 15 to 18 ft deep. For deeper holes, the production rates are low and the costs are high. Wagon drills may be used for holes 2 to $4\frac{1}{2}$ in. in diameter, with depths sometimes as great as 40 ft, although shallower depths are more desirable. Drifters are used to drill approximately horizontal holes, up to about 4 in. in diameter, in mining and tunneling operations. Jackhammers, wagon drills, and drifters are operated by compressed air, which actuates the hammer that produces

FIGURE 7.15 | Truck-mounted drill equipped with a dust collector.

FIGURE 7.14 | Gasoline-engine-operated hand drill.

the percussion that disintegrates the rock and blows it out of the holes. Replaceable bits, which are attached to the bottoms of the hollow drill steels, are commonly used.

A churn drill disintegrates the rock by repeated blows from a heavy steel bit which is suspended from a wire rope. Holes in excess of 12 in. in diameter can be drilled several hundred feet deep with this equipment. Water, which is placed in a hole during the drilling operation, will produce a slurry with the disintegrating rock. A bailer is used to remove the slurry.

Rotary drills can be used to drill holes 3 to 8 in. or more in diameter to depths in excess of 100 ft. Drilling is accomplished by a bit, which is attached to the lower end of a drill stem (see Fig. 7.16). Either water or compressed air can be used to remove the rock cuttings.

ANFO is frequently used as the explosive, although several other types of explosives are available. Dynamite is available in sticks of varying sizes, which are placed in the holes. The strength, which is specified as 40, 60, etc., percent, indicates the concentration of the explosive agent, which is nitroglycerin. The dynamite is usually exploded by a blasting cap, which is detonated by an electric current. At least one blasting cap is required for each hole. The charges in several holes can be shot at one time.

The amount of ANFO required to loosen rock will vary from about 0.25 to more than 1 lb/cy, depending on the type of rock, the spacing of holes, and the degree of breakage desired.

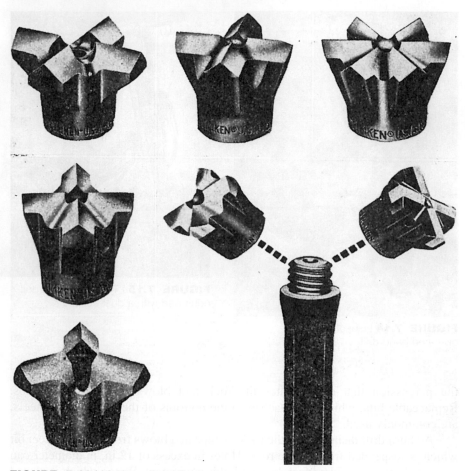

FIGURE 7.16 | Multiuse rock bits.

COST OF OPERATING A DRILL

The items of cost in operating a jackhammer or a wagon drill will include equipment and labor. The equipment cost will include the drill, drill steel, bits, air compressor, and hose. Since each of these items may have a different life, it should be priced separately.

Drill steel is purchased by size, length, and quality of steel used, with the cost of steel based on its weight. While the life of a drill steel will vary with the class of rock drilled and the conditions under which the steel is used, records from drilling projects show consumptions varying from $\frac{1}{20}$ to $\frac{1}{10}$ lb/cy of rock to be representative, with the higher consumption applicable to the harder rocks. Thus, if a drill steel for a wagon drill, 12 ft long, weighing 4.6 lb/ft, whose total weight is about 55 lb, is consumed at the rate of $\frac{1}{12}$ lb/cy of rock, this steel should drill enough hole to produce about 660 cy of rock.

Detachable bits are commonly used with jackhammers, drifters, and wagon drills. The depth of hole that can be drilled with a bit before it must be resharpened or discarded will vary considerably with the class of rock and the type of bit, with values ranging from less than a foot to as much as 100 ft or more for steel bits. Bits with carbide inserts will give much greater depths.

Rates of Drilling Holes

The rate of drilling rock will vary with several factors, including the type and hardness of the rock, type and size of drill used, depth of holes, spacing of holes, topography at the site, condition of the drilling equipment, etc. Although the rates given in Table 7.18 are based on observation, they should be used as a guide only. The rates include an allowance for lost time at the job.

Holes should be drilled 1 ft, or more, deeper than the desired effective depth of rock loosened. This is necessary because the rock usually will not break to the full depth of the holes over the entire area blasted.

TABLE 7.18 | Representative rates of drilling rock with various types of drills.

Size of hole, in.	Class of rock	Rate of drilling, ft/hr				
		Jack-hammer	Wagon drill	Churn drill	Rotary drill	Diamond drill
$1\frac{3}{4}$	Soft	15–20	30–45	—	—	5–8
	Medium	10–15	25–35	—	—	3–5
	Hard	5–10	15–30	—	—	2–4
$2\frac{3}{8}$	Soft	10–15	30–50	—	—	5–8
	Medium	7–10	20–35	—	—	3–5
	Hard	4–8	15–30	—	—	2–4
3	Soft	—	30–50	—	—	4–7
	Medium	—	15–30	—	—	3–5
	Hard	—	8–20	—	—	2–4
4	Soft	—	10–25	—	—	3–6
	Medium	—	5–15	—	—	2–4
	Hard	—	2–8	—	—	1–3
6	Soft	—	—	4–7	25–50	3–5
	Medium	—	—	2–5	10–25	2–4
	Hard	—	—	1–2	6–10	1–3

EXAMPLE 7.17

Estimate the cost of drilling and blasting limestone rock. An analysis of the job indicates that 3-in.-diameter holes will be drilled 14 ft deep at a spacing of 8 ft by 8 ft. Two medium-size wagon drills will be used for the drilling at a rate of 24 ft/hr each. It is anticipated that 1.0 lb of ANFO will be required to loosen each cubic yard of rock. Base all quantities and costs on operating 1 hour.

Job requirements:

Diameter of holes: 3 in.

Depth of holes: 14 ft

Effective depth of holes: 13 ft

Spacing of holes: 8 ft × 8 ft

Equipment and production rates:

Two medium-size wagon drills will be used, each with 50 ft of $1\frac{1}{2}$-in. hose and connections

Estimated rate of drilling: 24 ft/hr

Total production rate: 2 drills × 24 ft/hr = 48 ft/hr

Effective depth of hole drilled: 48 ft/hr × 13/14 = 44.6 ft/hr

Volume of rock produced: (44.6 ft × 8 ft × 8 ft)/(27cf/cy) = 106 cy/hr

Number of holes drilled per hour: (48 ft/hr)/(14 ft/hole) = 3.44 holes/hr

Materials for drilling and blasting:

ANFO required: 1.0 lb/cy of rock

Unit cost of explosive: $1.75/lb

Cost of bits: $24.10 each

Number of times sharpened: 3

Cost of sharpening bit: $3.15 each time

Depth of hole before sharpening bit: 36 ft

Drill steel consumed: $\frac{1}{15}$ lb/cy of rock

The estimated hourly costs for drilling and blasting are calculated as

Drill bit consumption and cost:

Original cost of bit: each = $24.10

Sharpening: 3 × $3.15   =   9.45

    Total cost per bit = $33.55

Number of uses per bit: original bit + 3 sharpenings = 4 uses/bit

Total depth of hole drilled by 1 bit: 4 uses × 36 ft = 144 ft/bit

Cost of bit per foot of hole: $33.55/bit × 144 ft/bit = $0.23/ft

Hourly cost of materials:

Cost of bits per hour: 48 hr/ft × $0.23/ft   = $11.04

Drill steel: 106 cy/hr × 1/15 lb/cy = 7.1 lb @ $1.82 =   12.92

ANFO: 106 cy/hr × 1 lb/cy = 106 cy @ $1.75/lb   =   185.50

Electric caps and wire: 3.44 holes/hr @ $1.45/hole =   4.99

                      Cost per hour for materials = $214.45/hr

Hourly cost of equipment:

Wagon drills: 1 hr × 2 drills @ $42.30/hr = $84.60

Air compressor: 1 hr @ $34.50/hr   =   34.50

Hose and connections: $1\frac{1}{2}$ in. diameter =   1.12

Utility truck: 1 hr @ $8.50/hr   =   8.50

Cost per hour for equipment = $128.72/hr

Hourly cost of labor:

Drill operators: 1 hr × 2 operators @ $30.00/hr = $60.00/hr

Laborers: 1 hr × 2 helpers @ $28.00/hr   =   56.00/hr

Detonator person: 1 hr @ $37.00/hr   =   37.00/hr

Helper: 1 hr × 1 helper @ $21.00/hr   =   21.00/hr

Foreman: 1 hr @ $32.00/hr   =   32.00/hr

Cost per hr for labor = $206.00/hr

Summary of costs:

Material   = $214.45/hr

Equipment =   128.72/hr

Labor   = $206.00/hr

Total = $549.17/hr

Cost per cubic yard of rock, $549.17/hr/106 cy/hr = $5.18/cy

PROBLEMS

7.1 A truck will haul an average load of 15 cy, heaped capacity, loose measure per load. Use average values of swell factors in Table 7.3 to determine the volume of the truck in bank measure when the material is sand, ordinary earth, dense clay, and well-blasted rock.

7.2 Earth has a unit weight of 105 lb/cf bank measure and a swell factor of 25%. The earth is loaded into a 15-ton empty weight truck that has a capacity of 18 cy loose measure. What is the total vertical weight of the truck plus payload? If the shrinkage factor is 15%, what is the volume of earth in the truck, in compacted measure?

7.3 A project requires 40,000 cy of compacted fill that must be completed in 32 working days. The fill material will be hauled to the site in 21-cy, loose measure, trucks. The cycle time for a truck is 40 min per trip. The swell factor for the earth is 25% and the shrinkage factor is 10%. Assume 8 hours per day and a 45-min effective hour. Determine the number of required trucks to complete the project in 32 days.

7.4 A project requires transporting 13,000 cy of sand using 12-cy dump trucks. A front-end loader will load trucks at a rate of 95 cy/hr. The trucks will transport the sand 7 miles at an average haul speed of 30 mi/hr, return speed of 40 mi/hr, and dump time of 3 minutes. Assume a 45-min effective hour for all equipment. Calculate the number of trucks required to balance the production rate of the loader and estimate the time required to transport the sand based on the calculated number of trucks.

7.5   Earth is hauled by trucks to a jobsite at a rate of 1,800 cy/day. Compaction of the soil will be accomplished by sheep's foot compactors that have 5 ft wide drums. The average speed of the compaction roller is 2 mi/hr. Six passes of the roller are required for each 6 in. lift to meet moisture/density requirements. Assume a 50-min effective hour and determine how many compactors are required to balance the production rate of the trucks.

7.6   Estimate the total direct cost and the cost per linear foot for excavating and back-filling a sewer trench 15 in. wide, 36 in. deep, and 8,500 ft long in firm clay using a 70 hp self-transporting trenching machine. Use the data shown in Table 7.7 and assume a 50-min effective hour to determine the production rate of the trencher. Assume the trencher will operate at 20% of the 5.0 mi/hr maximum forward speed of the trencher to backfill the trench.

   The cost of the trencher, with operator, is $75.00/hr. Assume one laborer at $21.00/hr will assist the trenching and backfilling operation. The cost of moving the trencher to and from the jobsite is $400.00. A foreman, with a pickup, at a cost of $38.00/hr will be used to supervise the operation.

7.7   A contractor is trying to decide whether to hand excavate ordinary common earth for a trench using laborers, or to rent a small backhoe excavator. The trench is 3-ft wide and 4-ft deep. The rental rate for the backhoe, with operator, is $95.00/hr. The backhoe can excavate 40 cy/hr.

   Use the hand excavation data from Table 7.5 to obtain the production rates for laborers to excavate the earth from the trench, spreading the loose earth back from the edge of the trench, and backfilling the trench. Assume 4 laborers will be used. The cost of each laborer is $21.00/hr. Calculate the minimum linear feet of trench necessary to economically justify bringing the backhoe to the job.

7.8   A backhoe with a $1\frac{1}{2}$-cy bucket will be used for excavating the basement of a building. The dimensions of the basement are 150 ft wide, 450 ft long, and 8 ft deep. The earth is hard, tough clay. The backhoe will load earth into 18-cy, loose measure, trucks that will haul the earth 10 miles to a waste dump. The average haul speed will be 40 mi/hr and return speed of 45 mi/hr. For each truck assume a fixed time of 5 min to dump a load and wait at the basement area to be loaded.

   The cost of the backhoe, with operator, is $105.00/hr and the cost of each truck, with driver, is $65.00/hr. Assume a 45-min effective hour for the loading and hauling operation. Estimate the total cost and cost per cubic yard, bank measure, based on using the adequate number of trucks to balance the production rate of trucks and the backhoe.

7.9   A 2-cy front shovel will be used to excavate 75,000 cy, bank measure, of earth from a pit that consists of a rock-earth mixture. The shovel can excavate the rock-earth mixture at the optimum depth of cut and at a 120° average angle of swing. The cost of the shovel, with operator, is $125.00/hr and it will cost $1,250 to transport the shovel to and from the job.

   The material will be loaded into 15-cy, loose measure, trucks and hauled 8 miles to the jobsite. The trucks will average 30 mi/hr loaded and 40 mi/hr empty. Assume 3 min to dump a truck load. The cost of each truck, with driver, is $55.00/hr.

   Estimate the total cost and the cost per cubic yard, bank measure, for excavating and hauling the earth. Assume the shovel and trucks will operate at a 45-min effective hour. Determine the economical number of trucks to balance the production rate of the shovel with the hauling rate of the trucks.

7.10 A $1\frac{1}{2}$-cy dragline will be used to excavate a ditch in dense clay. The ditch will be trapezoidal in cross-section, 18 ft wide at the bottom, 40 ft wide at the top, 8 ft deep. The ditch is 4,800 ft long. The excavated earth will be cast along the edge of the ditch. The angle of swing will be 145°. Assume a 50-min effective hour.

It will cost $5,200 to transport the dragline to the job, set it up for operation, and return it to the storage yard. The hourly cost of the dragline, with operator, is $125.00/hr. Estimate the total cost and cost per cubic yard, bank measure, for excavating the ditch.

7.11 A 15-cy, loose measure, self-loading scraper will be used to excavate and haul earth 1,800-ft for a highway project. The scraper will average 9 mi/hr loaded and 11 mi/hr empty. The estimated fixed time to load, accelerate, decelerate, dump, and maneuver the scraper is 7 min per trip. Assume a 50-min effective hour for the scraper.

The cost of the scraper, with operator, is $105.00/hr. Assume a swell factor of 25% and estimate the probable direct cost per cubic yard, bank measure, for excavating and hauling the earth.

7.12 Earth will be excavated and hauled by scrapers with an average capacity of 16-cy loose measure. Assume a swell factor of 25%. The distance from the borrow pit to the compaction area will average 1,700 ft. The scrapers can average 14 mi/hr loaded and 18 mi/hr empty. The fixed time for scrapers is 6 min per trip, which includes the time for loading, dumping, turning, waiting to load or dump, etc.

One dozer will be used to assist the scrapers in loading. It is expected that the average time for a dozer to serve a scraper will be about $2\frac{1}{2}$ min. For this project determine the number of scrapers that one dozer can serve.

The cost of a scraper, with operator, is $115.00/hr and the cost of a dozer, with operator, is $85.00/hr. Assume a 50-min effective hour and estimate the cost per cubic yard.

7.13 Estimate the total time and cost, and the cost per cubic yard, bank measure, for the second balance point, Sections IV and V, of the mass diagram in Figure 7.9. A self-loading scraper with a 15-cy, loose measure, capacity will be used for the excavating and hauling. Assume an average haul speed of 12 mi/hr and return speed of 20 mi/hr. Use a 3-min load time and 2-min dump time. Assume a 45-min effective hour and a swell factor of 8%. The hourly cost of the scraper, with operator, is $110.00/hr.

7.14 A compacted fill is being built with 2 sheep's foot compactors that each have a 5 ft wide drum, operating at 2.0 mi/hr. Six passes are required to obtain the 6-in. depth of compacted fill to meet the moisture/density specifications. A spread of 22-cy, loose measure, trucks will haul the fill dirt to the compacted area. Analysis of the soil conditions indicates a 22-cy, loose measure, truck is equivalent to 15.8 cy, compacted measure.

If the cycle time of a single truck is 15 min, how many trucks are required? If 130,000 cy of compacted fill are required using this compaction equipment, how many drums are required if the project is to be completed in 28 days?

8

Highways and Pavements

OPERATIONS INCLUDED

The chapter is organized into three parts: clearing land, concrete pavements, and asphalt pavements. Even though the coverage is limited to only a few of the methods used, the discussion and examples presented illustrate how estimates can be prepared for projects constructed by other methods.

CLEARING AND GRUBBING LAND

LAND-CLEARING OPERATIONS

Clearing land can be divided into several operations, depending on the type of vegetation to be removed, the type and condition of the soil and topography, the amount of clearing required, and the purpose for which the clearing is done as listed:

1. Complete removal of all trees and stumps, including tree roots
2. Removal of all vegetation above the surface of the ground only, leaving the stumps and roots in the ground
3. Disposal of the vegetation by stacking and burning it

Light Clearing

Light clearing includes removal of vegetation up to 2 in. in diameter. If the clearing involves only cutting at or above ground level, axes, machetes, brush hooks, and wheel-mounted circular saws can be used. Heavy-duty sickle mowers are available for removal of vegetation up to $1\frac{1}{2}$ in. in diameter. Tractor-mounted circular saws are also used.

When the vegetation must be knocked down to the ground, a bulldozer blade is commonly used. A small- to intermediate-size bulldozer is generally adequate. Also, an anchor chain drawn between two crawler tractors can be used to clear the land. Often light clearing includes incorporating the cleared vegetation into the soil. Disc plows and disc harrows are used for this purpose. The ground is worked with the plows and harrows to blend the cleared material with the soil.

Intermediate Clearing

For vegetation from 2 to 8 in. in diameter, a bulldozer blade, shearing blade, or angling blade can be used. For clearing the vegetation at the ground level, the equipment may include power chain saws, tractor-mounted circular saws, and single scissor-type tree shears. Hydraulic cylinders that cut the trees at ground level operate the scissor shear.

A bulldozer with a V blade is effective for shearing trees, stumps, and brush at ground level. The special V blade has a protruding stinger at its lead point, as illustrated in Fig. 8.1. The dozer moves at a steady speed as the blade slides along the surface of the ground, cutting vegetation flush with the surface and casting the removed material to the sides.

Large Clearing

For vegetation of 8 in. in diameter and larger, bulldozers with special blades are generally used. Before felling large trees, the bulldozers must excavate the earth from around the trees to cut the main roots. When stacking the felled trees and other vegetation, bulldozers transport earth into the piles of trees, which makes burning more difficult.

A variety of rakes are available that attach to the front of bulldozers for land clearing. They can penetrate the soil to remove small stumps, rocks, and roots. Figure 8.2 illustrates a tractor-mounted rake that can be used to group and pile

FIGURE 8.1 | Tractor-mounted V blade for clearing land.

FIGURE 8.2 | Tractor-mounted clearing rake.

trees, boulders, and other materials without transporting an excessive amount of earth. This can be a very effective machine for stacking materials in piles for burning.

RATES OF CLEARING LAND

The rates of clearing land will depend on the type, number, and size of trees. Other factors that can affect the rate of clearing include density of vegetation, type of soil, topography, rainfall, types of equipment used, skill of equipment operators, and requirements of the specifications governing the project. Prior to preparing a clearing estimate, the estimator should visit the project site and obtain information to evaluate these factors.

Clearing of Light Vegetation

Where only light vegetation is present, it is possible to clear the land at a constant speed of the equipment that is used for clearing. The production rate can be calculated by multiplying the speed of the tractor by the width of cut of the blade on the tractor. The base formula is

Production rate (acres/hr)

$$= \frac{\text{width of cut (ft)} \times 5{,}280 \text{ ft/mi} \times \text{speed (mi/hr)}}{43{,}560 \text{ sf/acre}} \quad \text{[8.1]}$$

The American Society of Agricultural Engineers' formula for estimating hourly production of land clearing at constant speed is based on an 82.5 percent efficiency, which is equivalent to a 49.5-min hour. With 82.5 percent efficiency, then Eq. [8.1] becomes:

$$\text{Production rate (acres/hr)} = \frac{\text{width of cut (ft)} \times \text{speed (mph)}}{10} \quad \text{[8.2]}$$

Width of cut is the effective working width of the equipment, measured perpendicular to the direction of tractor travel. The width of cut for clearing may not be the same as the rated width of the blade. The cut of an angle blade is not equivalent to the width of the blade. Even when working with a straight blade, it may not be the same as the blade width. Although the width of cut is sometimes estimated as a percentage of the rated width, it is better to measure the width of cut on the job.

EXAMPLE 8.1

A 165-hp crawler tractor will be used to clear small trees and brush from a 25-acre site. By operating in first gear, the tractor should be able to maintain a continuous forward speed of 0.9 mph. An angle-clearing blade will be used. From past experience the average effective width of cut will be 8 ft. Assume normal efficiency, and estimate the time required to clear the vegetation.

Quantity of work:

Clearing light vegetation = 25 acres

Production rate:

$$\text{Production rate (acres/hr)} = \frac{\text{width of cut (ft)} \times \text{speed (mph)}}{10}$$ [8.2]

$$= (8 \text{ ft}) \times (0.9 \text{ mph})/10$$

$$= 0.72 \text{ acres/hr}$$

Time for clearing:

Time = 25 acres/0.72 acres/hr = 34.7 hr, or about 35 hr

Estimating Time to Fell Trees

Most land clearing operations such as bulldozing, cutting, grubbing, raking, and piling are not performed at constant speed. Rome Industries developed Eq. [8.3] and Table 8.1 as a guide in estimating the required time for felling trees only, by using the tractor sizes in Table 8.1 with a K/G blade. The Rome Industry formula and tables of constants provide guidance for variable speed operations, but use of the results should be combined with field experience.

$$\text{Time (min/acre)} = H\left[A(B) + M_1N_1 + M_2N_2 + M_3N_3 + M_4N_4 + DF\right]$$

[8.3]

where

H = hardwood factor affecting total time

Hardwoods affect overall time as

75–100% hardwoods; add 30% to total time ($H = 1.3$)

25–75% hardwoods; no change ($H = 1.0$)

0–25% hardwoods; reduce total time 30% ($H = 0.7$)

TABLE 8.1 | Production factors for felling trees with Rome K/G blades.

Tractor, hp	Base minutes per acre, B*	Diameter range				
		1–2 ft, M_1	2–3 ft, M_2	3–4 ft, M_3	4–6 ft, M_4	>6 ft, F
165	34.41	0.7	3.4	6.8	—	—
215	23.48	0.5	1.7	3.6	10.2	3.3
335	18.22	0.2	1.3	2.2	6.0	1.8
460	15.79	0.1	0.4	1.3	3.0	1.0

*B is based on power-shift tractors working on reasonable level terrain, (10%) maximum grade, with good footing and no stones and an average mix of soft and hardwoods.

Source: Caterpillar, Inc.

A = tree density and presence of vines' effect on base time

Base time is affected by the density of material less than 1 ft in diameter and the presence of vines:

Dense: greater than 600 trees/acre; add 100% to base time ($A = 2.0$)

Medium: 400–600 trees/acre; no change ($A = 1.0$)

Light: less than 400 trees/acre; reduce base time 30% ($A = 0.7$)

Presence of heavy vines; add 100% to base time ($A = 2.0$)

B = base time for each tractor size per acre

M = minutes per tree in each diameter range

N = number of trees per acre in each diameter range, from field survey

D = sum of diameter in foot increments of all trees per acre above 6 ft in diameter at ground level, from field survey

F = minutes per foot of diameter for trees above 6 ft in diameter

When the job specification requires the removal of trees and grubbing of the roots and stumps greater than 1 ft in diameter in one operation, the total time per acre should be increased by 25 percent. When the specification requires the removal of stumps in a separate operation, the time per acre should be increased by 50 percent.

To develop the necessary input data for Eq. [8.3], the estimator must make a field survey of the area to be cleared and collect information on these items:

1. Density of vegetation less than 12 in. in diameter:

   Dense—600 trees/acre

   Medium—400 to 600 trees/acre

   Light—less than 400 trees/acre

2. Presence of hardwoods expressed in percent

3. Presence of heavy vines

4. Average number of trees per acre in each of these ground-level diameter size ranges:

   < 1 ft

   1–2 ft

   2–3 ft

   3–4 ft

   4–6 ft

5. Sum of diameter of all trees per acre above 6 ft in diameter at the ground level

---

EXAMPLE 8.2

Trees must be cleared for a highway project. The site is reasonably level terrain with firm ground and less than 25 percent hardwood. A 215-hp tractor equipped with a K/G blade will be used to fell trees. The highway specifications require the grubbing of tree stumps greater than 12 in. in diameter. Estimate the production rate to fell the trees. A field survey has been conducted and these data have been collected:

Average number of trees per acre = 700

Number of 1–2 ft dia. = 150 trees

Number of 2–3 ft dia. = 18 trees

Number of 3–4 ft dia. = 7 trees

Number of 4–6 ft dia. = 3 trees

Sum of diameter increments above 6 feet = none

Input data based on the field survey:

H = 0.7 based on percentage of hardwoods, < 25% hardwoods

A = 2.0 based on dense number of trees, > 600 trees/acre

Production factors from Table 8.1:

B = 23.48 for 215-hp tractor

M_1 = 0.5 for 1- to 2-ft-dia. trees

M_2 = 1.7 for 2- to 3-ft-dia. trees

M_3 = 3.6 for 3- to 4-ft-dia. trees

M_4 = 10.2 for 4- to 6-ft-dia. trees

F = 3.3 for diameter of increments above 6 ft

Production rate to fell trees:

$$\text{Time (min/ acre)} = H[A(B) + M_1N_1 + M_2N_2 + M_3N_3 + M_4N_4 + DF] \qquad \textbf{[8.3]}$$

$$= 0.7[2.0(23.48) + 0.5(150)$$

$$+ 1.7(18) + 3.6(7) + 10.2(3) + 0(3.3)]$$

$$= 208.4 \text{ min/acre}$$

Because the operation will include grubbing, the time must be increased by 25 percent.

60-min hour production rate = 1.25 × 208.4 min/acre = 260.5 min/acre

Converting to hours/acre = 260.5 min/acre × hr/60 min = 4.3 hr/acre

For a 45-min hour, production rate = 60/45 × 4.3 hr/acre = 5.8 hr/acre

Cost:

If the tractor with blade and operator cost $185/hr, the cost per acre of land cleared can be calculated as

Cost per acre: ($185/hr) × 5.8 hr/acre = $1,073/acre to fell trees

Estimating Time to Pile Trees

Usually the cut trees are piled for burning or piled in stacks so they can be easily picked up and hauled away. Rome Industries developed Eq. [8.4] and Table 8.2 as a guide to estimate the time for piling trees. The factors in Eq. [8.4] and Table 8.2

TABLE 8.2 | Production factors for piling-up in windrows.*

Tractor, hp	Base minutes per acre, B	1–2 ft, M_1	2–3 ft, M_2	3–4 ft, M_3	4–6 ft, M_4	>6 ft, F
		Diameter range				
165	63.56	0.5	1.0	4.2	—	—
215	50.61	0.4	0.7	2.5	5.0	—
335	44.94	0.1	0.5	1.8	3.6	0.9
460	39.27	0.08	0.1	1.2	2.1	0.3

*Can be used with most types of raking tools and angled shearing blades.
Windrows to be spaced approximately 200 ft apart.
Source: Caterpillar, Inc.

have the same definitions as when used previously in Eq. [8.3] and Table 8.1. Piling-up grubbed vegetation increases the total piling-up time by 25 percent.

$$\text{Time (min/acre)} = B + M_1N_1 + M_2N_2 + M_3N_3 + M_4N_4 + DF \qquad \textbf{[8.4]}$$

EXAMPLE 8.3

The vegetation cut in Example 8.2 is to be piled into a stack for burning. Estimate the cost for piling-up the cut vegetation.

$$\text{Time (min/acre)} = B + M_1N_1 + M_2N_2 + M_3N_3 + M_4N_4 + DF \qquad \textbf{[8.4]}$$
$$= 50.61 + 0.4(150) + 0.7(18) + 2.5(7) + 3.6(3) + \_\_(0)$$
$$= 151.5 \text{ min/acre}$$

Because the operation will include grubbing, the time must be increased by 25 percent,

60-min hour production rate = 1.25 × 151.5 min/acre = 189.4 min/acre

Converting to hours/acre = 189.4 min/acre × hr/60 min = 3.2 hr/acre

For a 45-min hour, production rate = 60/45 × 3.2 hr/acre = 4.2 hr/acre

Cost:

If the labor and equipment cost is $185/hr, the cost per acre for piling and stacking trees can be calculated.

Cost per acre, $185/hr × 4.2 hr/acre = $777/acre

DISPOSAL OF BRUSH

When brush is to be disposed of by burning, it should be piled in stacks and windrows, with a minimum amount of earth included. Felled trees and grubbed brush often have earth around their roots, which impedes burning. Shaking the rake while transporting the material to piles or into windrows will help remove the earth.

Because burning is usually necessary while the brush contains considerable moisture, it may be desirable to provide a continuous external source of fuel and heat to assist in burning the material. The burner illustrated in Fig. 8.3, which consists of a gasoline-engine-driven pump and a propeller, is capable of discharging the liquid fuel onto the pile.

FIGURE 8.3 | Burning brush with forced draft and fuel oil.

DEMOLITION

For remodels of buildings or replacement of infrastructures, it is necessary to remove a portion or all of an existing structure. The time and cost to perform this work is difficult to estimate because there are many factors that can affect the work. For example, removal of an existing concrete pavement depends on the thickness of the slab, amount of reinforcement, state of deterioration of the pavement, job conditions, type and size of demolition equipment, and skill of the operator.

CONCRETE PAVEMENTS

GENERAL INFORMATION

The cost of concrete pavement in place includes the cost of fine-grading the subgrade; side forms, if required; steel reinforcing, if required; aggregate; cement; mixing, placing, spreading, finishing, and curing concrete; expansion-joint material; saw cutting; and shaping the shoulders adjacent to the slab. For many paving operations, a slip-form paver is used to place the concrete, which eliminates the need for setting side forms. Some concrete pavements are designed for construction without reinforcing steel. If the subgrade is dry, it may be necessary to wet it before placing the concrete.

If the pavement is not uniformly thick, the average thickness can be determined to find the area of the cross section. This multiplied by the length will give the volume of concrete required, usually expressed in cubic yards. Payment for concrete pavement usually is at the agreed price per yard of surface area.

CONSTRUCTION METHODS USED

At least two methods are used to place concrete pavements: side forms or slip forms. The side-form method is used for small projects, such as city streets or parking lots of businesses. The slip-form method is commonly used for larger projects, such as long highway pavements or airport runways.

Using the side-form method, the subbase soil is brought to the specified density, grade, and shape and then side forms are set to control the thickness of the slab and to confine the concrete until it sets. After the concrete has cured sufficiently, the side forms are removed. The forms most commonly used are made of steel whose height is equal to the thickness of the concrete adjacent to the forms. Forms are manufactured in sections 10 ft long, with three holes per section for pins, which are driven into the ground to maintain alignment and stability. The top of the forms must be set to the exact elevation required. Forms usually are left in place 8 to 12 hr after the concrete is placed, after which they are removed, cleaned, oiled, and reused.

The slip-form method uses a slip-form paving machine to shape the slab. Two side forms, which confine the outer edges of the freshly placed concrete, are mounted on a self-propelled paver, which spreads, vibrates, and screeds the concrete to the specified thickness and surface shape as it moves along the job.

BATCHING AND HAULING CONCRETE

There are two types of concrete mixing operations: central-batch concrete and job-batch concrete. A central mix plant generally is used for projects that require large volumes of concrete, whereas the job-batch concrete is typically used for low-volume projects, such as city streets or parking lots. At remote locations or jobs requiring large quantities of concrete, generally concrete batch plants are set up at the jobsite. (See Fig. 8.4.) The concrete may be mixed completely in a stationary mixer at the concrete plant, or it may be batched at the plant and mixed in the concrete truck while en route to the job.

FIGURE 8.4 | Concrete batch plant at jobsite.

Courtesy: Sherwood Construction Company.

The term ready-mix concrete refers to concrete that is proportioned in a central location by a supplier and transported to the purchaser in a fresh state. Ready-mix concrete must be available within a reasonable distance from the project. The concrete is transported to the job in mixer trucks that range in capacity from 9 to 14 cy. The cost of ready-mix concrete depends on the strength of concrete and any additives that may be required in the mix design. Some ready-mix suppliers charge a mileage fee for concrete delivered in excess of a specified distance, such as over 12 mi. The concrete supplier sets the cost for ready-mix concrete.

For most highway projects, a batch plant is set up at the jobsite. Generally the plant is set up as close to the center of the project as possible. For a long paving project, more than one setup of the concrete plant may be justified to reduce the length of haul for the batched materials. A cost study should be made prior to locating the plant to determine if more than one location is desirable.

Bulk quantities of cement, sand, and aggregate are stored at the jobsite for the batching operation. The equipment for operating a batch plant will include bulk cement bins; overhead bins for aggregate storage, equipped with weight batchers; and a front-end loader or portable conveyors to handle aggregate. The cement is supplied to the jobsite in bulk quantities from cement transport trucks, each holding 20 T or more. Small quantities of cement can be purchased in paper bags, each containing 1 cf loose measure and weighing 94 lb net.

Most concrete batch plants have the capacity to mix about 8 to 12 cy per batch and can produce from 100 to 400 cy of concrete per hour. A small batch plant can mix about 20 cy/hr and is used for small jobs, less than 500 cy total. If the batch is fully mixed at the batch plant, it usually is transported to the job in dump trucks, or end-dump trailers for slip-form paving. For smaller projects or for fixed-side-form paving, many contractors load the batched concrete into mixer trucks, or trucks with open-top agitators, which transport the fresh concrete to the jobsite. Concrete that is not fully mixed at the batch plant is transported to the job in transit-mixer trucks.

PLACING CONCRETE PAVEMENTS

After the fresh concrete arrives at the jobsite, it must be moved to its final position without segregation of the mix and before it has achieved an initial set. For efficiency and economy, the production rate should be balanced between the batch plant, hauling units, and the paving equipment. For small paving projects, the concrete may be placed from the chute of the concrete truck into the hopper of the paving machine that spreads and places the concrete. Laborers hand finish the concrete and saw cut joints in the pavement.

For large paving projects, a paving train of equipment is used to place concrete pavements. A paving train consists of three types of equipment: concrete placer spreader, slip-form paver, and a tine and cure machine (see Fig. 8.5).

The concrete placer spreader is designed specifically for end-dump concrete delivery. The concrete is dumped from the delivery truck onto a conveyor that

FIGURE 8.5 I Concrete paving train: placer/spreader, paver, tine and cure machine.
Courtesy: Wittwer Paving, Inc.

discharges the concrete on grade in front of an auger that spreads the concrete evenly across the front of the paving machine. The unit can be set up for either left-side or right-side operation to allow trucks to pass by without interfering with the paving operation. Essentially, the concrete placer spreader machine feeds the concrete paver.

The slip-form paver follows the concrete placer spreader. The paver shapes the slab by two side forms that confine the outer edges of the freshly placed concrete. The paver vibrates and screeds the concrete to the specified thickness and surface shape as it moves along the job. After placement of the concrete by the paver, laborers provide an edge finish to both sides of the pavement and a straight edge finish to the surface of the pavement. Figure 8.6 shows a slip-form paving machine in operation.

The tine and cure machine follows the paver and performs two functions. It produces the desired texture on the surface of the pavement and it applies a curing compound. It provides a microtexture with a burlap drag, a macrotexture with transverse tines, and applies curing compound to the pavement. Figure 8.7 shows the tine and cure machine in operation.

For a high-production paving operation, a paving train consisting of two concrete placer spreader machines, one slip-form paver, and two tine and cure machines can produce up to 500 cy/hr.

FIGURE 8.6 | Slip-form paving machine in operation.

Courtesy: Sherwood Construction Company.

FIGURE 8.7 | Tine and cure machine.

Courtesy: Wittwer Paving, Inc.

CONCRETE PAVEMENT JOINTS

To reduce the danger of irregular and unsightly cracks across and along concrete pavements, it is common practice to install joints at regular intervals. Joints control the location of cracks, allow expansion and contraction of the concrete, and provide a method of separating concrete placements in the construction operation.

Three types of joints are installed in concrete pavements: construction, transverse, and longitudinal. Construction joints are installed where the concrete placement operation is stopped at the end of a working day. Transverse joints are at a right angle to the road centerline to allow for expansion and contraction of the concrete pavement. Longitudinal joints are installed to tie together adjacent lanes of the pavement. Transverse and longitudinal joints are sawed in the pavement after the concrete is placed. Figure 8.8 shows sections through several types of joints.

Prior to placing concrete, smooth steel bars called dowels are set at $\frac{1}{2}$ the concrete thickness at the location of transverse joints. The dowels are generally 1 in. in diameter and 24 in. long, spaced at 12 in. on centers along the transverse joint. For longitudinal joints, $\frac{5}{8}$- to $\frac{3}{4}$-in.-diameter deformed reinforcing bars are commonly used, placed at 24 in. on centers.

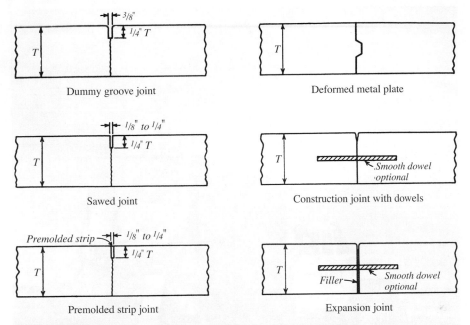

FIGURE 8.8 | Representative details of joints in concrete pavement.

Transverse joints are sawed across the pavement at spacings that vary from 15 to 20 ft. Longitudinal center joints are installed along the length of the pavement usually when the width exceeds 12 ft and the thickness is less than 12 in. The width of joints specified may vary from $\frac{1}{8}$ to $\frac{1}{4}$ in., and the depth is normally $\frac{1}{4}$ to $\frac{1}{3}$ of the concrete thickness. It is common practice to perform two saw-cutting operations. For example, for a 12-in.-thick pavement the first saw cut might be $\frac{1}{8}$ in. wide and 3 in. deep to control the location of the crack. After about 72 hr a second cut that is $\frac{1}{4}$ in. wide and 1 in. deep is made through the first cut. A backer-rod made of styrofoam is placed in the 1-in.-deep cut, then sealant material is placed.

Self-propelled or hand-pushed machines, using abrasive or diamond blades, are used for saw-cutting joints in concrete pavements, as illustrated in Fig. 8.9. Cutting speeds should vary from 2 to 10 ft/min, depending on the width and depth of the joint, type of blade used, and kind of aggregate used in the concrete mix. The cutting operation generally involves a crew of three workers, including the saw operator with a helper and a driver of the water truck that is used to cool the saw blade during the cutting operation.

Diamond-tip concrete saw blades are expensive and must be replaced after about every 10,000 to 20,000 lin ft of cut, depending on the depth and width of cut. A reasonable allowance should be made for breakage during the sawing operation.

The sawed joints are filled with a sealing compound, usually an asphaltic or rubber-based material, applied either hot or cold. Sealant material for concrete

FIGURE 8.9 | Saw-cutting joints in concrete pavements.

Courtesy: Wittwer Paving, Inc.

pavements is expensive and is available in 55-gal-capacity barrels. One gallon of joint sealant will fill about 125 lin ft of saw cut. Dowels, consisting of steel tie bars or rods, may be installed across all joints to transfer shear forces across the joint.

CURING CONCRETE PAVEMENT

Curing is normally accomplished with an impervious membrane-producing compound that is sprayed on the surface and sides soon after the concrete is placed. The membrane-producing compounds, frequently called curing compounds, will cover about 30 to 50 sy/gal.

Curing can also be accomplished by covering the fresh slab with Visqueen plastic film or waterproof paper to reduce the evaporation of initial water in the concrete. Burlap or cotton mat can also be used, but they must be kept wet for the specified time to reduce evaporation of water from the newly placed concrete.

EXAMPLE 8.4

Estimate the total cost and the cost per square yard, for bid purposes, for placing a concrete pavement 26 ft wide, of 10-in. average thickness, and 5.78 mi long. Smooth steel dowels will be placed at 12 in. along transverse joints and $\frac{5}{8}$-in. deformed steel bars will be placed at 24 in. along longitudinal joints. No reinforcing steel will be placed in the pavement.

A single longitudinal saw cut is required in the center of the pavement, and transverse saw cuts are required at 15 ft intervals for the full length of the pavement. Two saw cuts will be made: one will be 25 percent of the depth and $\frac{1}{8}$ in. wide and another will be 1 in. deep and $\frac{1}{4}$ in. wide.

The mix design of each cubic yard of concrete is given as:

Cement: 5.6 sacks

Sand: 1,438 lb

Gravel: 1,699 lb

Water: 39 gal

A portable central-mix plant will be set up near the midpoint of the project. The plant has a capacity of 8 cy/batch. Concrete trucks with a capacity of 8 cy each will haul the material an average of 2.5 mi at an average speed of 30 mph. A front-end loader and portable conveyor will be used to charge the aggregate into the bins.

Water will be obtained from a private pond, requiring a 3-in. gasoline-engine-operated pump having a capacity of 20,000 gal/hr. A royalty of $0.42/1,000 gal will be paid for the water. The water will be hauled an average distance of 3 mi by trucks whose capacities are 2,500 gal each. The average haul speed is estimated to be 25 mph. It will require 10 min to fill a truck and 10 min to empty a truck. The truck driver will operate the pump at the pond.

Quantity of materials:

Volume of concrete:

$$(5.78 \text{ mi} \times 5,280 \text{ ft/mi} \times 26 \text{ ft} \times 10/12 \text{ ft})/(27 \text{ cf/cy}) \quad = \quad 24,490 \text{ cy}$$

$$\text{Add 5\% waste} = \underline{\quad 1,225 \text{ cy}}$$

$$\text{Total} = 25,715 \text{ cy}$$

Square yards of pavement:

(5.78 mi × 5,280 ft/mi × 26 ft)/9 sf/sy = 88,164 sy

Calculated unit weight of the concrete:

Cement: 5.6 sacks/cy × 94 lb/sack   =   526.4 lb

Sand   =   1,438.0 lb

Gravel   =   1,699.0 lb

Water: 39 gal/cy × (1.0 gal/7.48 cf) × 62.4 lb/cf =   325.3 lb

Total = 3,988.7 lb/cy

Converting to lb/cf: 3,988.7 lb/cy × (1.0 cy/27 cy) = 147.7 lb/cf

Linear feet of saw cut joints:

Longitudinal along centerline: 5.78 mi × 5,280 ft/mi   = 30,518 lin ft

Transverse: 5.78 mi × 5,280 ft/mi × joint/15 ft × 26 ft/joint = 52,899 lin ft

Total saw cuts = 83,417 lin ft

Number of saw blades for $2\frac{1}{2}$-in.-deep, $\frac{1}{8}$-in.-wide cut:

(83,417 lin ft)/(12,500 lin ft/blade) = 6.7 blades, need 7 blades

Number of saw blades for 1-in.-deep, $\frac{1}{4}$-in.-wide cut:

(83,417 lin ft)/(18,500 lin ft/blade) = 4.5 blades, need 5 blades

Adding 2 blades for waste, 12 + 2 = 14 blades required

Transverse dowels and longitudinal tie bars:

Transverse: (52,899 ft)/(1.0 ft/dowel) = 52,899 smooth steel dowels

Longitudinal: (30,518 ft)/(2.0 ft/bar) = 15,259 deformed reinforcing bars

Bulk materials:

Cement: (25,715 cy × 526.4 lb/cy)/2,000 lb/T = 6,768 T

Sand: (25,715 cy × 1,438 lb/cy)/2,000 lb/T = 18,489 T

Gravel: (25,715 cy × 1,699 lb/cy)/2,000 lb/T = 21,845 T

Water: (25,715 cy × 39 gal/cy) = 1,002,885 gal

Curing compound: (88,164 sy/40 sy/gal) = 2,204 gal

Sealing compound: (83,417 lin ft)/(125 lin ft/gal) = 667 gal

Production rate of concrete plant:

An 8-cy batch plant will be used to batch and mix the concrete.

For each batch the quantity of material will be

Cement: 8 cy/batch × 5.6 sacks/cy × 94 lb/sack = 4,211 lb

Sand: 8 cy/batch × 1,438 lb/cy   = 11,504 lb

Gravel: 8 cy/batch × 1,699 lb/cy   = 13,592 lb

Water: 8 cy/batch × 39 gal/cy   = 312 gal

If the mixer discharges the entire load into one truck, the time per cycle can be calculated

Charging mixer   = 0.75 min

Mixing concrete   = 2.25 min

Discharging mixer = 0.50 min

Lost time   = <u>0.25 min</u> (average time of mechanical problems)

      Total cycle time = 3.75 min/batch

Number of batches per hour: (60 min/hr)/(3.75 min/batch) = 16 batches/hr

For a 45-min hour: 16 batches/hr × 8 cy/batch × 45/60 = 96 cy/hr

Time to complete the job: 25,715 cy/96 cy/hr = 267 hr

Rounding to a full 8-hr day, the time will be 272 hr ←

The time on the job is established by the 96 cy/hr production rate of the batch plant and the total quantity of concrete for the pavement. However, to keep the batch plant in operation to produce the 96 cy/hr there must be adequate production rate of water hauling trucks, concrete hauling trucks, and saw cutting crews to service the batch plant.

Production rate for transporting water:

Cycle time for water truck:

Filling tank: 10 min × (1.0 hr/60 min)   = 0.167 hr

Hauling water: 3 mi/25 mph   = 0.120 hr

Emptying tank: 10 min × (1.0 hr/60 min.) = 0.167 hr

Returning empty: 3 mi/25 mph   = <u>0.120 hr</u>

                 Total = 0.574 hr/trip

Water hauled per trip: 2,500 gal/trip × (1.0 trip/0.574 hr) = 4,355 gal/hr

For a 45-min hour: water hauled per trip = 4,355 gal/hr × 45/60 = 3,266 gal/hr

Quantity of water needed for concrete: 96 cy/hr × 39 gal/cy = 3,744 gal/hr

Number of water trucks needed: 3,744/3,266 = 1.1 trucks required

Use 2 water trucks to ensure continuous operation of batch plant.

Production rate for hauling concrete:

Cycle time for a transit mix concrete truck:

Load: 1.0 min × (1.0 hr/60 min)   = 0.017 hr

Haul: 2.5 mi/30 mph   = 0.083 hr

Dump: 1.5 min × (1.0 hr/60 min) = 0.025 hr

Return: 2.5 mi/30 mph   = <u>0.083 hr</u>

            Total = 0.208 hr/trip

Cubic yards hauled per trip: 8 cy × (1.0 trip/0.208 hr) = 38.5 cy/hr

For a 45-min hour, production rate = 38.5 cy/hr × 45/60 = 28.8 cy/hr

Number of trucks needed: 96/28.8 = 3.3 trucks required

Use 4 transit mix trucks to ensure a continuous paving operation.

Production rate for sawing joints:

Rate of sawing of $2\frac{1}{2}$-in.-deep joint $\frac{1}{8}$-in.-wide cut:

4 ft/min $\times$ 45 min/hr = 180 ft/hr

Time for sawing: (83,417 ft/180 ft/hr) = 463 hr

Rate of sawing of 1-in.-deep joint $\frac{1}{4}$-in.-wide cut:

7 ft/min $\times$ 45 min/hr = 315 ft/hr

Time for sawing: (83,417 ft/315 ft/hr) = <u>265 hr</u>

Total time for sawing = 728 hr

Number of saw crew required: 728 hr/272 hr = 2.7 needed, therefore use 3 crews

Costs to construct the pavement:

Material costs:

Cement: 6,768 tons @ $104.00/ton	=	$703,872
Sand: 18,489 tons @ $9.00/ton	=	166,401
Gravel: 21,845 tons @ $11.00/ton	=	240,295
Water: 1,002,885 gal @ $0.50/gal/1,000 gal	=	501
Transverse dowels: 52,899 dowels @ $2.50/dowel	=	132,248
Longitudinal tie bars: 15,259 bars @ 0.95/bar	=	14,496
Curing compound: 2,204 gal @ $8.75	=	19,285
Sealing compound: 667 gal @ $37.00/gal	=	<u>24,679</u>
Subtotal cost of materials	=	$1,301,777

Equipment costs:

Batch plant operation:

Portable batch plant: 272 hr @ $301.00/hr	=	$81,872
Conveyor: 272 hr @ $30.00/hr	=	8,160
Front-end loader: 272 hr @ 70.00/hr	=	19,040
Water trucks: 272 hr $\times$ 2 trucks @ $39.00/hr	=	21,216
Water pump: 272 hr @ $9.00/hr	=	2,448
Water tank for plant: 272 hr @ $7.00/hr	=	1,904

Pavement preparation operation:

Fine grader: 272 hr @ $95.00/hr	=	25,840
Earth roller, smooth-wheel: 272 hr @ $34.00/hr	=	9,248

Lay down paving operation:

Transit mix trucks: 272 hr $\times$ 4 trucks @ $70.00/hr	=	76,160
Concrete placer/spreader: 272 hr @ $240.00/hr	=	65,280
Paving machine: 272 hr @ $225.00/hr	=	61,200
Tine and cure machine: 272 hr @ $90.00/hr	=	24,480
Truck for paving foreman: 272 hr @ $7.50/hr	=	2,040

Pavement finishing operation:

Concrete saws: 272 hr × 3 saws @ $14.00/hr	=	11,424
Water truck for saws: 272 hr × 3 trucks @ $39.00/hr	=	31,824
Saw blades, diamond tips: 14 blades @ $750 each	=	10,500

Supporting equipment costs:

Generator: 272 hr @ $10.00/hr	=	2,720
Air compressor: 272 hr @ 9.00/hr	=	2,448
Surveying equipment: 272 hr @ $18.00/hr	=	4,896
Pickup trucks: 272 hr × 4 trucks @ $6.00/hr	=	6,528
Barricades, signs, small stools, etc.	=	2,500
Moving to and from project, excluding labor	=	15,000
Subtotal of equipment costs	=	$486,728

Labor costs:

Batch plant operation:

Batch plant operator: 272 hr @ $31.00/hr	=	$8,432
Batch plant laborers: 272 hr @ $20.00/hr	=	5,440
Front-end loader operator: 272 hr @ 28.00/hr	=	7,616
Water truck drivers: 272 hr × 2 drivers @ $25.00/hr	=	13,600
General laborer: 272 hr @ $19.00/hr	=	5,168

Pavement preparation operation:

Fine grader operator: 272 hr @ $27.00/hr	=	7,344
Roller compaction operator: 272 hr @ $25.00/hr	=	6,800
Dowel setters: 272 hr × 4 laborers @ $20.00/hr	=	21,760
Tie bar setters: 272 hr × 2 laborers @ $20.00/hr	=	10,880

Lay down paving operation:

Mixer truck drivers: 272 hr × 4 drivers @ $25.00/hr	=	27,200
Placer/spreader operator: 272 hr @ $26.00/hr	=	7,072
Paver operator: 272 hr @ $31.00/hr	=	8,432
Tine and cure machine operator: 272 hr @ $29.00/hr	=	7,888

Pavement finishing operation:

Edge finishers: 272 hr × 2 laborers @ $22.00/hr	=	11,968
Straight-edge finisher: 272 hr @ $22.00/hr	=	5,984
Saw operators: 272 hr × 3 operators @ $21.00/hr	=	17,136
Water truck for saws: 272 hr × 3 drivers @ $25.00/hr	=	20,400
Joint sealant laborers: 272 hr × 2 laborers @ $19.00/hr	=	10,336

Supporting labor costs:

General laborers: 272 hr × 3 laborers @ $19.00/hr	=	15,504
Surveyors: 272 hr × 2 surveyors @ $27.00/hr	=	14,688

Traffic control: 272 hr × 4 laborers @ $19.00/hr   =   20,672

Paving foreman: 272 hr @ $31.00/hr   =   8,432

Superintendent: 272 hr @ $35.00/hr   =   9,520

Subtotal of labor costs = $272,272

Taxes and insurance costs:

Material tax: 5% × $1,301,777   = $65,089

Social security tax: 7.65% × $272,272   =   20,829

Unemployment tax: 2.5% × $272,272   =   6,807

Worker's compensation insurance: 9% × $272,272 =   24,504

Public liability/property damage: 12% × $272,272   =   32,673

Subtotal of taxes = $149,902

Summary of direct costs:

Material cost: = $1,301,777

Equipment cost: =   486,728

Labor cost: =   272,272

Taxes/insurance: =   149,902

Direct costs = $2,210,679

Add-on costs:

Job overhead: 4% × $2,210,679 $88,427

Office overhead: 2% × $2,210,679   =   44,214

Contingency for risk: 3% × $2,210,679 =   66,320

Plant setup and field office: =   42,000

Subtotal of add-ons = $240,961

Total estimated cost:

Final estimate = direct costs + add-on costs

= $2,210,679 + $240,961

= $2,451,640

Bid amount:

Total estimated cost: = $2,451,640

Bonds: 1.2% × $2,451,640 =   29,420

Profit: 7% × $2,451,640   =   171,615

Bid estimate = $2,652,675

Unit costs:

Cost per cubic yard: $2,652,675/24,490 cy = $108.32/cy

Cost per square yard: $2,652,675/88,164 sy = $30.09/sy

Cost per mile: $2,652,675/5.78 mile = $458,940/mile

ASPHALT PAVEMENTS

Asphalt pavements are constructed by placing mixtures of mineral aggregates and asphaltic binders. The aggregate and asphalt binder are heated and mixed together in an asphalt plant to make the asphalt paving material, called *asphalt mix*.

The cost of hot-mix asphaltic-concrete pavements will include the initial expense of setting up a plant, the aggregates, liquid asphalt, burner fuel, electricity (if no generator is used), quality control, and all the labor and equipment costs associated with laying and compacting the mix. Additional costs can include site rent and a railroad spur for delivery of aggregates and liquid asphalt. The initial cost of setting up an asphalt plant can include clearing and grubbing and grading. It may also be necessary to stabilize the soil to minimize the expense of stock-pile loss. Because cost can range from $15,000 to $50,000 for this item, it is desirable to use one plant site per project whenever possible.

Some asphalt mixing plants are operated primarily as producers of multiple-type mixes that are available for sale at the plant to serve multiple small paving operations. Contractors purchase the asphalt mix at the plant and transport it to the paving machine at their jobsites. For larger projects, a portable asphalt plant is set up by the contractor at the jobsite to produce asphalt mixes for the specific job.

AGGREGATES

The load applied to an asphalt pavement is primarily carried by the aggregates in the mix. The aggregate portion of a mix accounts for 90 to 95 percent of the material by weight. Good aggregates and proper gradation of those aggregates are critical to the mix's performance.

The aggregates most commonly used include limestone, granite, sand, and other crushed stone products. Each asphalt pavement has its own unique job mix formula. The competitive bid environment for most asphalt paving contracts mandates that local materials be used due to the high cost of freight. The job-mix formula prescribes the types and amounts of aggregates to be used. To achieve the desired mix characteristics, such as density and stability, the aggregates are carefully proportioned ranging from the largest to the smallest particle sizes. The mix design specification designates the percentages passing and retained on screens with graded size openings.

Ideally, an aggregate gradation should be provided that enables the minimum amount of expensive binder (asphalt) to be used. The binder fills most of the voids between the aggregate particles as well as the voids in the particles. The larger-size aggregate, greater than a no. 8 sieve, used in an asphalt mix can be gravel or crushed stone. Crushed aggregate has angular rough surfaces, which interlock to produce higher strength asphalt mixes. Smooth surface aggregate, sometimes called *river run,* improves the workability of the mix but produces reduced strength compared to crushed aggregate. A mix using rough surface aggregates will require more binder than one using smooth aggregate. Both sand

and a small portion of some very fine material, less than a no. 200 sieve, referred to as mineral filler, can also be incorporated into the mix.

The asphalt plant provides the mechanism to achieve a controlled blend of the aggregates. The blending of different size aggregates to obtain a desired mix gradation is vital for a good-quality pavement. The amount of foreign matter, either soil or organic material, that may be present with the aggregate, will greatly reduce the strength of the asphalt pavement. A visual inspection can detect aggregate cleanness. Washing, wet screening, or other methods can be used to clean the aggregate.

For jobsite asphalt plants, aggregate is purchased from a rock quarry and transported to the jobsite asphalt plant, where it is stockpiled. Front-end loaders and portable conveyors move the material from stockpiles to the asphalt mixing plant. The price for purchasing aggregate is established by the aggregate supplier. The cost for hauling aggregate is discussed in Chapter 6.

ASPHALTS

The asphalts used in asphalt pavements can be divided into three categories: cutback asphalts, emulsified asphalts, and asphalt cements. Cutbacks and emulsions are used mostly for prime coats and stockpiled patching material. Asphalt cement is the type of asphalt most commonly used for main-line road construction. Liquid asphalt is rated with a performance (PG) system. For example, a PG64-22 designates the high and low temperature at which an asphalt binder would be expected to perform satisfactorily.

Cutback asphalt is produced by adding volatile products to asphalt cement. Emulsified asphalts are a mix of asphalt cement and water, using an emulsifying agent to control separation. Asphalt cement used as the binder in paving mixes is usually classified by the AASTHO standard penetration test, which consists of measuring the depth of penetration into a 77°F asphalt cement sample by a standard 100-gram weighted needle in a 5-sec time duration. The asphalt penetration grades are 40–50, 60–70, 85–100, 120–150, and 200–300.

The quantity of asphalt binder varies from 4 to 9 percent of the weight of the finished product. The cost of asphalt materials is highly dependent on the price of crude oil, which is subject to fluctuating changes. Therefore, it is necessary to obtain a firm quote from the supplier of asphalt material before preparing a cost.

ASPHALT PLANTS

Hot-mix asphalt is produced in plants that proportion the aggregates, dry them, and combine them with asphalt cement that has been heated from 275°F to 375°F. There are two types of asphalt plants: batch-type plants and continuous mixing drum-type plants. They differ only in the way the aggregates are proportioned and how the liquid asphalt is mixed with the aggregate.

Batch-type plants have been used for years in the asphalt paving industry. They are economical for operations that require production of several different mixes during a single production time frame. Batch-type plants are efficient because they can rapidly change the production mix with little wasted material.

Continuous mixing drum-type plants can produce higher volumes of output at faster rates than batch-type plants. Both types of plants require storage bins for aggregate, a burner for heating the materials, a mixing unit, and dust collectors to control pollution.

Batch-Type Asphalt Plants

The primary components of a batch-type asphalt plant, in the order of material flow, are cold bins, dryer unit, hot elevator, screens, and pugmill mixer. Cold feed systems usually consist of three or four open-top bins that are mounted together as a single unit. The individual bins can be fed from aggregate stockpiles by a front-end loader, clamshell, or a conveyor. A gate at the bottom of each bin controls the material flow and a feeder unit for metering the flow.

The dryer unit heats and dries the aggregate of the mix. The dryer has a burner that provides the heat energy for evaporating the moisture in the aggregate and then heats the aggregate to a discharge temperature of about 300°F.

The heated aggregate from the dryer is discharged into the hot elevator, which carries it to the screens at the top of the batch plant tower. The tower unit consists of vibrating screens that provide gradation control of the aggregate sizes. The heated aggregate sizes are fed into different hot bins.

Aggregate from the hot bins is dropped into a weight hopper that is located below the bins and above the pugmill. The aggregates are weighted cumulatively in the hopper, with the mineral filler added last. Once the correct proportions of aggregate are measured, the material is dropped into the pubmill. The liquid asphalt for the batch is either weighed or metered volumetrically and then is pumped through spray bars into the pubmill mixer. The mixer is sufficiently high to allow trucks to pass directly below for loading. Figure 8.10 shows a truck receiving a mix at a batch-type asphalt plant.

Continuous Mixing Drum-Type Plants

A continuous mixing drum-type plant has almost the same components as a batch-type plant. The primary unit in a drum-mix plant both dries the aggregate and performs the mixing function. As with a batch-mix plant, the cold feed system is the critical aggregate proportioning unit. The aggregates are fed into an inclined rotating drum mixer that provides continuous mixing. Usually the burner and the aggregate feed are both located at the upper end of the drum, which means the aggregates and hot gas flow are in the same direction. However, some plants have a counterflow.

As the material moves through the drum, paddles inside the drum lift the aggregates and then drop them through the flame and hot gases. Farther down the length of the drum, beyond the flame, the liquid asphalt is introduced and the drum paddles inside the drum provide the mixing mechanism. When liquid asphalt is combined with the aggregate for mixing, the temperature of the asphalt is about 300°F. The drum for a drum mix plant has a slope, approximately of $\frac{1}{2}$ to 1 in./ft of drum length. The drum rotates at a speed from 5 to 10 rpm. Figure 8.11 shows a drum-mix plant in operation.

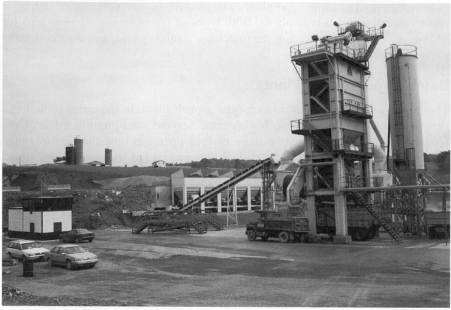

FIGURE 8.10 | Batch-type asphalt plant.
Courtesy: Astec Industries, Inc.

FIGURE 8.11 | Drum-mix asphalt plant.
Courtesy: Astec Industries, Inc.

TRANSPORTING AND LAYING ASPHALT MIXES

After asphalt mix is completed at the asphalt plant, it is discharged into trucks and hauled to the job, where it is spread in uniformly thick layers by a mechanical paver. It may be necessary to cover the material with a canvas or tarpaulin during transit to prevent excessive loss of heat between the mixing plant and the paver.

A specially designed distributor truck is used to apply an asphalt prime, tack, or seal coat for asphalt paving projects. These trucks must be able to apply the liquid asphalt to a surface at uniform rates. Prior to placement of an asphalt mix on a new base, a prime coat is applied to the base. Normal rates of application for prime coats vary between 0.20 and 0.60 gal/sy. The prime coat promotes adhesion between the base and the overlying asphalt mix course by coating the absorbent base material, which is gravel, crushed stone, or an earthen grade.

Tack coats are designed to create a bond between old existing pavements and new overlays. A tack coat acts as an adhesive to prevent the slippage of the two mats. The tack coat is a very thin uniform blanket of asphalt, usually 0.05 to 0.15 gal/sy of diluted emulsion.

Seal coats consist of an application of asphalt followed by a light covering of the fine aggregates, which is rolled in with pneumatic rollers. Application rates are normally from 0.10 to 0.20 gal/sy.

An asphalt paver consists of a tractor, either track- or rubber-tired, and a screed. Pavers can receive the mix directly into their hoppers from rear dump trucks (see Fig. 8.12). The paver pushes the truck forward during the unloading process.

FIGURE 8.12 | Rubber-tire asphalt paver receiving mix from a rear dump truck.

Courtesy: Astec Industries, Inc.

The tractor power unit has a receiving hopper in the front and a system of slat conveyors and augers to spread the asphalt evenly across the front of the screed. The screed controls the asphalt placement width and depth, and the initial finish and compaction of the material. Various sizes and models of asphalt pavers are available, with paving speeds that range from 125 to 300 ft/min. Hopper capacities are about 200 cf in size.

Continuous paving operations depend on balancing the production rates of the asphalt plant, hauling units, and paver operation. An adequate number of trucks must be available to haul the asphalt mix to the paver to ensure a continuous operation. There may be times when two trucks are at the paver at the same time and both are not needed, or there may be times when no truck is available at the paver. If a truck is not available, the paver must stop operations.

Several methods are used to prevent the problem of having too many trucks, or no trucks, at the paver. One method involves depositing the truckloads of asphalt mix in windrows parallel to and along the path of the paver. A windrow elevator is attached to the front of the paver, which picks up the asphalt mix from the windrow and deposits the material into the paver as it moves along the paving lane. This enables continuous operation of the paver without dependency on trucks. Another method involves placement of a silo at the asphalt plant. Asphalt mix material from the asphalt plant is deposited into the silo. Then, the asphalt mix is deposited from the silo into a truck and transported to the job when the paver needs the material.

Another method to help compensate for truck delays at the jobsite and to ensure continuous operation of the paver is to use a material transfer vehicle (see Fig. 8.13).

FIGURE 8.13 | Asphalt paver receiving mix from a material transfer vehicle.
Courtesy: Astec Industries, Inc.

ROLLER 1   ROLLER 2   PAVER

(a) Asphalt paving train with the mix deposited directly into the paver

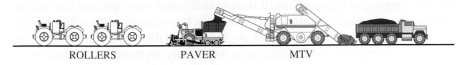

ROLLERS   PAVER   MTV

(b) Asphalt paving operation with a material transfer vehicle providing mix to the paver

FIGURE 8.14 | Methods of supplying asphalt mix material to the paver.

Courtesy: Astec Industries, Inc.

Material transfer vehicles are available with capacities of 25 to 30 T of asphalt mix. This machine operates ahead of the paver to feed the asphalt material into the paver. As the paver approaches the material transfer vehicle, it is refilled before the material transfer vehicle moves forward to receive the next truck discharge. An insert can be installed in the paver, allowing storage of 15 to 20 T of mix in the paver and providing a combined paver and material transfer vehicle storage capacity of 45 to 50 T. With storage of this capacity, the paver can operate continuously. When a material transfer vehicle is used, trucks can be stopped 100 to 200 ft ahead of the paver and dump their loads safely without moving.

Figure 8.14a shows a paving operation with trucks depositing the mix directly in the paver, and Fig. 8.14b shows a paving operation with a material transfer vehicle providing the asphalt mix to the paver.

COMPACTING ASPHALT CONCRETE MIXES

After each layer is deposited, it is compacted by a roller to the desired density. Three basic roller types are used to compact the asphalt-paving mixes: smooth-drum steel-wheel, pneumatic-tired, and smooth-drum steel-wheel vibrator rollers. After a fresh layer of hot mix asphalt concrete is placed, a steel-wheel roller is usually used for initial compaction. After the initial compaction, hairline cracks or marks will be left by the steel drum rollers. A pneumatic-tired roller is used to smooth the pavement surface to remove these marks or lines.

Steel-wheel rollers are typically 4 to 5 ft wide and weigh from 8 to 14 T with penetration depths from $\frac{1}{16}$ to $\frac{3}{4}$ in. Pneumatic-tired rollers weigh between 2,000 and 10,000 lb and have from 4 to 7 tires on the front of the roller and 3 to 6 tires on the rear of the rollers. The operating speed of compactors ranges from 1 to 4 mph.

EQUIPMENT FOR HOT-MIX ASPHALTIC-CONCRETE PAVEMENT

Considerable equipment is required for mixing and placing hot-mix asphaltic-concrete pavement. A front-end loader, clamshell, or conveyor is needed to handle the aggregate. Storage and feeder bins, elevators, aggregate drier, dust collector, aggregate screens, asphalt storage tanks, and a storage tank for fuel are required for the mixing operation. Hauling trucks, a distributor truck, a paving machine, and compaction equipment are required to distribute and compact the asphalt mix. If the subgrade or subbase is not already prepared, additional equipment will be required to do this work.

COST OF HOT-MIX ASPHALTIC-CONCRETE PAVEMENT

An asphalt plant is expensive to purchase and operate. Also, the cost of setting up and operating an asphalt plant for producing hot-mix asphaltic-concrete pavement is high. Consequently, a single setup is generally made for a given job. The cost will include the cost of aggregate, asphalt, moving the plant to the location, setting it up, operating the plant, hauling the material to the job, spreading, and rolling. The cost of operating the plant will include depreciation, interest, insurance and taxes, maintenance and repairs, and fuel. Additional costs may include the rental of land at the jobsite to stockpile material and operate the asphalt mixing plant.

EXAMPLE 8.5

Estimate the total direct cost and cost per ton for mixing and placing hot-mix asphaltic-concrete pavement 18 mi long, 24 ft wide, and 5 in. thick. The pavement consists of a 3-in.-thick base course and a 2-in.-thick wearing course.

A portable drum-mix plant with a 150 T/hr capacity will be used to provide continuous mix of the asphalt material. Assuming a 50-min effective hour, the production rate of the asphalt plant will be $50/60 \times 150 = 125$ T/hr.

The asphalt plant will be located at the middle of the projects. Dump trucks with 20-T capacity will haul the asphalt mix to the paver. Assume trucks will haul an average of 9 mi at 30 mph and return at 35 mph. Assume a 45-min effective hour for the hauling operation.

The pavement will be placed on a previously prepared subbase. Prior to placing the 3-in.-thick base, an asphalt prime coat will be applied to the subbase at the rate of 0.3 gal/sy. A 2-in.-thick wearing course will be placed on top of the base with no tack coat required.

The specifications require 6 percent asphalt cement content by weight measurement. The asphalt plant will require 60 percent coarse aggregate, larger than a no. 10 sieve, and 40 percent fine aggregate and filler. Assume that 5 percent of the coarse aggregate, 10 percent of the fine aggregate, and 2 percent of the asphalt will be lost through waste or other reasons.

The combined material will weigh about 3,600 lb/cy when compacted. The paving mixtures shall meet the requirements for grading and mix composition shown in the table.

Sieve size	Combined aggregate including filler, percent passing, by weight	
	Base course, 3 in. thickness	Wearing course, 2 in. thickness
2 in.	100	
$1\frac{1}{2}$ in.	95–100	
1 in.	—	100
$\frac{3}{4}$ in.	70–85	95–100
$\frac{1}{2}$ in.	—	75–90
No. 4	35–50	45–60
No. 10	25–37	35–47
No. 40	15–25	23–33
No. 80	6–16	16–24
No. 200	2–6	6–12
Asphalt cement, percentage of combined weight	6%	6%

Quantity of materials:

Total surface area = [18.0 mi × 5,280 ft/mi × 24 ft]/(9 sf/sy) = 253,440 sy

Material for 3-in. ($\frac{3}{12}$-ft) base course:

[(253,440 sy × $\frac{3}{12}$ ft × 3,600 lb/cy)/(3 ft/yd)]/[2,000 lb/T] = 38,016 T

Material for 2-in. ($\frac{2}{12}$-ft) wearing course:

[(253,440 sy × $\frac{2}{12}$ ft × 3,600 lb/cy)/(3 ft/yd)]/[2,000 lb/T] = <u>25,344 T</u>

Total aggregate weight = 63,360 T

Less 6% asphalt, effective weight of aggregate, 0.94 × 63,360 T = 59,558 T

Material quantities for costs:

Coarse aggregate (60% with 5% waste): 0.60 × 59,558 T × 1.05 = 37,522 T

Fine aggregate (40% with 10% waste): 0.40 × 59,558 T × 1.10 = 26,205 T

Asphalt cement (6% with 2% waste): 0.06 × 63,360 T × 1.02 = 3,878 T

Prime coat: 253,440 sy × 0.30 gal/sy = 76,032 gal

Production time for asphalt mix plant:

For a 50-min hour, the plant production rate: 50/60 × 150 T/hr = 125 T/hr

Time to complete the work: 63,360 T/125 T/hr = 507 hr

Rounding to a full 8-hour day: total time = 512 hr ←

Production time for transporting asphalt mix:

Cycle time for a truck:

Load: 20 T/125 T/hr	= 0.16 hr
Haul: 9 mi/30 mph	= 0.30 hr
Dump: 12 min × 1 hr/60 min	= 0.20 hr
Return: 9 mi/35 mph	= <u>0.26 hr</u>
Total cycle time	= 0.92 hr/trip

Quantity of asphalt mix per trip: 20 T/trip $\times$ 1.0 trip/0.92 hr = 21.7 T/hr

For a 45-min hour, asphalt mix hauled per trip = 21.7 T/hr $\times$ 45/60 = 16.3 T/hr

Required truck: plant production/truck production, 125/16.3 = 7.6 trucks required

Therefore, need 8 trucks with 20-T capacity to serve the paver

Using 8 trucks, the haul rate will be greater than the plant production rate

Costs to construct the pavement:

Material costs:

Asphalt cement: 3,878 tons @ $350.00/ton	=	$1,357,300
Coarse aggregate: 37,522 tons @ $11.00/ton	=	412,742
Fine aggregate: 26,205 tons @ $10.00/ton	=	262,050
Prime coat: 76,032 gal @ $1.20/gal	=	91,238
Subtotal cost of materials	=	$2,123,330

Equipment costs:

Batch plant operation:

Asphalt mix plant: 512 hr @ $212.00/hr	=	$108,544
Hot-oil heater fuel: 512 hr $\times$ 16 gal/hr @ $4.00gal	=	32,768
Fuel for drier: 512 hr $\times$ 360 gal/hr @ $4.00/gal	=	737,280
Fuel for mixer engine: 512 hr 4.6 gal/hr @ $4.00/gal	=	9,420
Fuel for drier engine: 512 hr $\times$ 8.2 gal/hr @ $4.00/gal	=	16,794
Lubricating oil: 512 hr $\times$ 0.5 quarts/hr @ 4.50/quart	=	1,152
Other lubricants and grease: 512 hr @ $1.50/hr	=	768
Front-end loader: 512 hr @ $70.00/hr	=	35,840
Portable conveyor: 512 hr @ $45.00/hr	=	23,040

Lay down paving operation:

Haul trucks: 512 hr $\times$ 8 trucks @ $35.00/hr	=	143,360
Asphalt distributor truck: 512 hr @ $45.00/hr	=	23,040
Asphalt paving machine: 512 hr @ $155.00/hr	=	79,360
Smooth-wheel roller: 512 hr $\times$ 2 rollers @ $25.00/hr	=	25,600
Pneumatic roller: 512 hr $\times$ 2 rollers @ $30.00/hr	=	30,720

Supporting equipment costs:

Service trucks: 512 hr $\times$ 2 trucks @ $7.50/hr	=	7,680
Surveying equipment: 512 hr @ $18.00/hr	=	9,216
Barricades, signs, small tools:	=	1,600
Pickup trucks: 512 hr $\times$ 4 trucks @ $6.00/hr	=	12,288
Moving to and from project, excluding labor	=	15,000
Subtotal of equipment costs	=	$1,313,470

Labor costs:

   Batch plant operation:

Batch plant operator: 512 hr @ $31.00/hr	=	$15,872
Batch plant laborer: 512 hr @ $20.00/hr	=	10,240
Front-end loader operator: 512 hr @ 28.00/hr	=	14,336
Mechanic: 512 hr @ $25.00/hr	=	12,800
Batch plant foreman: 512 hr @ $30.00/hr	=	15,360

   Lay down paving operation:

Haul truck drivers: 512 hr × 8 drivers @ $25.00/hr	=	102,400
Distributor operator: 512 hr @ $26.00/hr	=	13,312
Paver operator: 512 hr @ $31.00/hr	=	15,872
Smooth-wheel operators: 512 hr × 2 @ $29.00/hr	=	29,696
Paving foreman: 512 hr @ $31.00/hr	=	15,872

   Supporting labor costs:

General laborers: 512 hr × 3 laborers @ $19.00/hr	=	29,184
Surveyors: 512 hr × 2 surveyors @ $27.00/hr	=	27,648
Traffic control: 512 hr × 4 laborers @ $19.00/hr	=	38,912
Superintendent: 512 hr @ $35.00/hr	=	17,920
	Subtotal of labor costs =	$359,424

Summary of direct costs:

   Material cost: = $2,123,330

   Equipment cost: = 1,313,470

   Labor cost: = 359,424

      Direct costs = $3,796,224

Unit costs:

   Cost per cubic yard: $3,796,224/63,360 tons = $58.17/ton

   Cost per square yard: $3,796,224/253,440 sy = $14.54/sy

   Cost per mile: $3,796,224/18 miles = $204,758/mile

These costs do not include setup and takedown of the drum-mix plant or the indirect costs, such as taxes on labor and material, bonds and insurance, jobsite office, and general overhead. Also, contingency and profit are not included.

COMPUTER ESTIMATING OF HIGHWAY PROJECTS

The previous sections of this chapter presented methods, materials, and examples of estimating the cost of highway construction projects. To accurately estimate the cost the estimator must know the method of construction and the mix of labor and equipment that will be used on a particular job.

Estimating software programs are commercially available for estimating the cost of highway pavements. These programs store the contractor's equipment and labor rates, and indirect costs. They also store crew configurations and associated equipment spreads for specific operations such as shouldering, small concrete structures, main-line paving, etc. Since highway contracts are unit-price format with the estimated quantities provided, the estimator only has to select a crew from a menu in the software and determine the applicable production rate. Chapter 22 is devoted to computer estimating with emphasis on infrastructure type projects, such as highways.

PROBLEMS

8.1   A project requires clearing and grubbing 87 acres of light vegetation, small trees, and brush. A 250-hp dozer, operating with a 16-ft effective cutting blade, will perform the clearing. The dozer will operate in first gear at 2.3 mph during clearing.

It will cost $1,500 to transport the dozer to and from the job. The hourly cost of the dozer with operator is $130.00/hr. Assume a 45-min effective hour and estimate the total time and cost, and the cost per acre for clearing.

8.2   A highway project must be cleared of trees before the grading operation can be started. The site is reasonably level terrain with firm ground. A site survey reveals 75 to 100 percent hardwoods.

A 215-hp dozer will be used to fell trees. The cost of the dozer and operator is $130.00/hr. The cost of moving the dozer to the job and back to storage will be $1,500. Assume a 45-min effective hour and estimate the cost per acre to fell the trees based on the following tree count:

Average number of trees per acre = 900
Number of 1 to 2 ft diameter = 195
Number of 2 to 3 ft diameter = 85
Number of 3 to 4 ft diameter = 11

8.3   The trees in Problem 8.2 are to be piled into a stack for burning. A 165-hp dozer will be used to grub and stack the felled trees. Assume a 50-min effective hour for this operation. Estimate the cost of piling up the trees if the cost of the dozer with operator is $95.00/hr.

8.4   A paving operation is using a 200 cy/hr batch plant to pave a concrete highway, 19.7 miles long, 24 ft wide, and 9 in. thick. One longitudinal saw joint will be installed along the centerline of the pavement and transverse saw joints are required at 15 ft on centers. Water will be transported to the job for the concrete batch plant by water trucks, each with a 2,500 gal capacity. Preliminary estimates of production rates have been determined as:

Concrete batch plant = 200 cy/hr
Water needed by batch plant = 39 gal/cy
Transporting water = 0.75 hr/trip
Saw cutting joints = 250 lin ft/hr

For this project determine the number of water trucks and the number of saw cutting crews to ensure a continuous paving operation. What production rate, in square yards per day, is necessary for the slip-form paver?

8.5 Estimate the total cost and cost per square yard of concrete for bid purposes for furnishing materials, equipment, and labor for concrete of a pavement 24 ft wide, 9 in. thick, and 15.2 miles long.

Use the same job conditions, types of equipment, equipment costs, wage rates, material costs, and overhead costs as those in Example 8.4.

8.6 Estimate the total cost and cost per ton for furnishing materials, mixing, and placing a hot-mix asphaltic-concrete pavement 22.5 miles long, 26 ft wide, and 4 in. thick on a previously prepared base. The pavement will be placed in 2 layers, each 2 in. thick after compaction.

Use the same methods, materials costs, equipment costs, wage rates, and production rates as those used in Example 8.5, except change the asphalt cement to 7 percent of the combined weight of the concrete.

Foundations

TYPES OF FOUNDATIONS

Foundations that support structures include footings, piles, and drilled shaft foundations. Footings are shallow foundations that are constructed by excavating the soil from the ground and then installing reinforced concrete. Piles and drilled shaft foundations are deep foundations. A pile foundation is constructed by driving the pile directly into the ground, whereas a drilled shaft foundation is constructed by drilling a hole in the ground and installing reinforced concrete.

Footings are foundations placed at shallow depths, usually less than 5 ft. Footings may be placed at one location, called isolated or spot footings, to support a column that is placed on the footing. Some footings are continuous, such as a footing to support a wall of a building.

Pile foundations are driven into the ground by the hammer of pile-driving equipment. The piles may be timber, steel, or concrete. The driving depth depends on the magnitude of the load that will be placed on the pile from the structure above and the strength of the soil. Load-bearing piles resist loads by skin-friction or end bearing, or a combination of both. Piles are driven from 30 to 200 ft.

Drilled shaft foundations are installed by drilling holes to depths in the soils that have sufficient strength to support the loads from the structure above the foundation. For sandy soils, a steel casing may be required to stabilize the soil during the drilling operation. After the hole is drilled, a belling tool can be used to widen the bottom of the drilled hole, to increase the load-bearing capacity of the foundation. Reinforced concrete is installed after the shaft has been drilled and belled. Drilled shafts can be installed with diameters from 18 to 96 in. and at depths from 5 to 60 ft. Anchor bolts are installed at the top of the foundation to connect the foundation to the structure above, such as a steel column.

FOOTINGS

Construction of a footing includes excavating the soil to the required depth, erecting formwork, setting reinforcing steel, placing concrete, removing formwork, and backfilling soil above the footing to the surface of the ground.

A trench is excavated into the soil to construct a continuous footing. For some shallow continuous footings, the sides of the excavated soil may be stable enough to support itself without caving in. After the trench is excavated, reinforcing steel is installed and concrete is placed directly into the open trench without any formwork.

For some shallow continuous footings, the sides of the excavated soil are unstable. Thus, the trench is excavated wider than the footing, formwork is installed, and reinforced concrete is placed in the formwork. After the concrete has sufficiently cured, the formwork is removed and the sides of the footing are backfilled with soil. Chapter 10 presents the cost for formwork, reinforcing steel, and placing concrete.

Some projects require excavation into soil that is so unstable that the walls of the excavated earth must be supported to prevent them from caving into the pit or trench. For these situations, it will be necessary to install a system of shores, braces, and solid sheeting along the excavated walls to hold the earth in position. If groundwater is present, it may be necessary to install semi-watertight sheeting around the walls to exclude or reduce the flow of water into the pit. Another alternative is to pump the water out of the trench area where the footing is to be installed. Timber and steel are used for braces and sheeting.

SHEETING TRENCHES

If the earth is so unstable that it must be restrained for the full area of the wall of a trench, it will be necessary to install a *trench box,* as illustrated in Fig. 9.1. The sheeting of a *trench box* is typically constructed with 2-, 3-, or 4-in.-thick lumber, placed side by side or overlapping along the entire length of the trench. Wales, consisting of 4 × 6-in.-thick lumber, can be placed in a horizontal direction to provide additional support for the sheeting. The lumber may be rough sawed or S4S. Depending on the stability of the earth, it may be necessary to drive the sheeting and install some of the braces ahead of excavating. The sheeting may be driven with power equipment, such as a pneumatic hammer.

EXAMPLE 9.1

Estimate the cost of installing and removing solid sheeting and bracing for a trench box, as shown in Fig. 9.1. The trench is 100 ft long and 7 ft deep. The sheeting will be 2 × 12-in. S4S lumber, 8 ft long. Two horizontal rows of 4 × 6-in. wales will be placed on each side of the trench for the full length of the trench. Trench braces will be placed 4 ft apart along each row of wales. The sheeting will be driven one plank at a time by using an air compressor and a pneumatic hammer.

Quantity of materials:

Sheeting, actual dimension of 2 × 12 lumber is 1.5 in. × 11.25 in.

Number of pieces for 2 sides: [(100 ft)/(11.25 in./12 ft)] × 2 sides = 214 pieces

Quantity of sheeting, 214 pieces × [(2 × 12)/(1 × 12)] × 8 ft = 3,424 bf

Wales:

Number of pieces for 1 row, 100 ft/8 ft/wale = 12.5, therefore need 13 pieces

Quantity of lumber: 13 pieces × 2 rows × 2 sides × [(4 × 6)/(1 × 12)] × 8 ft = 832 bf

Trench braces:

Number of braces: (100 ft/4 ft/brace) 2 rows = 50 braces

Time to perform the job:

Two laborers installing sheeting: 214 pieces/4 pieces/hr = 53.5 hr

Two laborers installing wales and braces: (50 + 13) = 63 pieces/3 pieces/hr = 21.0 hr

Material cost:

Sheeting: 3,424 bf @ $0.73/bf = $2,499.52

Wales: 832 bf @ $0.92/bf   =   765.44

Braces: 50 @ $3.50 each   =   175.00

Total material cost = $3,439.96

Labor cost:

Laborers driving sheeting: 53.5 hr × 2 laborers @ $28.00/hr   = $2,996.00

Installing wales and braces: 21.0 hr × 2 laborers @ $22.00/hr =   924.00

Total labor cost = $3,920.00

Equipment cost:

Air compressor: 53.5 hr @ $8.75/hr   = $468.13

Air hammer, hose, etc.: 53.5 hr @ $1.75/hr =   93.63

Total equipment cost = $561.76

Summary of costs:

Material   = $3,439.96

Labor   =   3,920.00

Equipment =   561.76

Total cost = $7,921.72

Cost per linear foot of trench: $7,921.72/100 ft = $79.22/ft

Cost per square foot sheeting: $7,921.72/(100 ft × 7 ft × 2 sides) = $5.66/sf

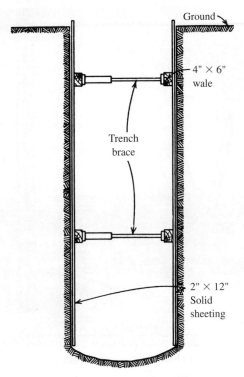

FIGURE 9.1 | Trench braces, wales, and solid sheeting.

PILE-DRIVING EQUIPMENT

Equipment used to drive piles usually consists of a truck-mounted or crawler-mounted crane, leads, a hammer, and a source of compressed air or steam to drive the hammer. When piles are driven into water, the driving rig is usually mounted on a barge. The actual driving of piles is accomplished with hammers. Several types are available, including the drop hammer, single-acting air or steam hammer, double-acting air or steam hammer, self-contained diesel-operated hammer, and vibratory hammer. Figure 9.2 illustrates a diesel-operated hammer driving a steel pile.

The size of a drop hammer is indicated by the weight of the hammer, whereas the size of an air, steam, or diesel hammer is indicated by the theoretical foot-pounds of energy delivered by each blow.

A drop hammer is a heavy metal weight that is lifted by a hoist line, then released and allowed to fall onto the top of the pile. Due to high dynamic forces, a pile cap is positioned between the hammer and the pile head. The pile cap provides a uniform distribution of the blow to the pile head and acts as a shock absorber. The cap contains a cushion block, which is commonly fabricated from wood.

FIGURE 9.2 | Diesel-operated hammer driving a steel pile.

Courtesy: Sherwood Construction Company.

Drop hammers are suitable for driving piles on remote projects that require only a few piles, or when the time of completion is not important. A drop hammer normally can deliver from four to eight blows per minute.

A single-acting air hammer has a free-falling weight, called a *ram,* that is lifted by steam or compressed air that flows to the underside of a piston, which is connected to the ram through a piston rod. When the piston reaches the top of the stroke, the steam or air pressure is released and the ram falls freely to strike the top of a pile. Whereas a drop hammer may strike 4 to 8 blows/min, a single-acting steam or air hammer will strike 40 to 60 blows/min when delivering the same energy per blow. A pile cap is used with a single-acting steam or air hammer. The cap is mated to the case of the hammer. This cap is commonly called an *anvil* or a *helmet.*

In the double-acting steam or air hammer, steam or air pressure is applied to the underside of the piston to raise the ram, then during the downward stroke, steam is applied to the top side of the piston to increase the energy per blow. Thus, with a given weight of ram, it is possible to attain a desired amount of energy per blow with a shorter stroke than the longer stroke single-acting hammer. The number of blows per minute will be approximately twice as great as for a single-acting hammer with the same energy rating. Double-acting

TABLE 9.1 I Recommended sizes of hammers for driving piles, in theoretical foot-pounds of energy per blow.

Length of piles, ft	Depth of penetration	Steel sheet*			Timber		Concrete		Steel		
		20	30	40	30	60	150	400	40	80	120
Driving through ordinary earth, moist clay, and loose gravel; normal frictional resistance											
25	$\frac{1}{2}$	2,000	2,000	3,600	3,600	7,000	7,500	15,000	3,600	7,000	7,500
	Full	3,600	3,600	6,000	3,600	7,000	7,500	15,000	4,000	7,500	7,500
50	$\frac{1}{2}$	6,000	6,000	7,000	7,000	7,500	15,000	20,000	7,000	7,500	12,000
	Full	7,000	7,000	7,500	7,500	12,000	15,000	20,000	7,500	12,000	15,000
75	$\frac{1}{2}$	—	7,000	7,500	—	15,000	—	30,000	7,500	15,000	15,000
	Full	—	—	12,000	—	15,000	—	30,000	12,000	15,000	20,000
Driving through stiff clay, compacted sand, and gravel; high frictional resistance											
25	$\frac{1}{2}$	3,600	3,600	3,600	7,500	7,500	7,500	15,000	5,000	9,000	12,000
	Full	3,600	7,000	7,000	7,500	7,500	12,000	15,000	7,000	10,000	12,000
50	$\frac{1}{2}$	7,000	7,500	7,500	12,000	12,000	15,000	25,000	9,000	15,000	15,000
	Full	—	7,500	7,500	—	15,000	—	30,000	12,000	15,000	20,000
75	$\frac{1}{2}$	—	7,500	12,000	—	15,000	—	36,000	12,000	20,000	25,000
	Full	—	—	15,000	—	20,000	—	50,000	15,000	20,000	30,000

*The indicated energy is based on driving two steel-sheet piles, simultaneously. When single piles are driven, use approximately two-thirds of the indicated energy.

hammers commonly deliver 95 to 300 blows/min. These hammers do not require cushion blocks. The ram strikes on an alloy steel anvil that fits on the pile head.

Table 9.1 gives recommended sizes of hammers for different types of sizes of piles and driving conditions. Table 9.2 gives information for various sizes and types of air or steam driven hammers.

A diesel pile-driving hammer is a self-contained driving unit that does not require an external source of energy such as a steam boiler or an air compressor. Thus, it is simpler and more easily moved from one location to another than a steam hammer. A complete unit consists of a vertical cylinder, a piston or ram, an anvil, fuel and lubrication oil tanks, a fuel pump, injectors, and a mechanical lubricator. After a hammer is placed on top of a pile, the combined piston and ram are lifted to the upper end of the stroke and released to start the unit operation. As the ram nears the end of the down stroke, it activates a fuel pump that injects the fuel into the combustion chamber between the ram and the anvil. The continued down stroke of the ram compresses the air and the fuel to ignition heat. The resulting explosion drives the pile downward and the ram upward to repeat its stroke. The energy per blow, which is controlled by the operator, can be varied over a wide range. Diesel hammers deliver about 40 to 50 blows/min. Typically, the diesel hammer is set in leads on the crane. Table 9.3 gives information on diesel hammers of a manufacturer of pile driving equipment.

TABLE 9.2 | Data on air or steam pile-driving hammers.

Hammer	Model	Ram weight, lb	Blows per minute	Rated energy, ft-lb
Vulcan single-action	2	3,000	70	7,260
	1	5,000	60	15,000
	0	7,500	50	24,325
	10	10,000	57	32,000
	14	14,000	59	42,000
	16	16,250	58	48,750
	20	20,000	59	60,000
	30	30,000	54	90,000
	340	40,000	60	120,000
	360	60,000	62	180,000
McKiernan-Terry single-action	S5	5,000	60	16,250
	S8	8,000	55	26,000
	S10	10,000	55	32,500
	S14	14,000	60	37,500
	S20	20,000	60	60,000
Vulcan differential action	30C	3,000	133	7,260
	50C	5,000	120	15,100
	80C	8,000	111	24,450
	100C	10,000	103	32,885
	140C	14,000	103	36,000
	200C	20,000	98	50,200
McKiernan-Terry double-action	9B3	1,600	145	8,750
	10B3	3,000	105	13,100
	11B3	5,000	95	19,150

TABLE 9.3 | Data on diesel hammers.

Hammer	Model	Piston weight, lb	Blows per minute	Rated energy, ft-lb
McKiernan-Terry single-action	DE10	1,000	40–50	8,800
	DE20	2,000	40–50	18,000
	DE30	2,800	40–50	25,200
	DE40	4,000	40–50	36,000
	DE50	5,000	40–50	45,000
	DE70	7,000	40–50	63,000

SHEET PILING

Sheet piles are used to form a continuous wall by installing sheets of steel that interlock. The cost of sheet piling in place will include the cost of the piling material, driving equipment, and labor. If the piling is to be salvaged, there will be an additional cost for extracting it. It is common practice to drive two piles simultaneously.

TABLE 9.4 | Properties of steel-sheet piles.

Type and section no.	Width, in.	Area, sq. in.	Weight lb/lin ft of pile	lb/sf of wall
Straight type				
PS31	19.69	14.96	50.9	31.0
PS27.5	19.69	13.27	45.1	27.5
Angle type				
PSA23	16	8.99	30.7	23.0
Z-type				
PZ22	22	11.86	40.3	22.0
PZ27	18	11.91	40.5	27.0
PZ35	22.64	19.41	66.0	35.0
PZ40	19.69	19.30	65.6	40.0

Sheet piling is manufactured in both flat and Z-section sheets that have interlocking longitudinal edges. The flat sections are designed to interlocking strength, which makes them suitable for construction of cellular structures. The Z-sections are designed for bending, which makes them more suitable for construction of retaining walls, bulkheads, cofferdams, and for supporting excavation.

Table 9.4 gives information for sheet piles manufactured by steel companies in the United States. Table 9.5 gives representative rates for driving steel-sheet piles when they are driven in pairs with the size hammer recommended in Table 9.1.

TABLE 9.5 | Representative number of steel-sheet piles driven per hour.

Length of pile, ft	Depth of penetration	Weight of pile, lb/lin ft 20	30	40
20	$\frac{1}{2}$	6	$5\frac{3}{4}$	$5\frac{1}{2}$
	Full	$5\frac{1}{2}$	$5\frac{1}{4}$	$5\frac{1}{4}$
25	$\frac{1}{2}$	5	$4\frac{3}{4}$	$4\frac{1}{2}$
	Full	$4\frac{1}{4}$	4	4
30	$\frac{1}{2}$	$4\frac{1}{2}$	$4\frac{1}{4}$	4
	Full	4	$3\frac{3}{4}$	$3\frac{1}{2}$
35	$\frac{1}{2}$	4	$3\frac{3}{4}$	$3\frac{1}{2}$
	Full	$3\frac{1}{2}$	$3\frac{1}{4}$	3
40	$\frac{1}{2}$	—	$3\frac{1}{4}$	3
	Full	—	3	$2\frac{3}{4}$
45	$\frac{1}{2}$	—	—	$2\frac{3}{4}$
	Full	—	—	$2\frac{1}{2}$
50	$\frac{1}{2}$	—	—	$2\frac{1}{2}$
	Full	—	—	$2\frac{1}{4}$

EXAMPLE 9.2

Estimate the cost of furnishing and driving steel-sheet piling for a cofferdam to enclose a rectangular area 60 by 100 ft. The piles will be 16 in. wide, 24 ft long, and weigh 30.7 lb/ft (see Table 9.4 section no. PSA23). The piles will be driven to full penetration into soil having normal frictional resistance.

A hammer delivering approximately 6,000 ft-lb of energy per blow should be used. A suitable steel cap should be placed on top of the piles to protect them from damage during driving. The hammer will be suspended from a crawler-mounted crane. A portable air compressor will supply air.

Quantity of material:

Perimeter of wall: (60 ft + 100 ft + 60 ft + 100 ft) = 320 ft

Number of 16-in.-wide sheet piles: 320 ft/(16/12 ft) = 240 piles
Corner sheet piles will be the same section, bent at right angles

Linear feet of driven sheet piling: 240 piles × 24 ft/pile = 5,760 ft
Weight of sheet piling: 5,760 ft × 30.7 lb/ft = 176,832 lb
Square feet of sheet piling: 320 ft × 24 ft = 7,680 sf

Time to drive piles:

From Table 9.5, for a 30 lb/ft sheet pile, 24 ft long, driven to full penetration
Number of steel-sheet pile driven per hour = 4 piles/hr
Linear feet of piles, based on pairs: 240 piles/4 piles/hr = 60 hr

Cost:

Material cost:
   Sheet piling: 176,832 lb @ $0.78/lb = $137,928.96
Equipment costs:
   Crane: 60 hr @ $85.00/hr   = $5,100.00
   Leads for hammer: 60 hr @ $14.30/hr   =   858.00
   Hammer: 60 hr @ $9.80/hr   =   588.00
   Air compressor: 60 hr @ $14.65/hr   =   879.00
   Other equipment and supplies: 60 hr @ $9.35/hr =   561.00
                        Total equipment cost = $7,986.00

Labor costs, allow 8 hr for crew to set up and take down equipment:
   Foreman: (60 hr + 8 hr) = 68 hr @ $34.00/hr   = $2,312.00
   Crane operator: 68 hr @ $31.00/hr   =   2,108.00
   General laborers: 68 hr × 3 laborers @ $25.00/hr =   5,100.00
                              Total labor cost = $9,520.00

Summary of costs:

Material	=	$137,928.96
Equipment	=	7,986.00
Labor	=	9,520.00
Total cost	=	$155,434.96

Cost per pile: $155,434.96/240 = $647.65/pile

Cost per linear foot of piling: $155,434.96/5,760 ft = $26.99/ft

Cost per square foot of sheet piling: $155,434.96/7,680 sf = $20.24/sf

These costs do not include taxes, insurance, overhead, contingency, or profit.

WOOD PILES

The cost of wood load-bearing piles in place will include the cost of the piles delivered to the job; the cost of moving the pile-driving equipment to the job, setting it up, taking it down, and moving out; the cost of equipment and labor driving the piles. Since the cost of moving in, setting up, taking down, and moving out is the same regardless of the number of piles driven, the cost per pile will be lower for a greater number of piles.

The cost of wood piles is usually based on the length, size, quality, treatment, and location of the job. Wood piles are tapered from an endpoint diameter of 6 to 8 in. to a butt diameter of 12 to 14 in. Lengths vary from 30 to 50 ft. For piles that are to be driven over 50 ft in depth, a splice of two sections of piles is required.

A preservative treatment of creosote or pentachlorophenol is generally applied to wood piles. Treatments reduce the tendency of decay but often increase the brittleness of wood, which increases the possibility of breakage during the driving process. A steel boot can be placed on the pile point to reduce the tendency of breakage during driving of the pile.

Because piles are sometimes broken during the driving process, a reasonable allowance for extra piles should be included in the estimate. If the contractor is paid for the total number of linear feet of piles driven, any piles that cannot be driven to full penetration will result in some wastage. When it is necessary to cut off the tops of piles to a fixed elevation, the estimate should include the cost of cutting.

The estimator should thoroughly evaluate each project to determine the number and length of piles, number of splices, desired steel boots for pile points, potential for breakage, and waste for cutting the tops of piles to match the elevation of the pile caps.

DRIVING WOOD PILES

A relatively fixed amount of energy is required to drive a pile of a given length into a given soil, regardless of the frequency of the blows. Consequently, a hammer delivering a given amount of energy per blow will reduce the driving time if it strikes more blows per minute. This is particularly true in driving piles into materials where skin friction and not point resistance is to be overcome, for the skin friction will not have as much time to develop between blows when the blows are struck more frequently.

Table 9.6 gives the approximate number of wood piles that should be driven per hour for various lengths in the driving conditions. The rates are based on using a hammer of the proper size.

TABLE 9.6 | Representative number of wood piles driven per hour, full penetration.

Length of pile, ft	Piles driven per hour	
	Normal friction	**High friction**
20	5	4
28	$4\frac{1}{4}$	3
32	$3\frac{1}{2}$	$2\frac{1}{2}$
36	3	$2\frac{1}{4}$
40	$2\frac{3}{4}$	2
50	$2\frac{1}{2}$	$1\frac{3}{4}$
60	2	$1\frac{1}{2}$

EXAMPLE 9.3

Estimate the cost of furnishing and driving 160 wood piles 36 ft long into a soil having normal frictional resistance. The piles will be purchased in 40-ft lengths and the tops will be cut to the required elevation after driving. The piles will be approximately 14 in. in diameter at the butt and 7 in. at the tip. The piles will be treated with creosote preservative at a rate of 2.5 lb/cf, therefore breakage should be considered.

The average weight of a pile will be about 37 lb/lin ft. Reference to Table 9.1 indicates that the hammer should deliver approximately 7,000 theoretical ft-lb of energy per blow. Use a Vulcan size 2 single-acting hammer.

Quantity of materials:

Number of piles to be driven   = 160 piles

Add 5% for breakage, 5% × 160 =   8 piles

Total = 168 piles

Number of linear feet of pile to be purchased: 168 ft × 40 ft = 6,720 ft

Number of linear feet of piles to be driven: 168 piles × 36 ft = 6,048 ft

Number of required boots for pile points = 168

Number of piles that must be cut at top = 160

Time to drive piles:

From Table 9.6, for 36-ft-long wood pile driven in normal friction to full penetration, the rate of driving will be 3 piles/hr.

Time to drive piles: (168 piles)/(3 piles/hr) = 56 hr ←

Cost for driving piles:

Material costs:

Treated piles: 6,720 ft @ $10.85/ft = $72,912.00

Pile points: 168 pile point @ $64.78 = <u>10,883.04</u>

Total material cost = $83,795.04

Equipment costs:

Crane: 56 hr @ $82.40/hr = $4,614.40

Leads for hammer: 56 hr @ $14.63/hr = 819.28

Hammer: 56 hr @ $10.80/hr = 604.80

Air compressor: 56 hr @ $14.65/hr = 820.40

Other equipment and supplies: 56 hr @ $9.35/hr = <u>523.60</u>

Total equipment cost = $7,382.48

Labor costs, allow 8 hr for the crew to set up and take down equipment:

Foreman: (56 hr + 8 hr) = 64 hr @ $34.00/hr = $2,176.00

Crane operator: 64 hr @ $31.00/hr = 1,984.00

General laborers: 64 hr × 3 laborers @ $24.00/hr = <u>4,608.00</u>

Total labor cost = $8,768.00

Summary of costs:

Material = $83,795.04

Equipment = 7,382.48

Labor = <u>$8,768.00</u>

Total cost = $99,945.52

Cost per driven pile: $99,945.52/160 piles = $624.66/pile

Cost per linear foot of driven pile: $99,945.00/6,048 ft = $16.53/ft

This cost does not include taxes, insurance, overhead, contingency, or profit.

PRESTRESSED CONCRETE PILES

Prestressed concrete piles are manufactured by companies that specialize in precast concrete. The piles may be round, square, or octagonal. The amount of steel, strength of concrete, and amount of prestressing depends on the specifications prepared by the designer. After the piles are cast, they are normally steam cured until they have reached sufficient strength to allow them to be removed from the forms.

The cost of concrete piles will depend on the length, diameter, and strength, and the distance from the fabrication yard of the supplier to the jobsite. The weight of concrete piles is high; therefore the cost of transporting the piles to the job can be significant.

Because of the varied conditions that can affect the costs, it is not possible to prepare a general estimate that will apply to a particular project without the information that must be obtained from a concrete supplier that is to furnish the piles. To estimate the cost, the estimator must obtain price quotes from the supplier.

CAST-IN-PLACE CONCRETE PILES

Cast-in-place piles are especially suitable for use on projects where the soil conditions are such that the depth of penetration is not known in advance and the depth varies among the piles driven.

Several methods of casting concrete piles in place are used. In general, they involve driving tapered steel shells or steel pipes, which are later filled with concrete. Although the shells are left in place, the pipes sometimes are withdrawn as the concrete is deposited.

The monotube pile is installed by driving a fluted, tapered steel shell to the desired penetration depth (see Fig. 9.3). The 8-in. tip is closed for the driving operation. If necessary, additional sections may be welded to the original tube as driving progresses to the required length. Any length of pile, up to approximately 125 ft, can be obtained by welding extensions to the shell. Table 9.7 provides data on monotube piles.

After driving is completed in a given area, the condition of the tubes is inspected for damage. If a tube is damaged, it can be removed and replaced. After approval by inspection, concrete is then poured into the tubes. No additional reinforcing steel is required. If steel anchor bolts are specified, they are placed into the tops of the pile before the concrete hardens.

Cost of Cast-in-Place Piles

In estimating the cost of cast-in-place concrete piles, it is necessary to determine the cost of the shells delivered to the project, the cost of equipment and labor required to drive the shells, and the cost of the concrete placed in the shells. All these costs, except the cost of driving the shells, are easy to obtain.

The rate of driving varies with the length of the piles, class of soil into which they are driven, type of driving equipment used, spacing of the piles, topography of the site, and weather conditions. If a project requires the driving of a substantial number of piles, it may be desirable to make subsoil tests, to obtain reliable information related to the difficulty of driving piles and the lengths of piles required.

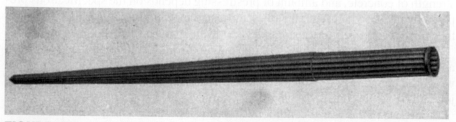

FIGURE 9.3 I Monotube pile.

TABLE 9.7 | Data on monotube piles (type, size, weight, and concrete volumes).

Type, taper per ft	Size, point diameter × butt diameter × length	Weight, lb per ft				Volume of concrete, cy
		9 ga*	7 ga	5 ga	3 ga	
F-type						
0.14 in./ft	$8\frac{1}{2}$ in. × 12 in. × 25 ft	17	20	24	28	0.43
0.14 in./ft	8 in. × 12 in. × 30 ft	16	20	23	27	0.55
0.14 in./ft	$8\frac{1}{2}$ in. × 14 in. × 40 ft	19	22	26	31	0.95
0.14 in./ft	8 in. × 16 in. × 60 ft	20	24	28	33	1.68
0.14 in./ft	8 in. × 18 in. × 75 ft	—	26	31	35	2.59
J-type						
0.24 in./ft	8 in. × 12 in. × 17 ft	17	20	23	27	0.32
0.24 in./ft	8 in. × 12 in. × 17 ft	18	22	26	30	0.58
0.24 in./ft	8 in. × 12 in. × 17 ft	18	24	28	32	0.95
0.24 in./ft	8 in. × 12 in. × 17 ft	—	26	30	35	1.37
Y-type						
0.40 in./ft	8 in. × 12 in. × 17 ft	17	20	224	28	0.18
0.40 in./ft	8 in. × 14 in. × 15 ft	19	22	26	30	0.34
0.40 in./ft	8 in. × 16 in. × 20 ft	20	24	28	33	0.56
0.40 in./ft	8 in. × 18 in. × 25 ft	—	26	31	35	0.86

Extensions (overall lengths 1 ft greater than indicated)						
Type	Diameter × length	9 ga	7 ga	5 ga	3 ga	(cy/ft)
N 12	12 in. × 12 in. × 20–40 ft	20	24	28	33	0.026
N 14	12 in. × 14 in. × 20–40 ft	24	29	34	41	0.035
N 16	12 in. × 16 in. × 20–40 ft	28	33	39	46	0.045
N 18	12 in. × 18 in. × 20–40 ft	—	38	44	52	0.058

* The abbreviation ga stands for gauge.

STEEL PILES

Steel piles are especially suited to projects where it is necessary to drive piles through considerable depths of poor soil to reach solid rock or another formation having high load-supporting properties. Standard rolled steel *HP*-sections or wide-flange beams are most frequently used, although fabricated sections are sometimes used for special conditions. Standard sections are less expensive than fabricated sections.

Driving operations are similar to those of other piles. Steel piles do not require the care in handling that must be observed in handling precast-concrete piles. A cap can be installed on the top of piles to reduce the potential for damage during the driving operation.

Table 9.8 gives data for various steel piles and Table 9.9 gives approximate rates for driving steel piles when one is using an appropriate size hammer.

TABLE 9.8 | Data on selected steel-pile sections.

Designation	Cross-sectional dimensions, in., depth × width	Area, sq. in.	Weight, lb/lin ft
HP14x117	14.21 × 14.89	34.4	117
14x102	14.04 × 14.79	30.0	102
14x89	13.83 × 14.70	26.1	89
14x73	13.01 × 14.59	21.4	73
HP12x84	13.15 × 13.21	24.6	84
12x74	12.95 × 13.11	21.8	74
12x63	12.75 × 13.00	18.4	63
12x53	12.54 × 12.90	15.5	53
HP10x57	9.99 × 10.25	16.8	57
10x42	9.70 × 10.08	12.4	42
HP8x36	8.02 × 8.155	10.6	36

TABLE 9.9 | Approximate number of steel piles driven to full penetration per hour.*

Length of pile, ft	Weight of pile, lb/lin ft			
	100	75	55	35
30	3	$3\frac{1}{4}$	$3\frac{1}{2}$	$3\frac{1}{2}$
35	$2\frac{1}{2}$	$2\frac{3}{4}$	3	3
40	$2\frac{1}{4}$	$2\frac{1}{4}$	$2\frac{1}{2}$	$2\frac{1}{2}$
45	$1\frac{1}{2}$	$1\frac{3}{4}$	$1\frac{3}{4}$	—
50	$1\frac{1}{2}$	$1\frac{1}{2}$	$1\frac{3}{4}$	—
55	$1\frac{1}{4}$	$1\frac{1}{2}$	—	—
60	1	1	—	—
70	$\frac{3}{4}$	—	—	—
80	$\frac{1}{2}$	—	—	—

*These rates are based on piles being delivered to the job in lengths up to 40 ft. For lengths greater than 40 ft, a field weld will be required for each pile.

EXAMPLE 9.4

Estimate the cost for bid purposes for furnishing and driving 180 steel piles, 40 ft long. The piles are designated as HP14x73. The piles will be driven to full penetration into soil having high frictional resistance.

A single-acting air-driven pile hammer will be used. Table 9.1 indicates that a hammer with 12,000 ft-lb of energy is required. Table 9.2 shows several models of pile driving equipment are available. For this project a 5,000-lb hammer that can deliver at least 15,000 ft-lb of energy will be used. The pile-driving hammer will be mounted on the leads of a crawler crane, which will be used to lift the piles into position, support the leads, and lift the hammer. The cost can be determined as

Quantity of materials:

Linear feet of piles: 180 piles $\times$ 40 ft = 7,200 ft

Total weight of piles: 7,200 ft $\times$ 73 lb/ft = 525,600 lb

Time to drive piles:

From Table 9.9 for a 75 lb/ft pile, 40 ft long the driving rate = $2\frac{1}{4}$ piles/hr

Time for driving piles: (180 piles)/($2\frac{1}{4}$ piles/hr) = 80 hr

Considering a 45-min hour, the total time will be 60/45 $\times$ 80 hr = 107 hr

Rounding to a full 8-hour day, the total time will be 112 hr $\leftarrow$

Cost to drive piles:

Materials:

Piles: 525,600 lb @ $0.32/lb	=	$168,192.00
Pile caps: 180 caps @ $87/cap =		15,660.00
Total material cost =		$183,852.00

Equipment costs:

Crane: 112 hr @ $82.40/hr	=	$9,928.80
Leads for hammer: 112 hr @ $14.63/hr	=	1,638.56
Hammer: 112 hr @ $10.80/hr	=	1,209.60
Air compressor: 112 hr @ $14.65/hr	=	1,640.80
Other equipment and supplies: 112 hr @ $9.35/hr =		1,047.20
Total equipment cost =		$15,464.96

Labor costs, allow 8 hours for crew to set up and take down equipment:

Foreman: (112 hr + 8 hr) = 120 hr @ $34.00/hr	=	$4,080.00
Crane operator: 120 hr @ $31.00/hr	=	3,720.00
General laborers: 120 hr $\times$ 3 laborers @ $24.00/hr =		8,640.00
Total labor cost =		$16,440.00

Taxes:

Worker's compensation: $4.37 per $100 $\times$ $16,440.00 =		$718.43
Social security tax: 7.65% $\times$ $16,440.00	=	1,257.66
Unemployment tax: 4% $\times$ $16,440.00	=	657.60
Material taxes: 5% $\times$ $183,852	=	9,192.60
Subtotal of taxes =		$11,826.29

Summary of direct costs:

Materials	=	183,852.00
Equipment	=	15,464.96
Labor	=	16,440.00
Taxes	=	11,826.29
Subtotal of direct costs =		$227,583.25

Office overhead: 2% × $227,583.25 = $4,551.67
Job overhead: 4% × $227,583.25   =   9,103.33
Contingency: 3% × $227,583.25   =   6,827.50
Subtotal = $20,482.50

Total direct costs   = $227,583.25
Overhead + contingency =   20,482.50
Subtotal = $248,065.75

Total estimated cost   = $248,065.75
Profit: 7% × $248,065.75 =   17,364.60
Subtotal = $265,430.35

Performance bond:
$265,430.35 @ $12.00 per $1,000 = $3,185.16

Sum of subtotal cost = $265,430.35
Performance bond   =   3,185.16
Bid price = $268,615.51 ←

Cost per driven pile: $268,615.51/180 piles = $1,492.30/pile
Cost per linear foot of driven pile: $268,615.51/7,200 ft = $37.31/ft

JETTING PILES INTO POSITION

If the formation into which the piles are driven contains considerable sand, the rate of driving can be increased by jetting with water. To accomplish this with concrete piles, steel pipes, $1\frac{1}{2}$ to 2 in. diameter, can be placed in the piles when they are cast. Water is injected into the pipe at a rate of 100 to 400 gal/min to jet a concrete pile into position. Required pressures will be 100 to 200 psi. For suitable soil conditions, this method will sink piles much more quickly than using pile-driving hammers.

DRILLED SHAFT FOUNDATIONS

Drilled-shaft foundations are installed by placing reinforced concrete in holes that have been drilled into the soil. (See Figure 9.4.) Truck-mounted rotary drilling rigs can drill holes with diameters from 12 to 96 in. Depths up to approximately 100 ft are possible; however, most drilled shafts range from 6 to 50 ft deep. The bottom of the drilled hole can be underreamed with a belling tool to provide an increased bearing capacity of the foundation.

Two types of belling tools are available to underream drilled shafts; one opens out from the top, as shown in Fig. 9.5a and one opens out from the bottom,

FIGURE 9.4 | Auger drilling hole for drilled shaft foundation.
Courtesy: Sherwood Construction Company.

as shown in Fig. 9.5b. Equation [9.1] can be used to determine the volume of the shaft for the underream of the type of hole illustrated in Fig. 9.5b.

$$V_c = \frac{\Pi (d_s)^2}{4}(D_f - h) + \frac{\Pi h}{12}(d_s^2 + d_s d_b + d_b^2) \qquad \textbf{[9.1]}$$

where

V_c = total volume of concrete in the shaft, cf
D_f = depth of foundation, ft
h = depth from the bottom of shaft to top of bell, ft
d_s = diameter of shaft, ft
d_b = diameter of bell, ft

Cohesive soils, such as dry clays, generally have sufficient strength to provide natural stability of the sides of the drilled shaft. However, if water is present the sides of the shaft drilled into clay may become unstable. Cohensionless soils such as sands are unstable in both the dry and wet condition. When drilling into unstable soils or where water may be present, it is usually necessary to

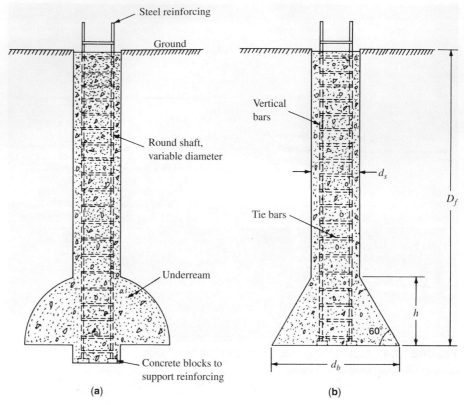

FIGURE 9.5 I Drilled hole for cast-in-place concrete piles; underream made by a belling tool that opens out (a) from the bottom and (b) from the top.

install cylindrical steel casing to hold the sides of the shaft until the reinforcing steel and concrete are placed (see Figure 9.6). The casing is pulled from the hole during placement of the concrete.

Table 9.10 provides representative rates of drilling various diameters of shafts into soils. Table 9.11 gives approximate rates for drilling into soft rock, such as shales. These rates can vary, depending on the moisture and hardness of the soil and the size of the rotary drilling rig. A thorough evaluation of the soil test borings is necessary to determine the rate of drilling.

Vertical reinforcing steel is tied together by horizontal ties or spiral hoops, and lowered into position in the drilled shaft. Lugs, attached to the reinforcing cage, hold them into position at the center of the hole until concrete is placed.

The cost of drilled-shaft foundations will include mobilization of equipment, drilling of the shaft, casing (if required), fabrication and placement of the reinforcing steel, and concrete material and placement.

FIGURE 9.6 | Steel casing for drilled shaft foundation.

Courtesy: Sherwood Construction Company.

TABLE 9.10 | Representative rates of drilling shafts into soils.

Diameter, in.	Rate of drilling in soils, ft/hr	
	No casing required	Casing required
18	24–26	19–21
24	22–25	15–17
30	18–20	10–12
36	15–17	7–8
48	12–13	6–7
60	10–12	5–6
72	9–11	4–5
84	8–10	3–4
96	7–8	2–3

TABLE 9.11 | Representative rates of drilling shafts into soft rock material.

Diameter, in.	Drilling in soft rock material, ft/hr
18	6.0–6.7
24	3.6–3.9
30	2.4–2.6
36	1.7–2.1
48	1.1–1.4
60	0.8–1.0
72	0.7–0.8
84	0.6–0.7
96	0.5–0.6

EXAMPLE 9.5

A 15-mi-long electric transmission line requires 64 steel truss towers to support the conductors and shield wires. There are four drilled shaft-foundations for each tower, one for each leg of the earth tower, similar to the type shown in Fig. 9.5b. Each foundation is to be installed with a 24-in. diameter at a depth of 9 ft in wet clay; therefore, casing will be required. A 42-in.-diameter bell is required for each foundation.

Reinforcing steel consists of eight vertical bars 9 ft long with $\frac{3}{4}$ in. diameters. Horizontal ties will be placed at 12 in. on centers. Tie bars will have $\frac{3}{8}$-in. diameters and will have a 3-in. overlap on their ends. A 3-in. clear cover is required on all reinforcing steel. Ready-mix concrete will be obtained from a local concrete supplier. The concrete will be delivered to the job and placed directly into the drilled holes.

Quantity of materials:

Number of drilled shafts, 64 structures $\times$ 4 foundations/structure = 256 drilled shafts

Total linear feet of drilling, 256 shafts $\times$ 9 ft/shaft = 2,304 lin ft

Concrete:

Volume of concrete per drilled shaft from Eq. [9.1]:

$$V_c = \frac{\Pi (d_s)^2}{4} (D_f - h) + \frac{\Pi h}{12} (d_s^2 + d_s d_b + d_b^2) \qquad \text{[9.1]}$$

$$= \frac{3.14 \times (2 \text{ ft})^2}{4} [9.0 \text{ ft} - 1.3 \text{ ft}]$$

$$+ \frac{3.14 \times 1.3 \text{ ft}}{12} [(2.0 \text{ ft})^2 + (2.0 \text{ ft} \times 3.5 \text{ ft}) + (3.5 \text{ ft})^2]$$

= 32.1 cf for each drilled shaft

Total volume of concrete: 256 shafts $\times$ (32.1 cf/shaft)/(27 cf/cy) = 304.4 cy

Add 3% waste = 　9.1 cy

Total volume = 313.5 cy

Reinforcing steel:

    Vertical bars ($\frac{3}{4}$ in. diameter):

        Total length: 256 shafts $\times$ 8 bars/shaft $\times$ 9 ft/bar = 18,432 lin ft

        Total weight: 18,432 lin ft @ 1.502 lb/ft = 27,685 lb

    Horizontal tie bars ($\frac{3}{8}$ in. diameter):

        Centerline diameter: [24 in. dia. $-$ (2 $\times$ 3 in. cover) $- \frac{3}{8}$ in.] = 17.625 in.

        Circumference: (3.14 $\times$ 17.625 in. + 3-in. lap)/(12 in./ft) = 4.86 ft

        Total length: 256 shafts $\times$ 9 bars/shaft $\times$ 4.86 ft/bar = 11,197 lin ft

        Total weight: 11,197 lin ft @ 0.376 lb/ft = 4,210 lb

Time to drill foundations:

    Rate of drilling, from Table 9.10 for 24-in. dia. with casing: rate = 16 ft/hr

    Time for drilling: 2,304 lin ft/16 ft/hr = 144 hr

    Considering 45-min. hour, the total time will be 60/45 $\times$ 144 hr = 192 hr

    Rounding to a full 8-hour day, the total time will be 192 hr $\leftarrow$

Cost of work:

    Material costs:

        Ready-mix concrete: 313.5 cy @ $115.00/cy   = $36,052.50

        Vertical bars in place: 27,685 lb @ $0.46/lb   = 12,735.10

        Horizontal tie bars in place: 4,210 lb @ $0.46/lb =   1,936.60

                         Total material cost = $50,724.20

    Equipment costs:

        Drilling rig: 192 hr @ $168.00/hr   = $32,256.00

        Casing rental: 2,304 lin ft @ $3.00/ft =   6,912.00

        Moving drill rig to and from job   =   2,500.00

               Total equipment cost = $41,668.00

    Labor costs:

        Drill operator: 192 hr @ $31.00/hr   = $5,952.00

        Laborers: 192 hr $\times$ 2 laborers @ $25.00/hr =   9,600.00

        Foreman: 192 hr @ $34.00/hr   =   6,528.00

                 Total labor cost = $22,080.00

    Summary of costs:

      Material   = $50,724.20

      Equipment =   41,668.00

      Labor   =   22,080.00

       Total cost = $114,472.20

    Cost per linear foot: $114,472.20/2,304 ft = $49.68/ft

    Cost per cubic yard: $114,472.20/304.4 cy = $376.06/cy

These costs do not include taxes, insurance, overhead, contingency, or profit.

PROBLEMS

9.1 Estimate the total direct cost and the cost per linear foot for furnishing and driving steel-sheet piling for construction of the basement of a building, 200 by 350 ft. The piles will be 22 in. wide, 20 ft long, and weigh 40.3 lb/lin ft (see Table 9.4, section no. PZ22). The piles will be driven to full penetration into soil having normal frictional resistance. Use the cost information in Example 9.2.

9.2 Estimate the total cost and cost per linear foot for furnishing and driving 108 creosote-treated wood piles to full penetration into soil having high frictional resistance. The piles will be 36 ft long, with 14-in. minimum diameter at the butt and 6-in. minimum diameter at the tip. The tip of each pile will be fitted with a metal point to facilitate driving. Use the cost information in Example 9.3.

9.3 Estimate the total direct cost and cost per linear foot for furnishing and driving the piles of 9.2 to full penetration into soil having normal frictional resistance.

9.4 Estimate the total direct cost and cost per linear foot for furnishing 75 steel piles, HP 14 × 73, into soil having high frictional resistance. The piles will be driven to full penetration at a depth of 45 ft. Use the cost information in Example 9.4.

9.5 Estimate the total cost and cost per cubic yard for installing 156 drilled-shaft foundations 18 ft deep in clay soil. The water table is 3 ft below the ground surface, therefore casing will be required. Each foundation will be installed, similar to the type shown in Figure 9.5(b), with a 36-in. diameter shaft and a 48-in. diameter underreamed bell. Vertical reinforcing steel consists of 16 bars with a $\frac{7}{8}$-in. diameter. Horizontal ties will be placed at 15-in. on centers with a 6-in. lap. Ties are $\frac{1}{2}$-in. diameter reinforcing steel. A 3-in. cover is required on all steel. Use the cost information in Example 9.5.

10

Concrete Structures

COST OF CONCRETE STRUCTURES

The items that govern the total costs of concrete structures include

1. Forms
2. Reinforcing steel
3. Concrete
4. Finishing, if required
5. Curing

Formwork for concrete structures is usually measured in square feet of contact of area (SFCA). The SFCA is the surface area that concrete will be in contact with the forms. For example, an 8-ft-high wall, 200 ft long, would have 3,200 SFCA, calculated as 8 ft × 200 ft × 2 sides = 3,200 SFCA. The unit of measure for reinforcing steel is pounds of weight, whereas the unit of measure for concrete is cubic yards of volume. The finishing and curing of concrete is measured in square feet.

When preparing an estimate, the cost of each of these items should be determined separately and a code number should be assigned to each item. The cost of an item should include all materials, equipment, and labor required for that item. Then, after all the costs are determined, the total cost can be divided by the SFCA. Following this procedure will simplify the preparation of an estimate of concrete structures.

FORMS FOR CONCRETE STRUCTURES

Concrete structures can be constructed in any shape for which it is possible to build forms. However, the cost of forms for complicated shapes is considerably greater than for simple shapes because of the extra material and labor costs required to build them, the reduced potential to reuse the forms, and the low salvage value after they are used.

Since the major cost of a concrete structure frequently results from the cost of forms, the designer of a structure should consider the effect that the shape of the structure will have on the cost of forms.

Forms for concrete are fabricated from lumber, plywood, steel, aluminum, and various composite materials, either separately or in combination. If the form material will be used only a few times, the lumber will usually be more economical than steel or aluminum. However, if the forms can be fabricated into panel sections or other shapes, such as round column forms, that will be used many times, then the greater number of uses obtainable with steel and aluminum may produce a lower cost per use than with lumber.

The material should be selected for the form that will give the lowest total cost of a structure, considering the cost of the forms plus the cost of finishing the surface of the concrete, if required, after the forms are removed. The use of lumber will usually leave form marks that may have to be removed at considerable cost, whereas the use of smooth-surface plywood or metal forms can reduce the cost of removing form marks. It is a better practice to spend extra money on good-quality form materials, than it is to spend money on labor to hand-finish concrete.

The cost of forms will include the cost of materials, such as lumber, nails, bolts, form ties, and the cost of labor making, erecting, removing, and cleaning the forms. Frequently there will be a cost of power equipment, such as saws and drills, and a small lifting crane for handling materials. If it is possible to reuse the forms, an appropriate allowance should be made for salvage value. If the forms are treated with oil prior to each use, the cost of the oil should be included.

MATERIALS FOR FORMS

Forms for concrete structures are generally fabricated from standard-dimension lumber and plywood sheets. Because the lumber must be cut and fastened for the particular shape of the concrete member, considerable waste often occurs. The cost of materials for forms should include an allowance for waste.

Plywood is used for sheathing of walls, columns, and floor slabs to achieve a smooth surface. Lumber is used to support the plywood. For wall and column forms, 2×4 and 2×6 dimension lumber is commonly used, whereas 4×4 and 4×6 lumber is typically used for floor slabs and bridge decks.

For large projects or jobs that have extremely high concrete pressures, steel forms are frequently used. The initial fabrication cost for steel is high, but the ability to reuse the forms many times makes them economical. Steel forms also have higher strength to resist the concrete pressure that is applied to the forms.

Plywood Used for Forms

Plywood sheets are made in various thicknesses with waterproof glue. Those commonly used are $\frac{3}{4}$, $\frac{7}{8}$, and 1 in. thick. They are available in sheets 4 ft wide and 8, 10, and 12 ft long. Other dimensions can be obtained on special order.

It is necessary to install the plywood with the outer layers perpendicular to the studs or joists to obtain maximum strength. Plywood and Plyform are priced by the sheet.

The plywood industry manufactures a special product, designated as Plyform, for use in forming concrete. Plyform is an exterior-type plywood limited to certain species of wood and grades of veneer to ensure high performance as a form material. The panels are sanded on both sides and oiled by the manufacturer. Plyform can be reused more times than plywood. A special high-density overlay (HDO) can be applied to the surface of Plyform by the manufacturer, which will allow the HDO Plyform to be used in excess of 100 uses.

Dimension Lumber Used for Forms

The net dimensions of lumber are given in Chapter 12. Because the net dimensions are less than nominal dimensions, it is necessary to use the net dimensions to determine the quantity of lumber required for formwork. Standard lengths are available in multiples of 2 ft. The common lengths of lumber are 8, 10, 12, 14, 16, and 18 ft. If an odd length, such as 9 ft, is required, it can be obtained by cutting an 18-ft-long board into two equal lengths or by cutting 1 ft off a 10-ft-long board, with a resulting waste.

Lumber is measured and priced by *board feet* (bf). A *board foot* is a piece of lumber whose nominal dimensions are 1 in. thick, 12 in. wide, and 1 ft long. Thus, a 2 × 4 piece of lumber is equivalent to 0.67 bf per foot of length, whereas a 2 × 8 piece of lumber is equivalent to 1.33 bf per foot of length.

There are many grades and species of lumber. The grade of lumber is classified as no. 1, 2, or 3. Grades 1 and 2 are commonly used as form material. Common species are Douglas Fir and Southern Pine. The price of lumber depends on grade, species, and size.

Nails Required for Forms

The quantity of nails required for forms usually will vary from 10 to 20 lb/1,000 bf of lumber for the first use of lumber and from 5 to 10 lb/1,000 bf for additional uses if the forms can be reused without completely refabricating them. The price of nails is based on pounds of weight.

There are three types of nails used for forms: common wire nails, box nails, and double-headed nails. The common wire nail is the most frequently used nail for fastening form members together. A box nail has a thinner shank than a common wire nail, which makes it more useful for built-in-place forms. Double-headed nails are frequently used in formwork when it is desirable to remove the nail easily, such as for stripping forms.

Form Oil

Form oil is applied to surfaces of the forms that will come in contact with concrete. It prevents the concrete from attaching to the form material and facilitates removal of the forms. Sometimes form oil is called bond-breaker. Form oil can be applied with mops, brushes, or pressure sprayers. Generally it is applied at a rate of 300 to 500 sf/gal.

Form Ties

To resist the internal pressure resulting from the concrete, form ties are placed between the forms, as shown in Fig. 10.1. Form ties serve two purposes: they hold the forms apart prior to placing the concrete, and they resist the lateral pressure of the freshly placed concrete. A tie clamp is used to secure both ends of the tie. Thus, two tie clamps are required for each tie. Sometimes cam-lock brackets are used in place of wedge tie clamps.

Many varieties of ties are available. The snap tie and coil tie are most commonly used. Both of these ties are designed to break off inside the concrete after the forms are removed. The coil tie has a plastic cone at each end of the tie that enables a cleaner break upon removal, which may be required for architectural finished concrete surfaces. The holes on the surface can be filled with concrete to obtain a smooth surface. Manufacturers specify the safe loads that can be resisted by their ties, which generally range from 3,000 to 5,000 lb.

A taper tie is available that can be reused. It consists of a tapered rod that is threaded at each end. Wing nuts or coil nuts with washers are attached at each end to secure the tie. Taper ties have larger strength than snap or coil ties. In ordering form ties it is necessary to specify the thickness of the wall, sheathing, studs, and wales.

Form Liners

Architectural concrete form liners can be used to create a virtually unlimited variety of textured or sculpted finishes on concrete surfaces. The surfaces can simulate wood grains, rough brick, blocks, or an irregular texture. Form liners may be rigid or flexible.

Polyvinyl chloride (PVC) sheets of rigid form liners are attached to the formwork prior to placing the concrete. Following placement and normal curing time, the formwork and liner are stripped, leaving a textured concrete surface. The sheets generally are 4 ft wide and 10 ft long with square edges to allow the placement of adjacent sheets for large areas of concrete. Sheets of rigid form liners are lightweight and easily stripped. A releasing agent should be applied to the form liner to ensure uniformity of the concrete surface and to protect the form liner for possible reuse. Flexible form liners are also available that are made of urethanes, which can be peeled away from the concrete surface, revealing the desired texture.

Form liners are expensive, often more than the cost of the plywood and lumber. The cost depends on the type of liner selected. The estimator must obtain a price quote from the concrete supplier who will provide the form liner for a particular job.

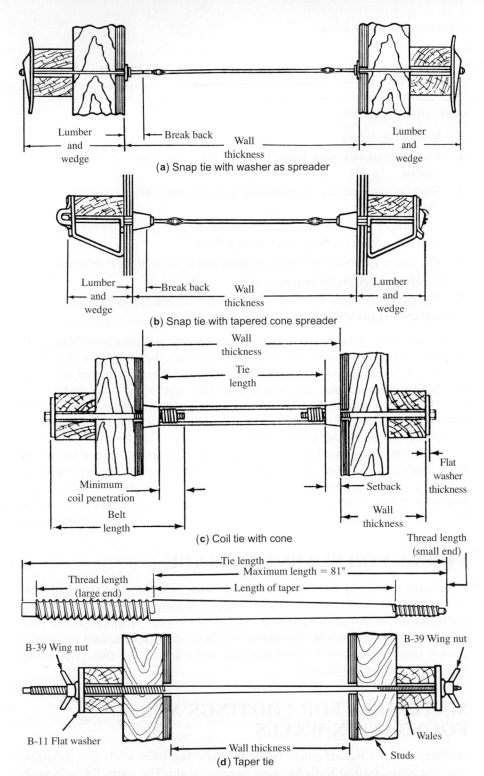

Lumber and wedge — **Break back** — **Wall thickness** — **Lumber and wedge**

(a) Snap tie with washer as spreader

Lumber and wedge — **Break back** — **Wall thickness** — **Lumber and wedge**

(b) Snap tie with tapered cone spreader

Wall thickness

Tie length

Flat washer thickness

Minimum coil penetration

Setback

Belt length

Wall thickness

(c) Coil tie with cone

Thread length (small end)

Tie length

Maximum length = 81"

Thread length (large end)

Length of taper

B-39 Wing nut

B-39 Wing nut

B-11 Flat washer

Wales

Wall thickness

Studs

(d) Taper tie

FIGURE 10.1 | Ties for concrete formwork: (a) snap tie with washer spreader, (b) snap tie with cone, (c) coil tie with cone, and (d) taper tie.

Courtesy: Dayton Superior Corporation.

LABOR REQUIRED TO BUILD FORMS

The factors that determine the amount of labor required to build forms for concrete structures include

1. Size of the forms.
2. Kind of materials used. Large sheets of plywood require less labor than lumber.
3. Shape of the structure. Complicated shapes require more labor than simple shapes.
4. Location of the forms. Forms built above the ground require more labor than forms built on the ground or on a floor.
5. The extent to which prefabricated form panels or sections can be used.
6. Rigidity of dimension requirements.
7. The extent to which the forms can be prefabricated in the shop and transported to the job.

If forms are prefabricated into panels or sections and then assembled, used, removed, and reused, it is desirable to estimate separately the labor required to make, assemble, remove, and clean them. Because making the forms is required only once, for additional uses it is only necessary to assemble, remove, and clean the forms.

The production rates given in this book are based on using power saws and other equipment as much as possible. If the fabricating is done with hand tools, the labor required should be increased.

The tables on production rates give the minimum and maximum number of labor-hours required to do a specified amount of work, for both carpenters and helpers. If union regulations specify the ratio of helpers to carpenters for certain locations, it may be necessary to transfer some of the helper time to carpenter time. The tables include an allowance for lost time by using a 45- to 50-min hour.

FORMS FOR SLABS ON GRADE

When concrete is placed directly on the ground, side forms of metal or constructed of 2-in. lumber are commonly used. The forms are placed around the perimeter of the slab and secured in position with wood stakes or metal pins. Either welded wire mesh or reinforcing steel is set in the forms, then concrete is placed, finished, and cured. Two carpenters and a laborer can place side forms at a rate of 400 to 500 lin ft/day.

MATERIALS FOR FOOTINGS AND FOUNDATION WALLS

Foundation walls include grade beam footings, basement walls for buildings, retaining walls, vertical walls for water reservoirs, etc. The forms for such walls typically consist of $\frac{3}{4}$-in.-thick plywood or Plyform sheathing, 2×4 or 2×6 studs, double 2 × 4 or 2 × 6 wales, and 2 × 4 or 2 × 6 braces. The sheathing, studs, and wales are erected on both sides of the wall, while the braces are spaced about 6 to 8 ft apart on one side of the wall only. Form ties are placed between wall forms to resist the lateral pressure against wall forms resulting from concrete.

Wall forms are frequently built as shown in Fig. 10.2. The studs are uniformly spaced to support the plywood sheathing. Studs should be spaced so that there will be a stud at the end of each plywood sheet. Wales are placed perpendicular to the studs and wall ties are used to secure the wall form together. To keep the damage to a minimum, a standard pattern for holes for wall ties should be adopted. The adoption of such a practice will reduce the need for drilling holes for subsequent uses. With proper care, plywood can be used at least 3 or 4 times. With proper care, HDO Plyform can be used 30 or 40 times.

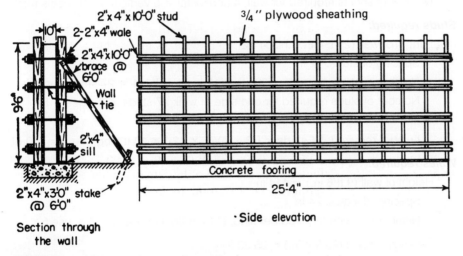

FIGURE 10.2 | Forms for concrete wall.

EXAMPLE 10.1

Estimate the quantity of material to form a concrete wall 9 ft 6 in. high and 25 ft 4 in. long. The sheathing will be $\frac{3}{4}$-in. plywood sheets, 4 ft wide by 8 ft long. The plywood sheathing will be placed with the 4 ft width in the vertical direction and the 8 ft length in the horizontal direction. Vertical studs will be 2 × 4 lumber placed at 12 in. on centers and the horizontal double 2 × 4 wales will be placed at 24 in. on centers. The tie spacing will be 24 in., center to center.

Because the height of the wall is not evenly divisible by 4 or 8 ft, there will be waste in the 4-ft by 8-ft plywood. Also, due to the height and length of the wall there will be waste in the studs and wales.

Prepare a list of materials and calculate the quantity of lumber per 100 sf of wall. Include all of the material to form the wall, including braces, scab splices, and stakes.

Plywood sheathing:

Height of wall (9 ft 6 in.) = 9.5 ft

Number of sheets in vertical direction: (9.5 ft)/(4 ft/sheet) = 2.4

Number of sheets of plywood for each 8-ft length of wall: 2.4 × 2 sides = 4.8 sheets

Cutting one sheet down the center, then 5 sheets are sufficient for vertical direction

Length of wall (25 ft 4 in.) = 25.33 ft

Number of sheets in horizontal direction: (25.33 ft)/(8 ft/sheet) = 3.2

Therefore use 4 sheets in the horizontal direction

Number of sheets required for wall, 4 horizontal × 5 vertical = 20 sheets total

Studs required:

Length of wall (25-ft 4-in.) = 304 in.

Spacing of studs: 12 in.

Number of studs per side: [(304 in.)/(12 in./stud)] + 1 = 27 studs per side

Number of studs required for one side of wall: 27 studs × 2 sides = 54 studs total

Lumber required: 54 studs × [(2 × 4)/(1 × 12)] × 10 ft long   = 360 bf

Wales required:

Height of wall (9 ft 6 in.) = 114 in.

Spacing of wales: 24 in.

Number of wales required per side: (114 in.)/(24 in./wale) = 4.8, use 5

Length of wall (25 ft 4 in.) = 25.33 ft

Use wales 14-ft-long lumber, end-to-end, to provide 28 ft in total length

Lumber for one row: {2 wales × [(2 × 4)/(1 × 12)] × 14 ft} × 2 = 37.3 bf

Lumber required for one side of wall: 5 rows × 37.3 bf = 186 bf

Lumber required for both sides of wall: 186 bf/side × 2 sides   = 372 bf

Sills required:

Top sill: 2 pieces × [(2 × 4)/(1 × 12)] × 14 ft long   =   19 bf

Bottom sill: 2 pieces × [(2 × 4)/(1 × 12)] × 14 ft long   =   19 bf

Scab splices required:

Number required per side: (2 splices/row) × (5 rows) = 10 splices

Number required for both sides: 2 sides × 10 splices = 20 splices total

Lumber: 20 splices × [(2 × 4)/(1 × 12)] × 2 ft long   =   27 bf

Braces required:

Braces on one side only, spaced at 6 ft with one brace at each end

Number of braces required: (25.3 ft)/(6 ft/brace) + 1 = 5 required

Lumber required: 5 braces × [(2 × 4)/(1 × 12)] × 10 ft long   = 33 bf

Stakes required:

Lumber required: 5 stakes × [(2 × 4)/(1 × 12)] × 3 ft long   = 10 bf

Total quantity of lumber = 840 bf

Nails required:

Pounds of nails: (10 lb/1,000 bf) × (840 bf) = 8.4 lb

Form ties required:

Number of ties per row of wales: (304 in.)/(24 in./tie) + 1 = 13.7, use 14 ties per row

Number of ties required: 5 rows × 14 ties/row = 70 ties total

Add 5% for waste   = 4

Total = 74 ties

Summary of materials:

Plywood = 20 sheets

Lumber = 840 bf

Nails = 8.4 lb

Ties = 74

Quantities of material per 100 SFCA:

Total square feet of contact area: 25.3 ft × 9.5 ft × 2 sides = 481 SFCA

Plywood: 20 sheets × 32 sf/sheet = 640 sf/4.81 SFCA = 133 sf/100 SFCA

Lumber: 840 bf/4.81 SFCA = 175 bf/100 SFCA

Ties: 74 ties/4.81 SFCA = 16 ties/100 SFCA

QUANTITIES OF MATERIALS AND LABOR-HOURS FOR WALL FORMS

Table 10.1 gives the approximate quantities of plywood, lumber, and form ties and the labor-hours required for 100 sf of concrete wall using plywood for various heights. For wall heights other than multiples of 2 ft, there will be wastage. Also door and window openings, wall columns, and irregular-shaped walls will usually cause waste and require more time to erect and remove the forms. The estimator should consider the extent of wastage and perform the calculations illustrated in Example 10.1 to obtain more accurate quantities of materials.

For walls up to about 12 ft high, 2 × 4 studs and double 2 × 4 wales are generally adequate. For greater heights, 2 × 6 studs and 2 × 6 wales may be

TABLE 10.1 I Quantities of plywood, lumber, and form ties and the labor-hours required for 100 SFCA of wall forms.*

Wall height, ft	Plywood, sf	Lumber, bf	Form ties	Labor-hours, making forms		Labor-hours erecting and removing forms	
				Carpenter	Helper	Carpenter	Helper
4	100–110	120–135	12	2.0–2.5	1.5–2.0	2.5–3.0	2.0–2.5
6	100–110	135–155	14	2.0–2.5	1.5–2.0	3.0–3.5	2.5–3.0
8	100–110	155–175	14	2.0–2.5	1.5–2.0	3.5–4.0	3.0–3.5
10	100–110	175–205	16	2.0–2.5	1.5–2.0	4.0–4.5	3.5–4.0
12	100–110	205–245	18	2.0–2.5	1.5–2.0	4.5–5.0	4.0–4.5
14	100–110	245–305	18	2.0–2.5	1.5–2.0	5.0–5.5	4.5–5.0
16	100–110	305–325	20	2.0–2.5	1.5–2.0	5.5–6.0	5.0–5.5

*These quantities are based on a rate of fill of approximately 4 ft/hr and at a temperature of 70°F using normal weight concrete. If the concrete is pumped, the quantities of material may be greater than shown.

necessary. Double rows of braces should be used for walls over 12 ft high, one near the middle and the other near the top of the wall.

Fabrication of concrete forms by a carpenter will range from 60 to 80 bf/hr, depending on the size of material that is fabricated. Generally a laborer will assist two or more carpenters during the fabrication process.

EXAMPLE 10.2

Estimate the cost for building, erecting, and removing concrete forms for a 12-ft high, 85-ft 6-in.-long wall for an industrial plant. The forms will be used three times. Thus, it will be necessary to build only 28.5 ft of the wall forms, but 85.5 ft of forms must be erected and removed.

Plyform sheets will be used for sheathing and dimension lumber will be used for studs and wales. A material supplier quotes these prices: $0.55/bf for lumber, $34/sheet for 4-ft by 8-ft Plyform, $2.31/lb for nails, and $14.00/gal for form oil. Carpenters cost $28/hr and laborers cost $21.00/hr.

Quantity of materials:

Forms to be erected and removed: (12 ft × 85.5 ft) × 2 sides = 2,052 SFCA

Forms to be built: 2,052 SFCA/3 uses = 684 SFCA

Use Table 10.1 median value for quantities of material:

Plyform sheathing: (684 SFCA) × (105 sf/100 SFCA) = 718.2 sf

718.2 sf/(32 sf/sheet) = 22.4 sheets, use 23 sheets of Plyform

Lumber: (684 SFCA) × (225 bf/100 SFCA) = 1,539 bf

Form ties: 2,052 SFCA × (18 ties/100 SFCA) = 370 ties

Form oil: (2,052 SFCA)/(400 SFCA/gal) = 5.1 gal

Nails for first use: (1,539 bf) × (15 lb/1,000 bf)   = 23.1 lb

Nails for 2 reuses: (1,539 bf) × 2 × (10 lb/1000 bf) = 30.8 lb

Total = 53.9 lb

Labor-hours required:

Reference Table 10.1 for labor-hours:

Making forms:

Carpenter: 684 SFCA × (2.25 hr/100 SFCA)	=	15.4 hr
Helper: 684 SFCA × (1.75 hr/100 SFCA)	=	12.0 hr

Erecting and removing forms:

Carpenter: 2,052 SFCA × (4.75 hr/100 SFCA)	=	97.5 hr
Helper: 2,052 SFCA × (4.25 hr/100 SFCA)	=	87.2 hr
Foreman based on 4 carpenters: (15.4 + 97.5)/4	=	28.2 hr

Cost of labor and material:

Labor costs:

Carpenter making forms: 15.4 hr @ $28.00/hr	=	$431.20
Helper making forms: 12.0 hr @ $21.00/hr	=	252.00
Carpenter erecting and removing: 97.5 hr @ $28.00/hr	=	2,730.00
Helper erecting and removing: 87.2 hr @ $21.00/hr	=	1,831.20
Foreman based on 4 carpenters: 28.2 hr @ $32.00/hr	=	902.40
Labor cost	=	$6,146.80

Material costs:

Plyform: 23 sheets @ $34.00/sheet	=	$782.00
Lumber: 1,539 bf @ $0.55/bf	=	846.45
Form ties: 370 ties @ $1.47/tie	=	543.90
Nails: 53.9 lb @ $2.31/lb	=	124.51
Form oil: 5.1 gal @ $14.00/gal	=	71.40
Material cost	=	$2,368.26

Summary of costs:

Total labor and material: $6,146.80 + $2,368.26 = $8,515.06

Cost/SFCA: $8,515.06/2,052 SFCA = $4.15/SFCA

These costs do not include reinforcing steel, concrete, taxes, overhead, or profit.

Considerable savings can be achieved by reusing forms, as illustrated in Examples 10.3 and 10.4. Example 10.3 shows the cost of formwork for a wall when the forms are used only one time and Example 10.4 shows the cost for the same wall forms, except the forms will be used three times.

EXAMPLE 10.3

Estimate the cost of formwork for a concrete wall 12 ft high and 126 ft long, assuming the forms will not be reused. Plyform sheets will be used for sheathing and dimension lumber will be used for studs and wales.

A material supplier quotes these prices: $0.55/bf for lumber, $34/sheet for 4-ft by 8-ft Plyform, $2.31/lb for nails, and $14.00/gal for form oil. Carpenters cost $28.00/hr and laborers cost $21.00/hr.

Quantity of material:

Total surface area of wall forms: 12 ft × 126 ft × 2 sides = 3,024 SFCA

Quantity of forms to be built, erected, and removed = 3,024 SFCA

Materials from Table 10.1:

Plyform sheathing: 3,024 SFCA × (105 sf/100 SFCA) = 3,175 sf

3,175 sf/(32 sf/sheet) = 99.2, use 100 sheets

Lumber: 3,024 SFCA × (225 bf/100 SFCA) = 6,804 bf

Form ties: 3,024 SFCA × (18 ties/100 SFCA) = 545

Nails: 6,804 bf × (15 lb/1,000 bf) = 102 lb

Form oil: (3,024 sf)/(400 sf/gal) = 7.6 gal

Labor cost:

Making forms:

Carpenter: 3,024 SFCA ×
(2.25 hr/100 bf) = 68.0 hr @ $28.00/hr   = $1,904.00

Helper: 3,024 SFCA ×
(1.75 hr/100 bf) = 52.9 hr @ $21.00/hr   =   1,110.90

Erecting and removing forms:

Carpenter: 3,024 SFCA ×
(4.75 hr/100 SFCA) = 143.6 hr @ $28.00/hr =   4,020.80

Helper: 3,024 SFCA ×
(4.25 hr/100 SFCA) = 128.5 hr @ $21.00/hr =   2,698.50

Foreman based on 4 carpenters:
(68.0 + 143.6)/4 = 52.9 hr @ $32.00/hr   =   1,692.80

Total labor cost = $11,427.00

Material cost:

Plywood: 100 sheets @ $34.00/sheet = $3,400.00

Lumber: 6,804 bf @ $0.55/bf   =   3,742.20

Ties: 545 @ $1.62/tie   =   882.90

Nails: 102 lb @ $2.31/lb   =   235.62

Form oil: 7.6 gal @ $14.00/gal   =   106.40

Total material cost = $8,367.12

Summary of costs:

Total labor and material cost: $11,427.00 + $8,367.12 = $19,794.12

Cost/SFCA = $19,794.12/3,024 SFCA = $6.55/SFCA

These costs do not include reinforcing steel, concrete, taxes, overhead, or profit.

EXAMPLE 10.4

Estimate the cost of formwork for the 12-ft-high 126-ft-long concrete wall of Example 10.3, assuming the forms will be used three times. Calculate the cost savings that can be achieved by reusing the forms.

Quantity of material:

Total surface area of wall forms: 12 ft × 126 ft × 2 sides = 3,024 SFCA

Quantity of forms to be built: 3,024 SFCA/3 uses = 1,008 SFCA

Quantity of forms to be erected and removed = 3,024 SFCA

Materials from Table 10.1:

Plyform sheathing: 1,008 SFCA × (105 sf/ 100 SFCA) = 1,058 sf

1,058 sf/(32 sf/sheet) = 33.1, use 34 sheets

Lumber: 1,008 SFCA × (225 bf/100 SFCA) = 2,268 bf

Ties: 3,024 SFCA × (18 ties/100 SFCA) = 545 ties

Nails: 2,268 bf × (10 lb/1,000 bf) = 23 lb

Form oil: (3,024 sf)/(400 sf/gal) = 7.6 gal

Labor cost:

Making forms:

Carpenter: 1,008 SFCA × (2.25 hr/100 bf) = 22.7 hr @ $28.00/hr	=	$635.60
Helper: 1,008 SFCA × (1.75 hr/100 bf) = 17.6 hr @ $21.00/hr	=	369.60

Erecting and removing forms:

Carpenter: 3,024 SFCA × (4.75 hr/100 SFCA) =143.6 hr @ $28.00/hr	=	4,020.80
Helper: 3,024 SFCA × (4.25 hr/100 SFCA) = 128.5 hr @ $21.00/hr	=	2,698.50
Foreman based on 4 carpenters: (22.7 + 143.6)/4 = 41.6 hr @ $32.00/hr	=	1,331.20
Total labor cost	=	$9,055.70

Material cost:

Plywood: 34 sheets @ $34.00/sheet =		$1,156.00
Lumber: 2,268 bf @ $0.55/bf	=	1,247.40
Ties: 545 @ $1.62/tie	=	882.90
Nails: 23 lb @ $2.31/lb	=	53.13
Form oil: 7.6 gal @ $14.00/gal	=	106.40
Total material cost =		$3,445.83

Summary of costs:

Total labor and material cost: $9,055.70 + $3,445.83 = $12,501.53

Cost/SFCA = $12,501.53/3,024 SFCA = $4.13/SFCA

These costs do not include reinforcing steel, concrete, taxes, overhead, or profit.

The $12,501.53 in Example 10.4 is significantly lower than the $19,794.12 in Example 10.3, which shows considerable cost savings can be achieved by reusing the forms. The cost difference is $19,794.12 − $12,501.53 = $7,292.59, which represents a savings of $7,292.59/$19,794.12 = 36.8%. Further cost savings would also be achieved due to lower taxes on material and labor, and overhead expenses.

PREFABRICATED FORM PANELS

Contractors usually make prefabricated form panels using 2×4 lumber for frames and $\frac{3}{4}$-in. plywood for sheathing. Because plywood sheathing is used it is a good practice to make panels 2 ft wide and 2, 4, 6, or 8 ft long. An assortment of lengths will permit considerable flexibility in fitting the length to variable-length walls. When the panels are erected, one on top of the other, the 2 ft width will permit the use of form ties at the top and bottom of adjacent panels, thus eliminating the need for holes through the plywood.

Figure 10.3 shows a method of constructing a typical panel, using $\frac{3}{4}$-in.-thick plywood for sheathing and 2×4 lumber for the frame. The total weight of this panel will be about 140 lb, which will require two people to handle it.

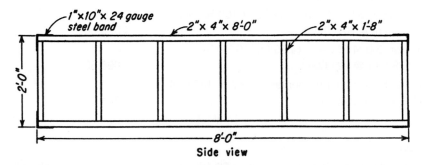

FIGURE 10.3 | Prefabricated panel form.

Figure 10.4 shows a typical set of forms for a foundation wall using prefabricated panels. The bottoms of the forms are held together by steel tie bands, which pass under the concrete wall and are nailed to the bottom members of the panel frames. The tops of the forms may be held together by lumber, as shown in Fig. 10.4, or with steel tie bands. The top and bottom edges of the panels may be grooved slightly to receive the form ties.

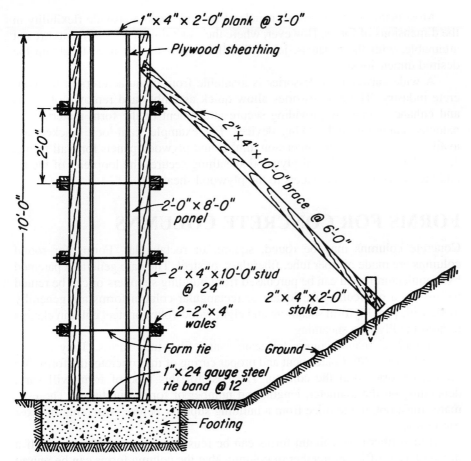

FIGURE 10.4 I Form for concrete wall.

COMMERCIAL PREFABRICATED FORMS

Plyform sheathing attached to steel or aluminum frames in various sizes can be purchased or rented from manufacturers who specialize in formwork materials. Forms consisting of steel sheets attached to steel frames can also be purchased or rented. All these forms can be used with straight or circular walls or for walls that vary in thickness either uniformly or in steps.

Proprietary panels, which are available in several types, are used frequently to construct forms, especially for concrete walls. Among the advantages of using these panels are many reuses, a reduction in the labor required to erect and dismantle the forms, a good fit, and a reduction in the quantity of additional lumber required for studs and wales.

Most manufacturers furnish panels of several sizes to provide flexibility in the dimensions of forms. However, where the exact dimensions of forms are not attainable, with these panels, job-built filler panels can be used to obtain the desired dimensions.

A wide variety of accessories is available from manufacturers in the concrete industry. These accessories allow quick assembly and removal of forms and enhance safety by providing secure attachment of the form materials by patented clamping and locking devices. For example, cam-lock brackets are available that can secure dimension lumber and plywood panels for wall forms. A cam-lock on each side of plywood sheathing secures the loop end of the tie and the horizontal wood walers to the plywood sheathing.

FORMS FOR CONCRETE COLUMNS

Concrete columns may be round, square, or rectangular. Forms for round columns are made of fiber tube, fiberglass, or steel. They are generally patented by manufacturers and can be purchased from building suppliers or can be rented for use on a particular job. Square or rectangular column forms are generally made with $\frac{3}{4}$-in. Plyform sheathing and eight wood yokes or steel column clamps to hold the Plyform sheathing.

Numerous manufacturers fabricate round fiber-tube forms. They can be used only once. After placement and proper curing of the concrete, the forms are pulled, or torn, from the column. The cost for this type of form will vary, depending on the diameter, length, and number of forms purchased. The estimator must obtain the price from a building supplier before estimating the cost for a project.

Round fiberglass column forms can be reused as many as four times. If a project has a sufficient number of columns that the column forms can be reused three or four times, then it may be economical to purchase fiberglass forms; otherwise, they can be rented. The estimator must determine the number of reuses before deciding whether it is more economical to purchase or rent round fiberglass column forms.

Round steel column forms are generally rented by the contractor for use on a particular project. Although they can be purchased and reused many times, there may be a substantial length of time between reuse, creating extra time for transporting, storing, and general handling of the forms.

Forms for square or rectangular concrete columns are generally made with $\frac{3}{4}$-in. Plyform sheathing and vertical wood battens (see Fig. 10.5). The wood battens are generally 2×4 or 2×6 lumber that is placed flat on the outside of the sheathing. The wood battens of column forms perform the same function as studs for wall forms. All steel square or rectangular forms are also available for purchase or rent from manufacturers.

Clamps are installed around the forms prior to filling them with concrete. The clamps are made of lumber or steel. The spacing of column clamps should

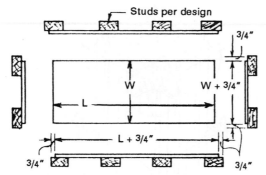

(a) Plan view of rectangular column form

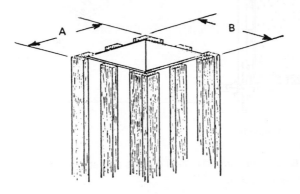

(b) Isometric view of square column form

FIGURE 10.5 | Column sheathing of plywood with vertical wood battens.

be such that the column forms will safely resist the maximum possible internal pressure of the concrete against the column forms.

For small columns, lumber is placed horizontally around all four sides of the column form. The 2 × 4 or 2 × 6 lumber overlaps the corners of the form and is secured by nailing the overlaps together, or by using a patented corner lock.

For larger column forms, steel clamps are commonly used, as illustrated in Figure 10.6. The steel clamps are adjustable for a wide range of sizes and can be reused several hundred times. Steel wedges are used to attach and secure the column clamps. The clamps automatically square the column as the wedge is tightened. Manufacturers provide recommendations for spacing of their column clamps. If it is expected that column clamps will be needed for many uses, it will be more economical to purchase or rent adjustable steel clamps.

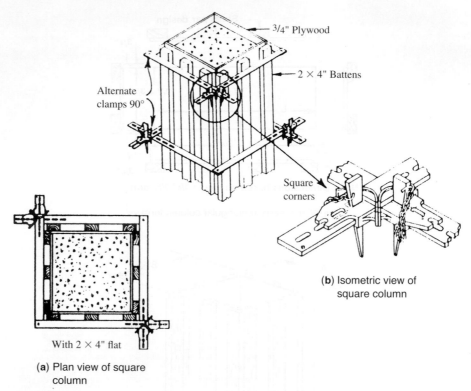

(b) Isometric view of
square column

(a) Plan view of square
column

FIGURE 10.6 | Column form with steel clamps and steel wedges.

Courtesy: Dayton Superior Corporation

MATERIAL REQUIRED FOR CONCRETE COLUMN FORMS

The sheathing or sides of column forms are typically $\frac{3}{4}$-in. Plyform, backed with 2×4 or 2×6 dimension lumber. The quantity of material will depend on the shape, cross section, length, and number of columns. Because Plyform sheets are available only in 4-ft widths, the estimator must determine the arrangement and number of sheets, including waste, necessary to build the forms for a particular job. Also, the number of board feet of lumber will depend on the size and spacing of the vertical wood battens.

The number of column clamps will depend on the height of the column and the pressure applied to the forms by the freshly placed concrete. Figure 10.7 provides the approximate number of column clamps for concrete columns. The quantity of nails for building wood column forms will vary from 7 to 12 lb per 1,000 bf of lumber.

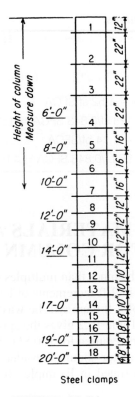

FIGURE 10.7
Approximate spacing of
column clamps.

EXAMPLE 10.5

Estimate the quantity of material for a 22-in. by 22-in. square column form. The column is 11 ft 6 in. high. The form will be constructed with $\frac{3}{4}$-in. Plyform sheathing (4 ft by 8 ft) and three 2 × 4 vertical wood battens on each side of the column form. The Plyform will be placed in the horizontal direction for maximum strength. Lumber will be purchased in 12-ft lengths.

Plyform sheathing:

Height of column: 11 ft 6 in. = 11.5 ft

Number of sheets in vertical direction:

(11.5 ft)/(4 ft/sheet) = 2.9, use 3 sheets

Perimeter of column: 22 in. × 4 sides = 88 in., or 7.3 ft

Number of sheets in horizontal direction:

(7.3 ft)/(8 ft/sheet) = 0.9,   use <u>1 sheet</u>

Total = 4 sheets

Total sf of sheathing: 4 sheets × (32 sf/sheet) = 128 sf

Lumber for vertical wood battens:

Quantity of lumber: 3 pieces × 4 sides × [(2 × 4)/(1 × 12)] × 12 ft = 96 bf

Summary of materials:

Total surface area: [(22 in.)/(12 in./ft)] × 11.5 ft × 4 sides = 84.3 SFCA

Plyform: 128 sf/84.3 SFCA = 1.52 sf/SFCA

Plyform per 100 SFCA: 128 sf/0.843 SFCA = 152 sf/100 SFCA

Lumber: 96 bf/84.3 SFCA = 1.14 bf/SFCA

Lumber per 100 SFCA: 96 bf/0.843 SFCA = 114 bf/100 SFCA

QUANTITIES OF MATERIALS AND LABOR-HOURS FOR COLUMN FORMS

For column heights and widths other than multiples of 2 ft, there will be waste because lumber is purchased in 2-ft increments of length. Likewise, for column widths other than multiples of 4 ft there will be waste because plywood is purchased in a 4-ft by 8-ft size. Table 10.2 gives the approximate quantities of plywood, lumber, and labor-hours required for 100 SFCA of concrete column forms for various heights. The estimator should consider the extent of wastage and perform the calculations illustrated in Example 10.5 to obtain more accurate quantities of materials.

TABLE 10.2 | Quantities of plywood, lumber, and labor-hours required for 100 SFCA of column forms.

Height, ft	Plywood, sf	Lumber, bf	Labor-hours, making forms		Labor-hours, erecting and removing forms	
			Carpenter	Helper	Carpenter	Helper
0–8	100–130	110–150	2.5–3.5	0.5–1.5	5.5–6.5	4.0–5.0
8–16	100–150	150–180	3.0–4.0	1.0–2.0	6.0–7.0	4.5–5.5

Column forms are usually prefabricated at the job prior to erecting in place. The lumber and sheathing are cut to the proper sizes. The four sides are assembled and fastened together with column clamps. After the forms are assembled, they are set into templates that have been accurately and securely located. It may be necessary to support the forms above the floor or column base until the column reinforcement is tied in place. Some contractors prefer to place the column reinforcing steel first and then assemble the forms around it.

If it is not necessary to rebuild the forms for each use, the estimator should separate the labor into two operations, making the forms and then erecting and removing them. For subsequent uses, if the sizes are not altered, only the labor for removing, cleaning, and erecting the forms should be considered.

EXAMPLE 10.6

Use the information in Table 10.2 to estimate the quantities of material and labor-hours to form 40 columns 30 in. by 30 in. by 14 ft 6 in. long. Assume the lumber will be used four times. The 4-ft by 8-ft plywood sheathing will cost $34.00/sheet, lumber will cost $0.55/bf. Steel column clamps will be rented at $4.50 each. The hourly rate for carpenters is $28.00/hr and helpers labor rate is $21.00/hr.

Quantity of materials:

Total surface area: 40 columns × (30 in./12 in./ft) × 4 sides × 14.5 ft
= 5,800 SFCA

Number of columns placed simultaneously: 40 columns/4 uses = 10 columns

Columns to be built: 5,800 SFCA/4 uses = 1,450 SFCA

Columns to be erected and removed = 5,800 SFCA

Since the 30 in. is not evenly divisible by a 4-ft × 8-ft sheet and the 14-ft 6-in. height is not compatible with a 2-ft increment, there will be waste for the plywood and lumber. Average values from Table 10.2 will be used for quantities.

From Table 10.2, the quantity will be:

Plywood: (125 sf/100 SFCA) × 1450 SFCA = 1,813 sf

   Number of sheets: (1,813 sf)/(32 sf/sheet) = 56.7, use 57 sheets

Lumber: (165 bf/100 SFCA) × 1450 SFCA = 2,393 bf

From Fig. 10.7, 10 columns × 10 clamps/column = 100 clamps

Form oil: (5,800 SFCA)/(400 gal/SFCA) = 14.5 gal

Material cost:

Plyform: 57 sheets @ $34.00/sheet = $1,938.00
Lumber: 2,393 bf @ $0.55/bf   =   1,316.15
Column clamps: 100 @ $4.50   =   450.00
Form oil: 14.5 gal @ $14.00/gal   =   203.00
                   Total = $3,907.15

Labor cost:

Making forms:

Carpenter: 1,450 SFCA ×
(3.5 hr/100 SFCA) = 50.8 hr @ $28.00/hr = $1,422.40

Helper: 1,450 SFCA ×
(1.5 hr/100 SFCA) = 21.8 hr @ $21.00/hr =   457.80
            Total cost making forms = $1,880.20

Erecting and removing forms:

Carpenter: 5,800 SFCA ×
(6.5 hr/100 SFCA) = 377.0 hr @ $28.00/hr = $10,556.00

Helper: 5,800 SFCA ×
(5.0 hr/100 SFCA) = 290.0 hr @ $21.00/hr =   6,090.00
        Total cost erecting and removing = $16,646.00

Foreman based on 4 carpenters: (50.8 hr + 377.0 hr)/4 = 106.9 hr @ $32/hr
= $3,420.80

Total labor cost = $1,880.20 + $16,646.00 + $3,420.80 = $21,947.00

Summary of costs:

Material = $3,907.15

Labor   =   21,947.00

   Total = $25,854.15

Cost/SFCA: $25,854.15/5,800 SFCA = $4.46/SFCA

Cost per column: $25,854.15/40 columns = $646.35/column

ECONOMY OF REUSING COLUMN FORMS

Considerable savings can be achieved by reusing concrete forms. For example, if the column forms in Example 10.6 are not used four times it would be necessary to build all of the 40 column forms at one time. Example 10.7 illustrates the economy of reusing forms.

EXAMPLE 10.7

Estimate the cost of the 40 columns in Example 10.6 assuming the forms will not be reused. Assume the column dimensions, material, and labor unit costs remain the same as given in Example 10.6.

Quantity of materials:

Total surface area: 40 columns × (30 in./12 in./ft) × 4 sides × 14.5 ft
= 5,800 SFCA

Number of columns placed simultaneously = 40 columns

Average values from Table 10.2:

Plywood: (125 sf/100 SFCA) × 5,800 SFCA = 7,250 sf

Number of sheets: (7,250 sf)/(32 sf/sheet) = 226.6, use 227 sheets

Lumber: (165 bf/100 SFCA) × 5,800 SFCA = 9,570 bf

From Fig. 10.7, 40 columns × 10 clamps/column = 400 clamps

Form oil: (5,800 SFCA)/(400 gal/SFCA) = 14.5 gal

Material cost:

Plyform: 227 sheets @ $34.00   = $7,718.00

Lumber: 9,570 bf @ $0.55/bf   =   5,263.50

Column clamps: 400 @ $4.50   =   1,800.00

Form oil: 14.5 gal @ $14.00/gal =   203.00

                 Total = $14,984.50

Labor cost:

Making forms:

Carpenter: 5,800 SFCA ×
(3.5 hr/100 SFCA) = 203 hr @ $28.00/hr = $5,684.00

Helper: 5,800 SFCA ×
(1.5 hr/100 SFCA) = 87 hr @ $21.00/hr   =   1,827.00

Total cost making forms = $7,511.00

Erecting and removing forms:

Carpenter: 5,800 SFCA ×
(6.5 hr/100 SFCA) = 377 hr @ $28.00/hr = $10,556.00

Helper: 5,800 SFCA ×
(5.0 hr/100 SFCA) = 290 hr @ $21.00/hr =   6,090.00

Total cost erecting and removing = $16,646.00

Foreman based on 4 carpenters: (203 hr + 377 hr)/4 = 145 hr @ $32.00/hr
= $4,640.00

Total labor cost = $7,511.00 + $16,646.00 + $4,640.00 = $28,797.00

Summary of costs:

Material = $14,984.50
Labor   =   28,797.00
Total = $43,781.50

Cost/SFCA: $43,781.50/5,800 SFCA = $7.55/SFCA
Cost per column: $43,781.50/40 columns = $1,095/column

As shown in Example 10.7 the cost of $43,781.50 is significantly higher than the $25,854.15 in Example 10.6, which shows considerable cost savings can be achieved by reusing the forms. When the forms are not reused, the cost increased from $4.46/SFCA to $7.55/SFCA.

The opportunity to reuse forms is reduced when multiple sizes of columns are required in a project. For example, due to reduced applied loads it is common practice to reduce the column size for the upper stories of multiple story buildings to save concrete cost. However, this prevents reuse of forms. Reducing the sizes of columns to save concrete cost may not be economical because different forms must be made, or the forms from the larger lower story columns must be remade, for use as forms of the upper story columns. Generally, the cost to remake forms is higher than the savings in concrete costs. Also, if the larger size column is used for the higher floors, in certain cases, it may be probable that the quantity of reinforcing steel required in a column will be less than would be required in a smaller column. When the design permits and coordinated with the structural designer, this practice could result in additional savings.

FIGURE 10.8 | Steel column forms with capitals.

Courtesy: EFCO Corp.

COLUMN HEADS, CAPITALS, AND DROP PANELS

Where the top of a column intersects the underside of a concrete floor slab, it is common practice to expand the width at the top of the column to provide more distribution of the load between the column and floor slab. This widening at the top of the column is called a column head.

A capital is a column head for round columns, as shown in Fig. 10.8. It is a cylindrical shaped conical section that tapers from a wide diameter at the bottom of the floor slab downward to the diameter of the concrete column. Column capitals are used when heavy floor loads must be distributed to the column below, to reduce punching shear through the concrete floor.

A drop panel is installed at the top of square or rectangular columns to distribute the transfer of load from the floor slab to the column. Forms for drop panels are built of lumber and plywood, consisting of flat surfaces, whereas column capitals consist of cylindrical conical surfaces. A drop panel can also be installed as a column head for round columns. However, installing rectangular drop panels above round columns requires complex forming to ensure good connections between straight and round surfaces, which greatly increases the cost of the formwork.

SHORES AND SCAFFOLDING

Vertical shores, or posts, and scaffolding are used to support concrete beams, girders, floor slabs, roof slabs, bridge decks, and other members until the concrete has gained sufficient strength to be self-supporting. Many types and sizes of shoring and scaffolding are available. They may be made from wood or steel, or a combination of the two materials. Aluminum shores and scaffolding are also available.

In general, shores are installed as single-member units that may be tied together at one or more intermediate points with horizontal and diagonal braces to give them greater stiffness and to increase their load-supporting capacities. They should be securely fastened at the bottom and top ends to prevent movement or displacement while they are in use.

Numerous patented shoring systems are available from companies that specialize in concrete accessories. Figure 10.9 is a shoring system that consists of either 4 × 4 or 4 × 6 wood posts, fastened by two special patented clamps. The bottom of one post rests on the supporting floor, while the second post is moved upward along the side of the lower one. The top post is raised to the desired

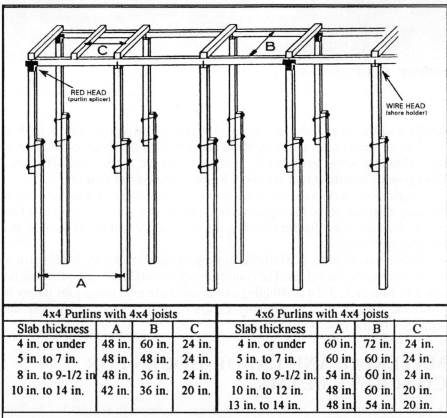

4x4 Purlins with 4x4 joists				4x6 Purlins with 4x4 joists			
Slab thickness	A	B	C	Slab thickness	A	B	C
4 in. or under	48 in.	60 in.	24 in.	4 in. or under	60 in.	72 in.	24 in.
5 in. to 7 in.	48 in.	48 in.	24 in.	5 in. to 7 in.	60 in.	60 in.	24 in.
8 in. to 9-1/2 in	48 in.	36 in.	24 in.	8 in. to 9-1/2 in.	54 in.	60 in.	24 in.
10 in. to 14 in.	42 in.	36 in.	20 in.	10 in. to 12 in.	48 in.	60 in.	20 in.
				13 in. to 14 in.	48 in.	54 in.	20 in.

Note: Douglas fir, construction grade joists and purlins, 3/4" plywood held strong way over joists with face grain parallel to span, maximum deflection *l*/360. Lace and brace as required. Shore height 12 ft and under.

FIGURE 10.9 | Ellis shoring system.

Source: Ellis Construction Specialties, Ltd.

FIGURE 10.10 | Formwork supported by tubular steel frames.

Courtesy: Dayton Superior Corporation.

height, and the two clamps automatically grip the two posts and hold them in position. Accessories are available to securely hold the shore to the horizontal purlins and screw jacks can be installed under the lower shore. The screw jacks provide adjustments of the shore height to the correct position.

Single-post steel shores are manufactured and available for purchase or rent by many manufacturers. Each post shore typically consists of two parts: a base post with a threaded collar for adjusting the shore height and a staff member that fits into the base post.

Tubular scaffolding is available to support high and heavy loads, such as elevated concrete floor slabs. The contractor often rents scaffolding for a particular job. However, if the scaffolding can be used often on many jobs it may be justifiable to purchase the scaffolding. Figure 10.10 shows forms for a floor slab supported by the tubular steel frames.

MATERIAL AND LABOR-HOURS FOR CONCRETE BEAMS

A typical example of forms for concrete beams is shown in Fig. 10.11. As illustrated in Fig. 10.11, several methods are used for constructing forms for concrete beams. In Fig. 10.11a, the beam bottom, or soffit, may be made from one or more pieces of 2-in.-thick lumber having the required width. The lumber should be cleated together on the underside at intervals of about 3 ft. The side forms are made of plywood panels or lumber, held together with 2 × 4 cleats. The ledgers

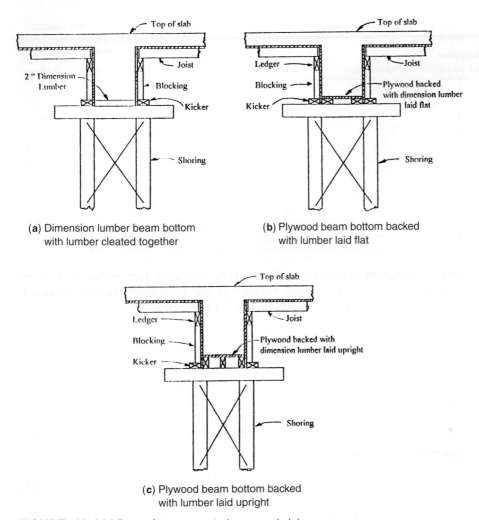

(a) Dimension lumber beam bottom
with lumber cleated together

(b) Plywood beam bottom backed
with lumber laid flat

(c) Plywood beam bottom backed
with lumber laid upright

FIGURE 10.11 | Forms for a concrete beam and slab.

attached to the vertical blocking carry the joist load to the shoring. The kickers at the bottom of the side form resist the lateral pressure from the concrete.

The beam bottom can also be plywood, backed by dimension lumber laid in the flat position, as illustrated in Fig. 10.11b. Plywood panels are placed on the top of the runners to provide a continuous smooth surface for the beam bottom. As illustrated in Fig. 10.11c, the beam bottom can also be plywood, backed with dimension lumber placed in the vertical direction to provide more resistance to bending and deflection. The side forms are made of plywood panels, held together by form ties. Side forms can also be made with lumber, held together with 2 × 4 cleats. The ledgers attached to the vertical 2 × 4 cleats assist in supporting the ends of the joists.

The quantities of material and labor-hours to build 100 SFCA of forms for concrete beams of various sizes are shown in Table 10.3. For concrete beam

TABLE 10.3 | Approximate quantities of material and labor-hours required for 100 SFCA of concrete beams.

Beam width, in.	Number of uses	Plywood, sf	Lumber, bf	Form ties	Labor-hours	
					Carpenter	Helper
12	1	100–110	260–290	12	13.5–14.5	3.0–4.0
	2	45–55	135–150	12	11.5–12.5	2.5–3.5
	3	35–45	105–115	12	9.5–10.5	2.0–3.0
	4	25–35	90–100	12	8.5–9.5	1.5–2.5
18	1	100–110	210–235	15	12.5–13.5	2.5–3.5
	2	50–60	120–135	15	10.5–11.5	2.0–3.0
	3	35–45	100–110	15	9.5–10.5	1.5–2.5
	4	25–35	75–85	15	9.0–10.0	1.5–2.5
24	1	100–110	190–210	20	11.5–12.5	2.5–3.5
	2	55–65	100–110	20	10.5–11.5	2.5–3.0
	3	35–45	75–85	20	9.5–10.5	2.0–2.5
	4	25–35	65–70	20	9.0–10.0	2.0–2.5

forms the amount of time for carpenters is significantly greater than for helpers, compared to walls and column because of the complexity of the formwork. Table 10.3 provides data on multiple uses of the forms.

EXAMPLE 10.8

Estimate the cost of forms for 15 concrete beams for a building. The beams are 12 in. wide, 19 in. deep, and 20 ft 6 in. long. The forms will be used three times. Use average values from Table 10.3 to determine the quantities of materials.

A material supplier quotes these prices: $0.55/bf for lumber, $2.31/lb for nails, and $14.00/gal for form oil. Carpenters cost $28.00/hr and laborers cost $21.00/hr.

Quantity of material:

$$\text{Surface area of forms: } 15 \text{ beams} \times \frac{12 \text{ in.} + 19 \text{ in.} + 19 \text{ in.}}{12 \text{ in./ft}} \times 20.5 \text{ ft} = 1{,}281 \text{ SFCA}$$

From Table 10.3 for 3 uses,

Quantity of lumber = 1,281 SFCA × (110 bf/100 SFCA) = 1,409 bf

Quantity of plywood: 1,281 SFCA × (40 sf/100 SFCA) = 512 sf

Number of sheets required: (512 sf)/(32 sf/sheet) = 16 sheets

Nails: (1,409 bf) × (12 lb/1,000 bf) = 16.9 lb

Form oil: (1,281 SFCA)/(400 SFCA/gal) = 3.2 gal

Form ties: 1,281 SFCA × 12 ties/100 SFCA = 154 ties

Labor-hours required:

Reference Table 10.3 for 3 uses,

Carpenter: 1,281 SFCA × (10 hr/100 SFCA) = 128.1 hr

Helper: 1,281 SFCA × (2.5 hr/100 SFCA) = 32.0 hr

Foreman based on 4 carpenters: 128.1 hr/4 = 32.0 hr

Cost of labor and materials:

Labor costs:

Carpenter: 128.1 hr @ $28.00/hr = $3,586.80

Helper: 32.0 hr @ $21.00/hr   =   672.00

Foreman: 32.0 hr @ $32.00/hr   =   1,024.00

Labor cost = $5,282.80

Material costs:

Lumber: 1,409 bf @ $0.55/bf   = $774.95

Plywood: 16 sheets @ $34.00/sheet =   544.00

Form ties: 154 ties @ $1.47/tie   =   226.38

Nails: 16.9 lb @ $2.31/lb   =   39.04

Form oil: 3.2 gal @ $14.00/gal   =   44.80

Material cost = $1,629.17

Summary of costs:

Total labor and material: $5,282.80 + $1,629.17 = $6,911.97

Cost/SFCA: $6,911.97/1,281 SFCA = $5.40/SFCA

FORMS FOR FLAT-SLAB CONCRETE FLOORS

The flat-slab-type concrete floor is sometimes used when the floor can be supported by the column heads, without beams or girders. Since there are no beam forms to assist in supporting the slab forms, it is necessary to classify all forms as slab forms. The cost of forms will vary with the thickness of the slab and the height of the slab above the lower floor. Figure 10.12 illustrates a method used in building forms for flat slabs.

The form builder has considerable freedom in selecting the size and spacing of joists and stringers. Joist sizes are usually 4×4 lumber, spaced at about 20 in. on centers. Stringers are usually 4×6 lumber spaced about 4 ft apart. Southern Pine and Douglas Fir, number 1 or 2 grades of lumber are common.

The decking is usually $\frac{3}{4}$-in.-, $\frac{7}{8}$-in.-, or 1-in.-thick plywood decking. If a smooth finish is required, high-density overlay (HDO) plywood is available. It also has the advantage of multiple reuses, up to 50 uses under favorable conditions.

Scaffolding or multiple single shores are commonly used to support the forming system, as shown in Figure 10.9. Commercial steel tower frames are available to support high and heavy loads. Tower frames are erected in sections, each consisting of four legs with heights of 8 or 10 ft.

Adjustable horizontal steel shores, consisting of two members per unit, are sometimes used to support the forms for concrete slabs, beams, and bridge decking. The members are designed to permit one member to telescope inside the

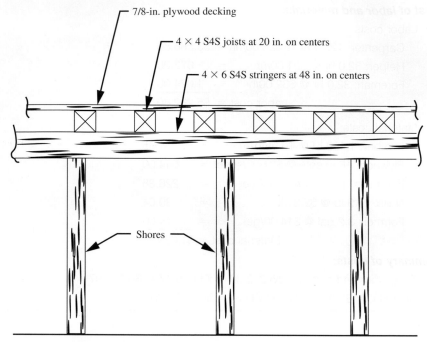

7/8-in. plywood decking

4 × 4 S4S joists at 20 in. on centers

4 × 6 S4S stringers at 48 in. on centers

Shores

FIGURE 10.12 | Typical slab forming system.

other, thereby providing adjustments in the lengths within the overall limits for a given shore. Horizontal shores are available for spans varying from 5 to 20 ft.

The quantity of material depends on the thickness of the slab and the height of the slab above the supporting floor. Table 10.4 gives the quantity of materials and number of labor-hours per 100 SFCA to build and remove forms for flat-slab concrete floors.

TABLE 10.4 | Approximate quantities of material and labor-hours required to build, erect, and remove 100 SFCA of flat-slab concrete floors.

Ceiling height, ft	Thickness of slab, in.	Plywood, sf	Lumber, bf	Labor-hours	
				Carpenter	Helper
10	6	100–105	150–155	6.0–6.5	2.0–2.5
	8	100–105	165–170	6.0–6.5	2.0–2.5
	10	100–105	180–190	6.0–6.5	2.0–2.5
12	6	100–105	150–155	6.5–7.0	2.5–3.0
	8	100–105	165–170	6.5–7.0	2.5–3.0
	10	100–105	180–190	6.5–7.0	2.5–3.0
14	6	100–105	150–155	7.0–7.5	3.0–3.5
	8	100–105	165–170	7.0–7.5	3.0–3.5
	10	100–105	180–190	7.0–7.5	3.0–3.5

EXAMPLE 10.9

Estimate the direct cost of furnishing, building, and removing forms for a flat-slab concrete floor whose size is 56 ft 8 in. by 88 ft 6 in. The ceiling height will be 11 ft 8 in. and the slab will be 6 in. thick.

Lumber will be 4 × 4 joists and 4 × 6 stringers, no. 2 Southern Pine, and the cost will be $0.78/bf. The $\frac{7}{8}$-in.-thick plywood will cost $37.00 for a 4 ft by 8 ft sheet. Nails cost $2.31/lb and form oil costs $14.00/gal. Assume carpenters cost $28.00/hr and laborers cost $21.00/hr.

Scaffolding, similar to that shown in Fig. 10.9, will be rented from a local building supplier at a rate of $0.45/sf of floor area. It will require 1 hr to set up and take down 100 sf of scaffolding.

Quantity of material:

Total surface area: 56 ft 8 in. × 88 ft 6 in. = 56.67 ft × 88.5 ft = 5,015 SFCA

From Table 10.4:

Lumber: (5,015 SFCA) × (152 bf/100 SFCA) = 7,623 bf
Plywood: (5,015 SFCA) × (102 sf/100 SFCA) = 5,115 sf
Sheets required: 5,115 sf/(32 sf/sheet) =   159.8, need 160 sheets

Nails: (7,623 bf) × (5 lb/1000 bf) = 38.1 lb
Form oil: (5,015 sf)/(400 sf/gal) = 12.5 gal

Material costs:

Lumber: 7,623 bf @ $0.78/bf	=	$5,945.94
Plywood: 157 sheets @ $37.00/sheet	=	5,809.00
Nails: 38.1 lb @ $2.31/lb	=	88.01
Form oil: 12.5 gal @ $14.00/gal	=	175.00
Material cost =		$12,017.95

Labor costs:

Carpenter: 5,015 SFCA ×
(6.7 hr/100 SFCA) = 336.0 hr @ $28.00/hr = $9,408.00

Laborer: 5,015 SFCA ×
(2.7 hr/100 SFCA) = 135.4 hr @ $21.00/hr =   2,843.40

Foreman based on 4 carpenters:
336 hr/4 = 84 hr @ $32.00/hr   =   2,688.00

Total labor = $14,939.40

Summary of form costs:

Total cost of material and labor: $12,017.95 + $14,939.40 = $26,957.35
Cost per SFCA: $26,957.35/5,015 SFCA = $5.38/SFCA

Cost of scaffolding:

Rental cost of scaffolding: 5,015 sf @ $0.45/sf = $2,256.75

Labor to set up and take down:
(5,015 sf) × (1 hr/100 sf) = 50.2 hr @ $21.00 =   1,054.20

Scaffolding cost = $3,310.95

Total cost:

Form costs + scaffolding cost = $26,957.35 + $3,310.95 = $30,268.30

Total cost per SFCA: $30,268.30/5,015 SFCA = $6.04/SFCA

PATENTED FORMS FOR FLOOR SLABS

Metal or fiberglass patented forms are available from commercial suppliers for concrete floor systems. Metal pans are placed end-to-end to provide one-way slab action, as illustrated in Fig. 10.13. Three-foot-long pans are available in depths of 6 in. to 24 in. and widths of 20 in. to 40 in. Metal pans are also available that are designed to be self-supporting for spans up to 12 ft.

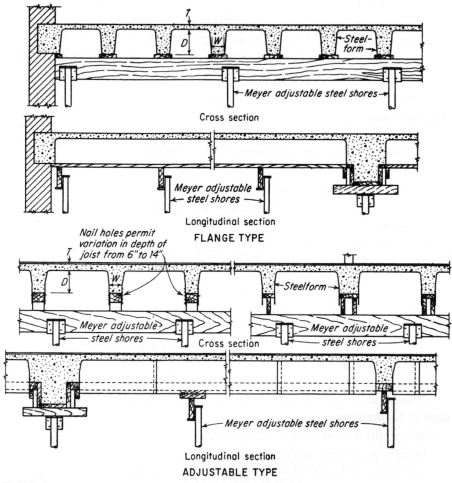

FIGURE 10.13 | Supporting forms for metal pans.

Fiberglass domes are available for forming a two-way floor system, sometimes called a waffle slab. Domes are available in depths of 12 in. to 24 in. and widths of 4 ft. Manufacturers of metal pan forms or domes furnish dimensions of their forms and other pertinent information.

When using commercial pans or domes, the total quantity of concrete is less than flat-slab floors, which reduces the cost of materials and also permits the use of smaller beams, columns, and footings. If the pans are used enough times, the cost per square foot of use, including the labor cost of installing and removing, can compare favorably with forms built entirely of lumber.

The structural designer specifies the particular system that is to be used. The structure is designed for a specific size and type of pan. Since the pans are designed for a specific job, contractors typically rent the pans at an agreed price per square foot per use, with a minimum number of uses.

MATERIAL AND LABOR-HOURS REQUIRED FOR METAL-PAN CONCRETE FLOORS

The lumber required for metal-pan construction consists of centering strips, stringers, shores, and braces. The metal pans are placed on centering strips that are spaced at a distance equal to the width of the pan plus the thickness of the joist between the pans. Stringers are typically 4 × 6 lumber, spaced about 4 ft on centers. Vertical shores or scaffolding are used to support the stringers. Table 10.5 provides the approximate quantities of material and labor-hours for metal-pan construction.

TABLE 10.5 | Approximate quantities of material and labor-hours required to build, erect, and remove 100 SFCA of metal-pan concrete floors.

| Ceiling height, ft | Floor thickness, in. | Lumber, bf | Labor-hours | |
			Carpenter	Helper
12	10	120	3.5–4.0	4.0–4.5
	12	130	4.0–4.5	4.5–5.0
	14	140	4.0–4.5	4.5–5.0
14	10	130	3.5–4.0	4.0–4.5
	12	140	4.0–4.5	4.5–5.0
	14	150	4.1–4.5	4.5–5.0
16	10	140	3.5–4.0	4.0–4.5
	12	150	4.0–4.5	4.5–5.0
	10	160	4.0–4.5	4.5–5.0

Note: The thickness of the floor equals the depth of the pan plus the slab thickness.

EXAMPLE 10.10

Estimate the cost of form lumber and metal pans in place and removed from a total floor area 39 ft 6 in. wide by 96 ft 6 in. long. The floor thickness will be 12 in. The joints will run one way only. The ceiling height will be 14 ft.

Rental cost of the metal pans is $4.75/sf and scaffolding will be rented at $0.45/sf. Assume 4 × 4 centering lumber and 4 × 6 stringers costs $0.78/bf. Wage rate for carpenters is $28.00/hr and $21.00/hr for laborers.

Quantity of materials:

Total floor area: 39.5 ft × 96.5 ft = 3,812 sf

From Table 10.5, lumber = 3,812 sf × (140 bf/100 sf) = 5,337 bf

Material costs:

Lumber: 5,337 bf @ $0.78/bf = $4,162.86

Nails: 26.7 lb @ $2.31/lb = 61.68

Form oil: 9.5 gal @ $14.00/gal = 133.00

Material cost = $4,357.54

Labor costs:

Carpenter: 3,812 SFCA × (4.2 hr/100 SFCA) = 160.1 hr @ $28.00/hr = $4,482.80

Laborer: 3,812 SFCA × (4.7 hr/100 SFCA) = 179.2 hr @ $21.00/hr = 3,763.20

Foreman based on 4 carpenters: 160.1 hr/4 = 40 hr @ $32.00/hr = 1,280.00

Total labor = $9,526.00

Rental costs:

Rental cost of metal-pan forms: 3,812 sf @ $4.75/sf = $18,107.00

Rental cost of scaffolding: 3,812 sf @ $0.45/sf = 1,715.40

Labor to set up and take down scaffolding = 850.00

Total cost = $20,672.40

Summary of costs:

Material costs = $4,357.54

Labor costs = 9,526.00

Rental costs = 20,672.40

Total costs = $34,555.94

Total cost per SFCA: $34,555.94/3,812 SFCA = $9.07/SFCA

CORRUGATED-STEEL FORMS

Corrugated-steel sheets are sometimes used as forms for concrete floor and roof slabs supported by steel joists, steel beams, and precast concrete joists. The sheets are available in several gauges, widths, and lengths to fit different requirements.

The sheets are laid directly on the supporting joists or beams with the corrugations perpendicular to the supports. Side laps should be one-half of a corrugation and end laps should be at least 3 in., made over supporting joists or beams. The sheets are fastened to the supports with special clips. The edges of adjacent sheets are fastened together with steel clips.

CELLULAR-STEEL FLOOR SYSTEMS

Cellular-steel panels are sometimes used for the forms and structural units of floor systems. This system is lightweight and acts as forms for concrete as well as providing raceways for utility services. Thus, the forming system combines the structural slab and electrical functions within the normal depth of the floor slab.

Each cellular unit has multiple cells that can be used for telephone, power, and computer cables. Cellular units are available in 24 in. and 36 in. widths. The panels are manufactured from roll-steel shapes and plates that are electrically welded together in various types. Openings in the cells provide routing for wire cable through the floor slab into the room area.

The panels are prefabricated by the manufacturer to fit any given floor condition and shipped to the project ready for installation without additional fabrication. If storage is necessary, the cellular units may be stacked on wood blocking clear off the ground and tilted longitudinally to ensure against entrapment of water. All deck bundles have labels or handwritten information clearly visible to identify the material and erection location.

After alignment, the cellular units are attached to the decking before the placement of concrete for the floor slab.

CONCRETE STAIRS

The construction of concrete stairways is complicated and presents a problem for the estimator. The costs per unit of volume or area varies greatly, depending on the length of the tread, the width of the tread, the height of the riser, the shape of the supporting floor for the shores, whether the treads are square or rounded, and whether the ends of the treads and risers are open or closed with curbs. Some stairs are straight-run, while others have an intermediate landing. Some stairs are completely in the open, while others have a wall on one or both sides. All of these conditions affect the cost of stairs.

The width of treads varies from 8 to 12 in., and the height of the risers varies from 6 to 8 in. for different stairs. Steel dowels should be set in the concrete floor and the beam supporting the landing prior to placing the concrete in the floor and landing. These dowels are tied to the reinforcing steel for the stairs as it is placed.

Lumber Required for Forms for Concrete Stairs

Figure 10.14 shows the details of one method of construction forms for concrete stairs. The height from floor to floor is 10 ft 6 in. With a riser height of 7 in., 18 risers and 17 treads will be required. The treads will be 12 in. wide, which gives a total horizontal length of 17 ft. The supporting floor is horizontal, which requires shores of variable lengths to support the forms.

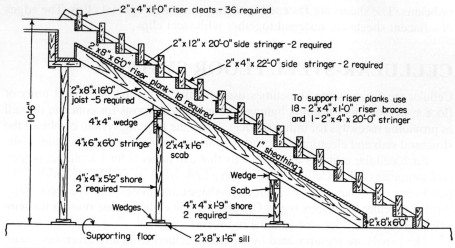

FIGURE 10.14 | Wood forms for concrete stairs.

EXAMPLE 10.11

Estimate the quantity of materials and cost for building the forms for the stairs illustrated in Fig. 10.14, assuming the stairs are 6 ft wide. Plywood will be used for sheathing under the stairs.

Quantity of material:

Diagonal side stringers:

2 pieces: 2 × 12 lumber × 20 ft long	=	80 bf
2 pieces: 2 × 4 lumber × 22 ft long	=	29 bf
Joists, 5 pieces: 2 × 8 lumber × 16 ft long	=	107 bf
Risers, 18 pieces: 2 × 8 lumber × 6 ft long	=	144 bf
Riser cleats, 36 pieces: 2 × 4 lumber × 1 ft long	=	24 bf
Riser stringer, 1 piece: 2 × 4 lumber × 20 ft long	=	13 bf
Riser braces, 18 pieces: 2 × 4 lumber × 1 ft long	=	12 bf
Horizontal stringers: 2 pieces: 4 × 6 lumber × 6 ft long	=	24 bf
Shores:		
2 pieces: 4 × 4 lumber × 6 ft long	=	16 bf
2 pieces: 4 × 4 lumber × 2 ft long	=	6 bf
Sills, 4 pieces: 2 × 8 lumber × 1 ft 6 in. long	=	8 bf
Wedges: 5 bf		5 bf
Headers, 1 piece: 1 × 8 lumber × 6 ft long	=	4 bf
Braces, 3 pieces: 1 × 6 lumber × 12 ft long	=	18 bf
	Total lumber =	490 bf

Plywood sheathing: 20 ft × 6 ft = 120 sf

   Sheets of plywood: 120 sf/(32 sf/sheet) = 3.75, use 4 sheets

Nails: 490 bf × (15 lb/1000 bf) = 7.4 lb

Costs of Materials:

Plywood: 4 sheets @ $34.00/sheet = $136.00

Lumber: 490 bf @ $0.55/bf   = 269.50

Nails: 7.4 lb @ $2.31/lb   =   17.09

                Total = $422.59

The cost just given is for a single use. If the forms can be removed and reused on stairs of the same design, the cost per use will be reduced by the number of times the forms will be used. Frequently, forms for stairs are supported by shores that rest on existing lower stairs. If this construction is used, the lengths of all shores will be increased to 9 ft for Fig. 10.14.

Labor Required to Build Forms for Concrete Stairs

Because of the irregularities of the forms for concrete stairs, it is difficult to estimate accurately the labor-hours of building forms. It is necessary to cut the shores to length, attach the 4 × 6 stringers, set the shores and stringers in place, and brace them securely. The 4 × 4 wedges are nailed to the tops of the stringers and the joists before the plywood sheathing is installed. The sheathing is cut and nailed to the joists.

The diagonal side stringers, 2 × 12 by 20 ft and 2 × 4 by 22 ft, are cleated together, and the positions for the 2 × 8 riser lumber are marked on the stringers. The stringers and the riser lumber are installed. Because of the length of the riser lumber, 6 ft, a center support should be used to hold them against deflection. A 2 × 4 riser stringer is placed at the center of the riser planks, and each riser plank is secured to it with a riser brace, 2 × 4 by 1 ft long.

Two carpenters should be used to build the forms for the stairs. Each carpenter should fabricate and install about 25 bf/hr of lumber.

REINFORCING STEEL

TYPES AND SOURCES OF REINFORCING STEEL

Reinforcing for concrete may consist of steel deformed bars or welded-wire fabric, used separately or together. The cost of reinforcing steel bars is estimated by the pound, hundredweight, or ton, while the cost of welded-wire fabric is estimated by the pound or square foot.

Usually reinforcing steel is cut to the required lengths and bent into the required shapes by steel suppliers prior to delivery to a job. Such shops are equipped with machines that will perform the fabricating operations more economically than when fabricating is performed on the job. Upon request, these shops will furnish quotations covering the supplying and fabricating of all reinforcing for a given project. Estimators frequently request such quotations before preparing estimates.

PROPERTIES OF REINFORCING BARS

Table 10.6 gives the sizes, areas, and weights of reinforcing bars. Reinforcing steel is available in strengths of 40 ksi, 50 ksi, and 75 ksi (ksi = kips per square inch, where kips = 1,000 lb). The deformed bars have ribbed projections rolled onto their surfaces to provide better bonding. The patterns differ with different manufacturers. The size and grades of reinforcing bars can be easily identified by bar identification marks that are placed on each bar. Three marks are on each bar. The top mark is a letter that represents the producing company, the middle mark is a number that gives the size of the bar in $\frac{1}{8}$ in., and the lower mark is a letter that identifies the type of steel.

TABLE 10.6 | Sizes, areas, and weights of reinforcing steel.

Bar no.	Diameter, in.	Area, sq. in.	Weight, lb/ft
3	$\frac{3}{8}$	0.11	0.376
4	$\frac{1}{2}$	0.20	0.668
5	$\frac{5}{8}$	0.31	1.043
6	$\frac{3}{4}$	0.44	1.502
7	$\frac{7}{8}$	0.60	2.044
8	1.0	0.79	2.670
9	1.128	1.00	3.400
10	1.270	1.27	4.303
11	1.410	1.56	5.313
14	1.693	2.25	7.650
18	2.257	4.00	13.600

When reinforced concrete is exposed to de-icing salts or seawater, epoxy-coated reinforcing bars are used. Such bars are more expensive than regular reinforcing steel. Also, the epoxy-coated bars must be carefully handled to prevent breaking off any of the coating during placement on the job.

ESTIMATING THE QUANTITY OF REINFORCING STEEL

When the reinforcing steel consists of bars of different sizes and lengths, each size and length should be listed separately. Such a list is sometimes called a *rebar schedule*. A form such as the one shown in Table 10.7 will simplify the listing and reduce the potential of errors. Each size and length should be

TABLE 10.7 | Rebar schedule for quantity of reinforcing steel.

Bar mark	No. required	Bar size	Length	Weight, lb/ft	Total weight, lb
A	120	4	30 ft 0 in.	0.068	2,405
B	56	4	20 ft 0 in.	0.668	749
C	116	5	24 ft 0 in.	1.043	2,900
D	42	6	12 ft 4 in.	1.502	780
E	36	6	14 ft 8 in.	1.502	794
F	28	6	19 ft 8 in.	1.502	826
G	16	7	22 ft 3 in.	2.044	604
H	72	7	22 ft 3 in.	2.044	3,280
I	84	7	18 ft 8 in.	2.044	3,200
J	24	8	24 ft 0 in.	2.670	1,535
K	18	8	21 ft 6 in.	2.670	1,037

assigned a number or a letter of the alphabet. Such designations are called *piece marks* or *bar marks*.

For a particular project there may be many rebar schedules. For example, there may be a separate rebar schedule for drilled shaft foundations, grade beams, columns, beams, and floor slabs.

COST OF REINFORCING STEEL

To determine the cost of reinforcing steel it is necessary to first determine the weight of reinforcing steel based on the lengths and sizes of bars and the nominal weights given in Table 10.6. Reinforcing bars are usually available in stock lengths of 40 and 60 ft. If an estimator wishes to determine the approximate cost of reinforcing steel for a project, she or he should list each size bar separately and then determine the total weight by size, as illustrated in Table 10.7. The items that determine the cost of reinforcing steel delivered to a project include

1. Base cost of the bars at the fabricating shop
2. Cost of preparing shop drawings
3. Cost of handling, cutting, bending, etc.
4. Cost of shop overhead and profit
5. Cost of transporting from the shop to the job
6. Cost of specialties, such as spacers, saddles, chairs, ties, etc.

The base cost is the cost of the reinforcing steel in standard lengths, usually stock lengths of 40 and 60 ft, without cutting to length or bending. In addition to the base price cost of the reinforcing steel, an extra charge based on the sizes of the bars will be made. These extras are subject to change and should be verified before an estimate is prepared because the price of steel can vary substantially depending on current economic conditions. The base cost of reinforcing steel varies with the total quantity of steel purchased. As the quantity of steel increases, the cost per pound decreases.

Before fabricating reinforcing steel it is necessary for the fabricating shop to prepare drawings that show how the bars are to be fabricated. Such drawings are called *shop drawings.* The shop drawings show the cut-lengths, bending details, and listing of each piece of reinforcing steel. A charge is made for this service, based on the complexity of the drawings and the quantity of reinforcing steel. After the shop drawings are prepared, they are submitted for approval before starting the fabricating operation.

The cost for fabricating reinforcing steel varies with the sizes of the bars and the complexity of the operations. Each bar must be cut and bent into the required shape. Steel suppliers have considerable experience with their fabricating shops and have the ability to accurately determine the cost for preparing the steel for a particular job. Thus, the estimator must rely on the steel fabricator to obtain the cost of fabricated reinforcing steel.

Typically, the estimator supplies the structural drawings to the steel supplier from which the reinforcing steel will be purchased for a particular job. Then, the supplier will develop shop drawings and submit them to the estimator for approval. Upon approval, the supplier will provide a quote of the total cost for the entire lot, fabricated, and delivered to the job.

LABOR PLACING REINFORCING STEEL BARS

The rates at which workers will place reinforcing steel bars will vary with these factors:

1. Sizes and lengths of bars
2. Shapes of the bars
3. Complexity of the structure
4. Distance and height the steel must be carried
5. Allowable tolerance in spacing bars
6. Extent of tying required
7. Skill of the workers

Less time is required to place a ton of steel when the bars consist of large sizes and long lengths than when they are small sizes and short lengths. For example, a worker can set and tie a no. 5 bar about as fast as a no. 4 bar, but the weight of the no. 5 bar will be greater than the weight of the no. 4 bar. Likewise, a worker can place a 10-ft-long bar in about the same time as an 8-ft-long bar.

Straight bars can be placed more rapidly than bars with bends and end hooks. For example, it is much easier to handle and align the straight compression and tension steel than the stirrups in a concrete beam. If many straight bars are required, it may be possible to tie sections of the steel in place on the ground and then lift the rebar assembly in one section and place it in the forms with a small crane.

If the bars must be placed in complicated structures such as stairs, the rate of placing will be less than for simple structures such as walls, floors, etc.

Placement of reinforcing steel where round members intersect with flat members, corners, and other complicated connections greatly increases the time to set and tie the steel in place.

Reinforcing steel should be stored as near the structure as possible, preferably not more than 50 to 100 ft away, to reduce the time required to carry them. If steel must be carried to upper floors or parts of a structure, additional time will be required. For some projects the lay-down area for materials is limited, which requires storage of bulk material in a remote location on the job. When bulk shipments of material, such as reinforcing steel, are delivered to the job a special effort should be made to store the material so it can be easily identified and moved to the work area without having to restack the material.

Rigid tolerances on the spacing of reinforcing bars will reduce the rate of placing steel. The estimator should review the contract documents and appropriate standards that apply to a particular job to determine the dimensional tolerances that must be used for setting the steel.

More time is required to tie bars at every intersection than when little or no tying is required. The estimator must review the contract documents to determine the requirements that apply.

Laborers or skilled steel setters may place reinforcing steel. The cost per hour of skilled steel setters will be higher than that of laborers, but they should be able to better read the drawings, place the steel at a faster rate, and require little or no supervision. The labor-hours to place reinforcing steel will vary depending on these factors.

Table 10.8 gives representative rates of placing reinforcing steel bars. The rates are based on carrying the steel by hand not more than 100 ft from the storage area to the structure. To estimate the cost of placing reinforcing steel, the

TABLE 10.8 | Rates of placing reinforcing steel bars, hr/T.

Type of work	Bar size	
	$\frac{5}{8}$ and less	$\frac{3}{4}$ and over
Beams and girders	19–21	11–12
Columns	20–22	13–15
Elevated slabs	10–12	10–12
Footings	14–16	8–9
Slab on grade	13–15	13–15
Spirals and stirrups	12–14	12–14
Walls	10–11	8–9

Note 1: These rates are based on skilled steel setters. If common laborers are used, the rates should be increased by 25 to 30 percent.

Note 2: These rates are based on handling the reinforcing steel less than 100 ft from the area where it is to be installed.

Note 3: Based on the job conditions, it may be necessary to add 0.5 hr/T to unload and sort the reinforcing steel.

Note 4: If a crane is used to handle the reinforcing steel, the crane should be charged to the job at a rate of 0.5 to 1.5 hr/T.

estimator must determine the quantity of steel (in pounds or tons) for each type of work and then use the data in Table 10.8 to determine the labor-hours. The total labor cost can then be determined by multiplying the labor-hours by the labor rate.

WELDED-WIRE FABRIC

For certain types of concrete projects, such as sidewalks, pavements, floors, canal linings, etc., it may be more economical to use welded-wire fabric for reinforcing instead of steel bars. This fabric is made from cold-drawn steel wire, electrically welded at the intersections of longitudinal and transverse wires, to form rectangular or square patterns. It is available in flat sheets or rolls, the latter frequently being 60 in. wide and 150 ft long. It is usually priced by the square foot or roll, the price depending on the weight.

The quantity required will equal the total area to be reinforced, with 5 to 10 percent of the area added for side and end laps. The fabric to be used is designated by specifying the style, such as 412-610. This style designates a rectangular fabric with longitudinal wires spaced 4 in. apart and transverse wires spaced 12 in. apart, using no. 6 gauge longitudinal wires and no. 10 gauge transverse wires. The gauge number represents the diameter of the wire.

Table 10.9 gives the properties of representative styles of welded-wire fabric. Many other styles are available. Based on a width of 60 in. and a length of 150 ft, the weight of a roll of fabric will vary from 75 to 1,000 lb or more.

Labor Placing Welded-Wire Fabric

Fabric is placed by unrolling it over the area to be reinforced, cutting it to the required lengths, lapping the edges and ends, and tying it at frequent spacings. On large regular areas, a laborer should place it at a rate of 0.25 hr/100 sf, while for irregular areas that require cutting and fitting the rate of placing may be 0.5 hr/100 sf.

CONCRETE

COST OF CONCRETE

The cost of concrete includes the cost of sand, aggregate, cement, water, admixtures, mixing, transporting, and placing the concrete. Concrete mixed at the jobsite is rarely used, except in small quantities that do not require rigid quality. Ready-mix concrete is used for most concrete structures, rather than mixing the concrete at the jobsite. The cost of concrete delivered to the job will vary with the size of the job, the location, and the quality of concrete. The cost to place the concrete after it arrives at the job depends on the extent to which equipment is used.

TABLE 10.9 | Properties of representative styles of welded-wire fabric.

Style	Weight, lb per 100 sf	Spacing of wire, in.		Gauge number		Sectional area of wires, in.²/ft	
		Longi-tudinal	Trans-verse	Longi-tudinal	Trans-verse	Longi-tudinal	Trans-verse
44-1010	31	4	4	10	10	0.043	0.043
44-88	44	4	4	8	8	0.062	0.062
44-66	62	4	4	6	6	0.087	0.087
44-44	85	4	4	4	4	0.120	0.120
66-1010	21	6	6	10	10	0.029	0.029
66-88	30	6	6	8	8	0.041	0.041
66-66	42	6	6	6	6	0.058	0.058
66-44	58	6	6	4	4	0.080	0.080
66-22	78	6	6	2	2	0.108	0.108
48-1012	20	4	8	10	12	0.043	0.013
48-912	23	4	8	9	12	0.052	0.013
48-812	27	4	8	8	12	0.062	0.013
412-1012	19	4	12	10	12	0.043	0.009
412-812	25	4	12	8	12	0.062	0.009
412-610	36	4	12	6	10	0.087	0.014
412-49	49	4	12	4	9	0.120	0.017
412-48	51	4	12	4	8	0.120	0.021
416-812	25	4	16	8	12	0.062	0.007
416-610	35	4	16	6	10	0.087	0.011
416-49	48	4	16	4	9	0.120	0.013
416-28	64	4	16	2	8	0.162	0.015
612-66	32	6	12	6	6	0.058	0.029
612-44	44	6	12	4	4	0.080	0.040
612-22	59	6	12	2	2	0.108	0.054
612-14	61	6	12	1	4	0.126	0.040
612-03	72	6	12	0	3	0.148	0.047

For example, the concrete may be placed directly into the forms using laborers to distribute the concrete. The concrete may also be placed using a crane with a bucket attached, or pumped by a concrete pump truck.

QUANTITIES OF MATERIALS FOR CONCRETE

The estimator should determine the cubic yards of each class of concrete in the job. A reasonable amount should be included for waste, such as 10 percent for a small job or 5 percent for larger jobs.

Concrete structures are designed for concretes having specified strengths, usually expressed in pounds per square inch in compression 28 days after it is placed in the structure. To produce a concrete with a specified strength, it is common practice to employ a commercial laboratory to design the mix. Such a laboratory specifies the weight or volume of fine aggregate, coarse aggregate, cement, and water to produce a batch or cubic yard of concrete having the required strength.

Estimators seldom have the laboratory design information when estimating the cost of a project. Tables giving the approximate quantities of aggregate, cement, and water for different quantities of concrete are available. The information given in these tables is sufficiently accurate for estimating purposes.

Table 10.10 gives the approximate quantities of cement, water, and coarse and fine aggregates required to produce 1 cy of concrete having the indicated 28-day compressive strengths. Note that the aggregate is saturated surface-dry when weighed. If the stockpiles of aggregates contain surface moisture, as they usually do, the quantity of water added should be decreased by the amount of water present on the surface of the aggregate. The weights of the aggregate should be increased to produce net weights that correspond to those given in the table.

TABLE 10.10 | Quantities of cement, water, and aggregates required for 1 cy of concrete having the indicated 28-day compressive strength.

Sacks of cement	Water, gal	Weights of saturated surface-dry aggregate, lb		28-day compressive strength, psi
		Fine	Coarse	
1-in. coarse aggregate				
4.9	39.1	1,370	1,860	2,250
5.6	39.1	1,345	1,847	2,750
6.0	39.0	1,260	1,860	3,000
6.5	39.0	1,235	1,820	3,300
7.2	39.8	1,150	1,875	3,700
8.0	40.0	1,120	1,840	4,250
2-in. coarse aggregate				
4.5	36.0	1,350	1,980	2,250
5.1	35.6	1,275	1,980	2,750
5.5	35.7	1,265	1,980	3,000
6.0	36.0	1,200	1,980	3,300
6.7	36.8	1,140	2,010	3,700
7.4	36.8	1,110	2,000	4,250

LABOR AND EQUIPMENT PLACING CONCRETE

A typical concrete crew consists of a finisher and not more than five laborers to place, spread, vibrate, and finish the concrete surface. The laborers place and spread the concrete in the forms. While concrete is placed in the forms, vibrators operated by gasoline engines are used to consolidate the concrete. For flat-work concrete, laborers screed the top of the concrete to the required level. The finisher applies the required surface texture. The surface of the concrete can be broom finished, hand toweled, or machine toweled.

The labor-hours required to place concrete will vary with the rate of delivery, type of structure, and location of the structure. Depending on the type of work, the concrete may be placed below the ground, on the ground, or above the ground.

Deep foundations require placement of the concrete into holes drilled into the ground. The concrete is discharged from the truck into a tremie to prevent separation of the concrete mix as it is dropped to depths from 5 to 50 ft below the surface of the ground. The tremie is a flexible hose that extends from the ground to the bottom of the drilled shaft. The tremie is pulled as the concrete placement moves from the bottom to the top of the drilled shaft. Generally a crew of not more than two to four workers is required.

For shallow foundations, such as spread footing, grade beams, and small slabs on grades, it may be possible to discharge the concrete directly into the forms from the chute on the back of the delivery truck. If the job conditions are such that the delivery truck cannot place the concrete directly into the forms, hand buggies and wheelbarrows may be used to transport from 4 to 9 cy of concrete, provided there are smooth and rigid runways on which to operate. Hand buggies are safer than wheelbarrows because they have two wheels rather than one. Hand buggies and wheelbarrows are recommended only for distances less than 200 ft.

For large slabs on grade, where the concrete delivery truck cannot be placed sufficiently close to directly discharge the concrete in the forms, it can be desirable to pump the concrete, otherwise the workers using concrete buggies must transport the concrete.

For elevated concrete structures, such as columns, beams, girders, and floor slabs, etc., two methods can be used: crane and bucket or pumping. If concrete is placed using a crane with a bucket, it is desirable to have two buckets available for the crane. Concrete can be discharged from the concrete truck into one bucket on the ground while the crane transports the filled bucket for placement of concrete in the elevated forms.

For large volumes of concrete placement, or concrete that must be placed at high elevations, it may be more economical to use a pump to place the concrete directly into the forms. Pumps require a steady supply of concrete to be effective. Pumps can be mounted on trucks, trailers, or skids. The concrete is placed into a hopper of the pump truck, which pumps the concrete through rigid pipes or flexible tubes to the location of placement. Truck-mounted pump and boom combination is particularly effective and cost-effective in saving labor.

Generally, the contractor hires the services of a company that specializes in pumping concrete. Pumps are available in a variety of sizes, capable of delivering concrete at sustained rates of 10 to 150 cy/hr. Effective pumping ranges from 300 to 1,000 ft horizontally and from 100 to 300 ft vertically.

Belt conveyors are available for rapid movement of fresh concrete. They may be portable or self-contained conveyors. Some models are designed to work in series, to carry the concrete from one conveyor to another until it is discharged. Units are available that have the capability to side-discharge and spread the concrete, for example, to place fresh concrete on a bridge deck.

Table 10.11 gives approximate labor-hours for various types of work required to place concrete.

TABLE 10.11 | Labor-hours required to place concrete, hr/cy.

| | Method of handling | | |
Type of work	Direct chute	Crane bucket	Pumped
Continuous footings	0.3–0.4	0.6–0.8	0.4–0.5
Spread footings	0.6–0.7	1.0–1.3	0.6–0.8
Grade beams	0.3–0.4	0.5–0.7	0.3–0.4
Slabs on grade	0.3–0.4	0.5–0.6	0.4–0.5
Walls	0.4–0.5	0.7–0.9	0.5–0.6
Beams and girders	—	1.4–1.8	1.0–1.2
Columns	—	1.1–1.3	0.6–0.8
Elevated slabs	—	0.6–0.8	0.4–0.5

LIGHTWEIGHT CONCRETE

When the strength of concrete is not a primary factor, and when a reduction in the weight is desirable, lightweight aggregate, such as cinders, burned clay, vermiculite, pumice, or other materials, may be used instead of sand and gravel or crushed stone. Concrete made with this aggregate and Portland cement may weigh 40 to 100 lb/cf, depending on the aggregate and the amount of cement used, which is substantially lighter than 150 lb/cf for normal weight concrete. Also, lightweight concrete has better insulating properties than conventional concrete.

PERLITE CONCRETE AGGREGATE

While several lightweight aggregates are used, only perlite will be discussed to illustrate the properties and costs of lightweight concrete. This aggregate is a volcanic lava rock that has been expanded by heat to produce a material weighing approximately 8 lb/cf. Concrete made from this aggregate may be used for subfloors, fireproofing steel columns and beams, roof decks, concrete blocks, etc.

The materials used in making concrete include Portland cement, aggregate, water, and an air-entraining agent. Mixing can be accomplished in a drum-type concrete mixer, or transit-mixed concrete can be purchased in some localities. Perlite is generally available in bags containing 4 cf, which weigh about 32 lb.

When lightweight concrete is used for floors or roof decks, it is usually placed in thinner layers than when conventional concrete is used. The reduction in volume, without a corresponding reduction in the number of laborers, will result in a higher labor cost per cubic yard of concrete. The cost of labor required to mix and place this concrete for a floor fill or roof deck should be increased at least 50 percent over that required for slabs constructed with conventional concrete.

TILT-UP CONCRETE WALLS

Tilt-up construction is frequently used for walls of warehouses and similar type structures. First, the walls are cast in a horizontal position on the concrete floor slab, usually at ground level. After they have cured sufficiently, the walls are tilted into the vertical position.

The concrete floor on which the wall sections are to be cast should be cleaned and coated with a bond-breaker before the forms are placed. Side forms, equal in height to the thickness of the wall, are placed around the panels to be poured. Frames for window and door openings should be installed before concrete is placed in the forms. Conduits can be placed in the wall easily. Stiffeners should be attached to the side forms to maintain true alignment. The reinforcing steel is placed in position. Steel dowels projecting through holes drilled in the side forms, or weld plates cast in the sides of the wall, permit the walls to be secured in position after they are tilted up. Leveling bolts are usually installed in the bottom of the wall panels to permit plumbing adjustments of the wall after it is tilted up. After the concrete has cured properly, the forms are removed and the wall panels are tilted to final position as units of a wall.

After the walls are tilted up, they are aligned and braced. The panels are tied into a solid continuous structure by joining adjacent panels with weld plates. The walls can be attached to reinforced concrete columns with the projecting steel dowels. A nonshrinkage grout is placed under the wall panels and caulking is installed to seal gaps between adjacent wall panels.

The size of wall panels is limited to the lifting capacity of the equipment used to tilt the wall up. To reduce the danger of structural failure in the panels while they are being tilted into position, it is advisable to use strongback, or steel beams temporally bolted to the panels. For this purpose, bolts should be embedded in the concrete or mechanical inserts of adequate strength may be used. The panels can be picked up with a face-lift or an edge-lift. Multiple pick-up points can be used in a face-lift to reduce the danger of cracking a panel as it is tilted into place.

The advantages of this method of construction include low cost of forms, low cost of placing reinforcing steel, and low cost of placing concrete, since high hoisting into vertical forms is not necessary.

PROBLEMS

10.1 A concrete basement wall 10 ft high for a building has dimensions of 90 ft wide and 150 ft long. A single monolithic placement of the concrete into the walls is planned. Thus, the entire formwork system is to be built and erected before placement of the concrete, thereby preventing any reuse of the forms.

  The forms will be constructed using plywood and dimension lumber. Use average values from Table 10.1 and the unit cost of material and labor as shown in Example 10.3 to estimate the total cost and cost per SFCA for the formwork.

10.2 Estimate the cost of the formwork of Prob. 10.1, assuming the forms will be used three times. Calculate the cost savings that can be achieved by reusing the forms.

10.3 Construction of a building requires 30 in. square concrete columns. The columns will be 12 ft high. A total of 80 columns are required in the structure. All of the concrete will be placed in the columns at one time, thereby preventing any reuse of the forms.

  The forms will be constructed using plywood and dimension lumber. Use average values from Table 10.2 and the unit cost of material and labor as shown in Example 10.6 to estimate the total cost and cost per SFCA for the formwork.

10.4 Estimate the cost of formwork of Prob. 10.3, assuming the forms will be used four times. Calculate the cost savings that can be achieved by reusing the forms.

10.5 Estimate the cost of forms for 75 concrete beams for a building. The beams are 12 in. wide, 20 in. deep, and 22 ft long. The forms will be used four times. The formwork will be constructed with plywood and dimension lumber. Use average values from Table 10.3 and the unit cost of material and labor as shown in Example 10.8 to estimate the total cost and cost per SFCA for the formwork.

10.6 Estimate the direct cost of furnishing, building, and removing forms for a flat-slab concrete floor whose size is 250 ft by 400 ft. The ceiling height will be 10 ft and the slab will be 8 in. thick.

  The forming system will be as shown in Figure 10.12, using plywood and lumber, with 4×4 joists and 4×6 stringers. Use average values from Table 10.4 and the unit cost of material and labor as shown in Example 10.9 to estimate the total cost and cost per SFCA.

10.7 Estimate the total cost and cost per pound of reinforcing steel for a concrete slab on grade. The floor will be 55 ft long, 70 ft wide, and 6 in. thick. The reinforcing steel will be no. 4 bars ($\frac{1}{2}$ in. dia) spaced not more than 12 in. apart each way in the slab. The bars laid next to and parallel to the edges of the slab will be placed 3 in. from the edges of the slab. The ends of the bars will extend to within 2 in. from the edges of the slab.

  Skilled steel setters will be used at a cost of $29.00/hr. The cost of reinforcing steel delivered to the job will be $0.46/lb. Use average values from Table 10.8.

10.8 If the concrete slab in Prob. 10.7 is 9 in. thick, estimate the labor cost for placement of the concrete by direct chute from the concrete delivery truck. Use average values from Table 10.11 and assume the unit cost of labor is $28.00/hr.

11

Steel Structures

TYPES OF STEEL STRUCTURES

Steel is used to erect such structures as multistory buildings, auditoriums, gymnasiums, theaters, churches, industrial buildings, roof trusses, stadiums, bridges, towers, etc. In addition to steel structures, steel members are frequently used for columns, beams, roof purlins, decking, joists, studs, lintels, and for other purposes.

Due to its high strength, steel is ideal for long spans and tall structures. Steel consisting of standard shapes or fabricated plate girders is used for many multiple-span bridges. Trusses made of steel can span large arenas for entertainment or educational facilities. Galvanized steel angles are fabricated into space-truss towers to support conductors and shield wires for electrical transmission structures. Transmission towers often are 120 to 200 ft tall. Steel is also used to support water towers and other types of elevated storage units. Skyscrapers are typically constructed of steel.

MATERIALS USED IN STEEL STRUCTURES

Insofar as it is possible, steel structures should be constructed with members fabricated from standard shapes, such as *W* section beams, *C* channels, structural tees, angles, pipes, tubes, rods, and plates. Members made from standard rolled shapes are usually more economical than fabricated members. However, if standard shapes are not available in sufficient sizes to supply the required strength, it is necessary to fabricate the members from several parts, such as adding plates to beam flanges or stiffeners to beam webs. Large built-up plate girders can be fabricated by welding together steel plates to form flanges and webs of any size.

The unit of measure of steel is pounds, hundredweight (cwt), or tons. The grade of structural steel denotes its yield strength. For example, a drawing may show a steel beam as a *W*18 × 55 Grade 50. The *W* represents a wide flange section, the 18 denotes the approximate depth in inches, the 55 represents the weight in pounds per ft of length, and the Grade 50 denotes that the yield strength is 50,000 pounds (psi).

ESTIMATING THE WEIGHT OF STRUCTURAL STEEL

In estimating the weight of structural steel for a job, the total number of linear feet for each shape by size and weight must be determined from the drawings for a project. Structural-steel handbooks give the nominal weights of all sections. However, variations in weights amounting to 2 percent above or below the nominal weights are permissible and may occur. The purchaser is charged for the actual weight furnished, provided that the weight does not fall outside the permissible variation.

The weight of the details for connections should be estimated and priced separately if a detailed estimate is desirable. In estimating the weight of a steel plate or irregular shape, the weight of the rectangular plate from which the shape is cut should be used. Steel weighs 490 lb/cf.

CONNECTIONS FOR STRUCTURAL STEEL

In fabricating standard shapes to form the required members or in connecting the members into the structure, two types of connections are used: bolts and welds. There are advantages and disadvantages in each type of connection, which will be discussed in greater detail later in this chapter.

ESTIMATING THE COST OF STEEL STRUCTURES

In estimating the cost of structural steel for a job, a contractor will submit a set of plans and specifications for the structure to a commercial steel supplier for quotations. The steel supplier will make a quantity takeoff, including main members, detailed connections, and miscellaneous items. Then, the supplier will add fabricating costs for cutting, punching, drilling, welding, overhead, and profit as a basis for submitting a quotation to the general contractor. For some jobs, the steel supplier will paint the fabricated steel before delivery to the job. When the steel fabricator paints the steel, it is called shop paint. The cost of transporting the steel to the project must be added to the cost of the finished products at the fabrication shop.

General contractors who erect buildings and similar structures usually subcontract the erection of the steel to subcontractors who specialize in this work. This practice is justified because the erection of steel is a highly specialized operation, which should be performed by a contractor with suitable equipment and a well-trained erection crew. Because of these conditions, the general contractor can usually have the erection performed more economically by a subcontractor than with the general contractor's equipment and employees. The charge for the erection is generally based on an agreed price per ton of steel in place, including bolting and welding the connections.

When estimating the cost of structural steel in place, a contractor will include in the estimate the cost of the steel delivered to the project, the cost of erection, and the cost of field welding and painting as required. To these costs the contractor will add the contractor's cost for job overhead, general overhead, and profit.

ITEMS OF COST IN A STRUCTURAL-STEEL ESTIMATE

The items of cost that should be considered in preparing a comprehensive detailed estimate for a steel structure include:

1. Cost of standard structural shapes delivered to the steel supplier
2. Cost of preparing drawings for use by the shop in fabricating the steel
3. Cost of fabricating the steel shapes into finished members
4. Cost of transporting the steel to the job
5. Cost of erecting the steel, including equipment, labor, bolts, or welding
6. Cost of field painting the steel structure
7. Cost of job overhead, general overhead, insurance, taxes, and profit

The base cost of steel is subject to change due to economic conditions. Therefore, the cost of any one or all of these items may vary considerably with respect to time. Consequently, the cost of each item must be estimated for each particular project.

COST OF STANDARD SHAPED STRUCTURAL STEEL

Structural steel is hot rolled into standard shapes, sizes, and lengths at steel mills. Steel suppliers purchase bulk quantities of structural steel from the mills in standard shapes and lengths, usually 40 ft long. The cost of steel delivered by the mill to the steel supplier's fabrication shop depends on the shape, weight per foot, grade, and total quantity ordered. Large quantities of purchased steel generally are priced lower than small quantities of steel.

Typically, the construction contractor purchases the steel from a local steel supplier, not the steel mill. The cost to purchase steel from the steel supplier before it is fabricated is commonly called the base price of steel. Steel suppliers have shops that specialize in fabricating standard shaped steel into members to meet the specifications required for each particular job. Before starting the fabrication process, the steel supplier prepares shop drawings that are submitted for approval.

COST OF PREPARING SHOP DRAWINGS

In preparing plans for a steel structure, the engineer or architect does not furnish drawings in sufficient detail to permit the shop to fabricate the members without additional information. The steel supplier must prepare shop drawings before the steel can be fabricated.

Steel suppliers maintain engineering and drafting departments, which prepare shop drawings in sufficient detail to enable their shops to fabricate the members. These drawings show actual cut-lengths of each member, precise location of drilled or punched holes, details of welded assemblies, and all other dimensions necessary to fabricate the steel. Figure 11.1 is an example of the fabrication details for a steel beam with bolted connections. The $W16 \times 40$ wide flange beam has a cut-length of 18 ft $4\frac{1}{2}$ in. with three $\frac{3}{8}$-in.-thick stiffener plates and four drilled holes at each end. Each member in a structural steel project is identified with its unique designation, called a piece mark, and detailed as illustrated in this figure. After all the members are detailed in the shop drawings, a bill of material is prepared for the entire project. The bill of material lists each piece of steel, its size, type, length, and weight. The bill of material for the $W16 \times 40$ beam is shown in Fig. 11.1.

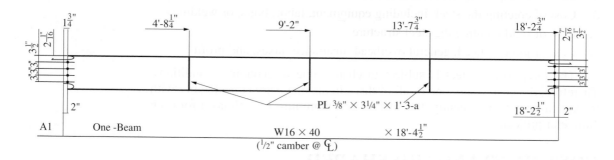

Bill of Material:

Item no.	Dwg. no.	Piece mark	Quant	Type and size	Notes	Length	Weight
1	1	A	1	$W16 \times 40 = A572$	beam	18 ft 4 $\frac{1}{2}$in.	735.00 lb
2	1	B	3	PL 3/8 in. $\times$ 3 $3\frac{1}{4}$	stiffener	1ft 3 in.	15.54 lb
Totals			4 piece marks			91.88 sf	750.54 lb

FIGURE 11.1 | Example of details for a bolted connection beam on a shop drawing.
Courtesy: W&W Steel Company.

Shop drawings are usually prepared on 24- by 36-in. size sheets. The number of shop drawings required depends on the size and complexity of a job. One job may require as few as one or two sheets of structural details, while another job may require more than 100 sheets. The cost of preparing the drawings is based on the complexity of the detailing and the number of sheets required. Since the total cost of the drawings is charged to the steel that will be furnished for a job, the cost per unit weight of steel will vary with the total cost of the drawings and the quantity of steel. If only one member is fabricated, the cost per member is high, whereas if a great many members are fabricated, the cost per member will be much lower.

After the shop drawings are prepared, they are provided to the contractor, who submits them to the designer for approval. The cost of preparing shop drawings is approximately 5 to 10 percent of the base price of steel.

COST OF FABRICATING STRUCTURAL STEEL

The cost of fabricating structural steel will vary considerably with the operations performed, sizes of members, and extent to which shop operations can be duplicated on similar members.

For bolted connections, the fabricating operations include cutting, punching, milling, planing, and marking each member. For welded connections, the fabricating operations include cutting, some punching for temporary bolt connections, milling, beveling, planing, and shop welding. It is usually more efficient and economical to perform welding in the shop, rather than in the field. Also, it is generally easier to control the quality of shop welds.

The cost per ton for lightweight roof trusses will be higher than the cost per ton for large-size beams fabricated from rolled sections. The cost per pound is significantly higher to fabricate angles and plates for space-truss towers, compared to beams and columns of buildings or plate girders for bridges. Thousands of members are required in space trusses, each of which must be cut to a precise length and holes must be punched in each end of each member so it can be bolted to other members in the completed structure.

The cost per ton or per hundredweight is based on the weight of the finished members, including details, cutting, punching, drilling, welding, etc. The time required to set up fabricating equipment is fairly constant, regardless of the number of operations performed. Therefore, when identical operation can be performed on many members, the cost per pound will be substantially lower compared to a shop operation on only a few members.

Specifications for structural steel frequently require the fabricator to apply a coat of paint after the fabricating is completed. A gallon of paint should cover about 400 sf of surface. Spray guns are generally used to apply the paint. A painter, using a spray gun, should be able to paint from 1 to 2 T/hr, depending on the size and shape of the steel members. The cost of shop painting structural steel is about 8 to 12 percent of the base price of steel.

Depending on the complexity and amount of fabrication required, the total cost of fabricating steel can range from 50 to 100 percent of the base price of steel.

COST OF TRANSPORTING STEEL TO THE JOB

Usually the steel supplier delivers the fabricated steel to the job. However, if the contractor is providing the delivery of the steel, the estimator should determine the freight or truck cost per ton or per hundredweight for the particular project to include the correct amount in the estimate.

The cost of transporting structural steel from the fabrication shop to the job will vary with the quantity of steel, method of transporting it, and the distance from the shop to the job. A truck should haul about 20 T/load.

COST OF FABRICATED STRUCTURAL STEEL DELIVERED TO A PROJECT

A complete list of material by size, shape, and length must be prepared to estimate the cost of fabricated steel. Example 11.1 illustrates the breakdown of costs for fabricating and delivering steel to a project.

EXAMPLE 11.1

Structural steel, consisting of beams, columns, angles, and plates, is used for a framed building. All members are to be fabricated at a shop for high-strength bolted connections. Wide flange beams will be used for beams and columns.

A list of members and details is given in the accompanying table. The small-size beams are used for short spans and larger-size beams will be used for the longer spans. The larger column sizes will be used for the lower floors, whereas the smaller column sizes are used for the upper floors. Angles are used as seats and clips for beam-to-column connections for joining together the structural members. Steel plates are used for column base plates and splicing of column members.

Based on current economic conditions, the base price of steel quoted from the supplier is $46.00/cwt for beams and columns, $52.00/cwt for plates, and $44.00/cwt for angles. The steel will be delivered by 20-T trucks from the steel supplier a distance of 134 mi to the job at a rate of $2.75/mi.

An illustrative breakdown of the base, fabrication, and delivery cost of the steel is shown in this example. The percentage values given in the preceding paragraphs of this book are used for fabrication, shop drawings, and painting.

Number of pieces	Description	Length each	Weight, lb/lin ft	Total weight, lb
18	Columns, $W10 \times 89$	24 ft 9 in.	89	39,649
12	Columns, $W10 \times 112$	24 ft 9 in.	112	33,264
18	Columns, $W10 \times 54$	21 ft 6 in.	54	20,898
12	Columns, $W10 \times 72$	21 ft 6 in.	72	18,576
Total weight of columns				112,387
30	16- $\times$ 16- $\times$ $1\frac{1}{2}$-in. base plates			3,267
120	6- $\times$ 18- $\times$ $\frac{3}{4}$-in. splice plates			2,763
120	∟ $3 \times 3 \times \frac{3}{8}$	0 ft 8 in.	7.2	575
420	∟ $2\frac{1}{2} \times 2 \times \frac{3}{16}$	0 ft 8 in.	2.75	577
Total weight of column details				7,182
68	Beams, $W14 \times 48$	17 ft 6 in.	48	57,120
54	Beams, $W14 \times 34$	19 ft 9 in.	34	36,261
24	Beams, $W12 \times 32$	16 ft 6 in.	32	12,672
18	Beams, $W12 \times 28$	22 ft 6 in.	28	11,340
Total weight of beams				117,393
492	∟ $3 \times 3 \times \frac{3}{8}$	1 ft 0 in.	7.2	3,542
164	∟ $3 \times 3 \times \frac{1}{4}$	0 ft 10 in.	4.9	670
Total weight of angles				4,212
Total weight of fabricated steel, lb				241,174
Total weight, tons				120.59

The total cost will be:

Base cost of steel:

Columns: 1,123.87 cwt @ $46.00/cwt = $51,698.02

Beams: 1,173.93 cwt @ $46.00/cwt = 54,000.78

Plates: 60.30 cwt @ $52.00/cwt = 3,135.60

Angles: 3 in., 47.87 cwt @ $44.00/cwt = 2,106.28

Angles: $2\frac{1}{2}$ in., 5.77 cwt @ $44.00/cwt = 253.88

Total base cost of steel = $111,194.56

Fabricating costs:

Columns: 1,123.87 cwt @ 85% $\times$ $46.00/cwt = $43,943.32

Beams: 1,173.93 cwt @ 85% $\times$ $46.00/cwt = 45,900.66

Plates: 60.30 cwt @ 95% $\times$ $52.00/cwt = 2,978.82

Angles: 3 in., 47.87 cwt @ 85% $\times$ $44.00/cwt = 1,790.34

Angles: $2\frac{1}{2}$ in., 5.77 cwt @ 90% $\times$ $44.00/cwt = 228.49

Total fabricating costs = $94,841.63

Shop painting:

7.5% $\times$ $94,841.63 = $7,113.12

Shop drawings:

10.0% $\times$ $94,841.63 = $9,484.16

Shipping steel to job:

Number of trucks needed: 120.59 T/(20 T/truck) = 6 trucks required

Total cost: 6 trucks $\times$ 134 mi @ $2.75/mi = $2,211.00

Subtotal of direct costs:

Base price of steel = $111,194.56

Fabricating costs = 94,841.63

Shop painting = 7,113.12

Shop drawings = 9,484.16

Shipping costs = 2,211.00

Total direct costs = $224,844.47

Overhead and profit:

12% $\times$ $224,844.47 = $26,981.34

Total cost of fabricated steel delivered to job:

$224,844.47 + $26,981.34 = $251,825.81

Summary of costs:

Cost per pound, $251,825.81/241,174 lb = $1.04/lb

Cost per hundredweight, $251,825.81/2,411.74 = $104.42/cwt

Cost per ton, $251,825.81/120.59 T = $2,088.28/T

ERECTING STRUCTURAL STEEL

Most general contractors subcontract the erection of structural steel for their projects. The subcontract may be assigned to a company that specializes in steel erection or it may be subcontracted to the steel supplier. Many steel suppliers have the crew and equipment and are capable of erecting the steel that they fabricate and deliver to the project.

When a structural-steel building is erected, the columns are erected first on the previously prepared concrete foundations, with anchor bolts already in place. The ends of the anchor bolts are usually threaded to allow installation of the column base plates, which have holes drilled to match the pattern of the anchor bolts. Nuts are installed on each threaded anchor bolt, one above and one below the column base plate, to adjust and plumb the column.

After the columns are erected, the beams are installed for the first tier of floors, usually two floors. The connections between the columns and the beams are temporarily bolted through holes, by using two or more bolts per connection. Horizontal bracing is necessary to stabilize the structural members during the erection process. After the tier of columns and beams is in place, it is necessary to plumb the structure before installing the permanent bolts. This operation is repeated for subsequent tiers until the erection of the structure is completed.

Equipment for Erecting Structural Steel

The equipment used for erecting steel structures depends on the type of structure, the size and height of the structure, and the location. Roof trusses are usually delivered to the job partly or completely assembled and hoisted directly from the delivery trucks into place by truck-mounted cranes. (See Figure 11.2.)

Several types of truck-mounted cranes are available. A hydraulic truck crane has a self-contained boom, which enables the unit to travel on public highways between projects under its own power without setup delays. A hydraulic truck crane is ideal for short-duration jobs, less than several days. For taller steel structures, a lattice-boom truck crane is sometimes used. The lattice-boom is cable suspended, and therefore acts as a compression member, not a bending member like the telescoping hydraulic boom. However, mobilization and demobilization of the lattice-boom crane require time. Sometimes a second crane is required to assemble and disassemble the lattice boom.

Multistory steel-frame buildings can be erected with truck-mounted cranes if the height is not excessive, usually about four stories. If a building is so tall that a truck-mounted crane cannot be used, the steel members may be placed with a tower crane. These are cranes that provide a high-lifting height with a good working radius and require limited space. Several types are available. Some units have a fixed vertical tower with a rotating horizontal boom truss. The climbing-frame-type crane is supported by the floors of the building that it is being used to construct.

Because cranes are used to hoist and move loads from one location to another, it is necessary to know the lifting capacity and working range of a crane selected

FIGURE 11.2 | Truck-mounted crane erecting steel truss.

to perform a given service. The lifting capacity of cranes for erecting steel structures typically varies from 20 to 100 T. However, lifting capacities up to 500 T are available. Individual manufacturers of equipment furnish specifications and information that describes the lifting capabilities and tipping loads of their equipment.

Steel erection contractors typically own the lifting equipment. However, for special jobs that may require an unusually heavy or high lift, the crane may be rented. The cranes can be rented per day, week, or month.

Bolting Structural Steel

When bolting structural steel, one worker will be required at each end of the member to install at least enough bolts to secure the member to the supporting structure. (See Figure 11.3.) One or two other workers may be required on the ground level to assist in attachment of slings and rigging the next structural member to be erected. As structural members are erected, adequate lateral bracing is necessary to ensure adequate rigidity and stability of the entire steel framing system.

After members are erected and plumbed, two ironworkers should be able to complete the permanent bolting, including application of the proper tightening torque. For safety, fall protection must be provided for ironworkers when they are working at high elevations.

FIGURE 11.3 I Iron workers erecting steel beam.

When a tower crane is used to hoist the members and place them into position, it is common practice to install enough bolts in the connections to permit the structure to be plumbed and braced as it is erected. If this procedure is followed, the rest of the permanent bolts will be installed later. Calibrated torque wrenches, selected to produce the desired tension in the bolts, can be used in tightening the nuts on the bolts. The crew size required for installing the bolts will vary with the number of bolts needed and the ease or difficulty in getting at the bolts.

Welding Structural Steel

In general, welded connections provide more rigidity than bolted connections. For welded steel structures, gasoline-operated welding machines are used, otherwise the general procedure and erection equipment are essentially the same as for a bolted structure. Welders, who are certified by laboratory tests, are used in producing high-quality welding of steel structures. The cost of field welding can be substantially reduced if the structure is designed for minimum overhead welding. Various types of structural welds are shown in Fig. 11.4.

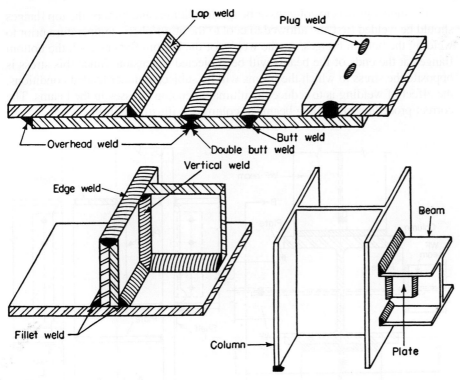

FIGURE 11.4 | Typical welds.

If a structure is designed for welded connections, it is possible to reduce the total weight of steel by as much as 5 percent, as compared to bolted connections. It is not necessary to punch bolt holes in members at points of critical stress for connection purposes. Where a few holes are required for temporarily bolting members in place, the holes can be punched at noncritical points, thus permitting the use of the full strength of a member at joints rather than a reduced net section through the member. Since the welded joints can be made as strong as the full sections of the members, it is possible to design beams as continuous members over several supports, thus reducing the critical bending-moment stresses. As a result of these conditions, lighter beams can be used in a structure, which will reduce the total weight of steel required. Also, smaller and less expensive foundation costs should result from reduction in deadweight of the structures.

After a member of the structure is plumbed or brought to the correct position, welding of the connections is started. To eliminate undesirable distortion of a structure, resulting from unequal heating at the connections, a definite pattern for welding should be established and rigidly followed. If beams are to be welded to opposite flanges of a column, the two welds should be performed simultaneously to eliminate unequal expansion of the two sides of the column, which would result from welding the beams separately.

If welding girders to columns or beams to columns and girders, the top flanges should be welded first and allowed to cool to the atmospheric temperature prior to welding the bottom flanges. As the welds for the bottom flanges cool, the bottom flanges at the ends of the beams will be subjected to tension. Since this stress is opposite the stress to which the beams will be subjected under loaded conditions, the effect of welding is to reduce the ultimate bending stresses in the beams. The correct procedure in welding beams to columns is illustrated in Fig. 11.5.

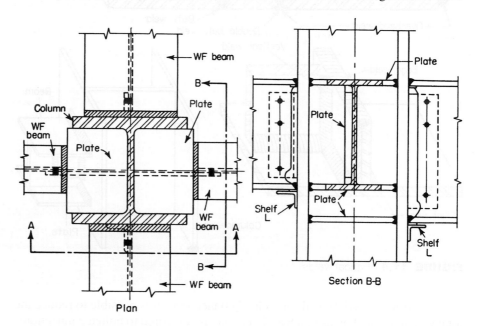

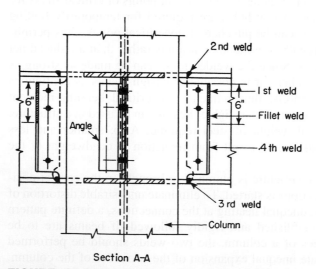

FIGURE 11.5 I Details of beam-to-column welds.

LABOR ERECTING STRUCTURAL STEEL

The cost of labor erecting structural steel will vary with the type of structure, the type of equipment used, sizes of members, climatic conditions, and the prevailing wage rates of the workers.

Construction laborers who erect structural steel are called ironworkers or structural steel workers. A common size of steel erection crew consists of four ironworkers, a crane operator, and foreman. However, crews for erecting steel may vary from two to eight ironworkers, depending on the size of structural members that must be handled and the height of the structure. For example, a high-rise multistory building may use a crew consisting of eight ironworkers, several welders, crane operator, hoist operator, and two foremen.

Table 11.1 gives the approximate number of crew-hours per 1,000 lin ft of member for erecting steel for various sizes and shapes of structural steel members that have been fabricated and delivered to the job. Wide-flange beams with deeper depths can span longer distances than shallower-depth beams. However, the time to set a 25-ft-long beam is about the same as the time to set a 22-ft-long beam. Therefore, the crew-hours per foot are less for erecting large-size beams, compared to smaller-size beams.

TABLE 11.1 | Approximate crew-hours per 1,000 lin ft for erecting various sizes and shapes of shop fabricated steel members.

Type of work	Crew-hours per 1,000 lin ft
Erecting steel beams, bolted connections	
Crew: 4 ironworkers, 1 crane operator, 1 foreman	
Equipment: 1 crane	
W-sections, 8 in. deep	15.6–17.3
W-sections, 10 in. deep	15.2–16.8
W-sections, 12 in. deep	12.6–13.8
W-sections, 14 in. deep	11.0–12.1
W-sections, 16 in. deep	10.5–11.6
Erecting steel beams, welded connections	
Crew: 5 ironworkers, 1 crane operator, 1 welder, 1 foreman	
Equipment: 1 crane, 1 welding machine	
W-sections, 18 in. deep	10.3–11.4
W-sections, 21 in. deep	9.3–10.2
W-sections, 24 in. deep	8.9–9.7
W-sections, 30 in. deep	8.6–9.4
W-sections, 36 in. deep	8.5–9.3
Erecting steel columns	
Crew: 4 ironworkers, 1 crane operator, 1 foreman	
Equipment: 1 crane	
W-sections, 8 in. deep	8.6–9.6
W-sections, 10 in. deep	8.9–9.9
W-sections, 12 in. deep	9.1–10.0
W-sections, 14 in. deep	9.3–10.3

Note: These crew-hours should be increased by 20 percent for erecting steel structures over six stories high.

If fabrication is required at the jobsite, the cost of fabricating and erecting the steel is substantially higher than when using shop fabricated steel. However, sometimes small members, such as angles, channels, or junior beams, are field fabricated. Table 11.2 gives the approximate crew-hours for field fabricating and erecting small-size steel members. The crew for performing this type of work consists of only one ironworker, one welder, and one foreman.

TABLE 11.2 | Approximate crew-hours per 1,000 lin ft for field fabricating and erecting of small members.

Type of work	Crew-hours per 1,000 lin ft
Field fabricating small members,	
Crew: 1 ironworker, 1 welder, 1 foreman	
Equipment: 1 welder	
Angles, $\frac{1}{2}$- to 1-in. legs	54–66
Angles, 2- to 3-in. legs	104–127
Channels, 2 in. to 5 in. deep	97–119
Channels, 6 in. to 8 in. deep	170–208
Junior beams, 3 in. to 5 in. deep	99–121
Junior beams, 6 in. to 8 in. deep	126–154

EXAMPLE 11.2

Estimate the cost of labor for erecting 7,931 lin ft of $W14 \times 53$ wide flange beams and 1,388 lin ft of $W12 \times 50$ columns. The structure is two stories high and bolted connections will be used. Use average values in Table 11.1 and assume these rates for labor and equipment: $31.00/hr for ironworkers, $36.00/hr for the crane operator, and $38.00/hr for the foreman. A 40-T crane will be used at a rate of $182.00/hr.

Quantity of materials:

Beams: 7,931 lin ft $\times$ 53 lb/ft = 420,343 lb

Columns: 1,388 lin ft $\times$ 50 lb/ft = <u>69,400 lb</u>

                    Total weight = 489,743 lb

Total hundredweight: 489.743 lb/(100 lb/cwt) = 4897.43 cwt

Total tons: 489,743 lb/(2,000 lb/T) = 244.87 T

Crew-hours required:

From Table 11.1 for bolted connections:

   Crew of 4 ironworkers, 1 crane operator, and 1 foreman

     For 14-in.-deep W-section beams: 11.55 crew-hours/1,000 lin ft

     For 12-in.-deep W-section columns: 9.55 crew-hours/1,000 lin ft

Crew-hours erecting structural steel:

Beams: 7,931 lin ft $\times$ (11.55 crew-hours)/(1,000/lin ft) = 91.6 crew-hours

Columns: 1,388 lin ft $\times$ (9.55 crew-hours)/(1,000 lin ft)= <u>13.3 crew-hours</u>

                           Total time = 104.9 crew-hours

Cost of erecting beams and columns:

Ironworkers: 104.9 crew-hours $\times$ 4 ironworkers @ \$31.00/hr = \$13,007.60

Crane operator: 104.9 crew-hours $\times$ 1 operator @ \$36.00/hr = 3,776.40

Foreman: 104.9 crew-hours $\times$ 1 foreman @ \$38.00/hr = 3,986.20

Crane: 104.9 crew-hours $\times$ 1 crane @ \$182.00/hr = 19,091.80

Total cost = \$39,862.00

Summary of costs:

Cost per pound: \$39,862.00/489,743 = \$0.08/lb

Cost per hundredweight: \$39,862.00/4,897.43 = \$8.14/cwt

Cost per ton: \$39,862.00/244.87 T = \$162.79/T

If the fabricated steel for these beams and columns cost \$2,088.28/T, then the total installed cost will be: \$2,088.28/T + \$162.79/T = \$2,251.07/T. This cost does not include the cost for lateral bracing, steel joists, field painting, tools, overhead, taxes, insurance, or profit.

FIELD PAINTING STRUCTURAL STEEL

Although structural steel is usually shop painted by the fabricator, it is sometimes necessary to perform limited field painting. The cost of applying coats of paint in the field to structural steel will vary with the type of structure, sizes of members to be painted, and access to steel members. The cost of painting a ton of steel for a roof truss will be considerably higher than for a steel frame building because of the greater area of steel per ton for the truss and also because of the difficulty of moving along the truss with the painting equipment.

Paint is usually applied with a spray gun, operated by an air compressor. One gallon of paint should cover about 400 sf. Two field coats are usually applied. Table 11.3 gives the approximate rates of applying a field coat of paint, by using a spray gun, to structural steel for various types of members and structures. The costs are based on painting 400 sf/gal. A painter should be able to field paint $\frac{3}{4}$ to $1\frac{1}{2}$ T/hr, depending on the structure.

TABLE 11.3 | Rate of applying a field coat of paint to steel structure, using a spray gun.

Member or structure	Square ft/ton
Beams	200–250
Girders	125–200
Columns	200–250
Roof trusses	275–350
Bridge structures	200–250

PROBLEM

11.1 The plan and elevation views of a five-story structural steel building are shown in the accompanying drawing. A column schedule is also given for the building. All beams will be $W14 \times 125$ with 20-ft span lengths. Estimate the cost of erecting the beams and columns using average rates from Table 11.1 and the type of crew and equipment, unit labor rates, and unit equipment rates as shown in Example 11.2.

Column Schedule	
Floor 1	$W10 \times 112$
Floor 2	$W10 \times 89$
Floor 3	$W10 \times 72$
Floor 4	$W10 \times 60$
Floor 5	$W10 \times 60$

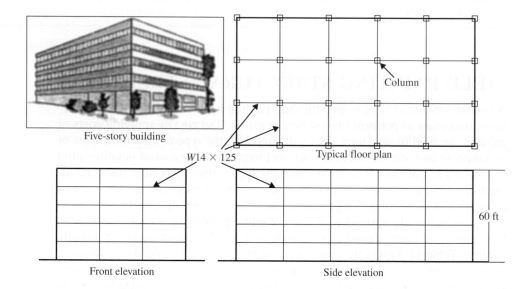

Five-story building

$W14 \times 125$

Typical floor plan

Column

Front elevation

Side elevation

60 ft

C H A P T E R 12

Carpentry

INTRODUCTION

Carpentry is classified as rough or finish. Rough carpentry involves the framing of structural components, such as floors, walls, window and door openings, and roof members. Finish carpentry involves installation of wood trim around doors and windows, cabinetry, moldings, and related work. Sometimes frame carpenters are called rough carpenters and finish carpenters are called trim carpenters.

Carpenters are used in both heavy and building type projects. They fabricate and install formwork for concrete structures, including foundations, box culverts, and bridge decks. They also frame and trim finished wood in residential and commercial building construction. Thus, carpenters are used in all types of construction.

CLASSIFICATION OF LUMBER

Lumber is classified by size, grade, and species. Lumber is measured and priced by *board feet* (bf). A *board foot* is a piece of lumber whose nominal dimensions are 1 in. thick, 12 in. wide, and 1 ft long. Thus, a 2×4 piece of lumber is equivalent to 0.67 bf per foot of length, whereas a 2×8 piece of lumber is equivalent to 1.33 bf per foot of length.

Lumber Sizes

The sizes of lumber, with the exception of certain specialties such as moldings, are designated by the nominal dimensions of the cross sections, which are the dimensions prior to finishing the lumber at the mill. After lumber is sawed lengthwise at the mill, it may be passed through one or more finishing operations, which will reduce its actual size to less than the identifying dimensions. Thus, a 2-in. by 4-in. plank (nominal size) will actually be $1\frac{1}{2}$ in. thick and $3\frac{1}{2}$ in. wide (actual size) after it is surfaced on all four sides (S4S). This lumber is called *dimension lumber* and it is designated by its nominal size as 2×4 S4S.

In addition to S4S dimension lumber, *dressed and matched* lumber is available. Dressed and matched lumber may be S2S *center match, tongue and groove,* or *shiplap.* For example, a 1-in. by 6-in. *tongue and groove* board will be $\frac{3}{4}$-in. by $5\frac{1}{8}$-in. actual size.

Because the finishing and dressing operations will result in a reduction in the actual widths of the lumber, a quantity estimate must include an allowance for this side wastage or shrinkage. For a plank nominally 6 in. wide, the net width will be $5\frac{1}{2}$ in. Thus, the side shrinkage will be $\frac{1}{2}$ in. The side shrinkage for quantity purposes will be $(0.5 \times 100)/5.5 = 9.1$ percent. To provide enough lumber for a required area, the quantity must be increased 9.1 percent to replace side shrinkage only.

Lumber is available in lengths that are multiples of 2 ft, with 20 ft as the common maximum length of an individual board. If other than standard lengths are needed, they must be cut from standard lengths. This operation may result in end waste. For example, if a 1×6 board 10 ft 6 in. long is needed, it must be cut from a board that is 12 ft 0 in. long with a 1-ft 6-in. end waste, unless the piece cut off can be used elsewhere. This will represent an end waste of $(1.5 \times 100)/10.5 = 14.3$ percent.

The combined side and end wastes for the cited examples will be:

Side waste   =   9.1%
End waste   = 14.3%
Total waste = 23.4%

Note that the side and end wastes are determined on the basis of the dimensions of lumber required instead of the nominal dimensions.

Table 12.1 lists the commonly used sizes of lumber, with the nominal and actual cross-sectional dimensions. In the table, the first dimension is the thickness and the second dimension is the width. Thinner members, 1 in. and 2 in., are often called lumber, whereas thicker, 3-in. to 12-in. members are referred to as timber.

Species and Grades of Lumber

Lumber is usually specified by species (such as pine, fir, oak, etc.), size, and grade. There are many species of wood. Southern pine, Douglas fir, hemlock, and western white spruce are common species for wood framing, such as studs, joists, girders, sills, and plates. Oak is frequently used for hardwood flooring. Cedar and redwood are used for exterior decking that will be exposed to the elements of weather. Examples of finishing woods include ash, birch, maple, mahogany, and walnut.

Lumber quality for structural framing wood is graded by a numbering system and the quality for appearance is graded by a lettering system:

Select grades: A, B, C, and D
Common grades: 1, 2, and 3

TABLE 12.1 | Dimensions and properties of lumber.

Nominal size, in.	Standard dressed size, in.	Area of section, in.^{2}	Moment of inertia, I	Section modulus, S
		Dimension lumber (S4S)		
1 × 4	$\frac{3}{4} \times 3\frac{1}{2}$	2.625	2.680	1.531
1 × 6	$\frac{3}{4} \times 5\frac{1}{2}$	4.125	10.398	3.781
1 × 8	$\frac{3}{4} \times 7\frac{1}{4}$	5.438	23.817	6.570
1 × 10	$\frac{3}{4} \times 9\frac{1}{4}$	6.938	49.446	10.695
1 × 12	$\frac{3}{4} \times 11\frac{1}{4}$	8.438	88.999	15.820
2 × 4	$1\frac{1}{2} \times 3\frac{1}{2}$	5.250	5.359	3.063
2 × 6	$1\frac{1}{2} \times 5\frac{1}{2}$	8.250	20.797	7.563
2 × 8	$1\frac{1}{2} \times 7\frac{1}{4}$	10.875	47.635	13.141
2 × 10	$1\frac{1}{2} \times 9\frac{1}{4}$	13.875	98.932	21.391
2 × 12	$1\frac{1}{2} \times 11\frac{1}{4}$	16.875	177.979	31.641
2 × 14	$1\frac{1}{2} \times 13\frac{1}{4}$	19.875	290.775	43.891
3 × 4	$2\frac{1}{2} \times 3\frac{1}{2}$	8.750	8.932	5.104
3 × 6	$2\frac{1}{2} \times 5\frac{1}{2}$	13.750	34.661	12.604
3 × 8	$2\frac{1}{2} \times 7\frac{1}{4}$	18.125	79.391	21.901
3 × 10	$2\frac{1}{2} \times 9\frac{1}{4}$	23.125	164.886	35.651
3 × 12	$2\frac{1}{2} \times 11\frac{1}{4}$	28.125	296.631	52.734
3 × 14	$2\frac{1}{2} \times 13\frac{1}{4}$	33.125	484.625	73.151
3 × 16	$2\frac{1}{2} \times 15\frac{1}{4}$	38.125	738.870	96.901
4 × 4	$3\frac{1}{2} \times 3\frac{1}{2}$	12.250	12.505	7.146
4 × 6	$3\frac{1}{2} \times 5\frac{1}{2}$	19.250	48.526	17.646
4 × 8	$3\frac{1}{2} \times 7\frac{1}{4}$	25.375	111.148	30.661
4 × 10	$3\frac{1}{2} \times 9\frac{1}{4}$	32.375	230.840	49.911
4 × 12	$3\frac{1}{2} \times 11\frac{1}{4}$	39.375	415.283	73.828
4 × 14	$3\frac{1}{2} \times 13\frac{1}{4}$	46.375	678.475	102.411
4 × 16	$3\frac{1}{2} \times 15\frac{1}{4}$	53.375	1,034.418	135.660
6 × 6	$5\frac{1}{2} \times 5\frac{1}{2}$	30.250	76.255	27.729
6 × 8	$5\frac{1}{2} \times 7\frac{1}{2}$	41.250	193.359	51.563
6 × 10	$5\frac{1}{2} \times 9\frac{1}{2}$	52.250	392.963	82.729
6 × 12	$5\frac{1}{2} \times 11\frac{1}{2}$	63.250	697.068	121.229
6 × 14	$5\frac{1}{2} \times 13\frac{1}{2}$	74.250	1,127.672	167.063
6 × 16	$5\frac{1}{2} \times 15\frac{1}{2}$	85.250	1,706.776	220.229
8 × 8	$7\frac{1}{2} \times 7\frac{1}{2}$	57.750	263.672	70.313
8 × 10	$7\frac{1}{2} \times 9\frac{1}{2}$	71.250	535.859	112.813
8 × 12	$7\frac{1}{2} \times 11\frac{1}{2}$	86.250	950.547	165.313
10 × 10	$9\frac{1}{2} \times 9\frac{1}{2}$	90.250	678.755	142.896
10 × 12	$9\frac{1}{2} \times 11\frac{1}{2}$	109.250	1,204.026	209.396
12 × 12	$11\frac{1}{2} \times 11\frac{1}{2}$	132.250	1,457.505	253.479

Dressed and matched*		Flooring		Shiplap	
S2S and CM		1 × 2	$\frac{3}{4} \times 1\frac{1}{8}$	1 × 4	$\frac{3}{4} \times 3\frac{1}{8}$
1 × 4	$\frac{3}{4} \times 3\frac{1}{8}$	1 × 3	$\frac{3}{4} \times 2\frac{1}{8}$	1 × 6	$\frac{3}{4} \times 5\frac{1}{8}$
1 × 6	$\frac{3}{4} \times 5\frac{1}{8}$	1 × 4	$\frac{3}{4} \times 3\frac{1}{8}$	1 × 8	$\frac{3}{4} \times 6\frac{7}{8}$
1 × 8	$\frac{3}{4} \times 6\frac{7}{8}$	1 × 6	$\frac{3}{4} \times 5\frac{1}{8}$	1 × 10	$\frac{3}{4} \times 8\frac{7}{8}$
1 × 10	$\frac{3}{4} \times 8\frac{7}{8}$			1 × 12	$\frac{3}{4} \times 10\frac{7}{8}$
1 × 12	$\frac{3}{4} \times 10\frac{7}{8}$				

*This lumber is also available in other thicknesses.

Select grades are used for trim, facings, moldings, etc., which may be finished in natural color or painted. Grade A is the best quality, and grade D is the poorest. Common grades are used for structural members such as sills, joists, studs, rafters, planks, etc. Grade 1 is the best quality, and grade 3 is the poorest. The lumber used for timber structures is generally designated as stress grade. For example, timber marked with a "P" is designated for posts and timber marked with a "B" is designated for beams.

PLYWOOD

Plywood is used for subfloors, siding, roof decks, and other similar purposes. Panels of plywood are available in 4-ft widths and 8-ft lengths with thicknesses from $\frac{1}{4}$ in. through $1\frac{1}{8}$ in. Larger sizes are available, such as 5 ft wide and 12 ft long.

Plywood is manufactured with odd numbers of wood plies, each placed at right angles to the adjacent ply, which accounts for the physical properties that make it efficient in resisting bending, shear, and deflection. Therefore, the position in which a panel is attached to the supporting members will determine its strength. For example, for a panel 4 ft wide and 8 ft long, the fibers of the surface plies are parallel to the 8-ft length. Such a panel, installed with the outer fibers perpendicular to the supporting members, such as studs, joists, or rafters, is stronger than it would be if the panel were attached with the outer fibers parallel to the supporting members.

Plywood is graded by the veneers of the plies. Veneer grade N or A is the highest grade level. Grade N is intended for a smooth surface and a natural finish, while grade A is intended for a smooth surface without a natural finish. Plywood graded as N or A has no knots or restricted patches. Veneer grade B has a solid surface, but may have small round knots, patches, and round plugs. Grade C has small knots, knotholes, and patches. Grade D has larger and more numerous knots, knotholes, and patches. Plywood can be purchased with different grades on opposite sides. For example, a plywood sheet may be $\frac{1}{2}$-in. AC, which designates a $\frac{1}{2}$-in.-thick panel with grade A on one side and grade C on the opposite.

The most commonly used thicknesses of plywood, in inches, are $\frac{1}{2}$, $\frac{5}{8}$, $\frac{3}{4}$, $\frac{7}{8}$, 1, and $1\frac{1}{8}$. Table 12.2 gives the thickness and approximate weight of 4-ft by 8-ft plywood panels.

TABLE 12.2 | Weights of 4-ft × 8-ft plywood panels.

Thickness, in.	Size, width × length	Approximate weight, lb
$\frac{1}{2}$	4 ft × 8 ft	48.0
$\frac{5}{8}$	4 ft × 8 ft	57.6
$\frac{3}{4}$	4 ft × 8 ft	70.4
$\frac{7}{8}$	4 ft × 8 ft	83.2
1	4 ft × 8 ft	96.0
$1\frac{1}{8}$	4 ft × 8 ft	105.6

COST OF LUMBER

Lumber is usually priced by the board foot or 1,000 bf, with the exception of certain types used for trim, moldings, etc., which may be priced by the linear foot. The cost varies with many factors, including:

1. Species of lumber: pine, fir, oak, birch, mahogany, redwood, etc.
2. Grade of lumber: select (A, B, C, or D) or common (1, 2, or 3)
3. Size of pieces: thickness, width, and length
4. Extent of milling required: rough, surfaced, or dressed and matched
5. Whether dried or green
6. Quantity purchased
7. Freight cost from mill to destination

The better grades of lumber cost more than the poorer grades. Select grades are more expensive than common grades. In the select grades, the costs decrease in the order A, B, C, and D, while the costs of common grades decrease in the order 1, 2, and 3.

The cost of lumber is higher for thick and wide boards than for thin and narrow ones. Lengths in excess of 20 to 24 ft may require special orders. Prices quoted may be for lengths through 20 ft, with increasingly higher prices charged for greater lengths.

Rough lumber is the product of the sawing operation, with no further surfacing operation. After the lumber is sawed, it may be given a smooth surface on one or more sides and edges, in which case it is designated *surfaced one edge* (S1E), *surfaced two sides* (S2S), or *surfaced on all edges and sides* (S4S). It may be surfaced on both sides with the edges matched with tongue and grooves, designated as dressed and matched, D and M. These operations reduce the thickness and width and increase the cost.

Freshly cut lumber contains considerable moisture that should be removed before the lumber is used, to prevent excessive shrinkage after it is placed in the structure. It usually is placed in a heated kiln to expel the excess moisture.

NAILS AND SPIKES

The sizes of common wire nails are designated by pennyweight, which specifies the diameter and length of the nail. For example, a common nail designated as 8d pennyweight is 0.131 in. in diameter and $2\frac{1}{2}$ in. long, whereas a 10d pennyweight nail is 0.148 in. in diameter and 3 in. long.

Common nails are used for attaching wood members together. Spikes are used to attach large-size members. Casing and finishing nails are used to attach members that will be treated with a stain or other type of surface that will be exposed. Shingle nails are used to attach shingles. The trade names, sizes, and weights of the more popular nails and spikes are given in Table 12.3. The approximate pounds of nails required for carpentry are given in Table 12.4.

TABLE 12.3 | Names, sizes, and numbers of nails and spikes per pound.

Size	Length, in.	Trade name					Roofing			
		Common	Spikes	Casing	Finishing	Shingle	Barbed	No. 8	No. 9	No. 10
	$\frac{3}{4}$	—	—	—	—	—	714	205	252	290
	$\frac{7}{8}$	—	—	—	—	—	469	179	219	253
2d	1	876	—	1,010	1,351	—	411	158	193	224
3d	$1\frac{1}{4}$	568	—	635	807	429	251	128	156	183
4d	$1\frac{1}{2}$	316	—	473	584	274	176	108	131	154
5d	$1\frac{3}{4}$	271	—	406	500	235	151	93	113	133
6d	2	181	—	236	309	204	103			
7d	$2\frac{1}{4}$	161	—	210	238					
8d	$2\frac{1}{2}$	106	—	145	189					
9d	$2\frac{3}{4}$	96	—	132	172					
10d	3	69	41	94	121					
12d	$3\frac{1}{4}$	63	38	87	113					
16d	$3\frac{1}{2}$	49	30	71	90					
20d	4	31	23	52	62					
30d	$4\frac{1}{2}$	24	17	46						
40d	5	18	13	35						
50d	$5\frac{1}{2}$	14	10							
60d	6	11	9							
	7	—	7							
	8	—	4							
	9	—	$3\frac{1}{2}$							
	10	—	3							
	12	—	$2\frac{1}{2}$							

Two methods are used for driving nails into wood, hand driven or by pneumatic nail guns. For rough carpentry framing, the use of nail guns can substantially reduce the time for driving nails, particularly for repetitive work such as roofing shingles.

BOLTS AND SCREWS

Although nails are the most commonly used fasteners for joining wood members, bolts are frequently used for timber structures. To attach timber members, bolts are available in sizes varying from $\frac{1}{4}$ in. to larger than 1 in. in diameter and varying from 1 in. to any desired length. The length does not include the head. Bolts with square heads and either square or hexagon nuts are used. Two washers should be used with every bolt, one for the head and one for the nut. If the head and the nut bear against steel plates, it is satisfactory to use stamped-steel washers; but if there are no steel plates under the heads and nuts, washers should be used to increase the bearing between the bolt and the timber member. Table 12.5 gives weights of bolts with square heads and hexagonal nuts in pounds per hundred.

Lag screws are sometimes used to securely fasten wood members. These screws are available in sizes from $\frac{1}{4}$-in. to 1-in. shank diameters and various

TABLE 12.4 | Approximate pounds of nails required for carpentry (includes 10% waste).

Size and type of lumber	Number of nails per support	Size and kind of nails	Pounds per 1,000 bf of lumber for nail spacing			
			12 in.	16 in.	20 in.	24 in.
General framing						
1 × 4	2	8d common	63	47	38	31
1 × 6	2	8d common	41	31	25	21
1 × 8	2	8d common	32	24	19	16
1 × 10	3	8d common	37	28	22	19
1 × 12	3	8d common	31	23	18	15
2 × 4	2	20d common	110	83	66	55
2 × 6	2	20d common	72	54	42	36
2 × 8	2	20d common	56	42	34	28
2 × 10	3	20d common	65	49	40	33
2 × 12	3	20d common	56	42	34	28
3 × 4	2	60d common	204	153	122	102
3 × 6	2	60d common	136	102	82	68
3 × 8	2	60d common	101	76	61	51
3 × 10	3	60d common	124	93	74	62
3 × 12	3	60d common	104	78	62	52
Stud framing						
2 × 4	—	16d common	15	11	9	7
2 × 6	—	16d common	11	8	6	5
2 × 8	—	16d common	11	8	7	6
Siding						
Bevel, $\frac{1}{2}$ × 4	1	6d finish	12	9		
Bevel, $\frac{1}{2}$ × 6	1	6d finish	8	6		
Bevel, $\frac{1}{2}$ × 8	1	6d finish	6	$4\frac{1}{2}$		
Drop, 1 × 4	2	8d casing	51	38		
Drop, 1 × 6	2	8d casing	33	25		
Drop, 1 × 8	2	8d casing	27	20		
Joist framing						
2 × 6		16d common	17			
2 × 8		16d common	10			
2 × 10		16d common	8			
2 × 12		16d common	7			
Bridging						
1 × 4	4 each	8d common	83			
Rafters						
2 × 4		16d common	18			
2 × 6		16d common	18			
2 × 8		16d common	15			
2 × 10		16d common	12			
Flooring						
1 × $2\frac{1}{4}$		3d finish	6			
Wood shingles						
$4\frac{1}{2}$ in. exposed	2	3d shingle	4.5			
5 in. exposed	2	3d shingle	4.1			
$5\frac{1}{2}$ in. exposed	2	3d shingle	3.8			
Asphalt shingles	2	$\frac{7}{8}$-in. barbed	4.2			

Note: If different sizes of common nails are used, make the following changes: increase the weight by 53 percent when substituting 10d for 8d nails; increase the weight by 58 percent when substituting 20d for 16d nails.

TABLE 12.5 | Weights of bolts with square heads and hexagonal nuts in pounds per 100.

Length under head, in.	\(\frac{1}{4}\)	\(\frac{5}{16}\)	\(\frac{3}{8}\)	\(\frac{7}{16}\)	\(\frac{1}{2}\)	\(\frac{5}{8}\)	\(\frac{3}{4}\)	\(\frac{7}{8}\)	1	\(1\frac{1}{8}\)	\(1\frac{1}{4}\)
					Diameter of bolt, in.						
1	2.7	5.0	7.2	11.2	14.9	28	43	—	—	—	—
\(1\frac{1}{4}\)	3.1	5.5	8.0	12.2	16.3	30	46	68	—	—	—
\(1\frac{1}{2}\)	3.4	6.1	8.8	13.3	17.7	32	49	73	103	144	190
\(1\frac{3}{4}\)	3.8	6.6	9.6	14.4	19.0	35	52	77	109	151	199
2	4.1	7.2	10.4	15.4	20.4	37	55	81	115	158	208
\(2\frac{1}{4}\)	4.5	7.7	11.1	16.5	21.8	39	58	85	120	165	216
\(2\frac{1}{2}\)	4.8	8.2	11.9	17.5	23.2	41	61	90	126	172	225
\(2\frac{3}{4}\)	5.2	8.8	12.7	18.6	24.6	43	64	94	131	179	234
3	5.5	9.3	13.5	19.7	26.0	45	68	98	137	187	242
\(3\frac{1}{4}\)	5.9	9.9	14.3	20.7	27.4	48	71	102	142	194	251
\(3\frac{1}{2}\)	6.2	10.4	15.1	21.8	28.8	50	74	107	148	201	260
\(3\frac{3}{4}\)	6.6	11.0	15.8	22.9	30.2	52	77	111	153	208	268
4	6.9	11.5	16.6	23.9	31.6	54	80	115	159	215	277
\(4\frac{1}{4}\)	7.3	12.0	17.4	25.0	33.0	56	83	119	165	222	286
\(4\frac{1}{2}\)	7.6	12.6	18.2	26.1	34.4	58	86	124	170	229	294
\(4\frac{3}{4}\)	8.0	13.1	19.0	27.1	35.7	61	89	128	176	236	303
5	8.3	13.7	19.8	28.2	37.1	63	93	132	181	243	312
\(5\frac{1}{4}\)	8.6	14.2	20.5	29.3	38.5	65	96	136	187	250	321
\(5\frac{1}{2}\)	9.0	14.8	21.3	30.3	39.9	67	99	141	192	257	329
\(5\frac{3}{4}\)	9.3	15.3	22.1	31.4	41.3	69	102	145	198	264	338
6	9.7	15.9	22.9	32.4	42.7	71	105	149	204	271	347
\(6\frac{1}{4}\)	10.0	16.4	23.7	33.5	44.1	74	108	153	209	278	355
\(6\frac{1}{2}\)	10.4	16.9	24.5	34.6	45.5	76	111	158	215	285	364
\(6\frac{3}{4}\)	10.7	17.5	25.2	35.6	46.9	78	114	162	220	292	373
7	11.1	18.0	26.0	36.7	48.3	80	118	166	226	299	381
\(7\frac{1}{4}\)	11.4	18.6	26.8	37.8	49.7	82	121	170	231	306	390
\(7\frac{1}{2}\)	11.8	19.1	27.6	38.8	51.1	84	124	175	237	313	399
\(7\frac{3}{4}\)	12.1	19.7	28.4	39.9	52.4	87	127	179	242	320	407
8	12.5	20.2	29.2	41.0	53.8	89	130	183	248	327	416
\(8\frac{1}{2}\)	—	21.3	30.7	43.1	56.6	93	136	192	259	341	434
9	—	22.4	32.3	45.2	59.4	98	143	200	270	356	451
\(9\frac{1}{2}\)	—	23.5	33.9	47.4	62.2	102	149	209	281	370	468
10	—	24.6	35.4	49.5	65.0	106	155	217	293	384	486
\(10\frac{1}{2}\)	—	—	37.0	51.6	67.8	111	161	226	304	398	503
11	—	—	38.6	53.7	70.5	115	168	234	315	412	520
\(11\frac{1}{2}\)	—	—	40.1	55.9	73.3	119	174	243	326	426	538
12	—	—	41.7	58.0	76.1	124	180	251	337	440	555
\(12\frac{1}{2}\)	—	—	—	60.1	78.9	128	186	260	348	454	573
13	—	—	—	62.3	81.7	132	193	268	359	468	590
\(13\frac{1}{2}\)	—	—	—	64.4	84.5	137	199	277	370	482	607
14	—	—	—	66.5	87.2	141	205	285	382	496	625
\(14\frac{1}{2}\)	—	—	—	—	90.0	145	211	294	393	510	642
15	—	—	—	—	92.8	150	218	302	404	525	660
\(15\frac{1}{2}\)	—	—	—	—	95.6	154	224	311	415	539	677
16	—	—	—	—	98.4	158	230	320	426	553	694
Per inch additional	1.4	2.2	3.1	4.3	5.6	8.7	12.5	17.0	22.3	28.2	34.8

lengths. Lag screws require prebored holes of the proper sizes. The lead hole for the shank should be of the same diameter as the shank. The diameter of the lead hole for the threaded part of the screw should vary from 40 to 50 percent of the shank diameter for lightweight species of wood, and 60 to 85 percent for dense hardwoods. During installation, some type of lubricant should be used on the lag screw or in the lead hole to facilitate insertion and to prevent damage to the lag screw. Lag screws should be turned, not driven, into the wood.

TIMBER CONNECTORS

The strength of joints between wood members can be increased substantially by using timber connectors. They are often used to securely connect wood members in prefabricated wood trusses. Several types are available, including split rings, shear plates, flat grids, flat clamping plates, and flanged clamping plates. Figure 12.1 illustrates various types of timber connectors. Table 12.6 gives weights and dimensions of timber connectors.

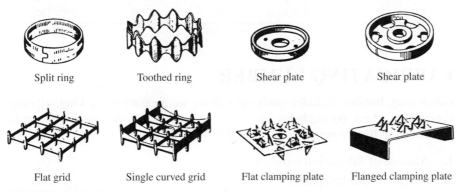

| Split ring | Toothed ring | Shear plate | Shear plate |

| Flat grid | Single curved grid | Flat clamping plate | Flanged clamping plate |

FIGURE 12.1 | Timber connectors.

Split rings, in sizes with $2\frac{1}{2}$ in. and 4 in. diameters, are used with wood-to-wood connections for heavy loads. It is necessary to drill a hole for the bolt and a groove for the ring. A complete connector unit includes a split ring, bolt, and two steel washers. Toothed rings in diameters of $2\frac{5}{8}$ in. and 4 in. are used for wood-to-wood connections in thin members with light loads. A complete unit includes a toothed ring, bolt, and two steel washers.

Spike grids and flat clamping plates, made of malleable metal, are used to joint two flat wood members, or connecting gusset plates for splicing the ends of multiple members in roof trusses. Metal plate connectors are available as hangers for connecting the end of joists to their supporting members. Toothed ring, spiked grid, and flat clamping plate types of connectors are installed by applying pressure against the assembly to force the connectors to penetrate the wood members.

TABLE 12.6 | Weights and dimensions of timber connectors.

Timber connector	Size, in.	Dimensions of metal, in.		Shipping weight per 100, pound
		Depth	**Thickness**	
Split rings	$2\frac{1}{2}$	0.75	0.163	28
	4	1.00	0.193	70
Toothed rings	2	0.94	0.061	9
	$2\frac{5}{8}$	0.94	0.061	12
	$3\frac{3}{8}$	0.94	0.061	15
	4	0.94	0.061	18
Shear plates				
Pressed steel	$2\frac{5}{8}$	0.375	0.169	35
Malleable iron	4	0.62	0.20	90
Spike grids				
Flat	$4\frac{1}{8} \times 4\frac{1}{8}$	1.00	—	48
Single curve	$4\frac{1}{8} \times 4\frac{1}{8}$	1.38	—	70
Circular	$3\frac{1}{8}$ diam.	—	—	26
Clamping plates				
Plain	$5\frac{1}{4} \times 5\frac{1}{4}$	—	0.077	59
Flanged	$5 \times 8\frac{1}{2}$	2.00	0.122	190

FABRICATING LUMBER

Fabricating lumber includes such operations as measuring, sawing, ripping, chamfering edges, boring holes, etc. The rates at which these operations can be performed will vary greatly with several factors, including:

1. Amount of fabrication required
2. Length of finished pieces
3. Size of pieces to be fabricated
4. Number of similar pieces fabricated with one machine setting
5. Care in sorting and storing lumber at the job
6. Clarity and amount of details on the drawing
7. Skill of the carpenter and quality of supervision

Cutting multiple pieces of lumber with table saws and radial arm saws requires considerably less time than when cutting individual members with hand saws. It requires considerably more time to fabricate rafters and stair stringers than it does to fabricate studs and joists.

Since a certain amount of time is required to handle and measure lumber, more time is required to fabricate a given quantity of lumber consisting of short lengths and small sections than when it consists of long lengths and large sections. Time is required to set a machine for a given operation. If only a few pieces are fabricated, after which the machine must be reset, the total time per piece will be higher than when many similar pieces are fabricated.

ROUGH CARPENTRY

HOUSE FRAMING

Houses can be framed by several methods, as illustrated in Fig. 12.2. For the balloon frame, the studs for the outside walls of a two-story building extend from

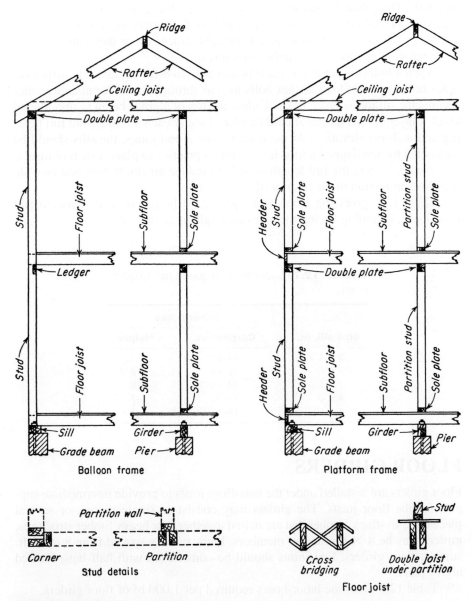

Balloon frame

Platform frame

Corner

Partition wall

Partition

Stud details

Cross bridging

Double joist under partition

Stud

Floor joist

FIGURE 12.2 | Typical wall section through a frame house.

the sill to the top plate on which the rafters rest. For the platform frame, the studs for the external walls are limited to one story in length. The lengths of studs for interior or partition walls for both types of frames are limited to one story.

SILLS

Sills are attached to grade beams, which are usually constructed of concrete or masonry. The grade beams may rest directly on the ground, or they may be placed on drilled concrete piers spaced at 4- to 6-ft intervals. The sills may be 2 × 4, 2 × 6, and 2 × 8 S4S lumber for light structures, or they can be 4 × 6 or larger members for heavy timber structures.

Anchor bolts are set in the grade beams at 6- to 8-ft intervals. The sills have holes bored to permit the anchor bolts to pass through them. A nut and washer secures the sill to the grade beam. Unless the top of a grade beam is smooth and level, a layer of stiff mortar should be placed under each sill to ensure full bearing and uniform elevation. At the corners and at end joints, the sills should be connected by half-lapped joints. It is common practice to place a strip of insulation material along the full length of sills to reduce air infiltration and provide heat-loss protection to the interior of the structure.

Table 12.7 gives the labor-hours per 1,000 bf required to fabricate and install sills, including boring the holes and tightening the nuts.

TABLE 12.7 | Labor-hours required per 1,000 bf of sills.

	Labor-hours	
Size sill, in.	Carpenters	Helper
2 × 4	24–30	8–10
2 × 6	18–22	6–8
2 × 8	14–18	5–6
4 × 6	15–20	5–7
4 × 10	14–19	5–7

FLOOR GIRDERS

Floor girders are installed under the first-floor joists to provide intermediate support for the floor joists. The girders may consist of single pieces, or several pieces of 2-in.-thick lumber that are nailed together. For heavy timber structures, girders may be 4 × 6 or larger members. Piers, usually spaced from 8 to 10 ft, support floor girders. End joints should be constructed with half-laps, located above piers.

Table 12.8 gives the labor-hours required per 1,000 bf of floor girders.

TABLE 12.8 | Labor-hours required per 1,000 bf of floor girders.

Number of girders and size, in.	Labor-hours	
	Carpenters	Helper
1—2 × 6	20–25	6–8
1—2 × 8	17–21	6–7
1—2 × 10	14–18	5–6
1—2 × 12	13–16	4–5
2—2 × 6	12–14	4–5
2—2 × 8	9–11	3–4
2—2 × 10	7–10	2–3
2—2 × 12	6–8	2–3
3—2 × 6	9–11	3–4
3—2 × 8	7–8	5–6
3—2 × 10	6–7	4–5
3—2 × 12	5–6	3–4
1—4 × 6	15–20	5–7
1—6 × 8	13–18	5–6
1—8 × 10	11–16	4–6

FLOOR AND CEILING JOISTS

Floor and ceiling joists are generally spaced 12, 16, 18, or 24 in. on centers, using 2 × 6, 2 × 8, 2 × 10, or 2 × 12 S4S lumber. The size and spacing depends on the span length of the joists. When the unsupported lengths exceed 8 ft, one row of cross bridging should be installed at spacings not greater than 8 ft. The bridging may be made of 1 × 4 lumber, with two 8d nails at each end of each strut. Joists may be 4 × 6 or larger members for heavy timber structures.

Table 12.9 gives labor-hours required per 1,000 bf of joists.

TABLE 12.9 | Labor-hours required per 1,000 bf of joists.

Size joist, in.	Labor-hours	
	Carpenters	Helper
2 × 4	13–16	4–5
2 × 6	11–13	4–5
2 × 8	9–12	4–5
2 × 10	8–11	4–5
2 × 12	7–10	4–5
Cross bridging per 100 sets		
1 × 3	5–6	1–2
2 × 3	5–6	1–2

EXAMPLE 12.1

Wood framing for a house requires 2 × 6 ceiling joists at 18 in. on centers. The structure is 35 ft wide and 90 ft long. The joists will bear on the two exterior walls and an interior wall that extends the full 90-ft length. A 1-ft overlap will be placed across the interior wall. Two rows of 1 × 3 bridging will be installed the full 90-ft length of the structure. Estimate the cost of material and labor for framing the joists, assuming $0.55/bf for lumber, $28.00/hr for carpenters, and $21.00/hr for laborers.

Quantities of material:

Joists:

Joist spacing = 18 in./(12 in./ft) = 1.5 ft

Number of joists = [length of structure/spacing of joists] + 1

= [(90 ft)/(1.5 ft/joist)] + 1 = 61 joists

Length of joists, assuming 18 ft/board: 61 joists × 2 boards × 18 ft
= 2,196 lin ft

Board feet of joists: 2,196 lin ft × [(2 × 6)/(1 × 12)] = 2,196 bf

Bridging:

Number of sets of bridging: 61 rows × 2 sets/row = 122 sets

Length of 1 × 3 bridging: 1.5-ft boards × 2/set × 122 sets = 366 lin ft

Board feet of bridging: 366 lin ft × [(1 × 3)/(1 × 12)] = 91.5 bf

Labor-hours required:

From Table 12.9 for framing joists and bridging:

Carpenters framing joists: 2,196 bf × 12 hr/1,000 bf = 26.4 hr

Carpenters framing sets: 122 sets × 5.5 hr/100 sets = 6.7 hr

Total carpenter hours = 33.1 hr

Laborers framing joists: 2,196 bf × 4.5 hr/1,000 bf   = 9.9 hr

Laborers framing bridging: 122 sets × 1.5 hr/100 sets = 1.8 hr

Total laborer hours = 11.7 hr

Cost of material and labor:

Joists: 2,196 bf @ $0.55/bf   = $1,207.80

Bridging: 91.5 bf @ $0.55/bf   =   50.33

Carpenters: 33.1 hr @ $28.00/hr =   926.80

Laborers: 11.7 hr @ $21.00/hr   =   245.70

Total cost = $2,430.63

Cost per board foot of lumber: $2,430.63/2,196 bf = $1.12/bf

Cost per linear foot of joist: $2,430.63/2,135 lin ft = $1.14/lin ft

In this example, the crew may consist of 3 carpenters and 1 laborer. The cost of nails, tools, job supervision, overhead, and profit are excluded.

STUDS FOR WALL FRAMING

Studs for one- and two-story houses usually are 2 × 4, spaced at 16 in. on centers, or 2 × 6 lumber, spaced at 24 in. on centers for exterior walls. The 2 × 6 studs permit installation of more insulation for reduced heat-loss and provide passages for plumbing pipes. Interior walls are generally 2 × 4 size studs.

At all corners and where partition walls frame into other walls, it is common practice to install three studs, as illustrated in Fig. 12.3. At all openings for windows and doors, two studs should be used, with one stud on each side of the opening cut to support the header over the opening.

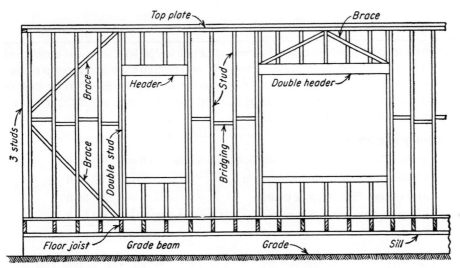

FIGURE 12.3 | Typical framing detail for studs.

Studs are generally purchased in standard 8-ft lengths. If the wall is taller than a standard 8-ft height, then it is necessary to cut all the studs for a wall of a room to the proper length. After the studs are cut to length, they are properly spaced and nailed to a bottom and top plate on the subfloor, then tilted into position. A second top plate can then be installed to join the sections or to tie partition walls to main walls. If 2 × 6 studs spaced at 24 in. on centers are used for exterior walls and the roof trusses spaced at 24 in. on centers are set directly on top of each stud, then the second top plate is sometimes omitted. Lateral rigidity for walls can be provided by plywood, diagonal steel tension straps, diagonal sheathing, or cut-in braces installed at each corner of the building.

Table 12.10 gives labor-hours required per 1,000 bf for wall framing consisting of studs with one bottom and two top plates.

TABLE 12.10 | Labor-hours required per 1,000 bf of stud wall framing with 1 bottom and 2 top plates.

Size studs, in.	Stud spacing, in.	Labor-hours	
		Carpenters	**Helper**
2 × 4	12	20–25	5–6
	16	22–26	5–6
	24	24–28	5–6
2 × 6	12	19–24	5–6
	16	21–25	5–6
	24	23–27	5–6

Note: These values apply to wall heights up to 10 ft. For walls over 10 ft high, increase these values by 25 percent.

EXAMPLE 12.2

The total length of walls of a wood-framed structure is 750 lin ft. The wall consists of 2 × 4 studs, 8 ft long, at 16 in. on centers with one bottom plate and two top plates. An evaluation of the drawings shows there will be 42 corners. Estimate the cost of material and labor, assuming lumber is $0.55/bf and labor rates of $28.00/hr for carpenters and $21.00/hr for laborers.

Quantity of material:

Studs: [750 lin ft/(16/12)] = 562.5, or 563 studs × 8 ft = 4,504 lin ft

Bottom plates: 750 lin ft × 1 board   =   750 lin ft

Top plates: 750 lin ft × 2 boards   = 1,500 lin ft

Corners: 42 corners × 3 boards = 126 boards × 8 ft = 1,008 lin ft

Total = 7,762 lin ft

Total board feet of lumber required: 7,762 lin ft × [(2 × 4)/(1 × 12)] = 5,175 bf

Labor-hours required:

From Table 12.10:

Carpenters: 5,175 bf × 24 hr/1,000 bf = 124.2 hr

Helpers: 5,175 bf × 5.5 hr/1,000 bf = 28.5 hr

Cost of material and labor:

Lumber: 5,175 bf @ $0.55/bf   = $2,846.25

Carpenter: 124.2 hr @ $28.00/hr =   3,477.60

Laborers: 28.5 hr @ $21.00/hr   =   598.50

Total = $6,922.35

Cost per board foot in place: $6,922.33/5,175 = $1.34/bf

Cost per linear foot of wall: $6,922.33/750 lin ft = $9.23/lin ft

For this example, the crew may consist of 4 carpenters and 1 laborer. The cost of nails, tools, job supervision, overhead, and profit are excluded.

FRAMING FOR WINDOW AND DOOR OPENINGS

Many sizes of windows are available, ranging from 2 ft to 5 ft in width and from 3 ft to 6 ft in height. Larger widths and heights are sometimes specified for a particular project. The size of a window is designated by its width and height. For example, a window that is 2 ft 6 in. wide and 5 ft 0 in. high is designated as a 2-6 5-0 window. Headers made of 2 × 6, 2 × 8, 2 × 10, or 2 × 12 lumber are framed above window openings. Each side of the window opening is framed with a trim stud that is the same size as the wall studs and a horizontal board, called a sill, which is framed along the bottom opening of the window. For windows greater than 6 ft wide, two trim studs are installed in each side of the window.

The standard height of doors is 6 ft 8 in. Common widths of door openings include 2 ft, 2 ft 8 in., 3 ft, 3 ft 6 in., and 4 ft. Larger heights and widths are sometimes specified for a particular project. The size of a door is designated by its width and height. For example, a door that is 3 ft 0 in. wide and 6 ft 8 in. high is designated as a 3-0 6-8 door. The headers above doors may be 2 × 6 lumber or larger. The vertical sides of door openings are called door jambs.

The quantity of lumber required to frame window and door openings depends on the width and height of the opening, which can be calculated from the drawings for a particular project. To calculate the quantity, some estimators do not deduct the size of the opening when calculating the lumber required to frame the stud walls, assuming the quantity of lumber required to frame the opening is approximately equal to the lumber that is required to frame the wall.

Table 12.11 gives the approximate labor-hours per 1,000 bf of lumber to frame openings for doors and windows.

TABLE 12.11 | Labor-hours required to frame openings for doors and windows in walls.

Type and size of opening	Labor-hours	
	Carpenters	**Helper**
Windows:		
Up to 5 ft wide	1.0–1.5	0.5–0.8
Over 5 ft wide	1.5–2.5	0.8–1.0
Doors		
Up to 7 ft high	0.8–1.0	0.5–0.8
Over 7 ft high	1.0–1.2	0.8–1.0

RAFTERS

One method of roof construction is called joist and rafter framing, as illustrated in Fig. 12.2. Rafters are the upper members of roofs. They are notched where they rest on the wall plates and are held in place by nailing them to the wall plates or by use of metal framing anchors. The ridge board is placed at the top and perpendicular to rafters. A horizontal collar beam is often placed below the collar ridge beam to assist in resisting wind load on the roof. For long span lengths, a diagonal knee brace is sometimes installed from the center of each rafter down to the interior wall.

The labor required to frame and erect rafters will vary considerably with the type of roof. The rise (vertical) over run (horizontal) dimension measures the steepness of a roof. Typical rise over run for roofs includes 3:12, 4:12, 5:12, and 6:12.

There are many configurations of roofs. Simple gable roofs are easier to frame than hip roofs. Double-pitch or gable roofs, with no dormers or gables, will require the least amount of labor, while hip roofs, with dormer or gables framing into the main roof, will require the greatest amount of labor.

Table 12.12 gives labor-hours required per 1,000 bf for rafters.

TABLE 12.12 | Labor-hours required per 1,000 bf of rafters.

Type of roof and size of rafter, in.	Labor-hours	
	Carpenters	Helper
Gable roofs—no dormers		
2 × 4	27–31	7–8
2 × 6	25–29	7–8
2 × 8	24–28	6–7
2 × 10	22–26	6–7
Hip roofs—no dormers		
2 × 4	30–34	8–9
2 × 6	29–33	8–9
2 × 8	28–32	7–8
2 × 10	26–30	7–8

EXAMPLE 12.3

Estimate the material and labor cost to frame rafters for a structure, 35 ft wide and 90 ft long. The slope (rise-over-run) of rafters is 5:12 and each rafter will extend 2 ft outside of each exterior wall. The 2 × 6 rafter spacing will be spaced at 18 in. on centers.

A 2 × 8 ridge board will extend the full 90 ft dimension of the roof. Horizontal collar ties of 1 × 6 lumber will be placed across each rafter at a distance of 2 ft below the peak of the rafters. Diagonal knee braces of 2 × 4 lumber will brace the midpoint of each rafter down to the center wall (see Fig. 12.4). Assume $0.55/bf for lumber, $28.00/hr for carpenters, and $21.00/hr for laborers.

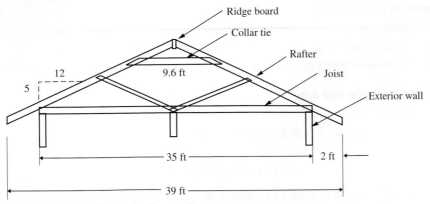

FIGURE 12.4 | Typical rafter section.

Quantities of material:

Rafters:

Total width of structure: 35 ft + 2 overhangs @ 2 ft = 39 ft total width

Diagonal dimension for a 5:12 slope: $\sqrt{(5)^2 + (12)^2} = 13$ ft

Length of rafter: 39 ft × (13/12) = 42.3 ft, therefore need two 22-ft-long boards

Number of rafters: [(length/spacing) + 1] = [90 ft/(18/12) + 1] = 61 rafters

Linear feet of rafter: 61 rafters × 42.3 ft/rafter = 2,580 lin ft

Collar ties:

Horizontal length at 2 ft below peak of rafters with a 5:12 slope:

[2 ft × (12/5)] × 2 sides = 9.6 ft, therefore need 10-ft-long boards

Ridge board:

Total length = 90 ft

Number of boards required using 18-ft lengths: (90 ft)/(18 ft/board)
= 5 boards

Knee braces:

Total length: 17.5 ft × (13/12) = 18.96 ft, therefore, need two 10-ft-long boards

Board feet of lumber:

Rafters: 61 rafters × 2 boards/rafter × 22 ft × [(2 × 6)/(1 × 12)] = 2,684 bf

Collar ties: 61 ties × 10 ft/tie × [(1 × 6)/(1 × 12)]   =   305 bf

Ridge board: 5 boards × 18 ft/board × [(2 × 8)/(1 × 12)]   =   120 bf

Knee braces:

61 rafters × 2 braces/rafter × 10 ft × [(2 × 4)/(1 × 12)]   =   813 bf

Total lumber = 3,922 bf

Labor-hours required:

Using average values from Table 12.12:

Carpenters: 3,922 bf × (27 hr/1,000 bf) = 105.9 hr

Laborers: 3,922 bf × (7.5 hr/1,000 bf) = 29.4 hr

Cost of material and labor:

Lumber: 3,922 bf @ $0.55/bf   = $2,157.10

Carpenters: 105.9 hr @ $28.00/hr =   2,965.20

Laborers: 29.4 hr @ $21.00/hr   =   617.40

Total cost = $5,739.70

Cost per board foot: $5,739.70/3,922 bf = $1.46/bf

Cost per linear foot of rafter: $5,739.70/2,580 lin ft = $2.22/lin ft

For this example, the crew may consist of 4 carpenters and 1 laborer. The cost of nails, tools, job supervision, overhead, and profit are excluded.

PREFABRICATED ROOF TRUSSES

Prefabricated roof trusses are sometimes used in place of the joist and rafter methods of framing that were discussed earlier. The trusses are prefabricated to prescribed span lengths and slopes at the shop of a supplier of wood materials. The supplier then delivers the roof trusses in bundles to the job where they are erected in place.

Prefabricated wood roof trusses are available in long span lengths, from 25 to 50 ft, with the entire roof load supported by the exterior walls. Each truss consists of a top and bottom chord, with a series of web members laced throughout the truss to stabilize the chord members. The top and bottom chords may be 2 × 4 or 2 × 6 S4S lumber and the web members are generally 2 × 4 sizes. Members are connected with metal plate connectors. The trusses are generally set at 24 in. on centers.

The joint connections of roof trusses are vulnerable to damage by excessive bending in the lateral direction. When they are delivered to the job, they must be stored in the horizontal direction with blocking placed at intervals of not more than 10 ft on a level surface and covered to minimize weather damage.

Short trusses can be erected with laborers by manually lifting them with the peak of the truss in the downward direction, then rotating the truss in the upright position and securely fastening them in place. Installing a board ramp from the ground to the top of the wall and pulling the truss up the ramp can reduce the amount of lifting. Longer trusses require a small crane or forklift with appropriate slings or spreader bars.

Proper lateral bracing of wood trusses is required during erection to ensure the safety of workers. The truss designer specifies permanent bracing, but the builder must provide temporary bracing until the permanent bracing is installed.

Proper bracing of the first end truss is especially important to obtain the correct alignment of the roof truss system.

The cost of prefabricated wood trusses is established by the supplier. Four carpenters and a small lifting crane can erect trusses for about 2,000 to 3,000 sf of floor area in one day.

ROOF DECKING

Roof deckings are usually constructed of $\frac{1}{2}$-, $\frac{5}{8}$-, or $\frac{3}{4}$-in.-thick sheets of plywood or oriented strand board (OSB). The thickness of the sheets will vary, depending on the required roof load and the spacing of the rafters. The decking is usually installed with a pneumatic nail gun. Table 12.13 gives the labor-hours required per 1,000 sf using plywood decking.

TABLE 12.13 I Labor-hours required per 1,000 sf of plywood or OSB roof decking.

Thickness of plywood, in.	Rafter spacing, in.	Labor-hours	
		Carpenters	Helper
$\frac{1}{2}$	12	10–12	2–3
	16	9–11	2–3
	18	8–10	2–3
$\frac{5}{8}$	16	11–13	2–3
	18	10–12	2–3
	20	9–11	2–3
$\frac{3}{4}$	16	12–14	2–3
	18	11–13	2–3
	24	10–12	2–3

WOOD SHINGLES

There are two types of wood shingles, red cedar and shakes. Red cedar shingles are available in lengths of 16 and 18 in., and in random widths. The shingles are about $\frac{3}{8}$ in. thick. These shingles are cut to relatively smooth surfaces and may be installed with an exposure of 5 or $7\frac{1}{2}$ in. on the roof. Prior to placing red cedar shingles, a 1 × 4 strip of lumber is nailed across the rafters of the roof. The spacing of the 1 × 4 lumber should correspond to the length of shingles exposed to the weather. Each shingle should be fastened to the 1 × 4 strips with $1\frac{3}{4}$- to 2-in.-long nails. Each shingle should be fastened with two nails.

Hand-split red cedar shakes have irregular surfaces and are available in 18- and 24-in. lengths with an $8\frac{1}{2}$- or 10-in. exposure on the roof. Two types of wood shakes are available, junior shakes and jumbo shakes. Junior shakes have thickness up to about $\frac{3}{4}$ in., whereas jumbo shakes have random thickness greater than $\frac{3}{4}$ in. These shingles are installed on a 30-lb organic felt underlayment.

Wood shingles are measured by the *square,* which is 100 sf. They are packaged in bundles, which will cover about 20 to 25 sf of roof, depending on the exposure. About 5 percent waste should be added for red cedar shingles and 10 percent waste for wood shakes.

In addition to using wood shingles on roofs, they are sometimes used for exterior wall siding. To estimate wood shingles on walls, the openings for doors and windows should be deducted from the total surface area of the walls.

Table 12.14 gives the labor-hours required to lay 100 sf of surface area for wood shingles.

TABLE 12.14 | Labor-hours required to install 100 sf of wood shingles.

Kind of work	Classification of labor	Length of exposure, in.			
		5	$7\frac{1}{2}$	$8\frac{1}{2}$	10
Shingle roofs, 16-in.	Carpenter	3.9	3.0	—	—
	Helper	0.5	0.4	—	—
Shingle roofs, 18-in.	Carpenter	2.9	2.2	—	—
	Helper	0.5	0.4	—	—
Shake roofs, 18-in.	Carpenter	—	—	4.4	4.0
	Helper	—	—	0.6	0.5
Shake roofs, 24-in.	Carpenter	—	—	3.5	3.2
	Helper	—	—	0.5	0.4
Shingle walls	Carpenter	4.3	3.2	3.0	2.9
	Helper	0.6	0.6	0.3	0.3

Note: Multiply the above values by 0.75 when pneumatic nail guns are used.

SUBFLOORS

Subfloors are usually constructed with plywood sheets. The thickness of the plywood will vary, depending on the required floor load and the spacing of the floor joists. A $\frac{5}{8}$- or $\frac{3}{4}$-in. thickness is common. An 8d common nail is placed at 6-in. spacing along the floor joists. It is a good practice to both glue and nail the plywood sheets to the floor joist to increase the strength of the subfloor and to reduce the tendency of a squeaking noise while someone is walking across the floor. When plywood is used for subflooring, the unit of measure is square foot. Table 12.15 gives labor-hours required per 1,000 sf using plywood subflooring.

TABLE 12.15 | Labor-hours required per 1,000 sf of plywood subflooring.

Thickness of plywood, in.	Joist spacing, in.	Labor-hours	
		Carpenters	Helper
$\frac{5}{8}$	12	10–11	2–3
	16	9–10	2–3
	18	8–9	2–3
$\frac{3}{4}$	16	11–12	2–3
	18	10–11	2–3
	24	9–10	2–3

EXTERIOR FINISH CARPENTRY

Finish carpentry involves placing final trim lumber to the rough carpentry. Trim carpenters usually perform their work without the assistance of laborers. They carefully measure, cut, and install the lumber with great accuracy because the lumber will be exposed after it is installed. The cost of finish carpentry will vary considerably with the species and grade of materials and the quality of the workmanship required.

After the walls and rafters are framed, trim boards are fabricated and installed to finish the exterior structure. Fascia boards are installed across the end of the rafters, and frieze boards are placed against the wall to provide support for the soffit, which is the horizontal member below the rafters that extends outward from the wall to the end of the roof, as illustrated in Fig. 12.5.

Interior finish carpentry is discussed later.

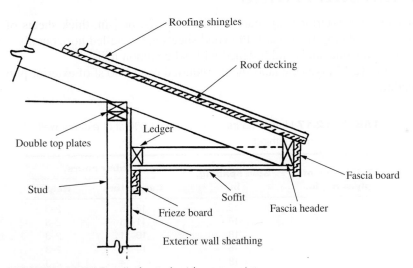

FIGURE 12.5 | Detail of exterior trim carpentry.

FASCIA, FRIEZE, AND CORNER BOARDS

The fascia, frieze, and corner boards are installed after the walls and rafters. The boards are usually 1-in.-thick S4S lumber. The quantity of lumber for these boards should be estimated by the 1,000 bf of lumber, but the labor-hours to place these boards are usually estimated by the 100 lin ft.

Table 12.16 gives the carpenter-hours to install 100 linear feet of fascia, frieze, and corner boards.

TABLE 12.16 | Carpenter-hours required per 100 lin ft of fascia, frieze, and corner boards.

Item	Carpenters
Fascia and frieze	4–5
Corner boards	3–4

SOFFITS

Soffits are measured by the square foot of area. Soffits may be $\frac{1}{4}$- or $\frac{1}{2}$-in. plywood, $\frac{3}{8}$-in. hardwood, or $\frac{5}{8}$-in.-thick wood fiber. The soffit is nailed to the ledger and fascia header. Typically, the soffits are installed by carpenters, without the help of laborers. It requires about 3.0 to 4.0 carpenter-hours to install 100 sf of soffit.

WALL SHEATHING

Sheathing for exterior walls may consist of $\frac{1}{2}$-, $\frac{5}{8}$-, or $\frac{3}{4}$-in.-thick sheets of plywood or insulating fiberboard. Plywood sheets are installed horizontally across the vertical wood studs and fastened with 8d common nails.

Table 12.17 gives the labor-hours required per 1,000 sf of exterior plywood sheathing.

TABLE 12.17 | Labor-hours required per 1,000 sf of exterior wall using plywood sheathing.

Thickness of plywood, in.	Stud spacing, in.	Labor-hours	
		Carpenters	Helper
$\frac{1}{2}$	12	13–15	2–3
	16	11–13	2–3
	18	10–11	2–3
$\frac{5}{8}$	16	14–16	2–3
	18	12–14	2–3
	20	11–13	2–3
$\frac{3}{4}$	16	15–17	2–3
	18	13–14	2–3
	24	12–14	2–3

AESTHETIC EXTERIOR SIDING

Several types of exterior wall treatment are used for aesthetic purposes, including board and batten siding, shiplap, and bevel siding. These types of exterior sidings are not commonly used, but occasionally they are selected for external siding of houses.

Board and batten siding usually consists of 1 × 10 or 1 × 12 S4S boards installed in the vertical direction with batts installed to cover the joints between boards. Underlayment plywood sheathing is usually installed before the board and batten siding is installed.

Rustic drop siding, which is available in shiplap patterns, and bevel siding are also sometimes used for aesthetic exterior wall siding. These sidings are placed in the horizontal direction across the studs.

Table 12.18 gives the approximate labor-hours per 1,000 bf for installing board and batten, shiplap, and bevel siding.

TABLE 12.18 | Approximate labor-hours required per 1,000 bf of aesthetic exterior wall siding.

Pattern and size, in.	Labor-hours	
	Carpenters	Helper
Board and battens		
1 × 10 S4S laid vertical	16–18	4–5
1 × 12 S4S laid vertical	15–17	4–5
Shiplap siding		
4-in. with $3\frac{1}{8}$-in. face	25–27	5–6
5-in. with $4\frac{1}{8}$-in. face	24–26	5–6
6-in. with $5\frac{1}{16}$-in. face	23–25	5–6
8-in. with $6\frac{7}{8}$-in. face	21–23	4–5
10-in. with $8\frac{7}{8}$-in. face	19–21	4–5
Bevel siding		
4-in. with $2\frac{3}{4}$-in. face	31–33	5–6
5-in. with $3\frac{3}{4}$-in. face	27–29	5–6
6-in. with $4\frac{3}{4}$-in. face	24–26	4–5
8-in. with $6\frac{3}{4}$-in. face	21–23	4–5

HEAVY TIMBER STRUCTURES

Heavy timber members are used extensively for temporary construction. The members are used for shoring, retainage walls, mats for outriggers of cranes, and numerous other applications. Also, some permanent structures are made entirely of heavy timber members.

Permanent structures made of heavy timber include churches, restaurants, auditoriums, gymnasiums, and other types of commercial facilities. The members, either glue laminated or sawn, are usually installed exposed and unfinished for a rustic appearance. Timber structures have much larger size wood than wood-framed structures, such as houses. Since the general public generally occupies these structures, architects and engineers design them, unlike residential wood structures that are designed by the builder who abides by local building codes and inspectors.

Columns for heavy timber structures are usually square in sizes from 8 to 12 in. The sides are chamfered or rounded and the bottoms rest on and are attached to the footings through fabricated metal column bases and bolts. Column caps may be all wood, all metal, or a combination of wood and metal with bolts and timber connectors to provide the transfer of loads and stresses from girders, floor beams, or trusses.

Floor beams are usually 4 or 6 in. thick and 8 to 10 in. high, with span lengths from 12 to 16 ft long. Girders are usually 8 to 12 in. thick and 12 to 24 in. high, with span lengths up to 20 ft. Beams and girders are attached with timber connectors or metal plates with bolts. Flooring for timber structures is usually constructed of S4S tongue and groove lumber.

To estimate the cost of timber structures, the estimator must perform a detailed material quantity takeoff, listing the number, size, and length of each member in the structure.

Fabricating timber members for heavy timber structures involves starting with commercial sizes of lumber and sawing them to the correct lengths and shapes, boring holes, chamfering edges, etc. Table 12.19 gives the approximate labor-hours per 1,000 board feet to fabricate and erect timbers. Because the members are larger and heavier than wood framed structures, such as houses, a small lifting crane is required to lift the members to the position for connecting. After the timber members are cut to the proper length they are attached by timber connectors, bolts, and plates.

TABLE 12.19 | Approximate labor-hours required per 1,000 bf to fabricate and erect heavy timber members.

Member	Labor-hours per 1,000 bf	
	Carpenters	**Helper**
Columns, 8 × 8 and larger	13–16	7–8
Girders, 8 × 12 and larger	6–8	6–7
Beams, 4 × 8 and larger	5–7	5–6

INTERIOR FINISH, MILLWORK, AND WALLBOARDS

INTERIOR FINISH CARPENTRY

Examples of interior finish carpentry include moldings around ceilings, floors, doors, and windows. Also included are hand railings, shelving, cabinets, vanities, fireplace mantels, louvers, shutters, etc. These items often are stained to expose the natural grains in the wood, which requires skilled carpenters.

Moldings are estimated by the linear foot, whereas doors, windows, cabinets, etc., are usually estimated by the unit. The cost of these items will vary considerably with the kind and grade of materials and the grade of workmanship required. The most dependable method of estimating the cost of interior finish and millwork is to list the quantity and cost of each item separately, together with the quantity and cost of the labor required to install each item.

LABOR-HOURS REQUIRED TO SET AND TRIM DOORS AND WINDOWS

Doors may be made of metal or wood. The time required to set doors will vary depending on the size of the opening and the quality of workmanship. Exterior doors are usually thicker and heavier than interior doors and require more time to fit and hang than interior doors. Solid wood is used for exterior doors and hollow wood core doors are often used for interior doors. Hollow wood doors are fabricated by placing external sheets of wood veneer over internal wood framing members. Hollow metal doors are also available. Prehung doors that are ready for installation when they are delivered to the job are available from suppliers.

The operations for setting solid wood doors include fitting the door to the opening in the frame, routing the jamb and door for butts, mortising the door for the lock, installing the butts and lock, and hanging the door.

Many types and sizes of windows are available. They may be fabricated of metal, wood, or vinyl. Prefabricated window frames are shipped from the supplier and installed in window openings that have been previously framed in the stud walls. Windows may be single-hung or double-hung. For single-hung windows, only the bottom half of the window can be raised and lowered, whereas both the top and bottom can be raised and lowered for double-hung windows.

Table 12.20 gives the carpenter-hours required to install various types of doors and windows. Trim wood is installed around the top and both sides of the door opening. Threshold is lumber placed between the bottom of the door and the floor. After the windows are installed, trim wood is placed around them. Table 12.21 gives the carpenter-hours to install trim around doors and windows.

TABLE 12.20 | Carpenter-hours required per opening to set doors and windows.

Item	Carpenters
Doors	
Wood, exterior	1.5–1.8
Wood, interior	1.2–1.5
Hollow core	0.8–0.9
Prehung, exterior	1.0–1.2
Prehung, interior	0.8–0.9
Metal frame	1.0–1.2
Metal door	0.9–1.1
Windows	
Metal	1.5–2.0
Wood	1.5–2.5
Vinyl	0.8–1.0

TABLE 12.21 | Carpenter-hours required per opening to install trim for doors and windows.

Item	Carpenters
Door trim	
Trim lumber	1.3–1.8
Threshold	0.3–0.5
Window trim	
Up to 4 ft wide	0.6–0.8
Over 4 ft wide	0.8–1.3

WOOD FURRING STRIPS

Wood furring strips are placed on masonry or concrete walls to permit installation of gypsum, prefinished plywood, or other type of finished interior wall materials. The labor-hours required to place wood furring strips and grounds is usually estimated by 100 lin ft. Table 12.22 gives the labor-hours to place 100 lin ft of furring strips.

TABLE 12.22 | Labor-hours required per 100 lf of wood furring strips.

Kind of work	Carpenters	Helper
Installed on masonry	1.5–3.5	0.4–0.6
Installed on concrete	2.5–4.5	0.5–0.8

GYPSUM WALLBOARDS

Gypsum wallboards, which are made by molding gypsum between two sheets of paper, are used on walls and ceilings of buildings. Gypsum wallboard is sometimes called sheetrock. It is manufactured in sheets, either $\frac{1}{2}$- or $\frac{5}{8}$-in. thick, 4-ft wide, and lengths of 8 or 10 ft. Gypsum wallboards are fastened to furring strips that are attached to concrete or masonry walls, or they may be installed directly against wood studs and ceiling joists.

Where gypsum wallboards are installed on wood studs and ceiling joists, the studs and joists should be placed at 12, 16, or 24 in. on centers because these dimensions are divisible into the 4-ft width. For walls, the boards are usually installed in vertical positions, with the length equal to the height of the wall. For ceilings, they may be installed perpendicular to or parallel with the joists. Some specifications require the use of nailing strips, such as 1 × 4 lumber, installed perpendicular to the ceiling joists on 12-in centers, to which the boards are fastened with nails.

Gypsum wallboards are fastened with sheetrock nails, size 5d or 6d with $\frac{3}{8}$- or $\frac{5}{16}$-in. heads, or by screws. Recommended spacing of fasteners is 8 in. for walls and 7 in. for ceiling installation. For studs or joists spaced at 12 in. on centers, the quantity of nails per 1,000 sf will be about 8 or 9 lb. For a 24-in. spacing, the quantity will be about 4 or 5 lb per 1,000 sf.

The joints between adjacent gypsum boards are filled with special cement, which is spread with a putty knife to cover a strip about $1\frac{1}{2}$-in. wide on each side of the joint. After the cement is applied, the joint is covered with a perforated fiber tape about 2-in. wide, and another layer of cement is applied and spread over the tape. After the cement dries, it can be sanded to a uniformly smooth surface to eliminate evidence of a joint for painting, or a textured surface can be spread over the entire surface of the gypsum wallboard.

The labor-hours required to install gypsum wallboard will vary considerably with the size and complexity of the area to be covered. For large wall or ceiling areas, which require little or no cutting and fitting of the boards, two carpenters should install a board 4-ft wide and 8-ft long in about 10 to 15 min. This is equivalent to 1 to $1\frac{1}{2}$ labor-hours per 100 sf. However, where an area contains numerous openings, it is necessary to mark the wallboards and cut them to fit the openings. For such areas, the labor-hours may be as high as 3 to 4 hr per 100 sf. Cutting is done by scoring one side of the board with a curved knife, which is drawn along a straightedge. A slight bending force will break the board along the scored line. Table 12.23 gives representative labor-hours required to install gypsum wallboard and perforated-tape joints.

TABLE 12.23 | Labor-hours required to install gypsum wallboard and perforated tape.

| | | Labor-hours | | |
| | | Spacing of studs or joists, in. | | |
Operation	**Unit of measure**	**12**	**16**	**24**
Install wallboard				
Large areas	100 sf	1.7	1.5	1.2
Medium-size rooms	100 sf	2.8	2.5	2.0
Small rooms	100 sf	4.0	3.5	3.0
Install perforated tape and sand the surface	100 lin ft	4.0	4.0	4.0

WALL PANELING

Prefinished plywood panels are used as covering for walls, ceilings, and wainscotings in buildings. Panels $\frac{1}{4}$- to $\frac{3}{4}$-in. thick, 4-ft wide, and 8-ft long are available. Prefinished plywood panels are available in many species, including birch, cherry, chestnut, mahogany, oak, pine, pecan, rosewood, and teak. Since the plywood is prefinished, special finishing nails are used to fasten panels to the framed walls.

The cost of prefinished plywood panels depends on the size and species and can be obtained from building suppliers. The cost of labor installing plywood panels will vary considerably with the size, shape, and complexity of the area to be covered. Where areas are large and plain, the waste of materials will be low, and labor can install them rapidly. However, where areas are small or irregular, with many openings, which require considerable cutting and fitting, the waste can be large and the cost of labor will be higher. The area to be covered should be carefully examined prior to preparing an estimate. Depending on the complexity of the area, it requires from 3 to 5 carpenter-hours per 100 sf to install prefinished plywood panels.

INTERIOR TRIM MOLDINGS

Baseboard moldings of pine or oak are installed around the perimeter of rooms at the floor level. Baseboards are usually $\frac{9}{16}$-in. thick and $3\frac{1}{2}$- or $4\frac{1}{2}$-in. wide. A $\frac{3}{4}$- by 1-in. base shoe is sometimes installed along the baseboards.

Cornice molding or crown moldings of pine are sometimes installed around the perimeter of walls at the ceiling level. The cornice molding is usually $\frac{9}{16}$-in. thick and $1\frac{3}{4}$- to $2\frac{1}{4}$-in. wide. Crown molding is usually $\frac{9}{16}$-in. thick and $3\frac{5}{8}$- to $4\frac{5}{8}$-in. wide. Many styles are available. Table 12.24 gives the approximate carpenter-hours per 100 lin ft to install interior trim moldings.

TABLE 12.24 | Carpenter-hours required per 100 lin ft of interior trim moldings.

Item	Carpenters
Baseboards	3.5–4.0
Baseboard shoes	3.0–3.5
Cornice molding	2.5–3.0
Crown molding	3.0–4.0

FINISHED WOOD FLOORS

The lumber used for finished wood floors includes pine, fir, maple, spruce, and oak. The most common hardwood floor is $\frac{23}{32}$-in.-thick oak in either $2\frac{1}{4}$- or $3\frac{1}{4}$-in. widths. Table 12.25 lists some of the sizes of wood flooring produced. Not all of the sizes listed may be available in all locations, especially hardwood flooring.

TABLE 12.25 | Dimensions of wood flooring.

Thickness,* in.		Width, in.		Addition for side waste, %
Nominal	Worked	Nominal	Face	
Softwood				
$\frac{1}{2}$	$\frac{7}{16}$	3	$2\frac{1}{8}$	41.2
$\frac{5}{8}$	$\frac{9}{16}$	4	$3\frac{1}{8}$	28.0
1	$\frac{3}{4}$	5	$4\frac{1}{8}$	21.2
$1\frac{1}{4}$	1	6	$5\frac{1}{8}$	17.1
$1\frac{1}{2}$	$1\frac{1}{4}$	6	$5\frac{1}{8}$	17.1
Hardwood				
1	$\frac{25}{32}$	$2\frac{1}{4}$	$1\frac{1}{2}$	50.0
1	$\frac{25}{32}$	$2\frac{3}{4}$	2	37.5
1	$\frac{25}{32}$	3	$2\frac{1}{4}$	33.3
1	$\frac{25}{32}$	4	$3\frac{1}{4}$	23.0

*Generally all widths of softwood flooring are available in each listed thickness. In some locations not all listed nominal widths of hardwood may be available. Also hardwood may be available in other thicknesses.

The percentages of side waste listed in Table 12.25 are determined by dividing the reduction in width by the net width. For example, if the nominal width is 4 in. and the face width is $3\frac{1}{4}$ in., then the loss in width is $\frac{3}{4}$ or 0.75 in. For these dimensions the table lists the percentage to be added for side waste. For example, $(0.75 \times 100)/3.25 = 23.1$ percent. The reason for using this procedure is that when the quantity of lumber required is determined, the area of the floor is known. If the quantity of side waste listed in Table 12.25 is added to the area of the floor for the particular nominal width of flooring selected, this allowance will be sufficient for the side waste. An additional amount may be required for end waste, resulting from sawing to shorter lengths.

Wood flooring is also available in prefinished strips, $\frac{3}{8}$ in. thick and widths of $2\frac{1}{4}$ and $3\frac{1}{4}$ in. These prefinished strips of tongue and groove wood are glued, rather than nailed, to either concrete or wood floors.

Parquet patterns of wood are also used for finished wood floors. This type of flooring is usually walnut, teak, or oak. Various patterns and sizes of blocks are available, about $\frac{5}{16}$-in. thick and in blocks of 9- or 12-in. square.

Table 12.26 gives the approximate labor-hours per 100 sf to install wood floors.

TABLE 12.26 | Labor-hours required per 100 sf to install wood flooring.

Type of flooring	Carpenters	Helper
Softwood lumber	3.0–4.0	0.4–0.6
Hardwood lumber	4.0–5.0	0.5–0.7
Prefinished strips	4.0–5.0	0.5–0.7
Parquet blocks	5.0–6.0	0.6–0.8

PROBLEMS

12.1 Estimate the total direct cost for furnishing and installing 108 floor joists size 2 × 8 lumber. The 16-ft-long joists will rest on wood sills at each end of the joist and at the midpoint of the joist. Use average values from Table 12.9, $28.00/hr for carpenters, $21.00/hr for laborers, and $0.55/bf for lumber costs.

12.2 Estimate the total direct cost and cost per sf for furnishing and installing plywood subflooring for a floor area of 2,240 sf. The 4-ft by 8-ft plywood, $\frac{3}{4}$-in. thick, will be placed on joists, spaced at 18 in. on centers. Use average values from Table 12.15, $34.00/sheet for the 4-ft by 8-ft plywood, $28.00/hr for carpenters, and $21.00/hr for helpers.

12.3 The total length of exterior walls of a residence is 650 lin ft. Studs, size 2 × 4 by 8 ft long, will be spaced not over 16 in. on centers. There will be 8 outside corners and 12 inside corners, each requiring three studs. The studs will rest on 2 × 4 plates and will be capped at the tops with two 2 × 4 plates for the full length of the walls.

Neglecting the door and window openings, prepare a list of materials for the studs and the bottom and top plates, showing the number of pieces of each length required. Estimate the total direct cost and cost per linear foot of wall for furnishing and installing the studs and plates. Assume $0.55/bf for lumber, $28.00/hr for carpenters, and $21.00/hr for laborers.

12.4 Estimate the total direct cost for furnishing and installing 2 × 8 rafters, spaced 24 in. on centers, for a rectangular building whose outside dimensions are 35 ft wide and 140 ft long. The framing for rafters is the same as shown in Figure 12.4, using one set of knee braces, collar tie, and ridge board. Use the same unit costs for material and labor as shown in Example 12.3.

12.5 The following is a list of windows and doors for a wood frame structure. Estimate the cost of labor for framing the openings, assuming $28.00/hr for carpenters and $21.00/hr for laborers.

Size	Number
Doors:	
2 ft 8 in. × 6 ft 8 in.	5 each
3 ft 0 in. × 6 ft 8 in.	14 each
3 ft 6 in. × 6 ft 8 in.	2 each
Windows:	
2 ft 0 in. × 4 ft 0 in.	4 each
2 ft 6 in. × 5 ft 0 in.	17 each
3 ft 0 in. × 5 ft 0 in.	8 each
4 ft 0 in. × 6 ft 0 in.	1 each

12.6 The following is a door and window schedule for a building. Use average values from Table 12.20 and estimate the labor cost to set the doors and windows, assuming carpenters cost $28.00/hr.

Doors:			
2 ft 8 in. × 6 ft 8 in.	5 each	Hollow core	Interior
2 ft 8 in. × 6 ft 8 in.	12 each	Hollow core	Interior
3 ft 0 in. × 6 ft 8 in.	14 each	Wood	Interior
3 ft 6 in. × 6 ft 8 in.	2 each	Wood	Exterior
Windows:			
2 ft 0 in. × 4 ft 0 in.	4 each	Metal	
2 ft 6 in. × 5 ft 0 in.	17 each	Wood	
3 ft 0 in. × 5 ft 0 in.	8 each	Wood	
4 ft 0 in. × 6 ft 0 in.	1 each	Vinyl	

12.7 Estimate the labor cost to install trim for the doors and windows in the following schedule, assuming carpenters cost $28.00/hr. Include threshold only for exterior doors.

Doors:			
2 ft 8 in. × 6 ft 8 in.	12 each	Hollow core	Interior
3 ft 0 in. × 6 ft 8 in.	14 each	Wood	Interior
3 ft 6 in. × 6 ft 8 in.	2 each	Wood	Exterior
Windows:			
2 ft 0 in. × 4 ft 0 in.	4 each	Wood	
3 ft 6 in. × 5 ft 0 in.	17 each	Wood	
4 ft 0 in. × 5 ft 0 in.	8 each	Wood	
5 ft 0 in. × 6 ft 0 in.	1 each	Wood	

12.8 Estimate the total cost and cost per square foot to furnish and install gypsum wallboard on the 8-ft high walls of a wood frame building. The structure has 125 lin ft of exterior walls and 480 lin ft of interior walls. A review of the construction drawings reveals that the structure has medium-size rooms and the stud spacing is 16 in. on centers. The gypsum board must be installed on one side of the exterior wall and both sides of the interior walls. Neglect the cost of installing perforated tape and sanding the surface. Assume that a 4-ft by 8-ft sheet of gypsum board costs $11.00 each and carpenters cost $28.00/hr.

12.9 Estimate the labor cost for installing gypsum wallboard for the ceilings of a structure consisting of 4,800 sf. A review of the drawings for the project shows that 10 percent of the rooms are large, 70 percent of the rooms are medium size, and 20 percent of the rooms are small size. Neglect the cost to install perforated tape and to sand the surface. Assume carpenters cost $28.00/hr.

CHAPTER 13

Roofing and Flashing

ROOFING MATERIALS

Roofing refers to the furnishing of materials and labor to install coverings for roofs of buildings. Several kinds of materials are used, including but not limited to:

1. Shingles
   a. Wood
   b. Asphalt
2. Slate
3. Clay tile
4. Built-up
5. Metal
   a. Copper
   b. Aluminum
   c. Galvanized steel
   d. Tin

AREA OF A ROOF

The unit of measure for roofing is the *square,* which is 100 sf. Roofing materials used to cover ridges and valleys are measured by linear feet, with the width specified.

The area to be roofed is the full length and width, measured along the slope of the roof, including eaves, overhangs, etc. Some estimators do not deduct for the areas of openings containing less than 100 sf. A minimum of 5 percent waste should be added to calculated quantities.

STEEPNESS OF ROOFS

Some estimators define the steepness of a roof by its pitch. Others define steepness by the vertical rise over the horizontal run, measured in inches. These terms and their values are

Pitch	Rise in./ft
$\frac{1}{8}$	3
$\frac{1}{4}$	6
$\frac{1}{3}$	8
$\frac{1}{2}$	12
$\frac{2}{3}$	16

Rise over run, vertical inches divided by horizontal inches	Diagonal slope dimension, inches
3:12	12.3693
4:12	12.6491
5:12	13.0000
6:12	13.4164
7:12	13.8924
8:12	14.4222
9:12	15.0000
10:12	16.6205
12:12	16.9706

ROOFING FELT

Most specifications require installation of an asphalt felt underlayment prior to placing roofing shingles, such as asphalt, slate, shakes, and tile. Two weights of asphalt felt are used, either 15 or 30 lb/square. This material is sold in rolls containing 108, 216, and 432 sf, which will cover 100, 200, and 400 sf, respectively, with a 2-in. lap. Galvanized nails about $\frac{3}{4}$-in. long, with $\frac{7}{16}$-in. heads, spaced about 6 in. apart along the edges, are used to hold the felt in place.

The labor-hours required to lay asphalt felt underlayment should be about 0.5 labor-hours/square.

ROOFING SHINGLES

WOOD SHINGLES

Wood shingles, both red cedar and shakes, are discussed in Chapter 12 under Carpentry.

ASPHALT SHINGLES

Asphalt shingles are manufactured in several styles, colors, and sizes, including organic asphalt, fiberglass asphalt, hexagon, etc. The shingles are fastened to wood decking with galvanized roofing nails, 1- to $1\frac{1}{2}$-in. long, starting at the lower edge of the roof. Asphalt starting strips should be laid along the eaves of a roof prior to laying the first row of shingles, or the first row can be doubled, as illustrated in Fig. 13.1.

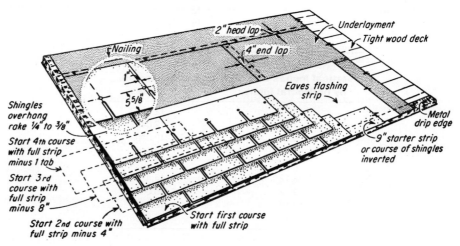

FIGURE 13.1 | Three-tab strip asphalt shingles.

The calculated net area of a roof should be increased by at least 5 percent for gable roofs and 10 percent for hip roofs. If the roof has numerous valleys or dormers, additional waste should be added to the net area.

Table 13.1 gives the net quantities of asphalt shingles of various styles required to cover one square. The cost of asphalt shingles will vary with the style, materials, and weight per square for the shingles selected. The quoted price per square is based on furnishing enough shingles to cover 100 sf of roof in the manner indicated by the manufacturer, with no allowance for waste included.

TABLE 13.1 | Quantities of asphalt shingles and roofing nails per square.

Style of shingle	Size, in.	Number per square	Length exposed, in.	Number of nails per shingle	Nails per square, lb
Asphalt strip, 3 tab, 235 lb	12 × 36	80	5	4	1.0
Fiberglass asphalt, 340 lb	12 × 36	80	5	4	1.0
Hexagon strip, 170 lb	12 × 36	86	5	6	1.5

Labor Required to Lay Asphalt Shingles

Asphalt shingles may be laid by carpenters or experienced roofers (see Figure 13.2). The latter should lay shingles more rapidly. The rate of laying shingles will be lower for simple areas than for irregular areas. Hips, gables, dormers, etc., require the cutting of shingles for correct fit, which will reduce the production rates. Also, the rate of laying shingles on steep roofs will be lower than installing them on roofs that have a low pitch.

FIGURE 13.2 I Laborers installing asphalt shingles.

Table 13.2 gives the labor-hours required to lay a square of asphalt shingles. The lower rates should be used for gable roofs and the higher rates for roofs with valleys, hips, dormers, etc. The rate of laying ridge shingles is about 0.5 labor-hours per 100 lin ft. The use of pneumatic nail guns will greatly reduce the time for installing asphalt shingles.

TABLE 13.2 I Labor-hours required to lay a square of asphalt shingles, using carpenters.

Style of shingle	Carpenter	Helper
Individual	1.7–2.5	0.7–0.9
Strip, 3 tab	1.1–1.8	0.6–0.8
Hexagonal strip	1.4–2.0	0.6–0.8
Double coverage	1.5–2.5	0.7–0.9

EXAMPLE 13.1

Estimate the cost of material and labor for installing asphalt shingles for the structure shown in the plan view in Fig. 13.3. As shown in the plan view there are three gables with the same slope of 5:12, rise over run. The dimensions shown are for the outside walls of the building. A 2-ft overhang is required, therefore the end of the roof extends 2 ft on either side of the outside walls.

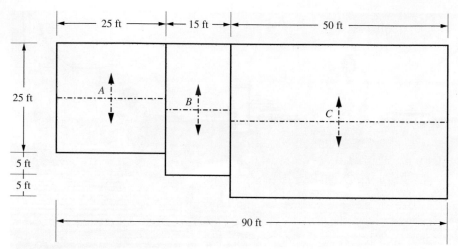

FIGURE 13.3 I Plan view of roof.

A 15-lb felt underlayment will be installed prior to installing three-tab, 235-lb asphalt shingles. Assume material costs of $3.40/square for the felt, $55.00/square for shingles, and $1.45/lb for nails. The cost of labor includes $28.00/hr for carpenters and $21.00/hr for laborers.

Quantities of materials:

   Slope lengths (see Fig. 13.4):

   Section *A:* 29 ft × (13/12) = 31.42 ft

   Section *B:* 34 ft × (13/12) = 36.83 ft

   Section *C:* 39 ft × (13/12) = 42.25 ft

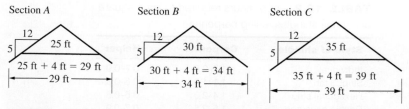

FIGURE 13.4 I Cross sections through roof.

Surface area of roofing:

    Section *A:* 31.42 ft × 25 ft =   785.5 sf

    Section *B:* 36.83 ft × 15 ft =   552.5 sf

    Section *C:* 42.25 ft × 50 ft = 2,112.5 sf

                          Total = 3,450.5 sf, or 34.5 squares, net

Required materials, assuming 5% waste:

    Felt underlayment: 1.05 × 34.5 squares = 36.2 squares required

    Asphalt shingles: 1.05 × 34.5 squares = 36.2 squares required

    Nails: 36.2 squares × 1 lb/square = 36.2 lb

Labor required:

    Simple gable roof with no valleys, use lower values from Table 14.2:

    Carpenters: 36.2 squares × 1.1 hr/square = 39.8 hr

    Laborers: 36.2 squares × 0.6 hr/square = 21.7 hr

    Labor-hours to install felt: 36.2 squares × 0.5 hr/square = 18.1 hr

Cost of material and labor:

Asphalt shingles: 36.2 squares @ $55.00/square	=	$1,991.00
Felt underlayment: 36.2 squares @ $3.40/square	=	123.08
Nails: 36.2 lb @ $1.45/lb	=	52.59
Carpenters for shingles: 39.8 hr @ $28.00/hr	=	1,114.40
Laborers for shingles: 21.7 hr @ $21.00/hr	=	455.70
Laborers for roofing felt: 18.1 hr @ $21.00/hr	=	380.10
	Total cost =	$4,116.77

Cost per square, $4,116.77/34.5 squares = $119.32/square

SLATE ROOFING

Slate for roofing is made by splitting slate blocks into pieces having the desired thicknesses, length, and width. The slate is usually $\frac{3}{16}$- or $\frac{1}{4}$-in. thick and up to 16 in. in widths and lengths. Colors include black, green, gray, red, and purple. Slate is heavy material, weighing from 170 to 180 lb/cf. For a 3-in. head lap, $\frac{3}{16}$-in.-thick slate will weigh 700 lb/square and $\frac{1}{4}$-in.-thick slate will weigh about 900 lb/square.

    Slate is priced by the square, with sufficient pieces furnished to cover a square with a 3-in. head lap over the second course under the given course. Thus, for slates 16 in. long, the length exposed to weather will be $(16 - 3)/2 = 6.5$ in. This same calculation will apply to any length for which the head lap is 3 in. If the slates are 8-in. wide, the area covered will be $8 \times 6.5 = 52$ sq. in. The number of slates required to cover a square will be $(100 \times 144)/52 = 277$. For other sizes the number required to cover a square can be determined in the same manner. An area should be increased 10 to 25 percent to allow for waste.

Laying Slate Roofing

Roofing slate is laid on asphalt felt, weighing 30 lb/square. The first course of slate should be doubled, with the lower end laid on a wood strip to give the slate the proper cant for the succeeding courses. Joints should be staggered. Each slate should be fastened with two nails driven through prepunched holes. Hips and valleys will require edge mitering of adjacent slates.

The labor operations required for slate roofing will include laying the felt and installing the slates. Production rates will vary considerably with the sizes and thicknesses of slate, slope of the roof, kind of pattern specified, and complexities of the areas. Table 13.3 gives the labor-hours to apply a square of asphalt felt and slate. The lower rates should be used for plain roofs, such as gable roofs, and the higher rates for steep roofs with valleys, hips, and dormers.

TABLE 13.3 | Labor-hours required to apply a square of asphalt felt and roofing slate.

Slate size, in.	Slater	Helper
12 × 8	5.0–7.0	2.5–3.5
16 × 8	4.0–6.0	2.0–3.0
16 × 12	3.5–5.2	1.8–2.7
16 × 16	3.2–5.0	1.6–2.5

CLAY TILE ROOFING

Clay tile roofing is installed similarly to slate roofing. A 30-lb asphalt felt underlayment is placed over the roof area before the clay tile is installed. The tiles are nailed to the roof decking with galvanized nails. The clay tiles are more uniform in shape than slates and therefore are easier to install. However, the tiles are brittle and are subject to breakage, which can reduce the production rate.

The rate of placing the tiles will depend on the steepness and complexity of the roof. The labor-hours required to lay clay tile roofing should be about 4.5 to 7.0 labor-hours per square for the roofer and 2.0 to 3.5 labor-hours per square for the helpers.

BUILT-UP ROOFING

Building up roofing consists of applying alternate layers of roofing felt and hot pitch or asphalt over the area to be covered, with gravel, crushed stone, or slag applied uniformly over the top layer of pitch or asphalt. This type of roofing may be applied to wood sheathing, concrete, poured gypsum, precast concrete tiles, precast gypsum blocks, book tile, and approved insulation.

The quality of built-up roofing is designated by specifying the weight and number of plies of felt, the weight and number of applications of pitch or asphalt, and the weight of gravel or slag used. The unit of area is a square.

FELT FOR BUILT-UP ROOFING

Pitch-impregnated felt should be used with pitch cement and asphalt-impregnated felt with asphalt. Felt weighing 15 or 30 lb/square may be used, with the 15-lb felt being more commonly specified. Felt is available in rolls 36-in. wide with gross areas of 108, 216, and 432 sf.

Fiberglass felt is sometimes used in place of organic felt for built-up roofing. Fiberglass, type IV, is available in 540-lb rolls.

PITCH AND ASPHALT

Pitch should be applied at a temperature not exceeding 400°F and asphalt at a temperature not exceeding 450°F. Applications are made with a mop to the specified thickness or weight. Pitch and asphalt will weigh about 10 lb/gal. Pitch is purchased in 200-lb cartons, and asphalt is purchased in 100-lb cartons.

GRAVEL AND SLAG

After the final layer of pitch or asphalt is applied, and while it is still hot, gravel or slag is spread uniformly over the area. The aggregate should be $\frac{1}{4}$ to $\frac{5}{8}$ in. in size and thoroughly dry. Application rates are about 400 lb/square for gravel and 300 lb/square for slag.

LAYING BUILT-UP ROOFING ON WOOD DECKING

While specifications covering the roofing laid on wood decking will vary, this method is representative of common practice.

1. Over the entire surface lay two plies of 15-lb asphalt felt, lapping each sheet 19 in. over the preceding one and turning these felts up not less than 4 in. along all vertical surfaces. Nail as often as necessary to secure, until the remaining felt is laid.

2. Over the entire surface embed in asphalt two plies of asphalt felt, lapping each sheet 19 in. over the preceding one, rolling each sheet immediately behind the mop to ensure a uniform coating of hot asphalt, so that in no place shall felt touch felt. Each sheet shall be nailed 6 in. from the back edge at intervals of 24 in. These felts shall be cut off at the angles of the roof deck and all walls or vertical surfaces.

3. Over the entire surface spread a uniform coating of asphalt into which, while hot, embed not less than 400 lb of gravel or 300 lb of slag per 100 sf of area. Gravel or slag must be approximately $\frac{1}{4}$ to $\frac{5}{8}$ in. in size, dry and free from dirt. If the roofing is applied during cold weather or the gravel or slag is damp, it should be heated and dried immediately before application.

Not less than these quantities of materials should be used for each 100 sf of roof area:

Material	Weight, lb	
	Gravel	**Slag**
Four plies of 15-lb asphalt felt	60	60
Roofing asphalt	100	100
Gravel or slag	400	300
Total weight	560	460

The application previously specified is designated as four-ply roofing, with two plies dry and two plies mopped.

Sometimes specifications require that the decking first be covered with a single thickness of sheathing paper, weighing not less than 5 or 6 lb/square, with the edges of the sheets lapped at least 1 in.

LAYING BUILT-UP ROOFING ON CONCRETE

A method of laying built-up roofing on concrete is shown in Fig. 13.5. Before built-up roofing is applied on a concrete deck, the concrete should be cleaned and dried, after which a coat of concrete primer should be applied cold at a rate of about 1 gal/square. After the primer has dried, embed in hot asphalt two, three, four, or five plies of asphalt felt, lapping each sheet enough to give the required number of plies. Roll each sheet immediately behind the mop to ensure a uniform coating of hot asphalt, so that in no place shall felt touch felt. Over the entire area spread a uniform coating of hot asphalt into which, while hot, embed not less than 400 lb of dry gravel or 300 lb of dry slag per square.

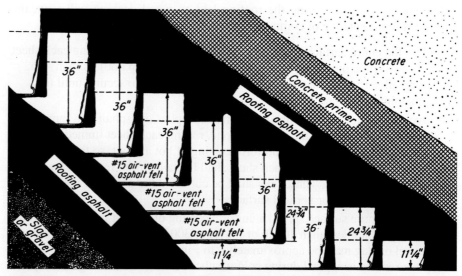

FIGURE 13.5 | Method of laying built-up roofing on concrete.

For a three-ply roofing the quantities of materials given in this table might be used per square.

Material	Weight, lb Gravel	Slag
Concrete primer	10	10
Three plies of 15-lb asphalt felt	45	45
Roofing asphalt	125	125
Gravel or slag	400	300
Total weight	580	480

LABOR LAYING BUILT-UP ROOFING

The operations required to lay built-up roofing will vary with the type of roofing specified. If the time required to perform each operation is estimated, the sum of these times will give the total time for a roof or for completing a square. If a building is more than two or three stories high, additional time should be allowed for hoisting materials.

A typical crew for laying roofing on buildings up to three stories high would include one person each tending the kettle, handling hot asphalt or pitch, laying felt, rolling felt, and mopping, with a foreman supervising all operations. On some jobs one person may be able to tend the kettle and handle the asphalt or pitch.

Table 13.4 gives the labor-hours per square required to perform each operation and to complete all operations for various types of built-up roofing. Use the lower values for large plain roofs and the higher values for roofs with irregular areas.

TABLE 13.4 | Labor-hours required to perform operations and lay a square of built-up roofing.

Operation	Labor-hours*
Apply sheathing paper on wood deck	0.10–0.15
Apply primer on concrete	0.10–0.15
Lay 1 ply of roofing felt	0.10–0.15
Apply asphalt with mop	0.10–0.15
Apply asphalt and gravel	0.50–0.75
Apply 2-ply roofing on wood deck	1.3–2.0
Apply 3-ply roofing on wood deck	1.5–2.2
Apply 4-ply roofing on wood deck	1.7–2.5
Apply 5-ply roofing on wood deck	1.8–2.7
Apply 3-ply roofing on concrete	1.6–2.3
Apply 4-ply roofing on concrete	1.8–2.6
Apply 5-ply roofing on concrete	1.9–2.8

*The labor-hours for the last eight applications include three workers on the roof and two workers heating and supplying hot asphalt.

FLASHING

Flashing is installed to prevent water from passing into or through areas such as valleys and hips on roofs, where roofs meet walls, or where openings are cut through roofs such as the chimney shown in Fig. 13.6. Materials used for flashing include sheets of copper, tin, galvanized steel, aluminum, lead, and sometimes mopped layers of roofing felt.

Flashing is usually measured by the linear foot for widths up to 12 in. and by the square foot for widths greater than 12 in.

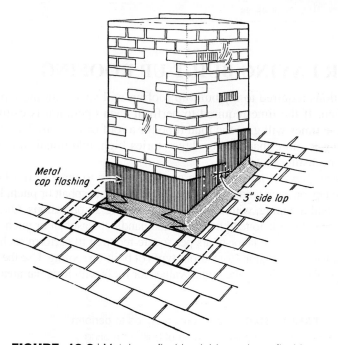

FIGURE 13.6 | Metal cap flashing laid over base flashing.

METAL FLASHING

When metal flashing requires nails or other metal devices to hold it in place, it is essential that the fastener and the flashing be of the same metal; otherwise, galvanic action will soon destroy one of the metals.

FLASHING ROOFS AT WALLS

Where a roof and a parapet wall join, it is common practice to extend the layers of built-up roofing 4 to 8 in. up the wall, with mopping applied to the wall and the felt, with no metal flashing underneath the felt. A metal counter flashing, whose

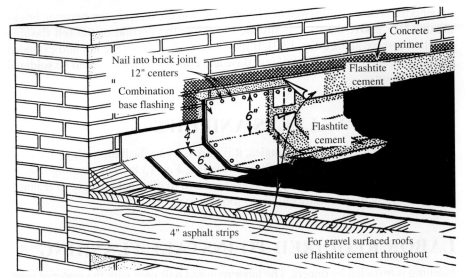

Concrete primer

Nail into brick joint 12" centers

Combination base flashing

Flashtite cement

6"

4"

6"

Flashtite cement

4" asphalt strips

For gravel surfaced roofs use flashtite cement throughout

FIGURE 13.7 | Method of flashing metal roof at parapet wall.

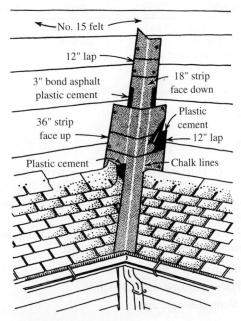

No. 15 felt

12" lap

3" bond asphalt plastic cement

36" strip face up

Plastic cement

18" strip face down

Plastic cement

12" lap

Chalk lines

FIGURE 13.8 | Use of roll roofing for valley flashing.

upper edge is bent and inserted in a raggle or slot in the mortar joint between bricks about 12 in. above the roof, is installed to cover the portion of the felt flashing attached to the wall (see Fig. 13.7). If metal is installed under the felt and extended up the wall, it is called *base flashing*. Base and counter flashing should be soldered at all end joints.

FLASHING VALLEYS AND HIPS

When roofs are covered with shingles, it is necessary to flash the valleys and hips, usually with metal flashing 10 to 30 in. wide (see Fig. 13.8). The flashing may be soldered at end joints, or it may be lapped enough to eliminate danger of leakage. The quantity is measured by the linear foot.

LABOR REQUIRED TO INSTALL FLASHING

Flashing is installed by tinners. The labor required will vary with the material used, type of flashing required, and specifications covering the installation. Table 13.5 gives the labor-hours required to install flashing.

TABLE 13.5 | Labor-hours required to install flashing.

Class of work	Unit	Labor-hours
Metal base around parapet wall	100 lin ft	5.0–6.0
Metal counter flashing around walls	100 lin ft	5.0–6.0
Metal along roof and wood walls	100 lin ft	4.0–5.0
Metal valleys and hips	100 lin ft	5.0–6.0
Metal shingles along chimneys	100 each	8.0–10.0

PROBLEMS

13.1 Estimate the total cost and the cost per square for furnishing and laying three-tab asphalt shingles on a wood roof decking whose area is 4,800 sf. The roof will have numerous valleys, hips, and dormers. Use the unit cost of materials and labor as shown in Example 13.1.

13.2 Estimate the labor cost for installing flashing for a building, assuming $27.00/hr for roofers. Quantity takeoffs from the plans reveal

Metal base around parapet wall: 870 lin ft

Metal counter flashing around walls: 870 lin ft

Metal shingles along chimney: 2 each

14

Masonry

MASONRY UNITS

Masonry units that are commonly used for construction include brick, concrete masonry units, and stone, either natural or artificial. They are bonded by suitable mortar. Masonry units are available in several sizes, grades, and textures. Plans and specifications for a particular job designate the type of masonry unit, including its size, grade, and texture. The specifications also define the kind of mortar and thickness of joints.

Members of the construction craft that install masonry units are called masons or bricklayers. Sometimes bricklayers are called brick masons. A helper usually assists masons or bricklayers. Masonry construction is labor intensive. Although the masonry units may be placed beside the masons, they must pick up each individual unit, properly spread the mortar, and fit the units in place one at a time.

ESTIMATING THE COST OF MASONRY

In estimating the cost of a structure to be constructed entirely or partly of masonry units, the estimator should determine separately the quantity and cost of each kind of unit required. For example, a project may require both brick veneer and concrete masonry units. The concrete masonry units may be separated into bond beams, columns, lintels, etc. An appropriate allowance should be made for waste, resulting primarily from breakage.

The quantities of masonry units and mortar should include an allowance for waste. The labor-hours required to install masonry depend on the type of unit, pattern of the masonry, and the height the masonry must be placed. The construction equipment required to install masonry units may include mortar mixer, scaffolding, and a forklift for handling the masonry units.

MORTAR

Mortar for masonry units is made by mixing portland cement with lime or masonry cement, and sand. The quantity of each ingredient may vary to produce a mortar suitable for a particular job.

Mortar is designated as ASTM types M, S, N, and O. The M type is high-strength, suitable for masonry subjected to high compressive loads, severe frost action, high lateral loads from earth pressures, and structures below grade. The S type is used for structures requiring high flexural bond strength, but subject only to normal compressive loads. The N type mortar is used in medium-strength mortar for general use above grade, specifically for exterior walls such as masonry veneers applied to frame construction. Type O is used in low-strength mortar for interior non-load-bearing partitions and for fireproofing.

The lime used for making masonry mortar is hydrated lime, which is calcium hydroxide. Lime, which sets only on contact with air, gives the mortar workability, water retentivity, and elasticity. Hydrated lime can be purchased in paper bags weighing 50 lb and containing 1 cf, or it can be purchased in bulk quantities.

Sand acts as a filler and also contributes to the strength of the mix. If fine sand is used, the workability of the mortar will be much better than if coarse sand is used. The sand should be screened and washed at the rock quarry pit. The price of sand may be quoted by cubic yard or ton.

Colored mortar is often required. The desired color is obtained by adding a color pigment to the mortar mix. The cost of the color pigment is usually minimal, but adding color to the mortar does require additional labor-hours to ensure consistency of the color during multiple mixing operations during a particular job.

Mortar properties may be specified by strength or by proportions. Table 14.1 gives the quantities of materials for 1 cy of various types of masonry mortar.

TABLE 14.1 | Quantities of materials for 1 cy of concrete masonry mortar.

		Cementitious materials		
Mortar type	Sand, dry, loose, cy	Portland cement, bag or cf	Masonry cement, bag or cf	Hydrated lime, bag or cf
M	1.0 cy	4.5	4.5	None
	1.0 cy	7.5	None	2.0
S	1.0 cy	3.0	6.0	None
	1.0 cy	6.0	None	3.0
N	1.0 cy	None	9.0	None
	1.0 cy	4.5	None	4.5
O	1.0 cy	None	9.0	None
	1.0 cy	3.0	None	6.0

Note: Cementitious materials are usually 1 cf volume per bag.

BRICKS

SIZES AND QUANTITIES OF BRICKS

Clay bricks are manufactured in many shapes, sizes, colors, and textures. The costs, which vary considerably, are usually based on 1,000 units, either at the supplier or delivered to the job.

The estimator must determine the quantity of bricks for each type. The quantity of bricks required for a job depends on the area of wall to be covered, plus a reasonable amount of waste. The estimator should deduct the openings, such as doors and windows, from the total gross area to be covered.

Brick may be designated as nonmodular or modular. Nonmodular bricks are solid bricks and are generally used for load-bearing walls. The standard size of a nonmodular brick is $2\frac{1}{4}$ in. high, $3\frac{3}{4}$ in. deep, and 8 in. long. An oversize nonmodular brick is available with the dimensions of $2\frac{3}{4}$ in. by $3\frac{3}{4}$ in. by 8 in.

Modular bricks have three or more holes in them, equally spaced along their centerlines. They are commonly used for brick veneer walls. Table 14.2 gives the names and nominal size of modular bricks. The listed dimensions of modular bricks are nominal, and are equal to the manufactured or specified dimension plus the thickness of the mortar joint with which the unit is designed to be laid. For example, the manufactured length of a unit whose nominal length is 12 in. would be $11\frac{1}{2}$ in. if the unit were designed to be laid with $\frac{1}{2}$-in. joints, or $11\frac{5}{8}$ in. for $\frac{3}{8}$-in. joints.

TABLE 14.2 | Types, nominal dimensions, and quantities of modular brick per square foot of wall area.

Name	Nominal dimensions, in.			Quantity of brick, bricks/sf
	Height	Depth	Length	
Standard	$2\frac{2}{3}$	4	8	6.75
Engineer	$3\frac{1}{5}$	4	8	5.63
Economy	4	4	8	4.50
Roman	2	4	12	6.00
Norman	$2\frac{2}{3}$	4	12	4.50
Norwegian	$3\frac{1}{5}$	4	12	3.75
SCR	$2\frac{2}{3}$	6	12	4.50

PATTERN BONDS

Masonry units may be oriented in various positions for construction of walls. Bricks laid with the long dimension placed in the horizontal direction are called *stretchers*. When the bricks are laid with the long dimension placed in the vertical direction, they are called *soldiers*. Bricks laid with their ends (height and depth dimensions) exposed to the exterior are called *headers*.

The arrangement of masonry units is called *pattern bond.* The common pattern bonds are *running bond, common bond, English bond, Flemish bond, basket pattern,* and *stack bond.* Figure 14.1 illustrates these pattern bonds.

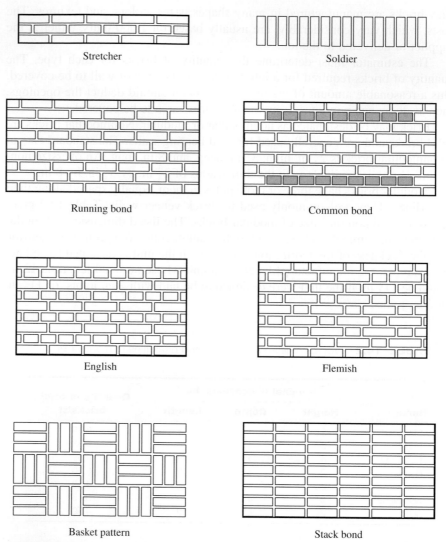

Stretcher

Soldier

Running bond

Common bond

English

Flemish

Basket pattern

Stack bond

FIGURE 14.1 | Brick patterns for masonry units.

Running bond uses only stretcher courses with head joints centered over stretchers in the course below. Common bond is similar to the running bond, except that a header course is repeated about every sixth course. English bond is made up of alternated courses of headers and stretchers, with headers centered on stretchers. Flemish bond alternates stretchers and headers in each course with headers centered over stretchers in the course below. Stack bond provides no interlocking between adjacent masonry units and is used for its architectural effect. Horizontal reinforcement should be used with stack bond to provide lateral bonding.

TYPES OF JOINTS FOR BRICK MASONRY

Several types of mortar joints are specified for brick masonry that can affect the rate of laying bricks. The more common types are shown in Fig. 14.2.

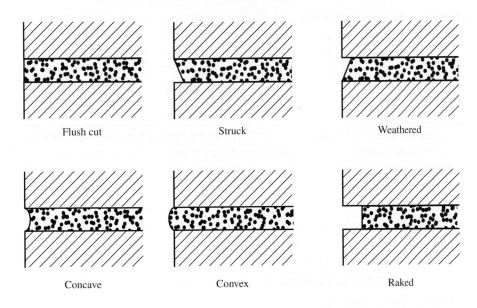

Flush cut   Struck   Weathered

Concave   Convex   Raked

Types of mortar joints

FIGURE 14.2 | Mortar joints for masonry.

Flush-cut joints are made by passing the trowel across the surface of the bricks and removing any excess mortar. This operation requires very little time.

Struck and weathered joints are made with a trowel after the mortar has gained some stiffness. This operation requires more time than flush-cut joints.

Concave joints are made by tooling the mortar with a wood or nonstaining metal rod before the mortar gains final set.

Convex joints are made by compressing the mortar during the process of laying the brick.

Raked joints are made by removing the mortar to a depth of $\frac{1}{4}$ to $\frac{3}{8}$ in., by using a special tool.

ESTIMATING MORTAR FOR BRICKS

The quantity of brick required per square foot of wall will depend on the size of brick and the size of horizontal and vertical mortar joints. The nominal sizes of modular bricks are given in Table 14.2. The actual manufactured sizes of these bricks will be less than the nominal dimensions given in the table. For example, the Norman brick may be manufactured with the dimensions of $2\frac{1}{4}$ by $3\frac{5}{8}$ by $11\frac{5}{8}$ for use with $\frac{3}{8}$-in. mortar, or with the dimensions of $2\frac{1}{4}$ by $3\frac{1}{2}$ by $11\frac{1}{2}$ for use with $\frac{1}{2}$-in. mortar.

The quantity of mortar required for a single brick can be calculated with the following equation.

$$\text{Volume per brick} = (L + H + t) \times t \times D \qquad \textbf{[14.1]}$$

where t = thickness of mortar

L = length of brick

H = height of brick

D = depth of brick

EXAMPLE 14.1

Calculate the quantity of mortar per 1,000 bricks for Roman type modular brick for construction of a brick veneer wall using running bond and $\frac{1}{2}$-in. thick mortar. Table 14.2 gives the nominal dimensions of this modular brick as 2 in. × 4 in. × 12 in. for the height (H), depth (D), and length (L), respectively. When this brick is purchased from the manufacturer for $\frac{1}{2}$-in. mortar, the actual size will be $1\frac{1}{2}$ in. × $3\frac{1}{2}$ in. × $11\frac{1}{2}$ in.

Quantity of mortar for a single brick, using Eq. [14.1]:

Volume $= (L + H + t) \times t \times D$

$\qquad = (11\frac{1}{2}$ in. $+ 1\frac{1}{2}$ in. $+ \frac{1}{2}$ in.$) \times \frac{1}{2}$ in. $\times 3\frac{1}{2}$ in.

$\qquad = 23.625$ cubic inches of mortar per brick

Converting to cubic feet:

Volume $= 23.625$ cubic inches/(1,728 cubic inches/cubic foot)

$\qquad = 0.0136719$ cf of mortar per brick

Volume, in cubic feet, of mortar per 1,000 bricks:

Volume $= (0.0136719$ cf/brick$) \times 1,000$ bricks

$\qquad = 13.671875$ cf/1,000 bricks

Volume, in cubic yards, of mortar per 1,000 bricks:

Volume $= (13.671875$ cf/1,000 bricks$)/(27$ cf/cy$)$

$\qquad = 0.5063657$ cy/1,000 bricks

QUANTITY OF MORTAR FOR BRICK VENEER WALLS

As illustrated in Example 14.1, the quantity of mortar required for brick masonry will vary with the type of brick and the thickness of the mortar joints. It is difficult to accurately estimate the quantity of mortar that will be wasted in laying bricks. Most estimators add 20 to 25 percent for mortar waste.

Table 14.3 gives the approximate cubic yards of mortar per 1,000 bricks. No waste factor is given for the amounts shown.

TABLE 14.3 | Approximate cubic yards of mortar per 1,000 bricks.

Name	Thickness of mortar	
	$\frac{3}{8}$ in.	$\frac{1}{2}$ in.
Standard	0.300	0.382
Engineer	0.319	0.404
Economy	0.341	0.433
Roman	0.397	0.506
Norman	0.419	0.533
Norwegian	0.433	0.552
SCR	0.648	0.837

Note: No allowance for waste is included.

ACCESSORIES FOR BRICK VENEER WALLS

Metal ties are used to anchor the brick veneer to wood framed walls. Wall ties made of 14- to 24-gauge hot-dipped galvanized steel in widths of $1\frac{1}{4}$ in. and lengths of $3\frac{1}{2}$ in. with 2-in. bends are used for this purpose. One end of the wall tie is nailed to the wood stud wall and the other end is embedded in the mortar joints. The wall ties are generally spaced about every 16 in. in the horizontal direction and 24 in. in the vertical direction.

If the brick veneer is placed as face brick on the outside of a concrete masonry unit wall, the wall ties are flat with one end embedded in the mortar of the brick veneer and the other end embedded in the mortar of the concrete masonry unit wall.

Column and beam flange straps are used to anchor brick veneer to structural steel framing. One end of the strap hooks around the flanges of the beam or column and the other end is embedded in the mortar of the brick veneer or concrete masonry unit.

CLEANING BRICK MASONRY

Brick masonry can be cleaned by several methods, including bucket and brush by hand, high-pressure water cleaning, or sandblasting. Hand cleaning is the most popular method due to its simplicity and readily available cleaning material. The cleaning material may consist of detergent or soap solutions, a mixture of 10 percent muriatic acid and water, or propriety compounds.

High-pressure water cleaning is sometimes used because it is faster and thereby reduces the cost of labor. A portable spraying unit on wheels is used for this purpose. Dry sandblasting has been used for many years to clean brick masonry, particularly on restoration work. Sandblasting requires a qualified operator to prevent scarring of the face of brick units and mortar joints.

SOLID BRICK WALLS

A brick veneer wall using a single thickness of modular bricks is generally constructed for aesthetic purposes and is considered a non-load-bearing wall. Structural load-bearing walls may be constructed with solid nonmodular bricks, usually with standard size of solid bricks, size $2\frac{1}{4}$ in. by $3\frac{3}{4}$ in. by 8 in.

Solid brick walls may be designated as 4, 8, 12, and 16 in. thick. The actual thickness will vary with the number of bricks of thickness and the thickness of the vertical mortar joints between rows of bricks. Figure 14.3 illustrates a two-brick-thick wall with common bond. A three-brick-thick wall is referred to as a 12-in. wall; but if $\frac{1}{2}$-in. mortar joints are used, the actual thickness will be $12\frac{1}{4}$ in. In a similar manner, a four-brick-thick wall will actually be $16\frac{1}{2}$ in. thick.

Table 14.4 gives the number of standard bricks ($2\frac{1}{4}$ in. by $3\frac{3}{4}$ in. by 8 in.) for various thicknesses of solid brick walls. Table 14.5 gives the cubic yards of mortar required per 1,000 bricks of size $2\frac{1}{4}$ in. by $3\frac{3}{4}$ in. by 8 in. for full joints, using common bond, with no allowance included for waste.

Buildings constructed with solid brick walls are frequently erected with a 4-in.-thick veneer of face bricks, while the balance of the wall thickness is obtained with common bricks. If a common bond is used, it is customary to lay face bricks as headers every sixth course for bond purposes. Since a header course will require twice as many face bricks as a stretcher course, the total number of face bricks required must be increased over the number that would be required for a uniform 4-in. thickness. This is equivalent to one extra stretcher course in six courses, amounting to an increase of $16\frac{2}{3}$ percent in the number of face bricks required. The number of common bricks may be reduced in an amount equal to the increase in the number of face bricks.

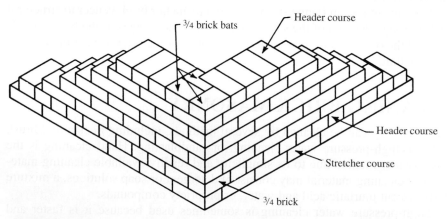

FIGURE 14.3 | Solid brick wall laid with common bond.

TABLE 14.4 | Number of $2\frac{1}{4}$-in. by $3\frac{3}{4}$-in. by 8-in. solid nonmodular bricks per 100 sf of wall area.

Nominal thickness of wall, in.	Thickness of horizontal mortar joint, in.				
	$\frac{1}{4}$	$\frac{3}{8}$	$\frac{1}{2}$	$\frac{5}{8}$	$\frac{3}{4}$
End joints $\frac{1}{4}$ in. thick					
4	698	665	635	608	582
8	1,396	1,330	1,270	1,216	1,164
12	2,095	1,995	1,905	1,824	1,746
16	2,792	2,660	2,540	2,432	2,328
End joints $\frac{1}{2}$ in. thick					
4	677	645	615	588	564
8	1,354	1,290	1,230	1,176	1,128
12	2,031	1,935	1,845	1,764	1,692
16	2,708	2,580	2,460	2,352	2,256

TABLE 14.5 | Cubic yards of mortar required per 1,000 standard-size ($2\frac{1}{4}$-in. by $3\frac{3}{4}$-in. by 8-in.) bricks for full joints, using common bond (no allowance included for waste).

Nominal thickness of wall, in.	Thickness of horizontal mortar joint, in.				
	$\frac{1}{4}$	$\frac{3}{8}$	$\frac{1}{2}$	$\frac{5}{8}$	$\frac{3}{4}$
Vertical joints $\frac{1}{4}$ in. thick					
4	0.211	0.291	0.376	0.458	0.542
8	0.301	0.384	0.474	0.562	0.649
12	0.346	0.432	0.523	0.614	0.703
16	0.429	0.519	0.616	0.709	0.805
Vertical joints $\frac{1}{2}$ in. thick					
4	0.262	0.348	0.433	0.518	0.627
8	0.365	0.456	0.546	0.637	0.751
12	0.414	0.506	0.600	0.693	0.809
16	0.433	0.520	0.621	0.715	0.832

LABOR LAYING BRICKS

The labor-hours required to lay bricks vary with a number of factors, such as the quality of work, type of bricks, kind of mortar used, shape of the walls, kind of bond pattern used, and weather conditions.

If walls are irregular in shape with frequent openings, pilasters, or other changes in shape, the labor requirements will be greater than for long, straight walls. If the joints must be tooled, more labor will be required than if the joints are simple cut flush with a trowel. For some jobs, more than one bond pattern is used, which increases the labor required to lay bricks.

Table 14.6 gives the approximate labor-hours to lay 1,000 bricks for modular brick veneer walls and for nonmodular solid brick walls. The indicated time includes mixing the mortar and laying the bricks.

TABLE 14.6 | Approximate labor-hours to lay 1,000 bricks for modular brick veneer walls and solid brick walls.

Type of wall	Hours per 1,000 bricks	
	Bricklayers	**Helper**
Brick veneer running bond using modular bricks	15–17	10–11
Solid brick walls common bond using $2\frac{1}{4}$-in. × $3\frac{3}{4}$-in. × 8-in. bricks	10–11	9–10

Note: (1) For other than running bond pattern brick veneer increase by 7 percent; (2) for cavity walls, the labor-hours should be increased by 15 percent; and (3) for curved walls, the labor-hours should be increased by 30 percent.

EXAMPLE 14.2

Estimate the total direct cost of material, labor, and equipment for constructing the walls of a rectangular building 116 ft long, 68 ft wide, with brick veneer walls 10 ft high. The total area of openings for doors and windows will be 468 sf.

The walls will be 4 in. thick, using Roman type bricks laid in common bond with $\frac{1}{2}$-in. mortar joints on all sides. A header course will be placed every sixth course between the stretcher courses. Wall ties will be placed every 16 in. in the horizontal direction and every 24 in. in the vertical direction.

Material costs will be $615/1,000 bricks, $83/cy for mortar, and $4.25/100 wall ties. Labor costs include $29.00/hr for bricklayers and $22.00/hr for helpers. The mixer for mortar will cost $9.50/hr. Scaffolding rental will be $1,800 for the job.

Quantity of materials:

Gross outside area:

[(116 ft × 10 ft) × 2 walls] + [(68 ft × 10 ft) × 2 walls]   = 3,680 sf

Deduct for four corners: [(4 in.)/(12 in./ft)] × 10 ft × 4 corners =   −13 sf

Deduct for doors and windows   = −468 sf

Net area = 3,199 sf

Bricks required:

From Table 14.2 for Roman type modular brick

Stretcher bricks: [(1 brick)/(2 in. × 12 in.)] × 144 sq. in./sf = 6.0 bricks/sf

Header bricks: [(1 brick)/(2 in. × 4 in.)] × 144 sq. in./sf = 18 bricks/sf

For common bond, every sixth course of bricks will be headers and the other courses of bricks will be stretchers. Therefore, the number of bricks can be calculated as:

For stretchers: 3,199 sf $\times \frac{5}{6} \times$ 6.0 bricks/sf = 15,995 stretcher bricks

For headers: 3,199 sf $\times \frac{1}{6} \times$ 18.0 bricks/sf = <u>9,597 header bricks</u>

Total = 25,592 bricks

Add 5% for waste = <u>1,280 bricks</u>

Total bricks = 26,872 bricks

Mortar required:

From Table 14.3 for Roman brick with $\frac{1}{2}$-in. mortar joints = 0.506 cy/1,000 brick

Quantity of mortar required: 26,872 bricks $\times$ 0.506 cy/1,000 bricks = 13.6 cy

Add 25% waste = <u>3.4 cy</u>

Total mortar = 17.0 cy

Wall ties required:

Spacing of metal wall ties: 16 in. $\times$ 24 in. = 384 sq. in./tie

Number of ties per square foot of wall:

144 sq. in./384 sq. in./tie = 0.375 ties/sf

Number of ties: 3,199 sf $\times$ 0.375 ties/sf = 1,200 ties

Add 5% waste = <u>+ 60 ties</u>

Total wall ties = 1,260 ties

Labor-hours required:

From Table 14.6 using average values for brick veneer walls with common bond:

Bricklayers: 26,872 bricks $\times$ 16 hr/1,000 bricks $\times$ 1.07 for common bond = 460 hr

Helpers: 26,872 bricks $\times$ 10.5 hr/1,000 bricks $\times$ 1.07 for common bond = 302 hr

Equipment required:

Time for mortar mixer assuming 3 bricklayers: 460 hr/3 = 153 hr

Rental cost of scaffolding = $1,800

Cost of material, labor, and equipment:

Materials:

Bricks: 26,872 bricks @ $615/1,000 bricks = $16,526.28

Mortar: 17.0 cy @ $83/cy   = 1,411.00

Wall ties: 1,200 ties @ $4.25/100 ties   = 51.00

Labor:

Bricklayers: 460 hr @ $29.00/hr   = 13,340.00

Helpers: 302 hr @ $22.00/hr   = 6,644.00

Equipment:

Mortar mixer: 153 hr @ $9.50/hr   = 1,453.60

Scaffolding rental   = <u>1,800.00</u>

Total cost = $41,225.88

Cost per brick = $41,225.88/25,592 bricks = $1.61/brick

Cost per 1,000 bricks: $41,225.88/25.592 = $1,610.89/1,000 bricks

Cost per square feet: $41,225.88/3,199 sf = $12.89/sf

This cost does not include indirect costs of taxes, insurance, contingency, or profit.

CONCRETE MASONRY UNITS

Concrete masonry units (CMUs) are concrete blocks that are manufactured in various sizes from Portland cement, sand, and gravel, or cement and lightweight aggregates. The concrete blocks may be solid or hollow. Exterior walls of hollow CMUs are usually filled with insulation material, such as granular vermiculite, to provide protection for heat loss or gain in the building. CMUs, either solid or hollow, can be placed to form a load-bearing wall or a non-load-bearing wall.

Table 14.7 gives the actual sizes, approximate weights, and quantities of mortar required for joints for the more popular sizes of CMUs. The most common nominal size of a CMU is 8 in. × 8 in. × 16 in. The actual sizes of blocks usually will be $\frac{3}{8}$ in. less than the nominal sizes. When $\frac{3}{8}$-in.-thick mortar joints are used, the dimensions occupied by blocks will equal the nominal sizes.

The quantities given in Table 14.7 are based on 100 sf of wall area, with no allowance for waste or breakage of the blocks. The quantities of mortar given do not include any allowance for waste, which frequently will amount to 20 to 50 percent of the net amount required.

TABLE 14.7 | Sizes, weights, and quantities of mortar for concrete blocks, using $\frac{3}{8}$-in. joints.*

Actual size: thickness, height, length, in.	Approximate weight per block, lb		Quantities per 100 sf of wall area	
	Standard	Light-weight	Number of blocks	Mortar, cy
$3\frac{5}{8} \times 4\frac{7}{8} \times 11\frac{5}{8}$	11–13	8–10	240	0.15
$5\frac{5}{8} \times 4\frac{7}{8} \times 11\frac{5}{8}$	17–19	12–14	240	0.16
$7\frac{5}{8} \times 4\frac{7}{8} \times 11\frac{5}{8}$	22–24	14–16	240	0.17
$3\frac{5}{8} \times 7\frac{5}{8} \times 11\frac{5}{8}$	17–19	12–14	150	0.11
$5\frac{5}{8} \times 7\frac{5}{8} \times 11\frac{5}{8}$	26–28	17–19	150	0.12
$7\frac{5}{8} \times 7\frac{5}{8} \times 11\frac{5}{8}$	33–35	21–23	150	0.13
$5\frac{5}{8} \times 3\frac{5}{8} \times 15\frac{5}{8}$	17–19	11–13	225	0.18
$7\frac{5}{8} \times 3\frac{5}{8} \times 15\frac{5}{8}$	22–24	14–16	225	0.19
$3\frac{5}{8} \times 7\frac{5}{8} \times 15\frac{5}{8}$	23–25	16–18	113	0.10
$5\frac{5}{8} \times 7\frac{5}{8} \times 15\frac{5}{8}$	35–37	24–27	113	0.11
$7\frac{5}{8} \times 7\frac{5}{8} \times 15\frac{5}{8}$	45–47	28–32	113	0.12

*These quantities do not include any allowance for waste. Add 2 to 5 percent for blocks and 20 to 50 percent for mortar.

LABOR LAYING CONCRETE MASONRY UNITS

Concrete masonry units are laid by masons. Joints are made by spreading mortar along the inside and outside horizontal and vertical edges. Joints may be cut smooth with a steel trowel, or they may be tooled as for brick. The joints are more resistant to the infiltration of moisture when they are tooled, because the tooling increases the density of the mortar.

A *bond beam* is a continuously reinforced horizontal beam of masonry designed to provide additional strength and to prevent cracks in masonry walls. Bond beams are often placed about every sixth course of CMU walls. Special U-shaped CMUs are placed with their open sides facing upward. Generally two no. 5 reinforcing bars are placed horizontally in the U-shaped CMU and then filled with concrete. A mason and a helper working together should be able to install about 15 lin ft of bond beam per hour, including laying the U-shaped CMUs, setting the reinforcing steel, and placing the concrete.

Additional reinforcement is obtained by placing latticed smooth wire reinforcing steel in the horizontal mortar joints. *Lintels* are short beams of wood, steel, stone, or reinforced masonry used to span openings in masonry walls.

Expansion or *control joints* are placed at intervals to permit differential movement of wall sections caused by shrinkage of concrete due to fluctuations in temperature. Expansion joints may be placed at door and window openings, or at columns and pilasters. Expansion joints are also placed at offsets in walls. Control joints are grooves placed in masonry to control shrinkage cracking.

Table 14.8 gives the labor-hours required to lay 1,000 concrete masonry units of various sizes. The time given for masons includes laying the units and tooling the joints, if required. The rates provide for different classes of work. The time given for laborers includes supplying mortar and CMUs for the masons.

EXAMPLE 14.3

Estimate the total direct cost and cost per 1,000 CMUs for furnishing and laying standard-weight concrete masonry units for a rectangular building 120 ft long, 80 ft wide, and 14 ft high. The total area of the openings will be 516 sf.

The CMUs will be 8 in. by 8 in. by 16 in. and mortar joints will be $\frac{3}{8}$-in. thick. Three rows of bond beams are required, reinforced with two no. 5 reinforcing steel bars. The manufacturer's literature shows 0.22 cf of concrete is required to fill each linear foot of the U-shaped CMUs that will be used for the bond beams.

Material costs will be $2.20/CMU and $95/cy for mortar. Material costs for the bond beam will be $0.46/lb for reinforcing steel and $115/cy for the concrete. Labor costs include $29.00/hr for masons, $22.00/hr for helpers, and $25.00/hr for the mixer operator. Equipment for mixing the mortar will cost $9.50/hr. Scaffolding rental will be $2,500 for the job.

TABLE 14.8 | Labor-hours required to handle and lay 1,000 concrete blocks.*

Nominal-size block: thickness, height, length, in.	Labor-hours per 1,000 blocks	
	Mason	Helper
Standard blocks		
4 × 5 × 12	33–38	33–38
6 × 5 × 12	38–44	38–44
8 × 5 × 12	44–50	44–50
4 × 8 × 12	38–44	38–44
6 × 8 × 12	44–50	44–50
8 × 8 × 12	50–55	50–55
6 × 4 × 16	38–44	38–44
8 × 4 × 16	44–50	44–50
4 × 8 × 16	38–44	38–44
6 × 8 × 16	44–50	44–50
8 × 8 × 16	52–57	52–57
Lightweight blocks		
4 × 5 × 12	30–35	30–35
6 × 5 × 12	35–40	35–40
8 × 5 × 12	40–45	40–45
4 × 8 × 12	35–40	35–40
6 × 8 × 12	40–45	40–45
8 × 8 × 12	45–50	45–50
6 × 4 × 16	35–40	35–40
8 × 4 × 16	40–45	40–45
4 × 8 × 16	35–40	35–40
6 × 8 × 16	40–45	40–45
8 × 8 × 16	47–52	47–52

*Add the cost of a hoisting engineer if required.

Quantity of materials:

Outside perimeter of wall: (120 ft + 80 ft) × 2 walls = 400 lin ft

Gross outside area:

[(120 ft × 14 ft) × 2 walls] + [(80 ft × 14 ft) × 2 walls]   = 5,600 sf

Deduct for four corners: [(8 in.)/(12 in./ft)] × 14 ft × 4 corners =   −37 sf

Deduct for doors and windows   = −516 sf

Net area = 5,047 sf

CMUs required:

From Table 14.7, number of concrete masonry units = 113 CMUs/100 sf

Number of CMUs required: 5,047 sf × (113 CMUs/100 sf) = 5,703 CMUs

Add 5% waste = +285 CMUs

Total = 5,988 CMUs

Mortar required:

From Table 14.7 for 8-in. by 8-in. by 16-in. CMUs = 0.12 cy/100 sf

Quantity of mortar required: 5,047 sf × (0.12 cy/100 sf) = 6.1 cy

Add 30% waste = 1.8 cy

Total mortar = 7.9 cy

Material for bond beams:

Concrete: 3 rows × 400 lin ft × 0.22 cf/lin ft × (1.0 cy/27 cf) = 9.8 cy

Add 15% waste = 1.5 cy

Total = 11.3 cy

Weight of no. 5 bars: 3 rows × 400 lin ft × 2 bars × 1.043 lb/ft = 2,503 lb

Add for lap splicing:

3 rows × [(400 lin ft)/(20 lin ft/lap)] 2 bars × 1 ft/lap × 1.043 lb/ft = 125 lb

Total = 2,628 lb

Labor-hours required:

Labor installing CMUs in wall:

Table 14.8 using average values for 8-in. by 8-in. by 16-in. CMUs:

Masons installing CMUs: 5,988 CMUs × 54.5 hr/1,000 CMUs = 326 hr

Masons installing bond beams: (400 lin ft/15 lin ft/hr) × 3 rows = 80 hr

Total = 406 hr

Helpers: 5,988 CMUs × 54.5 hr/1,000 bricks = 326 hr

Helpers installing bond beams: (400 lin ft/15 lin ft/hr) × 3 rows = 80 hr

Total = 406 hr

Mixer operator assuming 3 masons: 406 hr/3 = 135 hr

Equipment required:

Time for mortar mixer assuming 3 masons: 406 hr/3 = 135 hr

Rental cost of scaffolding = $2,500

Cost of materials, labor, and equipment:

Materials:

Bricks: 5,988 CMUs @ $2.20/CMU = $13,173.60

Mortar: 7.9 cy @ $95.00/cy = 750.50

Reinforcing: 2,566 lb @ $0.46/lb = 1,180.32

Concrete: 11.3 cy @ $115.00/cy = 1,299.50

Labor:

Bricklayers: 406 hr @ $29.00/hr = 11,774.00

Helpers: 406 hr @ $22.00/hr = 8,932.00

Mixer operator: 135 hr @ $25.00/hr = 3,375.00

Equipment:

Mortar mixer: 135 hr @ $9.50/hr = 1,282.50

Scaffolding rental = 2,500.00

Total cost = $44,267.46

Cost per CMU = $44,267.46/5,703 CMU = $7.76/CMU

Cost per square feet: $44,267.46/5,047 sf = $8.77/sf

This cost does not include lifting equipment if required or the indirect costs of taxes, insurance, contingency, or profit.

STONE MASONRY

Several kinds of stone, both natural and artificial, are used in structures such as buildings, walls, piers, etc. Natural stones used for construction include sandstone, limestone, dolomite, slate, granite, marble, etc. Artificial limestone is available in many areas.

Each kind of stone and work should be estimated separately. The cost of stone in place can be estimated by the cubic yard, ton, cubic foot, square foot, or linear foot. Because of the various methods of pricing stonework, an estimator should be very careful to use the correct method in preparing the estimate.

BONDS FOR STONE MASONRY

Figure 14.4 illustrates the more common bonds for stone masonry. Rubble masonry is formed of stones of irregular shapes that are laid in regular courses or at random with mortar joints. Ashlar masonry is formed of stones cut with rectangular faces. The stones may be laid in courses or at random with mortar joints.

MORTAR FOR STONE MASONRY

The mortar used for setting stones may be similar to that used for brick masonry. Sometimes special nonstaining white or stone-set cement may be specified instead of gray Portland cement. Hydrated lime is usually added to improve the working properties of the mortar.

The quantity of mortar required for joints will vary considerably with the type of bond, the thickness of the joints, and the size of stones used. Table 14.9 gives representative quantities of mortar required per cubic yard of stone.

WEIGHTS OF STONE

Table 14.10 gives the ranges in weights of stones commonly used for masonry. The given weights are expressed in pounds per net cubic foot of volume.

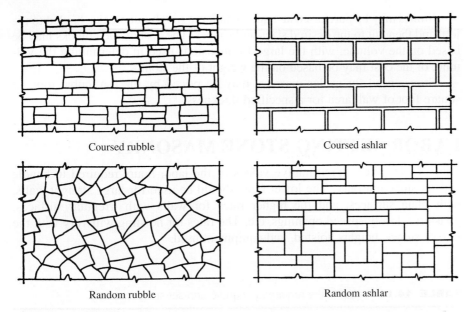

Coursed rubble

Coursed ashlar

Random rubble

Random ashlar

FIGURE 14.4 | Bonds for stone masonry.

TABLE 14.9 | Quantities of mortar required per cubic yard of stone masonry.

Type of bond	Quantity of mortar, cf
Coursed rubble	6.5–8.5
Random rubble	7.5–9.5
Cobblestone	6.5–9.5
Coursed ashlar, $\frac{1}{4}$-in. joints	1.5–2.0
Random ashlar, $\frac{1}{4}$-in. joints	2.0–2.5
Coursed ashlar, $\frac{1}{2}$-in. joints	3.0–4.0
Random ashlar, $\frac{1}{2}$-in. joints	4.0–5.0

TABLE 14.10 | Weights of building stones.

Stone	Weight, lb/cf
Dolomite	155–175
Granite	165–175
Limestone	150–175
Marble	165–175
Sandstone	140–160
Slate	160–180

COST OF STONE

The cost of stone varies so much with the kind of stone, extent of cutting done at the quarry, and location of use that no estimate that requires accurate pricing should be made without obtaining current prices for the particular stone. The cost of freight to the destination must be added to the cost at the source to determine the cost at the job.

Stones suitable for rubble masonry may be priced by the ton for the specified kind of stone and sizes of pieces.

Stones may be purchased at a quarry in large rough-cut blocks, hauled to the job, and then cut to the desired sizes and shapes. The cost of such blocks may be based on the volume, with the largest dimensions used in determining the volume; or the cost may be based on the weight of the stone.

Stones used for ashlar masonry may be priced by the ton, cubic foot, or square foot of wall area for a specified thickness.

LABOR SETTING STONE MASONRY

Table 14.11 gives representative values of the labor-hours required to handle and set stone masonry. The lower rates should be used for large straight walls, with plain surfaces, and the higher rates for walls with irregular surfaces, pilasters, closely spaced openings, etc. The rates for hoisting will vary with the size of stones, heights hoisted, and equipment used.

TABLE 14.11 | Labor-hours required to handle and set stone masonry.

| | Labor-hours required | | | |
| | Per cy | | Per cf | |
Operation	Skilled	Helper	Skilled	Helper
Unloading stones from truck	0.2–0.4	0.8–1.6	0.01–0.02	0.03–0.06
Rough squaring	4.0–10.0	—	0.15–0.37	
Smoothing beds	6.0–12.0	—	0.22–0.45	
Setting stone by hand				
Rubble	2.0–4.0	3.0–6.0	0.08–0.15	0.11–0.22
Rough squared	3.0–6.0	4.0–9.0	0.11–0.22	0.15–0.33
Ashlar, 4- to 6-in. veneer	10.0–15.0	10.0–15.0	0.37–0.55	0.37–0.55
Setting stones by using a hand derrick				
Heavy cut stone, ashlar, cornices, copings, etc.	2.0–3.0	12.0–18.0	0.08–0.11	0.45–0.67
Light cut stone, ashlar, sills, lintels, cornices, etc.	2.5–4.0	12.5–20.0	0.09–0.15	0.46–0.75
Setting stones by using a power crane				
Heavy cut stone				
Crane operator	1.0–2.0	—	0.04–0.08	
Stone setter	1.0–2.0	—	0.04–0.08	
Helpers	—	7.0–14.0	—	0.26–0.52
Medium cut stone				
Crane operator	1.5–3.0	—	0.06–0.11	
Stone setter	1.5–3.0	—	0.06–0.11	
Helpers	—	9.0–18.0	—	0.33–0.67
Pointing cut stone				
Heavy stone	0.5–0.8	0.2–0.4	0.02–0.03	0.01–0.02
Veneer, 4 to 6 in. thick	1.0–1.5	0.2–0.8	0.04–0.06	0.02–0.03
Cleaning stone				
Stone setter, 1.5 hr/100 sf				
Helper, 0.75 hr/100 sf				

EXAMPLE 14.4

Estimate the cost per 100 sf for furnishing and setting stone in random ashlar bond for a building. The stone will be sawed on the top and bottom beds and on the front and back faces. It will be furnished in random lengths to be sawed to the desired lengths at the job. The thickness will be 4 in. The stone will be set with $\frac{1}{2}$-in. mortar joints. The stone will be hoisted with a hand derrick.

The stone, which will weigh 150 lb/cf, will be delivered to the job at a cost of $326.00/T. Allow 10 percent for waste.

The quantity of stone required for 100 sf will be

$$\frac{100 \times 4 \times 1.1}{12} = 36.7 \text{ cf}$$

The weight of the stone will be 36.7 × 150/2,000 = 2.75 T

The cost will be:

Stone, f.o.b. job: 2.75 T @ $326.00/T	=	$896.50
Cutting end joints: 36.7 cf @ $2.46	=	90.28
Mortar: 36.7/27 = 1.36 cy of stone,		
1.36 cy × 4.5 cf/cy = 6.1 cf @ $3.06	=	18.67
Metal wall ties: 100 sf ÷ 2 sf/tie = 50 @ $0.92	=	46.00
Stone setters, 36.7 cf:		
36.7 cf × 0.12 hr/cf = 4.4 hr @ $31.00/hr	=	136.40
Helpers, 36.7 cf:		
36.7 cf × 0.6 hr/cf = 22 hr @ $24.00/hr	=	528.00
Stone setter pointing stone, 36.7 cf:		
36.7 cf × 0.05 hr/cf = 1.8 hr @ $31.00/hr	=	55.80
Helper pointing stone, 36.7 cf:		
36.7 cf × 0.03 hr/cf = 1.1 hr @ $24.00/hr	=	26.40
Stone setter cleaning stone, 100 sf:		
100 sf × 1.5 hr/100 sf = 1.5 hr @ $31.00/hr	=	46.50
Helper cleaning stone, 100 sf:		
100 sf × 0.75 hr per 100 sf = 0.75 hr @ $24.00/hr =		18.00
Foreman, based on 2 stone setters: 3.9 hr @ $34.00/hr =		132.60
Equipment, saws, derrick, scaffolds, etc.		250.00
Total cost	=	$2,245.15
Cost per sf: $2,245.15/100	=	$22.45

PROBLEMS

14.1 Estimate the total direct cost and cost per 1,000 bricks for construction of walls of a rectangular building, 60 ft wide and 150 ft long, with brick veneer walls 8 ft 9 in. high. The total area of openings for doors and windows will be

324 sf. The walls will be 4 in. thick, using Norman type bricks laid in running bond with $\frac{1}{2}$-in. mortar joints on all sides. Wall ties will be placed every 16 in. in the horizontal direction and every 24 in. in the vertical direction.

Material cost will be $586/1,000 for bricks, $95/cy for mortar, and $4.25/100 for wall ties. Labor costs include $29.00/hr for bricklayers and $22.00/hr for helpers. The mixer for mortar will cost $9.50/hr. Scaffolding rental will be $1,800 for the job.

14.2 Estimate the total direct cost and cost per 1,000 bricks for construction of walls of a rectangular building, 80 ft wide and 320 ft long, with brick veneer walls 10 ft high. The total area of openings for doors and windows will be 560 sf. The walls will be 4 in. thick, using Norwegian type bricks laid in common bond with $\frac{1}{2}$-in. mortar joints on all sides. A header course will be placed every fifth course between the stretcher courses. Wall ties will be placed every 16 in. in the horizontal direction and every 24 in. in the vertical direction.

Material cost will be $612/1,000 for bricks, $95/cy for mortar, and $4.25/100 for wall ties. Labor costs include $29.00/hr for bricklayers and $22.00/hr for helpers. The mixer for mortar will cost $9.50/hr. Scaffolding rental will be $1,800 for the job.

14.3 Estimate the total direct cost and cost per 1,000 CMUs for furnishing and laying standard-weight concrete masonry units for a rectangular building 185 ft long, 95 ft wide, and 12 ft high. The total area of the openings will be 725 sf.

The CMUs will be 8 in. by 8 in. by 16 in. and mortar joints will be $\frac{3}{8}$ in. thick. Two rows of bond beams are required, reinforced with two no. 5 reinforcing steel bars. The manufacturer's literature shows 0.33 cf of concrete is required to fill each linear foot of the U-shaped CMUs that will be used for the bond beams.

Use the unit costs for material, labor, and equipment shown in Example 14.3.

15

Floor Systems and Finishes

This chapter describes several types of floor systems that can be used as substitutes for concrete-beam-and-slab, concrete slab only, or pan-and-joist concrete floors. An examination of the cost developed for each of the floor systems and a comparison with the cost of an all-concrete floor will reveal that reductions in the cost frequently can be affected through the choice of the floor system.

Although these analyses are made for floor systems only, the results should demonstrate that similar analyses for other parts of structures may also permit the selection of methods and materials that will produce reductions in the costs. There are many types of finishes that are applied to floors. Several of the more popular types will be discussed in the last part of this chapter.

FLOOR SYSTEMS

STEEL-JOIST SYSTEM

Open-web steel joists (Fig. 15.1) are fabricated in the shop into the form of a Warren truss by an arc-welding process. The chord members consist of T sections, angles, or bars. The web is made of a single round bar, or angle, that is welded to the top and bottom chord members. A steel bearing seat is welded to each end of a joist to provide proper bearing area. The bearing seats are designed to permit the ends of the joist to rest on masonry walls or structural-steel beams.

Bridging

Typical bridging for steel joists includes horizontal angle struts that are welded between adjacent joists. If cross bridging is required, angle struts are fabricated that are bolted or welded to the top chord of one joist and the bottom chord of

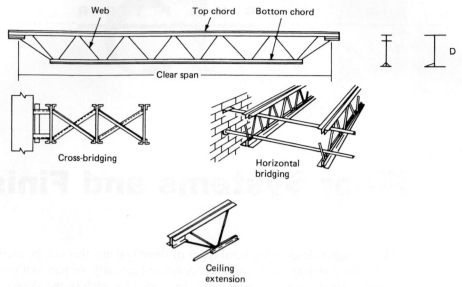

FIGURE 15.1 | Steel joist.

an adjacent joist. Two struts are installed for each line of bridging to give a cross-bridging effect.

Cross bridging should be installed in accordance with the specifications of the designer or the manufacturer of the joists, which will be approximately as shown in Table 15.1. These are common bridging requirements for light joists. Longer spans can be obtained with heavier chord sizes with the above lines of bridging.

TABLE 15.1 | Cross bridging for steel joists.

	Span, ft	
Lines of bridging	**K series joists**	**LH series joists**
Row near center	Up to 16	Up to 22
Rows at $\frac{1}{3}$ points	16–24	22–33
Rows at $\frac{1}{4}$ points	24–28	33–44

Metal Decking

To support the concrete floor or roof, a continuous layer of ribbed metal decking is installed over the joists, with ribs perpendicular to the joists. The decking is fastened to the top chords of the joists by welding or screw fasteners, spaced not over 15 in. apart along the joists. Metal decking is fabricated under various trade names by several manufacturers. It is designated by shape, size, and gauge. Table 15.2 provides a representative example of metal decking.

TABLE 15.2 | Data for form decking.

Joist spacing, in.	Size of rib, in.	Gauge	Weight, lb/sf
Up to 30	$\frac{9}{16}$	28	0.86
30–36	$\frac{9}{16}$	26	1.01
36–60	1	26	1.06

Floor decking should be lapped at least 2 in. beyond the center of supporting joists at the end of sheets and should be securely fastened to the next sheet.

Joist End Supports

Each end of the joist is secured to either a load-bearing wall, or a structural beam and column framing system. If the joist is supported by a load-bearing masonry or concrete wall, the joist is welded to a steel bearing plate or clip angle that is anchored in the wall. For structural beam and column framing systems, the ends of the joist are welded directly to the beams and columns.

Size and Dimensions of Steel Joists

Steel joists are available in a great many sizes and lengths for varying loads and spans. Standard open-web joists, which are designed in accordance with the Steel Joist Institute standard specifications, are available for spans varying from 5 to 60 ft for K series joists. Spans up to 144 ft are available in the DLH series joists. Joists can be installed with any desired spacing from 18 to 72 in. or more, provided that the maximum safe load is not exceeded. Table 15.3 provides representative samples of steel joists.

Cost of Steel Joists

The cost of steel joists varies so much with the number and sizes of the joists, accessories required, and location of the job that a table of costs is of little value to an estimator. Before preparing an estimate for a particular job, the estimator should submit the plans and specifications to a representative of a manufacturer for a quotation. Care should be taken to include in the quotation all necessary accessories such as bridging and extensions. The quotation should specify whether the prices are f.o.b. the shop or the job. If the prices are f.o.b. the shop, the estimator must add the cost of transporting the materials to the job.

TABLE 15.3 | Representative dimensions of joists.

Type	Depth, in.	Span length, ft	Approximate weight, lb/lin ft
K series			
12K3	12	12–24	5.0
14K3	14	14–28	6.0
16K4	16	16–32	7.0
18K6	18	18–36	8.5
20K6	20	20–40	9.0
22K7	22	22–44	10.0
24K8	24	24–48	11.5
26K8	26	26–52	12.0
28K9	28	28–56	13.0
30K10	30	30–60	15.0
LH series			
18LH04	18	25–36	12.0
20LH06	20	25–40	15.0
24LH08	24	33–48	18.0
28LH09	28	41–56	21.0
32LH11	32	49–64	24.0
36LH12	36	57–72	25.0
44LH13	44	73–88	30.0
48LH15	48	81–96	36.0
DLH series			
52DLH13	52	89–104	34.0
56DLH14	56	97–112	39.0
60DLH14	60	105–120	40.0
64DLH16	64	113–128	46.0
68DLH16	68	121–136	49.0
72DLH17	72	129–144	56.0

Labor Erecting Steel Joists

The labor cost of erecting steel joists and accessories is usually estimated by the ton. To arrive at a reasonable unit price per ton, it is necessary to determine the probable rate at which the joists will be erected. The rate will vary with the size and length of the joists, method of supporting them, type of end connections, type of bridging, height that they must be lifted, and complexity of the floor area.

In some locations, union regulations require that all labor erecting steel joists must be performed by union mechanics, while in other locations helpers are permitted to assist in the erection.

Table 15.4 gives the approximate labor-hours required to erect 1 T of steel joists, including the installation of bridging and accessories. If helpers are not permitted to assist the ironworkers, the time shown for helpers should be added to that shown for the ironworkers.

TABLE 15.4 | Labor-hours required to erect 1 T of steel joists.

Length of span, ft	hr/T	
	Ironworker	**Helper***
Irregular construction, small areas		
6–10	6.5	3.25
10–16	6.0	3.0
16–24	5.5	2.75
24–30	5.0	2.5
Regular construction, large areas		
16–24	4.5	2.25
24–32	4.0	2.00

*For each floor above the first floor, add 1.5 helper-hours if the joists are carried up by hand.

Labor Installing Metal Decking

Sheets of metal decking are laid perpendicular to the joists and secured with welds or screw fasteners. Two skilled ironworkers working together should install 180 to 240 sf/hr of decking, depending on the complexity of the floor area. This is equivalent to 7.5 to 11 labor-hours per 1,000 sf.

Labor Placing Welded-Wire Fabric

Welded-wire fabric is frequently used to reinforce the concrete slab placed on steel joists. Table 10.9 gives the properties of representative samples of welded-wire fabric. Many sizes and weights are manufactured. An experienced worker should place approximately 400 sf/hr of welded-wire fabric on straight-run jobs. If cutting and fitting are necessary, the rate will be reduced. Since the fabric is furnished in rolls containing approximately 750 sf, it may be necessary to use mechanical equipment to hoist it to floors above the ground level.

Concrete for Slabs

The subject of mixing and placing concrete for the rough floor has been discussed in Chapter 10. Since the information given in Chapter 10 was developed primarily for floors having a greater thickness than is generally used with floors supported by steel joists, the amounts of labor required to haul, spread, and screed the concrete for joist-supported floors should be increased to provide for the additional time needed. An increase of approximately 25 percent for the operations affected should be adequate.

EXAMPLE 15.1

Estimate the total direct cost of steel joists, metal decking, and a $2\frac{1}{2}$-in.-thick light-weight concrete slab for a floor area 48 ft wide and 72 ft long. The floor will be divided into four bays, each 18 ft wide and 48 ft long as shown in Fig. 15.2. The floor is one story above the ground level.

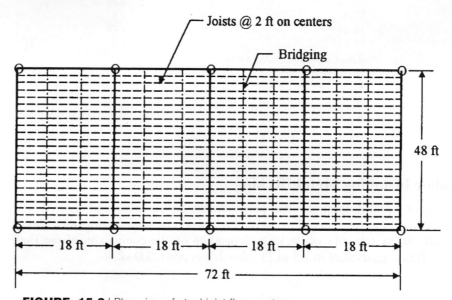

FIGURE 15.2 | Plan view of steel joist floor system.

The steel joists will be type 16K4, 18 ft long, spaced 2 ft on centers, with two rows of bridging per joist. The joists will be welded to the supporting steel members by two 1-in.-long welds at each end of each joist.

The joists will be covered with $\frac{9}{16}$-in. 28-gauge corrugated-metal decking weighing 0.86 psf. The decking will be fastened to the joists with $\frac{1}{2}$-in. welds spaced not over 15 in. apart.

The concrete will be reinforced with a 6 × 6-10/10 welded-wire fabric. The top surface of the concrete will be screeded to the required thickness, but will not be finished under this estimate.

Quantities of materials:

Floor area: 48 ft × 72 ft = 3,456 sf

Concrete:

Cubic feet of concrete: 3,456 sf × [2.5 in./(12 in./ft)] = 720 cf

Cubic yards of concrete: 720 cf/(27 cf/cy) = 26.7 cy

Add 10% waste, 10% × 26.7 cy = 2.7 cy

Total = 29.4 cy

Reinforcing steel:

    Square feet of welded-wire fabric = 3,456 sf

        Add 5% waste, 5% × 3,456 sf =   <u>173 sf</u>

                     Total area = 3,629 sf

Metal decking:

    Weight of decking: 3,456 sf × 0.86 lb/sf = 2,972 lb

         Add 2% waste, 2% × 2,972 lb =   <u>59 lb</u>

                    Total weight = 3,031 lb

Joists:

    Number of joists per bay: 48 ft/2 ft/joist − 2 = 22 joists/bay

    Total number of joists: 22 joists/bay × 4 bays = 88 joists total

    Total linear feet of joists: 88 joists × 18 ft/joist = 1,584 lin ft

    Weight of 16K4 joists: 88 joists × 18 ft @ 7 lb/ft = 11,088 lb

    Converting to tons: 11,088 lb/(2,000 lb/T) = 5.5 T

Bridging:

    Linear feet of struts: 48 ft/row × 2 rows/bay × 4 bays = 384 lin ft

Cost of labor and materials:

Joists and bridging:

Joists delivered to job, 18 ft long: 88 joists @ $64.26 each =		$5,654.88
Bridging delivered to job, 2 ft long: 384 struts @ $0.72	=	276.48
Ironworkers: 5.5 T × 4.5 hr/T = 24.8 hr @ $27.00/hr	=	669.60
Laborers: 5.5 T × 2.25 hr/T = 12.4 hr @ $21.00/hr	=	260.40
Small truck-mounted crane: 24.8 hr @ $48.50/hr	=	1,202.80
Crane operator: 24.8 hr @ $29.00/hr	=	719.20
Portable welding machine: 24.8 hr @ $5.75/hr	=	142.60
Electrodes: 5 lb @ $0.65	=	3.25
Welder: 24.8 hr @ $31.00/hr	=	768.80
Foreman: 24.8 hr @ $34.00/hr	=	<u>834.20</u>
	Total for joists =	$10,532.21

    Joist cost per linear foot, $9,744.89/1,584 lin ft = $6.15/lin ft

Metal decking:

Metal decking: 3,031 lb @ $0.60/lb	=	$1,818.60
Ironworkers: 3,456 sf × (9 hr/1,000 sf) = 31.1 hr @ $27.00/hr	=	839.70
Small truck-mounted crane: 31.1 hr @ $48.50/hr	=	1,510.35
Crane operator: 31.1 hr @ $29.00/hr	=	901.90
Portable welding machine: 31.1 hr @ $5.75/hr	=	178.83
Electrodes: 12 lb @ $0.57	=	6.84
Welder: 31.1 hr @ $31.00/hr	=	<u>963.10</u>
	Total for decking =	$6,219.32

Decking cost per square foot: $6,220.32/3,456 sf = $1.80/sf

Reinforcing Steel:

Welded-wire fabric: 3,629 sf @ $23.22/100 sf =	$842.65
Tie wire: 4 lb @ $0.98/lb =	3.92
Ironworkers: 3,629 sf × 1.0 hr/400 sf = 9.1 hr @ $27.00/hr =	245.70
Small truck-mounted crane: 9.1 hr @ $48.50/hr =	441.35
Crane operator: 9.1 hr @ $29.00/hr =	263.90
Total for reinforcing steel =	$1,797.52

Reinforcing steel per square foot: $1,797.52/3,456 sf = $0.52/sf

Concrete slab:

Concrete: ready-mix delivered to job, 29.4 cy @ $115/cy =	$3,381.00
Concrete pump truck and operator: 1 day @ $975/day =	975.00
Laborers: 8 hr × 4 laborers @ $22.00/hr =	704.00
Total for concrete =	$5,060.00

Concrete cost per cubic yard: $5,060.00/26.7 cy = $189.51/cy

Summary of costs:

Joists =	$10,532.21
Metal decking =	6,220.32
Reinforcing steel =	1,797.52
Concrete =	5,060.00
Total cost =	$23,610.05

Cost per square foot in place: $23,610.05/3,456 sf = $6.83/sf

COMBINED CORRUGATED-STEEL FORMS AND REINFORCEMENT FOR FLOOR SYSTEM

Description

A complete floor system that is suitable for floors supporting light to heavy loads is constructed with a deep corrugated-steel combined form and reinforcing unit and concrete. The steel sheets used in constructing this system are sold under the various manufacturers' trade names. The corrugated sheets are fabricated from high-strength steel, varying in thickness from 16 to 22 gauge with lengths up to 40 ft. The deep corrugated-steel sheets serve as longitudinal reinforcing for positive bending moments. Embossments on the vertical ribs transfer positive shear from the concrete to the steel. Welded-wire fabric is installed in the slab for temperature and shrinkage crack control. Conventional reinforced-concrete design procedures are followed for simple and continuous spans.

Table 15.5 provides an illustrative example of data for corrugated sheets used for this type of floor system. The quantities of concrete, in cubic yards per square foot of floor area for various thicknesses of floor slab, will vary depending on the manufacturer. The thickness of slab shown in Table 15.5 is measured from the top of the slab to the bottom of the corrugation.

TABLE 15.5 | Data for corrugated sheets.

Gauge	Cover, in.	Form depth, in.	Slab depth, in.	Volume of concrete, cf/sf
22	36	1.5	3.5	0.210
22	36	2.0	4.0	0.253
20	36	2.0	4.5	0.294
20	24	2.0	5.0	0.336
20	24	3.0	5.5	0.333
20	24	3.0	6.0	0.375

Source: Vulcraft Division of Nucor Corporation.

Installing Corrugated Sheets

The sheets are equally suited to concrete or steel-frame construction. They should be fastened to the supports by welds, screws, bolts, or other means. Conventional negative reinforcing steel is installed over the supports. Conduit for electric wires is laid on the sheets with openings made through them. Provisions for large openings, such as for stairs, should be prefabricated by the manufacturer.

To avoid excessive form stresses and deflection while supporting the wet concrete between permanent supporting members, one or two lines of temporary supports should be installed until the concrete has gained strength. The spacing of the temporary supports should conform to the recommendations of the manufacturer.

Labor Installing Corrugated Sheets

The labor required to install corrugated sheets will vary somewhat with the sizes and weights of the sheets, types of supports, and complexity of the floor. Based on a floor with 18-ft 0-in. by 24-ft 0-in. bays, with supporting steel beams spaced 8 ft 0 in. apart, requiring 22-gauge 1.25-in.-deep corrugations, the rates of placing should be about as follows:

Operation	Labor-hours per 1,000 sf
Sorting and arranging sheets	2.5
Laying and fastening decking	5.0
Installing welded-wire fabric	3.0
Installing negative reinforcing bars	8.0
Installing and removing temporary supports	11.0

EXAMPLE 15.2

Estimate the total direct cost and cost per square foot for materials and labor for a bay 18 ft wide and 24 ft long using corrugated forms and reinforcement and a concrete slab.

The corrugated sheets will be 22 gauge with 1.5-in.-deep corrugations, with 6 × 6-10/10 welded-wire fabric for shrinkage steel. The permanent supporting members will be steel beams, spaced at 8 ft on centers, parallel to the 18-ft side of the bay, as shown in Fig. 15.3.

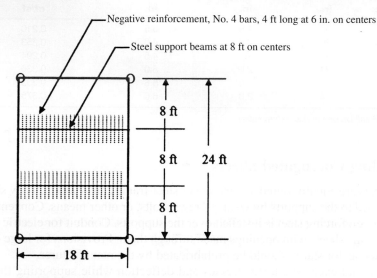

FIGURE 15.3 | Plan view of floor system.

The concrete will have a maximum depth of 3.5 in. Negative reinforcing, installed over each permanent beam, will consist of no. 4 bars, 4 ft long, spaced at 6 in. on centers.

Temporary intermediate support for the corrugated sheets will consist of one 2 × 8 wood stringer and adjustable shores, spaced 4 ft 6 in. apart. The stringers will be installed parallel with the 18-ft sides of the bay.

Quantities of materials:

Floor area: 18 ft × 24 ft = 432 sf

Corrugated sheets:

Square feet of surface area   =   432.0 sf

Add 1% waste, 432 sf × 1% =   4.3 sf

Total =   436.3 sf

Reinforcing steel:

Negative steel: [18 ft/(0.5 ft/bar) + 1] × 2 rows = 74 pieces 4 ft long

Weight of negative rebar: 74 pieces × 4 ft × 0.668 lb/ft = 197.8 lb

Square feet of negative reinforcing: 18 ft × 4 ft × 2 rows = 144 sf

Welded-wire fabric: 432 sf + 5% waste = 454 sf

Concrete:

> From Table 16.5, volume of concrete = 0.210 cf/sf × 432 sf = 90.7 cf
> Converting to cubic yards of concrete: 90.7 cf/(27 cf/cy) = 3.4 cy
> Add 10% waste: 10% × 3.4 cy = <u>0.3 cy</u>
> Total = 3.7 cy

Labor installing materials:

Sorting and arranging sheets: 432 sf × 7.5 hr/1,000 sf   = 3.2 hr
Steel setters installing welded-wire fabric: 454 sf × 3.0 hr/1,000 sf = 1.4 hr
Ironworker placing no. 4 negative rebar: 144 sf × 8 hr/1,000 sf   = 1.2 hr
Concrete workers placing concrete: 3.7 cy/(10 cy/hr)   = 0.4 hr
Installing and removing temporary supports:
432 sf × 11 hr/1,000 sf   = <u>4.8 hr</u>
Total = 11.0 hr

Cost of material, labor, and equipment:

Materials:

Corrugated sheets: 436.3 sf @ $2.60/sf   = $1,134.38
Welded-wire fabric: 454 sf @ $0.22/sf   =   99.88
No. 4 negative steel: 197.8 lb @ $0.47/lb =   92.97
Highchairs, tie wire: 432 sf @ $0.21/sf   =   90.72
Concrete: 3.7 cy @ $115.00/cy   =   <u>425.50</u>
Total material cost = $1,843.45

Labor:

Sorting and arranging sheets: 3.2 hr @ $21.00/hr   = $67.20
Steel setters installing welded-wire fabric: 1.4 hr @ $25.00/hr =   35.00
Ironworker placing no. 4 negative rebar: 1.2 hr @ $27.00/hr =   32.40
Laborers placing concrete: 6 laborers 0.4 hr @ $21.00/hr   =   50.40
Laborers installing/removing supports: 4.8 hr @ $21.00/hr   = 100.80
Crane operator: 11.0 hr @ $29.00/hr   = 319.00
Foreman: 11.0 hr @ $32.00/hr   = <u>352.00</u>
Total labor cost = $956.80

Equipment:

Small hydraulic crane: 11.0 hr @ $48.50/hr = $533.50
Concrete pump truck: 1/2 day @ $950/day =   <u>475.00</u>
Total equipment cost = $1,008.50

Summary of costs:

Materials   = $1,843.45
Labor   =   956.80
Equipment = <u>1,008.50</u>
Total cost = $3,808.75
Cost per square foot in place, $3,808.75/432 sf = $8.82/sf

FLOOR FINISHES

Many types of finishes are applied to floors. Several of the more popular types will be discussed in this section.

CONCRETE-FLOOR FINISHES

The methods of estimating the cost of materials, equipment, and labor for concrete subfloors were discussed in the preceding pages of this chapter. The costs included all operations through the screeding of the rough concrete floors but did not include finishing the floors. These costs will be discussed in this section.

For most concrete floors, it is necessary to add other materials to the top of the subfloors to produce finished floors of the desired types. The types of finishes most commonly used are *monolithic topping* and *separate topping,* to either of which may be added coloring pigments or hardening compounds.

Monolithic Topping

One of the best and most economical finishes for concrete floors is monolithic topping. After the concrete for the floor has been screeded to the desired level and has set sufficiently, it is floated with a wood float or a power-floating machine until all visible water disappears. Then the surface of the concrete is dusted with dry cement or a mixture of cement and sand, in approximately equal quantities. After the cement or mixture is applied to the floor, it is lightly floated with a wood float and then finished with a steel trowel.

Materials Required for a Monolithic Topping
The thickness of the monolithic topping may vary from $\frac{1}{16}$ to $\frac{1}{8}$ in. Table 15.6 gives the quantities of materials required for finishing 100 sf of concrete floor by using various proportions and thicknesses.

TABLE 15.6 | Quantities of materials for 100 sf of monolithic topping.

Thickness, in.	Proportion		Cement, sacks	Sand, cf
	Cement	Sand		
$\frac{1}{16}$	1	1	0.25	0.25
$\frac{1}{8}$	1	1	0.50	0.50

Labor Finishing Concrete Floors Using Monolithic Topping
In estimating the amount of labor required to finish a concrete floor, the estimator must consider several factors that can affect the total time required. After the concrete is placed and screeded, the finishers must wait until the

concrete is ready for finishing. The length of wait may vary from as little as 1 hr under favorable conditions to as much as 6 hr under unfavorable conditions. The factors that affect the time are the amount of excess water in the concrete, the temperature of the concrete, and the temperature of the atmosphere. If the concrete is free of excess water and the weather is warm, finishing may be started soon after the concrete is placed. If the concrete contains excess water and the weather is cold and damp, finishing may be delayed several hours.

If concrete-placing operations are stopped at the normal end of a day, the concrete finishers may have to work well into the night to finish the job. This requires payment of overtime wages at prevailing rates.

Table 15.7 gives the labor-hours required to finish 100 sf of monolithic concrete floor under various conditions.

TABLE 15.7 I Labor-hours required to finish 100 sf of monolithic concrete floor.

Classification	Conditions		
	Favorable	**Average**	**Unfavorable**
Cement finisher	0.75	1.25	1.75
Helper	0.75	1.25	1.75

EXAMPLE 15.3

Estimate the cost of materials and labor required to hand-finish 5,000 sf of floor area with $\frac{1}{8}$-in. thick monolithic topping. Assume average working conditions. The cost of cement is $9.49/sack and sand is $1.15/cf. The hourly rate of cement finishers is $29.00/hr and $22.00/hr for helpers.

The cost will be:

Cement, from Table 15.6 for $\frac{1}{8}$-in. monolithic topping:

    5,000 sf × 0.5 sacks per 100 sf = 25 sacks @ $9.49 = $237.25

Sand, from Table 15.6 for $\frac{1}{8}$-in. monolithic topping:

    5,000 sf × 0.5 cf per 100 sf = 25 cf @ $1.15   =   28.75

Cement finishers, from Table 15.7 for average conditions:

    5,000 sf × 1.25 hr/100 sf = 62.5 hr @ $29.00/hr   =   1,812.50

Helpers, from Table 15.7 for average conditions

    5,000 sf × 1.25 hr/100 sf = 62.5 hr @ $22.00/hr   =   1,375.00

                                    Total cost = $3,453.50

Cost per sf: $3,453.50/5,000 sf = $0.69/sf

Separate Concrete Topping

A separate concrete topping may be applied immediately after the subfloor is placed, or it may be applied several days or weeks later. The latter method is not considered as satisfactory as the first. The surface of the subfloor should be left rough to ensure a bond with the topping. The topping may vary from $\frac{3}{4}$ to 1 in. thick. Concrete for the topping is usually mixed in the proportions 1 part cement, 1 part fine aggregate, and $1\frac{1}{2}$ to 2 parts coarse aggregate, up to $\frac{3}{8}$-in. maximum size, with about 5 gal of water per sack of cement.

After the concrete topping has set sufficiently, it is floated with a wood float and finished with a steel trowel. It may be desirable to dust the surface with cement during the finishing operation.

Materials Required for a Separate Topping

The quantities of materials required for a separate topping will vary with the thickness of the topping and the proportions of the mix. Table 15.8 gives quantities of materials required for 100 sf of floor with various thicknesses and proportions.

TABLE 15.8 | Quantities of materials for 100 sf of separate topping.

Thickness, in.	Proportions by volume			Cement, sacks	Sand, cy	Gravel, cy
	Cement	Sand	Gravel			
$\frac{1}{2}$	1	1	$1\frac{1}{2}$	1.75	0.07	0.10
	1	1	2	1.62	0.06	0.12
	1	$1\frac{1}{2}$	3	1.18	0.07	0.13
$\frac{3}{4}$	1	1	$1\frac{1}{2}$	2.62	0.10	0.15
	1	1	2	2.42	0.09	0.18
	1	$1\frac{1}{2}$	3	1.70	0.10	0.18
1	1	1	$1\frac{1}{2}$	3.50	0.13	0.20
	1	1	2	3.24	0.12	0.23
	1	$1\frac{1}{2}$	3	2.27	0.13	0.25

Labor Mixing, Placing, and Finishing Separate Topping

The concrete for a separate topping should be mixed with a small concrete mixer. A finisher and a helper finishing 80 sf/hr will require 0.25 cy/hr of topping for a 1-in.-thick layer. A crew of three finishers and three helpers should finish 240 sf/hr under average conditions. This will require 0.75 cy/hr of 1-in.-thick topping. A concrete mixer with three laborers can mix and place the topping. If the topping is placed above the ground floor, a hoisting engineer will be needed on a part-time basis.

Table 15.9 gives approximate labor-hours required to mix, place, and finish 100 sf of topping under average conditions.

TABLE 15.9 | Labor-hours required to mix, place, and finish 100 sf of separate topping.

Thickness, in.	Finishers per crew	Labor-hours*			
		Finisher	**Helper**	**Laborer**	**Foreman**
$\frac{1}{2}$	1	1.25	1.25	1.25	0
	2	1.25	1.25	1.25	0.63
	3	1.25	1.25	0.80	0.42
$\frac{3}{4}$	1	1.25	1.25	1.25	0
	2	1.25	1.25	1.25	0.63
	3	1.25	1.25	0.80	0.42
1	1	1.25	1.25	1.25	0
	2	1.25	1.25	1.25	0.63
	3	1.25	1.25	1.25	0.42
	4	1.25	1.25	1.00	0.31

*Add time for a hoisting engineer if one is required.

Labor Finishing Concrete Floors with a Power Machine

Gasoline-engine- or electric-motor-driven power machines can be used to float and finish concrete floors. One manufacturer, the Whiteman Manufacturing Company, makes two sizes, 46- and 35-in. ring diameter. Each machine is supplied with three trowels, one set for floating and one for finishing the surface. The trowels are easily interchanged. The machines rotate at 75 to 100 rpm.

On an average job the 46-in.-diameter machine should cover 3,000 to 4,000 sf/hr one time over, including time for delays. The 35-in.-diameter machine should cover 2,400 to 3,000 sf/hr. For most jobs it is necessary to go over the surface three to four times during the floating and the same during the finishing operation.

In operating a power machine, a crew of two to three will be required—one operator and one or two finishers—to hand-finish inaccessible areas.

Table 15.10 gives approximate labor-hours required to finish 100 sf of surface area, based on going over the surface four times during each operation.

TABLE 15.10 | Approximate labor-hours required to finish 100 sf of surface area with power finishers.

Operation	Machine operator	Finisher
46-in.-diameter machine, 750 sf/hr		
Floating	0.133	0.266
Finishing	0.133	0.266
Total time	0.266	0.532
35-in.-diameter machine, 600 sf/hr		
Floating	0.167	0.333
Finishing	0.167	0.333
Total time	0.334	0.666

TERRAZZO FLOORS

A terrazzo floor is obtained by applying a mixture of marble chips or granules, Portland cement, and water laid on an existing concrete or wood floor. White cement is frequently used. The thickness of the terrazzo topping usually varies from $\frac{1}{2}$ to $\frac{3}{4}$ in. Several methods are used to install the topping on an existing floor.

Terrazzo Topping Bonded to a Concrete Floor

If terrazzo is to be placed on and bonded to a concrete floor, the concrete surface should be cleaned and moistened. Then a layer of underbed not less than $1\frac{1}{4}$ in. thick—made by mixing one part Portland cement, four parts sand, by volume, and enough water to produce a stiff mortar—is spread uniformly over the concrete surface.

While the underbed is still plastic, brass or other metal strips are installed in the mortar to produce squares having the specified dimensions.

The terrazzo topping is made by dry-mixing about 200 lb of granulated marble and 100 lb of cement, or in other proportions as specified. Water is added to produce a reasonably plastic mix. This mix is placed on the underbed inside the metal strips, after which it is rolled to increase the density, then hand-troweled to bring the top surface flush with the tops of the metal strips.

After the surface has hardened sufficiently, it is rubbed with a machine-powered coarse carborundum stone. The surface is then covered with a thin layer of grout made with white cement. At the time of final cleaning, this coating of grout can be removed with a machine-powered fine-grain carborundum stone.

Terrazzo Placed on a Wood Floor

When terrazzo is placed on a wood floor, it is common practice to cover the floor with roofing felt and galvanized wire netting, such as no. 14 gauge 2-in. mesh, which is nailed to the floor. A concrete underbed not less than $1\frac{1}{4}$ in. thick, as described earlier, is installed to a uniform depth over the floor. Metal strips and terrazzo topping are installed and finished, as described previously.

Labor Required to Place Terrazzo Floors

The labor rates required to place terrazzo floors will vary considerably with the sizes of the areas placed. The rates for large areas will be less than those for small areas. If terrazzo bases are required, additional labor should be allowed for placing and finishing the bases.

Table 15.11 gives representative labor-hours for the several operations required in placing terrazzo floors. Use the lower rates for large areas and the higher rates for small areas.

TABLE 15.11 | Labor-hours required to place terrazzo floors.

Operation	Labor-hours
Cleaning concrete floor, per 100 sf	0.75–1.25
Placing roofing felt on wood floors, per 100 sf	0.3–0.5
Placing netting on wood floor, per 100 sf	0.5–0.7
Mixing and placing $1\frac{1}{4}$-in.-thick underbed, per 100 sf:	
Mechanic	1.0–1.5
Helper	2.5–3.5
Mixing and placing $\frac{3}{4}$-in.-thick terrazzo topping and metal strips, per 100 sf:*	
Mechanic	2.0–3.0
Helper	2.5–3.0
Finishing terrazzo topping, per 100 sf	7.0–8.0
Mixing and placing 100 lin ft of 6-in. terrazzo cove base:	
Mechanic	14.0–18.0
Helper	14.0–18.0
Finishing 100 lin ft of 6-in. terrazzo cove base	8.0–10.0

*These rates are for metal strips placed to form 5-ft 0-in. squares. For other size squares, apply the following factors: 4 ft 0 in., 1.10; 3 ft 0 in., 1.20; 2 ft 0 in., 1.30.

EXAMPLE 15.4

Estimate the cost of 100 sf of terrazzo floor placed on a concrete slab, with a $1\frac{1}{4}$-in.-concrete underbed, $\frac{3}{4}$-in.-thick terrazzo topping, and brass strips to form 4-ft 0-in. squares. Use white cement for the topping.

The quantities will be:

Underbed: $100 \times 1.25/12 = 10.4$ cf

Topping: $100 \times 0.75/12 = 6.25$ cf

The cost will be:

Sand, including waste: 12 cf @ $1.15	=	$13.80
Gray cement: 3 sacks @ $9.49	=	28.47
Marble: 625 lb @ $0.52	=	325.00
White cement, including grout: 3.5 sacks @ $26.00	=	91.00
Brass strips, including waste: 60 lin ft @ $5.25	=	315.00
Labor cleaning concrete floor: 1 hr @ $21.00	=	21.00
Mechanic placing underbed: 1.25 hr @ $29.00	=	36.25
Helper placing underbed: 3.0 hr @ $21.00	=	63.00
Mechanic placing topping and strips: 2.75 hr @ $29.00	=	79.75
Helper placing topping and strips: 3.0 hr @ $21.00	=	63.00
Helper finishing terrazzo topping: 7 hr @ $21.00	=	147.00
Finisher and sundry equipment: 7 hr @ $15.00	=	105.00
	Total cost =	$1,288.27

Total cost sf, $1,288.27/100 sf = $12.88/sf

VINYL TILE

Vinyl tiles are available in sizes 6 by 6 in., 9 by 9 in., and 12 by 12 in., and in thicknesses of $\frac{1}{16}$ and $\frac{1}{8}$ in. The cost of tile will vary with the thickness, color, and quality.

When the quantity of tile required for a given floor area is estimated, an allowance should be included for waste due to cutting at the edges of the floor and to irregularities in the locations of the walls around the floor. Wastage may amount to 2 to 15 percent of the net area of a floor.

Laying Vinyl Tile on a Concrete Floor

Before tile is laid, the floor should be cleaned, after which a primer may be applied at a rate of 200 to 300 sf/gal. An adhesive cement is then applied at a rate of approximately 200 sf/gal; then the tiles are laid and rolled with a smooth-wheel roller.

Laying Vinyl Tile on a Wood Floor

Prior to the laying of the tile, the floor should be finished with a sanding machine or by some other approved method. A layer of felt is bonded to the floor with a linoleum paste, applied at a rate of approximately 150 sf/gal, and rolled thoroughly. Then an adhesive cement or emulsion is applied at a rate of approximately 150 sf/gal. The tiles are then laid and rolled.

Labor Laying Vinyl Tile

The labor required to lay vinyl tile will vary considerably with the size and shape of the floor covered. If specified color patterns are required, the amount of labor will be greater than that for single colors. Table 15.12 gives representative labor-hours for the operations required in laying vinyl tile.

TABLE 15.12 | Labor-hours required to lay 100 sf of vinyl tile.

Type of floor	Size tile, in.	Labor-hours
Concrete	12 × 12	1.5–2.0
	9 × 9	2.0–2.5
	6 × 6	2.8–3.3
Wood	12 × 12	1.8–2.2
	9 × 9	2.3–2.8
	6 × 6	2.0–3.5

EXAMPLE 15.5

Estimate the cost of materials and labor required to lay 100 sf of vinyl tile $\frac{1}{8}$ in. thick, using 12- × 12-in. squares, on a concrete floor.

The cost will be:

Prime coat: 0.5 gal @ $8.70	=	$4.35
Adhesive cement: 0.5 gal @ $14.10	=	7.05
Tile, including waste: 105 sf @ $4.85	=	509.25
Tile setter, 1.8 hr @ $28.00	=	50.40
	Total cost =	$571.05

Cost per sf, $571.05/100 sf = $5.71/sf

PROBLEMS

15.1 A concrete slab for the second-story floor of a building will be 58 ft 6 in. wide, 136 ft 0 in. long, and 6 in. thick. The slab will be finished with a monolithic topping $\frac{1}{8}$ in. thick, using hand finishers. Assume average working conditions. Estimate the total cost and the cost per square foot for the materials and the labor for the topping. Assume that the unit costs for materials and labor will be the same as those used in Example 15.3.

15.2 Estimate the total direct cost and the cost per square foot for furnishing the materials and labor required for a 4,480 sf terrazzo floor placed on a concrete slab, with a $1\frac{1}{4}$-in.-thick underbed, $\frac{3}{4}$-in.-thick terrazzo topping, and brass strips to form 4-ft 0-in. squares. Use gray cement and marble chips for the topping.

The floor consists of large rooms and corridors at approximately ground level. Use the unit costs for materials and labor from Example 15.4.

15.3 Estimate the total direct cost and the cost per square foot for furnishing and laying $\frac{1}{8}$-in.-thick vinyl tile, size 12- by 12-in. squares, on a concrete floor whose area is 2,360 sf. Use the unit costs for materials and labor from Example 15.5.

16

Painting

Painting is the covering of surfaces of wood, plaster, masonry, metal, and other materials with a compound for protection or for the improvement of the appearance of the surface painted. Many kinds of paint are used, some of which will be described briefly.

Since various practices are used in determining the area of the surfaces to be painted, it is desirable to state what methods are used. Some estimators make no deductions for the areas of openings except those that are quite large, whereas other estimators deduct all areas that are not painted. The latter method seems to be more accurate. Practices vary in estimating the areas for trim, windows, doors, metal specialties, masonry, etc.

Paints may be acrylic latex or enamel. Acrylic latex paint is considered a waterbase and is often preferred because the paintbrushes and rollers can be cleaned with water. Enamel paint is an oil base and the paintbrushes and rollers must be cleaned with a petroleum product, such as turpentine. Enamel paint is considered more durable than acrylic latex.

Paints can be applied in one, two, three, or more coats, with sufficient time allowed between successive coats to permit the prior coat to dry thoroughly. Paint can be applied with a brush, roller, or spray gun. The first coat, which is usually called the *prime coat,* should fill the pores of the surface, if such exist, and bond securely to the surface to serve as a base for the other coats.

The cost of painting includes materials, labor, and sometimes equipment, especially when the paint is applied with a spray gun. Because the cost of the paint is relatively small compared with the other costs, it is not good economy to purchase cheap paint.

MATERIALS

It is beyond the scope of this book to discuss all the materials used for paints, but those most commonly used will be discussed briefly.

Ready-mixed paints. These paints, which are mixed by the manufacturers, can be purchased with all the ingredients combined into a finished product.

Colored pigments. Colored pigments are added to paints by the manufacturer or at the job to produce the desired color.

Turpentine. Turpentine, which is obtained by distilling the gum from pine trees, can be used to thin oil paints, especially the priming coat, to obtain better penetration into the wood.

Varnish. Varnish, which is a solution of gums or resins in linseed oil, turpentine, alcohol, or other vehicles, is applied to produce hard transparent surfaces. Spar is a special varnish that is used on surfaces that may be exposed to water for long periods.

Shellac. Shellac is a liquid consisting of a resinous secretion from several trees dissolved in alcohol. It can be applied to knots and other resinous areas of lumber prior to painting to prevent bleeding of the resinous substance through the paint.

Stains. Stains, which are liquids of different tints, are applied to the surfaces of wood to produce desired color and texture effects. The vehicle may be oil or water.

Putty. Putty, which is a mixture of powdered chalk or commercial whiting and raw linseed oil, is used to fill cracks and joints and to cover the heads of countersunk nails. It should be applied following the application of the priming coat of paint. If it is applied directly on wood, the wood will absorb most of the oil before the putty hardens, and the putty will not adhere to the wood.

COVERING CAPACITY OF PAINTS

The covering capacity of paint is generally expressed as the number of square feet of area covered per gallon of one coat, which will vary with several factors, including:

1. Type of material painted: wood, gypsum wallboard, concrete, metal
2. Texture and porosity of the surface
3. Type of paint: enamel, latex, varnish, shellac, stain
4. Extent to which paint is spread as it is applied
5. Extent to which a thinner is added to the paint
6. Quality of the paint
7. Temperature of air. Thinner coats are possible during warm weather, resulting in greater coverage.

Table 16.1 gives representative values for the covering capacities of various paints, varnishes, and stains when applied to different surfaces.

TABLE 16.1 | Surface area covered by 1 gal of paint, stain, and varnish.

Surface and Material	Surface area covered, sf/gal
Applied on wood	
Paint, enamel or latex	350–450
Varnish, flat	500–600
Varnish, glossy	400–450
Shellac	600–700
Oil-base stain on wood	500–600
Oil-base paint on floors	400–450
Stain on wood shingles	125–225
Applied on gypsum wallboard	
Prime coat paint	400–450
Paint, enamel or latex	350–450
Applied on plaster and stucco	
Sealer	600–700
Flat-finish paint	500–550
Oil-base paint on stucco, first coat	150–160
Oil-base paint on stucco, second coat	350–375
Applied on metal	
Prime coat	400–450
Paint, enamel or latex	350–450

PREPARING A SURFACE FOR PAINTING

The operations required to apply a complete paint coverage will vary with the kind of surface to be painted, number of coats to be applied, and kind of paint used.

When new wood surfaces are being painted, it may be necessary to cover all resinous areas with shellac before applying the priming coat of paint. After this coat is applied, all joints, cracks, and nail holes can be filled with putty, which should be sanded smooth before the second coat of paint is applied.

It may be necessary to apply a filler to the surfaces of certain woods, such as oak, before they are painted. The surfaces may be sanded before and after applying the filler and sometimes following the application of each coat of paint or varnish, except the last coat.

Before paint is applied to plaster surfaces, it may be necessary to apply a sealer to close the pores and neutralize the alkali in the plaster.

The surfaces of new metal may be covered with a thin film of oil that must be removed with warm water and soap prior to applying the priming coat of paint.

Sometimes it is necessary to place masking tape, such as a strip of kraft paper with glue on one side, over areas adjacent to surfaces to be painted. This is especially true when paint is applied with a spray gun.

LABOR APPLYING PAINT

The labor required to apply paint can be expressed in hours per 100 sf of surface area; per 100 lin ft for trim, cornices, posts, rails, etc.; or sometimes per opening for windows and doors. If the surfaces to be painted are properly prepared, and the paint is applied with first-grade workmanship, more time will be required than when an inferior job is permitted. An estimator must consider the kinds of surfaces to be painted and the requirements of the specifications before she or he can prepare an accurate estimate for a job.

These factors should be considered in preparing an estimate for painting:

1. Treatment of the surface prior to painting, removing old paint, sanding, filling, etc.
2. Kind of surface, wood, plaster, masonry, etc.
3. Size of area, large, flat, small, irregular.
4. Height of area above the floor or ground.
5. Kind of paint. Some flows more easily than others.
6. Temperature of air. Warm air thins paint and permits it to flow more easily.
7. Method of applying paint, with brushes or spray guns. Spraying may be five times as fast as brushing.

Table 16.2 gives representative labor-hours required to apply paint. The lower values should be used when the paint is applied under favorable conditions and the higher values when it is applied under unfavorable conditions or when first-grade workmanship is required.

Since the heavy pigment will settle to the bottom of a paint container, it may be necessary for a painter to spend 5 to 15 min/gal mixing paint before it can be used.

EQUIPMENT REQUIRED FOR PAINTING

The equipment required for painting with brushes will include brushes, ladders, sawhorses, scaffolds, and foot boards. Inside latex paints can be applied with rollers. Where painting is done with spray guns, it will be necessary to provide one or more small air compressors, paint tanks, hoses, and spray guns, in addition to the equipment listed for brush painting.

COST OF PAINTING

Example 16.1 illustrates a method of determining the cost of furnishing materials and labor in painting a surface.

TABLE 16.2 | Approximate labor-hours required to apply a coat of paint.

Operation	Unit	Hours per unit
Exterior walls and siding		
Brush painting	100 sf	1.0–1.2
Roll painting	100 sf	0.5–0.6
Spray painting	100 sf	0.4–0.5
Exterior doors and windows		
Door, both sides, and frame	Each	0.8–0.9
Window, outside, frame and trim	Each	0.6–0.7
Exterior trim and molding		
Brush painting fascia and molding	100 linear ft	1.2–1.3
Spray painting fascia and molding	100 linear ft	0.4–0.6
Interior walls		
Brush painting	100 sf	0.9–1.1
Roll painting	100 sf	0.5–0.6
Spray painting	100 sf	0.4–0.5
Interior ceilings		
Brush painting	100 sf	1.1–1.6
Roll painting	100 sf	0.8–0.9
Spray painting	100 sf	0.5–0.6
Interior doors and windows		
Door, both sides, and trim	Each	0.7–0.8
Window, interior side, and trim	Each	0.3–0.4
Interior floors		
Brush painting	100 sf	0.7–0.8
Roll or spray painting	100 sf	0.3–0.4
Interior trim, molding, and base boards		
Painting or staining	100 linear ft	0.5–0.8
Cabinets and casework		
Painting, staining, or varnishing	100 sf	1.2–2.5
Structural steel members		
Spray painting beams and columns	Tons	0.20–0.25
Spray painting steel trusses	Tons	0.25–0.35

EXAMPLE 16.1

A building consists of 286 lin ft of exterior walls, 944 lin ft of interior walls, 2 exterior doors, and 8 interior doors. Exterior walls are 9 ft high and of wood shingle siding that will be sprayed with two coats of stain. Interior walls are 8-ft-high gypsum wallboard that will be roller painted with a prime coat and one finish coat of latex paint. All doors are metal and require a prime coat and one finish coat of enamel. Exterior doors are 3 ft wide and 7 ft high and interior doors are 2 ft 8 in. wide and 6 ft 8 in. high.

The cost of stain is $32.00/gal, primer is $18.00/gal, latex paint is $35.00/gal, enamel is $29.00/gal, and a painter costs $26.00/hr. Estimate the cost of material and labor for staining and painting.

Quantity of work:

 Exterior walls = 286 lin ft × 9 ft = 2,574 sf

 Interior walls = 944 lin ft × 8 ft = 7,552 sf

 Exterior doors = 2 doors × 3 ft × 7 ft × 2 sides = 84 sf

 Interior doors = 8 doors × 2.67 ft × 6.67 ft × 2 sides = 285 sf

Quantity of stain and paint, using average values from Table 16.1:

 Stain required,

  Exterior walls = [2,574 sf/175 sf/gal] × (2 coats) = 29.4 gal; use 30 gal

 Prime paint required,

  Interior walls = 7,552 sf/425 sf/gal = 17.8 gal

  Interior doors = 285 sf/425 sf/gal = 0.7

  Exterior doors = 84 sf/425 sf/gal = 0.2

            Total = 18.7 gal; use 19 gal

 Enamel paint required,

  Interior doors = 285 sf/400 sf/gal = 0.7 gal

  Exterior doors = 84 sf/400 sf/gal = 0.2

            Total = 0.9 gal; use 1 gal

 Latex paint required,

  Interior walls = 7,552 sf/400 sf/gal = 18.9 gal; use 19 gal

Time for staining and painting, using average values from Table 16.2:

 Exterior walls spraying stain = 2,574 sf × 2 coats × 0.45 hr/100 sf = 23.2 hr

 Exterior doors painting = 2 doors × 2 coats × 0.85 hr/door   = 3.4

 Interior walls rolling prime coat = 7,552 sf × 0.55 hr/100 sf = 41.5

 Interior walls rolling finish coat = 7,552 sf × 0.55 hr/100 sf = 41.5

 Interior doors painting = 8 doors × 2 coats × 0.75 hr/door   = 12.0

                Total time = 121.6 hr

  Rounding for 8-hr day = 121.6 hr/8 hr/day

           = 15.2 days; use 16 days @ 8 hr/day = 128 hr

Cost of material and labor:

 Material cost:

  Stain, 30 gal @ $32.00/gal   = $960.00

  Primer, 19 gal @ $18.00/gal = 342.00

  Latex, 19 gal @ $35.00/gal   = 665.00

  Enamel, 1 gal @ $29.00/gal   = 29.00

 Labor cost, 128 hr @ $26.00/hr = 3,328.00

        Total = $5,324.00

The above costs do not include an air compressor and hoses for spray guns, brushes and rollers, ladders, or scaffolding that may be used. Also, taxes on labor and equipment, overhead, and profit are not included.

PROBLEMS

16.1   Estimate the total direct cost of furnishing and roll painting a prime coat and one finish coat of enamel paint on 3,700 sf of exterior wood surface. The cost of the primer is $18.00/gal, the paint is $27.00/gal, and the painter cost is $26.00/hr

16.2   Estimate the total direct cost of furnishing and spray painting two coats of latex wall paint on 3,600 sf of interior gypsum wallboard. The cost of paint is $23.00/gal and the cost of the painter is $26.00/hr.

16.3   Estimate the total direct cost of furnishing and brush painting a glossy varnish on 1,800 sf of wood flooring. The cost of varnish is $28.00/gal and the painter cost is $26.00/hr.

16.4   Estimate the total direct cost of furnishing and applying a sealer and two coats of oil-base paint on a stucco wall that is 180 ft long and 10 ft high. The prime coat will be rolled and the two finish coats will be sprayed. The cost of the sealer is $15.00/gal, the oil-base paint is $24.00/gal, and the painter cost is $26.00/hr.

16.5   An industrial building has 84 interior metal doors that will require a prime coat and one coat of enamel. The doors will be brush painted. Assume 0.5 gal per door for the prime coat and the paint. The cost of the primer is $18.00/gal, the enamel paint is $31.00/gal, and the painter cost is $26.00/hr.

16.6   A commercial building has 4,200 lin ft of wood baseboard that requires one coat of flat varnish. The varnish cost is $33.00/gal and the painter cost is $26.00/hr. Estimate the total direct cost of furnishing and applying the varnish.

17

Plumbing

Although general contractors usually subcontract the furnishing of materials and the installation of plumbing in a building, they should have a reasonably good knowledge of the costs of plumbing. The fact that general contractors do not prepare detailed estimates for plumbing does not eliminate the need for estimating; it simply transfers the preparation to another party.

Plumbing involves the furnishing of materials, equipment, and labor to distribute water, wastewater, gas, and electrical conduits throughout a building; the furnishing and installation of fixtures; and the removal of the water and waste from the building. Water usually is obtained from a water main, and the waste is discharged into a sanitary sewer line. Materials include various types of pipe, fittings, valves, and fixtures, which will be more fully described later. Consumable supplies must be included in an estimate.

PLUMBING REQUIREMENTS

Some plans and specifications clearly define the types, grades, sizes, and quantities and furnish other information required for a complete plumbing installation, while other plans and specifications furnish limited information and place on the plumbing contractor the responsibility for determining what is needed to satisfy the owner and the local plumbing ordinance.

All cities have ordinances that require that plumbing installations conform with the plumbing code for the city in which the project will be constructed. A contractor must obtain a plumbing permit prior to starting an installation, and the work is checked by a plumbing inspector for compliance with the code. Although plumbing codes in different localities are similar, there will be variations that make it necessary for a plumbing contractor to be fully cognizant of the requirements of the code that will apply to a given project.

PLUMBING CODE

Some of but not all the requirements of the plumbing code for a major city are given here. The information is intended to serve as a guide in demonstrating the requirements of city codes for plumbing. When preparing an estimate of the cost of furnishing and installing items for plumbing services, the estimator should use the appropriate code for the area in which the project will be constructed.

Permit fee   A fee, which is generally paid by the plumbing contractors and included in their estimate, is charged by a city for issuing a plumbing permit. This fee may vary with the type and size of the project and with the location.

Plumbing-fixture facilities   Table 17.1 illustrates the minimum requirements for plumbing-fixture facilities for a representative city. The requirements may differ in other cities.

TABLE 17.1 | Minimum requirement for plumbing-fixture facilities (one fixture for each designated group).

Type of building	Water closet	Urinal	Lavatory	Drinking fountain	Shower	Bathtub	Kitchen sink
Dwellings and apartment houses	Each family	—	Each family	—	Choice of 1 per family		Each family
Places of employment, such as mercantile and office buildings, workshops, and factories where 5 or more persons work	25 males 20 females	25 males	15 persons	75 persons			
Foundries, mines, and places where exposed to dirty or skin-irritating materials where 5 or more persons work	25 males 20 females	25 males	5 persons	75 persons	15 males 15 females		
Schools	20 males 15 females	25 males	20 persons	75 persons			
Dormitories	10 males 8 females	25 males	6 persons	50 persons	8 males 10 females	40* males 35* females	

* Half may be additional showers.

Minimum-size trap and outlet   Table 17.2 gives the minimum-size trap and outlet permitted for the indicated fixture.

Fixture-unit values   Table 17.2 gives the value of each fixture unit for determining the relative load factors of different kinds of plumbing fixtures and estimating the total load carried by soil and waste pipe.

TABLE 17.2 | Fixture-unit values and minimum-size traps and outlets required.

Fixture	Fixture-unit value as load factor	Minimum-size trap and outlet connection, in.
Bathroom group consisting of water closet, lavatory, and bathtub or shower stall and		
Tank with water closet	6	3
Flush-valve water closet	8	3
Bathtub with or without overhead shower	2	$1\frac{1}{2}$
Bidet	3	$1\frac{1}{2}$
Combination sink and tray	3	$1\frac{1}{2}$
Combination sink and tray with food disposal unit	4	$1\frac{1}{2}$
Dental unit or cuspidor	$\frac{1}{2}$	$1\frac{1}{4}$
Dental lavatory	1	$1\frac{1}{4}$
Drinking fountain	$\frac{1}{2}$	1
Dishwasher, domestic	2	$1\frac{1}{2}$
Floor drains	1	2
Kitchen sink, domestic	2	$1\frac{1}{2}$
Kitchen sink, domestic, with food disposal unit	3	$1\frac{1}{2}$
Lavatory	2	$1\frac{1}{2}$
Lavatory, barber, beauty parlor	2	$1\frac{1}{2}$
Lavatory, surgeon's	2	$1\frac{1}{2}$
Laundry tray, 1- or 2-compartment	2	$1\frac{1}{2}$
Shower stall, domestic	2	2
Showers, group, per head	3	Varies
Sinks:		
Surgeon's	3	$1\frac{1}{2}$
Flushing rim, with valve	8	3
Service, standard trap	3	3
Service, with P trap	2	2
Pot, scullery, etc.	4	$1\frac{1}{2}$
Urinal, pedestal, syphon jet blowout	8	3
Urinal, wall lip	4	$1\frac{1}{2}$
Urinal, stall, washout	4	2
Urinal trough, each 2-ft section	2	$1\frac{1}{2}$
Wash sink, circular or multiple, each set of faucets	2	$1\frac{1}{2}$
Water closet:		
Tank-operated	4	3
Valve-operated	8	3

PIPING USED FOR PLUMBING

Pipes used in plumbing are fabricated from several materials; including plastic, copper tubing, black and galvanized steel, and cast iron pipe. Pressure pipe for water lines usually consists of PVC plastic, copper tubing, or black and galvanized steel.

Non-pressure pipe for drainage lines are plastic, designated as DWV pipe, or cast iron steel. Most vent pipes are fabricated of plastic. Each of these types of pipes is discussed in subsequent sections of this chapter.

STEEL PIPE

Black and galvanized steel pipes are available in standard weights for plumbing water lines. Sizes are fabricated from $\frac{1}{2}$ in. to 12 in. in diameter. Table 17.3 gives the dimensions and weights of black and galvanized pipe used for water pipe.

Galvanized pipe is steel with a zinc coating to resist corrosion. This pipe is used primarily in commercial and industrial buildings, with little use in residential construction. Steel pipe is strong and has an average life span of 40 years for galvanized steel.

TABLE 17.3 | Dimensions and weights of black and galvanized standard steel pipe.

Size, in.	Diameter, in. External	Diameter, in. Internal	Thickness, in.	Internal area, sq. in.	Weight, lb/lin ft	Threads per in.
$\frac{1}{8}$	0.405	0.269	0.068	0.057	0.24	27
$\frac{1}{4}$	0.540	0.364	0.088	0.104	0.42	18
$\frac{3}{8}$	0.675	0.493	0.091	0.191	0.56	18
$\frac{1}{2}$	0.840	0.622	0.109	0.304	0.84	14
$\frac{3}{4}$	1.050	0.824	0.113	0.533	1.12	14
1	1.315	1.049	0.133	0.861	1.67	$11\frac{1}{2}$
$1\frac{1}{4}$	1.660	1.380	0.140	1.496	2.25	$11\frac{1}{2}$
$1\frac{1}{2}$	1.900	1.610	0.145	2.036	2.68	$11\frac{1}{2}$
2	2.375	2.067	0.154	3.356	3.61	$11\frac{1}{2}$
$2\frac{1}{2}$	2.875	2.467	0.203	4.780	5.74	8
3	3.500	3.066	0.217	7.383	7.54	8
$3\frac{1}{2}$	4.000	3.548	0.226	9.886	9.00	8
4	4.500	4.026	0.237	12.730	10.67	8
5	5.563	5.045	0.259	19.985	14.50	8
6	6.625	6.065	0.280	28.886	18.76	8
7	7.625	7.023	0.301	38.734	23.27	8
8	8.625	7.981	0.322	50.021	28.18	8
9	9.625	8.937	0.344	62.72	33.70	8
10	10.750	10.018	0.336	78.82	40.07	8
12	12.750	12.000	0.375	113.09	48.99	8

COPPER PIPE

Copper tubing is frequently used instead of steel pipes for water lines. There are two types of copper tubing, Type K and Type L. Type K tubing has thicker walls than Type L tubing and is typically used for underground burial. Type L is used for most purposes. Table 17.4 gives standard sizes of Type K and L copper tubing. The common sizes used of Type K tubing are $\frac{1}{4}$ in. to 2 in., whereas Type L

TABLE 17.4 I Dimensions and weights of copper tubing used in plumbing.

Nominal size, in.	Outside diameter, in.	Type K		Type L	
		Wall thickness, in.	Weight, lb/lin ft	Wall thickness, in.	Weight, lb/ft
$\frac{1}{2}$	0.625	0.049	0.344	0.040	0.285
$\frac{5}{8}$	0.750	0.049	0.418	0.042	0.362
$\frac{3}{4}$	0.875	0.065	0.641	0.045	0.455
1	1.125	0.065	0.839	0.050	0.655
$1\frac{1}{4}$	1.375	0.065	1.040	0.055	0.884
$1\frac{1}{2}$	1.625	0.072	1.360	0.060	1.140
2	2.125	0.083	2.060	0.070	1.750

common sizes range from $\frac{1}{4}$ in. to 8 in. The printing on the pipe is color coded: red for Type K and green for Type L.

There are two tempers for copper tubing, drawn temper (sometimes called hard) and annealed temper (sometimes called soft). Flare fittings may be installed on soft tubing, but not on hard tubing. Both types can be soldered or brazed. Hard temper copper is available in 50-ft. lengths and soft temper copper is available in 60-ft, 100-ft, and 200-ft coils.

Another type of copper tubing is DWV (drain waste vent), which is available only in hard temper, and fabricated in sizes from $\frac{1}{4}$ in. to 8 in. DWV pipe is non-pressurized for waste drains and vents. It is available in 20-ft lengths. Copper tubing designated as ACR is used for air conditioning and refrigeration. It is sized by the actual outside diameter, with diameters from $\frac{1}{8}$ in. to $4\frac{1}{8}$ in. in diameter and 20-ft lengths in hard temper and 50-ft lengths in soft temper.

PVC PLASTIC WATER PIPE

Polyvinyl chloride (PVC) plastic pipe is used extensively in plumbing for water, drainage, and vent lines. It is fabricated as pressure pipe with the ability to withstand pressures of 150 to 200 psi, which is adequate for plumbing of water lines in buildings. PVC pipe is also fabricated as non-pressure pipe for use in drainage and vent lines.

PVC is lightweight and highly resistant to corrosion and has low resistance to the flows of water. Workers can easily lift and transport pipes and fittings without the use of lifting equipment. PVC can be cut easily with hand tools and installed rapidly and economically. The joint between a pipe and fitting is made by applying a coat of solvent adhesive around the end of the pipe and then inserting the pipe into the fitting. Table 17.5 gives representative sizes and costs of PVC plastic water pipe used in plumbing buildings. Table 17.6 gives representative costs of fittings for PVC plastic water pipe.

TABLE 17.5 |
Representative costs
of PVC plastic water
pipe.

Pipe size, in.	Cost per lin ft
$\frac{1}{2}$	$0.51
$\frac{3}{4}$	0.59
1	0.93
$1\frac{1}{4}$	1.13
$1\frac{1}{2}$	1.32
2	1.50

TABLE 17.6 | Representative costs of fittings for white PVC water pipe.

Fitting	Price each by size, in.					
	$\frac{1}{2}$	$\frac{3}{4}$	1	$1\frac{1}{4}$	$1\frac{1}{2}$	2
90° elbow	$0.24	$0.24	$0.40	$0.71	$0.76	$1.19
45° elbow	0.34	0.53	0.61	0.85	1.08	1.39
Tee	0.26	0.29	0.54	0.83	1.00	1.47
Coupling	0.12	0.17	0.33	0.44	0.46	0.72
Adapter	0.17	0.19	0.36	0.45	0.59	0.76
End cap	0.18	0.21	0.34	0.46	0.53	0.61

INDOOR CPVC PLASTIC WATER PIPE

This pipe, made of chlorinated polyvinyl chloride (CPVC), can be used inside a building for hot or cold water. It will withstand internal water pressures up to 100 lb/sq. in. or more, at temperatures up to 180°F. Joints between pipes and fittings are made by applying coatings of solvent cements to the ends of the pipes and then inserting them into the fittings.

LABOR INSTALLING PLASTIC WATER PIPE

Installing plastic water pipe involves cutting standard lengths of pipe to the desired lengths, if necessary using a hacksaw, applying a solvent cement to the ends, and then inserting the ends into plastic fittings. This forms a solid joint, which should be free of any leakage.

Connections between plastic pipes and standard steel pipes can be made by using plastic female or male adapters, which are threaded at one end to match the threads of the steel pipe.

Table 17.7 gives representative times required to perform various operations in installing plastic pipe.

TABLE 17.7 | Representative hours required to install plastic water pipe and fittings.

Operation	Size of pipe, in.				
	$\frac{1}{2}$	$\frac{3}{4}$	1	$1\frac{1}{4}$	$1\frac{1}{2}$
Cut pipe	0.08	0.10	0.11	0.12	0.13
Apply cement and join pipe and fittings*	0.05	0.05	0.06	0.07	0.08
Join plastic pipe to steel pipe with adapter	0.20	0.20	0.25	0.25	0.30

*This is the time required for making a joint. A coupling and an elbow require two joints, and a tee requires three joints.

SOIL, WASTE, AND VENT PIPES

The pipes that convey discharge liquids from plumbing fixtures to sewer drain pipes are called soil and waste pipes. Pipes that receive the discharge from water closets (toilets) are called *soil pipes*, while the pipes that receive the discharge from other fixtures (sinks and lavatories) are called *waste pipes*. Pipes that provide ventilation to prevent siphoning of water from drain traps are called *vent pipes*.

Soil and waste pipes are non-pressurized and carry discharged liquids by gravity flow. PVC plastic pipe is used extensively for soil and waste pipe in residential construction and used to a large extent in commercial buildings. Waste and vent pipe are almost always installed as PVC plastic pipe, designated as DWV (drain-waste-vent).

Cast iron pipes are more commonly used for soil and waste pipes in industrial buildings. Table 17.8 gives the weights of cast iron pipe. The joints are usually sealed with rubberized seals. However, joints may also be made with lead and oakum. The oakum is wrapped around the spigot end of the pipe. A runner is wrapped around the pipe adjacent to the hub, with an opening at the top, into which molten lead is poured to fill the joint in one operation. After the lead solidifies, the runner is removed and the lead heavily caulked. Table 17.9 gives approximate quantities of lead and oakum required for a joint of cast iron soil pipe.

TABLE 17.8 | Approximate weights of soil pipe, lb/lin ft.

Size, in.	Standard, single hub	Extra heavy	
		Single hub	Double hub
2	3.6	5.0	5.2
3	5.2	9.0	9.4
4	7.0	12.0	12.6
5	9.0	15.0	15.6
6	11.0	19.0	20.0
8	17.0	30.0	31.4
10	23.0	43.0	45.0
12	33.0	54.0	57.0

TABLE 17.9 | Approximate quantities of lead and oakum required for cast-iron soil pipe.

Size pipe, in.	Weight per joint, lb	
	Lead	Oakum
2	1.5	0.13
3	2.5	0.16
4	3.5	0.19
5	4.3	0.22
6	5.0	0.25
8	7.0	0.38
10	9.0	0.50
12	11.0	0.70

HOUSE DRAIN PIPE

The cast-iron soil pipe must extend a specified distance outside a building, depending on the plumbing code. The balance of the drain pipe extending to the sanitary sewer main may be PVC plastic, ductile iron, vitrified clay, or concrete sewer pipe.

FITTINGS

Fittings for black and galvanized-steel pipe should be malleable iron. Fittings for copper and brass pipe and tubing should be copper and brass, respectively, threaded or sweat type. Fittings for plastic pipe should be of the same material as the pipe. If different types of plastic are used, the solvent cement will not be effective on both types of materials. Fittings for cast-iron pipe should be cast iron.

Adapters are available that can be used in joining one type of pipe or fitting to another type.

VALVES

Valves are installed to shut off the flow of water. Several types are used, including globe, gate, check, sill cocks, drain, etc. They are usually made from bronze and brass.

TRAPS

Traps are installed below the outlets from fixtures to retain water as a seal to prevent sewer gases from entering a building. Vent pipes installed into the drain pipes below the traps prevent the water in the traps from being siphoned out.

ROUGHING IN PLUMBING

Roughing in includes the installation of all water pipes from the meter into and through a building, soil pipe, waste pipe, drains, vents, traps, plugs, cleanouts, etc., but does not include the installation of plumbing fixtures, which is called *finish plumbing*.

ESTIMATING THE COST OF ROUGHING IN PLUMBING

In preparing a detailed estimate covering the cost of furnishing materials, equipment, and labor for roughing in the plumbing, a comprehensive list of items required should be used as a check and for establishing all costs. Each item should be listed separately by description, grade, size, quantity, and cost, both unit and total. If there are other costs such as tapping fees, permits, or removing and replacing pavement, then they must be included in an estimate. Table 17.10 may be used as a guide in preparing a list for estimating purposes.

TABLE 17.10 | Checklist for estimating the cost of roughing in plumbing.

Description	Quantify	Unit cost	Total cost
Permits			
Tapping fees			
Removing and replacing pavement			
Water meter and box, if required			
Steel pipe			
Copper pipe and tubing			
Pipe fittings			
Valves			
Soil pipe			
Water pipe			
Vent pipe and stacks			
Traps, cleanouts			
Cast-iron fillings			
Floor drains			
Roof flashings			
Vitrified-clay sewer pipe and fittings			
Concrete sewer pipe and fittings			
Catch basins and covers			
Manholes and covers			
Lead, solder, oakum, gasoline, etc.			
Cement and sand			
Brackets, hangers, supports, etc.			
Other items			

COST OF MATERIALS FOR ROUGH PLUMBING

While the costs of materials for rough plumbing will vary with grades, quantities, locations, and time, the costs given in the tables in this book can be used as a guide in preparing an approximate estimate. The actual costs that will apply should be determined and used in preparing an estimate for bid purposes.

COST OF LEAD, OAKUM, AND SOLDER

Ingot or pig lead used for joints with cast-iron pipe will cost about $1.05 per pound. Oakum will cost about $4.75 per pound. Wiring solder used to sweat joints for copper pipe and tubing will cost about $4.20 per pound.

PLASTIC DRAINAGE PIPE AND FITTINGS

Building codes permit the use of plastic drainage pipe and fittings, such as ABS-DWV ASTM D2661-11, or later, for use in draining liquids from buildings. These pipes are generally available in sizes $1\frac{1}{4}$ in. in diameter and larger and in lengths of 10 or 20 ft or more. Joints are made by applying a solvent cement to the spigot end of a pipe before inserting it into a fitting. Adapters are available to permit this pipe to be connected into cast-iron pipe.

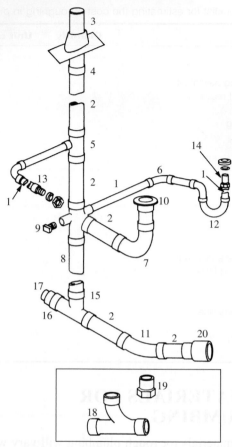

FIGURE 17.1 | Representative piping and fittings for a drain system.

Figure 17.1 illustrates an assembly of pipes and fittings representing a typical drainage system for a home or other building. Table 17.11 lists the items illustrated in Fig. 17.1, together with representative costs of the items.

LABOR REQUIRED TO ROUGH IN PLUMBING

Plumbers frequently work in teams of two, a plumber and a helper. However, since this arrangement is not always followed, the labor time given in the tables is expressed in labor-hours, which includes the combined time for a plumber and a helper.

The operations required to install water pipe will include cutting and threading the pipe and screwing it together with the appropriate fittings. Cutting and threading may be done with hand or power tools. The time required will vary with the size of the pipe, tools used, and working conditions at the job. If power tools can be moved along with the work, a minimum amount of time will be required; but if the pipe must be carried some distance to the tools, a great deal of time will

TABLE 17.11 | Representative costs of PVC drainage waste pipe and fittings.

Number	Item	Size, in.	Unit of measure	Unit cost
1	PVC pipe	$1\frac{1}{2}$	10 lin ft	$4.88
	PVC pipe	2	10 lin ft	6.67
2	PVC pipe	3	10 lin ft	9.39
3	Roof flashing, neoprene	$1\frac{1}{2}$	Each	8.30
		2	Each	10.40
		3	Each	12.48
4	Coupling	$1\frac{1}{2}$	Each	0.69
		2	Each	1.23
		3	Each	1.74
5	Sanitary tee	$1\frac{1}{2}$	Each	1.09
		2	Each	2.60
		3	Each	3.76
6	90° elbow	$1\frac{1}{2}$	Each	1.59
		2	Each	2.50
7	90° elbow	3	Each	4.90
8	Sanitary tee with two $1\frac{1}{2}$-in. side outlets	3	Each	3.37
9	Slip plug	$1\frac{1}{2}$	Each	0.75
10	Reducing closet flange	4×3	Each	4.50
11	45° elbow	$1\frac{1}{2}$	Each	0.80
		2	Each	1.96
		3	Each	2.76
12	P-trap with union	$1\frac{1}{2}$	Each	3.36
		2	Each	8.48
13	Trap adapter, $1\frac{1}{2}$-in. pipe to $1\frac{1}{4}$-in. waste	$1\frac{1}{4}$	Each	1.28
14	Trap adapter, $1\frac{1}{2}$-in. pipe to $1\frac{1}{2}$-in. waste	$1\frac{1}{2}$	Each	1.62
15	Wye branch	$1\frac{1}{2}$	Each	2.32
		2	Each	3.64
		3	Each	5.19
16	Cleanout adapter	$1\frac{1}{2}$	Each	0.70
		2	Each	1.18
		3	Each	2.75
17	Threaded cleanout plug	$1\frac{1}{2}$	Each	0.25
		2	Each	0.32
		3	Each	0.47
18	Combination Y and $\frac{1}{8}$ bend	3	Each	4.82
19	Male iron pipe adapter	$1\frac{1}{2}$	Each	0.78
		2	Each	1.35
20	Adapter, plastic to iron hub	$1\frac{1}{2} \times 2$	Each	1.87
		2×2	Each	2.14
	Adapter, plastic to plastic hub	3×4	Each	3.89

be consumed in walking. If the pipe is suspended from hangers, additional time will be required to install the hangers. Some specifications require that all pipe cut on the job be reamed to remove burrs. This will consume additional time.

Copper tubing is furnished in coils up to 60 ft long in all sizes to 2 in. Since the tubing is longer than steel pipe and changes in direction are obtained by bending the tubing, fewer fittings are required and the labor time will be less than for steel pipe. Joints usually are made by sweating the tubing into fittings, using solder and a blowtorch. Some cutting of the tubing with a hacksaw will be necessary.

Cast-iron soil, waste, and vent pipes are joined with molten lead and oakum joints. Many of these joints can be made with the pipe or fittings standing in an upright position, which will reduce the time required. Line joints are made by pouring the lead with the pipe in place, using a runner to hold the lead in the joint. After the lead has solidified, it must be caulked. Time should be estimated by the number of joints.

Vitrified-clay sewer pipe is manufactured in lengths varying from $2\frac{1}{2}$ to 4 ft. Joints are made with an O-ring seal, oakum and cement mortar, or a heated asphaltic joint compound.

Many plumbing contractors add a percentage of the cost of materials to cover the cost of labor required to rough in plumbing. The rates used vary from 40 to 80 percent of the cost of materials. Although this is a simple operation, it does not necessarily produce an accurate estimate.

Tables 17.12 through 17.20 provide representative costs of the various types of pipes and fittings that are commonly used to rough in plumbing.

TABLE 17.12 | Representative cost of standard-weight steel pipe.

Pipe size, in.	Cost per linear foot	
	Black	Galvanized
$\frac{1}{2}$	$1.28	$1.52
$\frac{3}{4}$	1.54	1.76
1	2.16	2.46
$1\frac{1}{4}$	2.75	3.20
$1\frac{1}{2}$	3.16	3.71
2	4.13	4.85

TABLE 17.13 | Representative cost of malleable fittings for steel pipe per fitting.

Fitting	Size, in.					
	$\frac{1}{2}$	$\frac{3}{4}$	1	$1\frac{1}{4}$	$1\frac{1}{2}$	2
Black						
Coupling	$1.60	$1.84	$2.81	$3.54	$4.52	$6.78
Elbow	1.29	2.15	2.35	2.87	5.29	8.82
Tee	1.48	2.20	3.84	5.87	7.64	12.50
Reducer	0.79	2.08	2.93	4.39	5.44	6.81
Union	5.18	6.37	8.00	10.43	13.04	16.54
Cap	0.94	1.69	1.82	2.55	3.26	4.35
Plug	1.04	1.24	1.32	1.64	2.20	3.23
Galvanized						
Coupling	$1.91	$2.32	$3.88	$5.74	$6.91	$11.61
Elbow	1.40	1.68	3.12	4.75	6.41	10.90
Tee	1.76	2.84	4.54	7.23	9.34	15.29
Reducer	1.91	2.54	3.43	5.92	7.29	9.92
Union	8.40	7.30	9.16	12.45	16.42	19.51
Cap	1.40	2.04	2.23	3.27	4.23	5.19
Plug	1.75	1.92	2.11	3.50	4.58	5.77

TABLE 17.14 |
Cost of copper pipe and tubing per linear foot.

Size, in.	Type L
$\frac{1}{2}$	$5.29
$\frac{3}{4}$	$8.50
1	12.61
$1\frac{1}{4}$	17.95
$1\frac{1}{2}$	23.17
2	36.60

TABLE 17.15 | Cost of copper solder fillings, per fitting.

Fittings	Size, in.					
	$\frac{1}{2}$	$\frac{3}{4}$	1	$1\frac{1}{4}$	$1\frac{1}{2}$	2
Couplings	$0.29	$0.57	$1.16	$2.18	$2.89	$4.86
Elbows	0.36	0.81	2.03	3.06	4.80	9.65
Tees	0.65	1.58	4.85	7.11	10.95	17.10
Reducers	0.61	1.02	2.41	2.96	5.04	7.88
Bushings	0.91	1.21	2.55	3.53	4.00	10.61
Caps	0.27	0.52	1.23	1.65	2.40	4.73
Adapters	3.79	4.86	6.16	19.05	23.17	37.42
Unions	4.43	5.22	8.77	15.63	20.61	35.10

TABLE 17.16 | Cost of bronze threaded valves, each.

Valves	Size, in.					
	$\frac{1}{2}$	$\frac{3}{4}$	1	$1\frac{1}{4}$	$1\frac{1}{2}$	2
Gate	$76.69	$94.22	$128.84	$189.22	$221.42	$281.43
Globe	129.83	151.01	217.65	318.28	362.73	537.07
Check	83.32	105.52	145.01	199.12	228.38	339.83
Ball	50.02	71.02	97.66	170.05	217.02	353.24

TABLE 17.17 | Cost of extra heavy cast-iron soil pipe per joint.

Pipe size, in.	Single hub, 10 linear ft	Double hub, 5 linear ft
2	$157.40	$138.90
3	217.10	182.50
4	284.30	226.70
5	376.80	310.80
6	424.10	365.20

TABLE 17.18 | Cost of single-hub fittings for cast-iron soil pipe, per fitting.

Fitting	Size, in.				
	2	3	4	6	8
Gasket	$15.82	$20.80	$24.52	$33.98	$42.43
$\frac{1}{4}$ bend	26.10	51.80	75.50	119.30	125.60
$\frac{1}{8}$ bend	25.91	41.40	62.70	88.90	94.20
Sanitary tee	47.50	117.50	124.40	238.20	240.70
45° Y	45.41	112.50	136.70	226.60	288.90
P-trap	59.70	100.20	150.40	243.57	385.50
SV clean out	22.10	39.46	38.20	71.04	110.08
Brass plug	18.22	18.66	33.38	86.24	155.96

TABLE 17.19 | Cost of vitrified clay sewer pipe, per linear ft.

Pipe size, in.	ASTM C-700	ASTM C-425 rubber seal
4	$3.15	$2.84
6	4.84	3.92
8	7.50	6.96
10	11.40	11.17
12	15.43	15.20
15	27.67	25.12
18	41.09	40.41

TABLE 17.20 | Cost of vitrified clay sewer pipe fittings with rubber seals, per fitting.

Pipe size, in.	Wyes and tees	$\frac{1}{8}$, $\frac{1}{4}$, $\frac{1}{2}$ bends	Cleanout wyes	Stoppers
4	$10.13	$9.68	$18.32	$2.63
6	23.18	16.65	29.64	3.52
8	29.70	26.55	39.53	5.10
10	44.13	61.35	55.95	6.75
12	71.25	66.20	98.14	9.38

Table 17.21 gives representative time in labor-hours required to rough in plumbing. The values are based on an analysis of the rates reported by a substantial number of plumbing contractors whose individual rates varied considerably. The rates given in the table include the combined time for plumbers and helpers. Thus, if a plumber and a helper work together, one-half of the labor-hours should be assigned to each person.

FINISH PLUMBING

Finish plumbing includes fixtures such as lavatories, bathtubs, shower stalls, water closets, sinks, urinals, water heaters, etc., and the accessories that usually are supplied with the fixtures. The prices of finish plumbing vary a great deal with the type, size, and quality specified. No book of this kind can give a complete listing of fixtures and prices. An estimator should list each item required for a building and then refer to a current catalog for prices.

Table 17.22 gives representative prices for plumbing fixtures in the medium grade range; these should be used as a guide only.

TABLE 17.21 | Labor-hours required for roughing in plumbing.

Class of work	Labor-hours
Hand cut, thread, and install steel pipe, per joint:	
$\frac{1}{2}$- and $\frac{3}{4}$-in. pipe	0.6–0.7
1- and $1\frac{1}{4}$-in. pipe	0.8–0.9
$1\frac{1}{2}$- and 2-in. pipe	1.3–1.5
$2\frac{1}{2}$- and 3-in. pipe	1.8–2.4
4-in. pipe	2.2–2.5
Machine cut, thread, and install steel pipe, per joint:	
$\frac{1}{2}$- and $\frac{3}{4}$-in. pipe	0.4–0.7
1- and $1\frac{1}{4}$-in. pipe	0.5–0.6
$1\frac{1}{2}$- and 2-in. pipe	0.7–0.8
$2\frac{1}{2}$- and 3-in. pipe	0.9–1.0
4-in. pipe	1.2–1.4
Install copper tubing, per joint:	
$\frac{1}{2}$- and $\frac{3}{4}$-in. tubing	0.3–0.4
1- and $1\frac{1}{4}$-in. tubing	0.4–0.5
$1\frac{1}{2}$- and 2-in. tubing	0.5–0.6
$2\frac{1}{2}$- and 3-in. tubing	0.6–0.7
4-in. tubing	0.7–0.8
Install cast-iron soil pipe and fittings, per joint:	
2-in. diameter	0.3–0.4
3-in. diameter	0.4–0.5
4-in. diameter	0.5–0.6
6-in. diameter	0.7–0.9
8-in. diameter	1.0–1.3
Install vitrified-clay or concrete pipe, per joint:	
4-in. diameter	0.2–0.3
6-in. diameter	0.2–0.3
8-in. diameter	0.3–0.4
10-in. diameter	0.5–0.6
12-in. diameter	0.7–0.9
Rough in for fixtures:	
Bathtub	10–18
Bathtub with shower	16–24
Floor drain	4–6
Grease trap	5–10
Kitchen sink, single	8–14
Kitchen sink, double	10–16
Laundry tub, two-compartment	8–12
Lavatory	10–15
Shower with stall	12–18
Slop sink	8–12
Urinal, pedestal type	8–12
Urinal, with stall	10–14
Urinal, wall type	7–10
Water closet	9–18
Water heater, 30- to 50-gal automatic	10–12
Water heater, 50- to 100-gal automatic	12–15

TABLE 17.22 | Representative costs of plumbing fixtures and accessories.

Fixture	Cost per unit
Bathtub, cast iron roll top, claw-foot legs	$1,400.00
Bathtub, acrylic	500.00
Bathtub faucet	120.00
Showers, one piece fiberglass	800.00
Shower faucet	230.00
Bathroom sink, porcelain, self-rimming	150.00
Bathroom faucet	230.00
Kitchen sink, vitreous china, self-rimming	1,200.00
Kitchen sink, enamel on cast iron, under mount	600.00
Kitchen sink, stainless steel, self-rimming	170.00
Kitchen faucet	350.00
Lavatory, porcelain on cast iron, 18 × 24 in.	200.00
Urinal stall, vitreous china	230.00
Water closet, vitreous china	300.00
Garbage disposal unit, with $\frac{3}{4}$ hp motor	180.00

LABOR REQUIRED TO INSTALL FIXTURES

The labor required to install plumbing fixtures will vary with the kind and quality of fixture, the kind of accessories used, the type of building, and to some extent the building code. The simplest fixtures will require the least time, and the more elaborate fixtures will generally require the most time.

TABLE 17.23 | Labor-hours required to install plumbing fixtures and accessories.

Fixtures	Labor-hours
Bathtub, leg type	10–16
Bathtub, flat bottom, no shower	14–20
Bathtub, flat bottom, with shower	18–26
Kitchen sink, enamel, single	6–10
Kitchen sink, enamel, double	8–12
Laundry tub, double	10–14
Lavatory, wall type	5–8
Lavatory, pedestal type	6–9
Urinal, wall type	8–12
Urinal, pedestal type	9–13
Urinal with stall	16–20
Water closet	6–9
Garbage disposal unit	6–8
Drinking fountain	5–8

A lavatory to be attached to the wall will be delivered to a job packed in a crate, complete with wall brackets, faucets, pop-up stopper, and other accessories, unassembled. It is necessary to remove it from the crate, attach the brackets, then attach and level the lavatory, connect the hot- and cold-water lines, and connect the trap and drain into the waste pipe. Additional time will be required to install legs and towel bars.

More time is required to install a combination bathtub with a shower than to install a bathtub only. Also considerably more time is required to install a built-in bathtub, especially in a bathroom with tile floors and walls, than to install a tub supported on legs. All factors that affect the rates of installation must be considered in preparing an estimate.

Table 17.23 gives representative labor-hours required to install plumbing fixtures, based on reports furnished by a substantial number of plumbing contractors. Use the lower rates for simple fixtures installed under favorable conditions and the higher rates for more complicated fixtures installed under more difficult conditions.

PROBLEMS

For these problems use the national average wage rates and the costs of materials given in this book. Use a plumber and a helper as a team to perform the work.

17.1 Estimate the direct cost of furnishing and installing 286 lin ft of 1-in. galvanized steel pipe, whose average length per joint prior to cutting will be 20 ft 0 in. It will be necessary to cut and thread the pipe for installing eight galvanized tees, six galvanized elbows, four galvanized unions, and three brass gate valves. The pipe will be cut and threaded by hand.

17.2 Estimate the direct cost of furnishing and installing the pipe, fittings, and valves of Prob. 17.1 when the pipe is cut and threaded by machines.

17.3 Estimate the cost of furnishing and installing 468 lin ft of $\frac{3}{4}$-in. type-L soft-copper tube, including 14 couplings, 26 tees, 16 elbows, and 4 unions, using sweated joints.

17.4 Estimate the cost of furnishing and installing 256 lin ft of 4-in. single-hub standard cast-iron pipe, including six wyes, eight standard quarter bends, and three P traps, all standard fittings, under average conditions. All joints will be made with oakum and lead.

17.5 Estimate the cost of furnishing and installing 280 lin ft of 2-in.-diameter PVC drain pipe supplied in 20-ft lengths. The following fittings will be required, with each requiring that the pipe be cut: six sanitary tees and five 90° elbows. The cost of a 1-pt can of solvent cement will be $4.20. This is enough cement to make 50 joints with $1\frac{1}{2}$-in. pipe.

18

Electric Wiring

Electric wiring is generally installed by a subcontractor who specializes in this type of work, with all materials and labor furnished under the contract. This procedure does not eliminate the need for estimating a job; it simply transfers the operation from the general to the electrical contractor.

Approximate estimates, and sometimes estimates for bid purposes, are prepared by counting the number of outlets required for a job and then applying a unit price to each outlet to arrive at the total cost. While this method will permit estimates to be prepared quickly, it will not always give an accurate cost for a job, because conditions may vary a great deal between two jobs. When this method is used, each switch, plug, and fixture served will count as an outlet.

FACTORS THAT AFFECT THE COST OF WIRING

The cost of wiring a building will be affected by the type, size, and arrangement of the building and the materials used in constructing the building. Wiring for a frame building may be attached to the framing members or run through holes drilled in the members. Wiring for a building constructed of masonry may be placed in conduit attached to or concealed within the walls. Wiring for concrete structures is placed in conduit installed in the concrete, which will require the installation of the conduit prior to placing the concrete.

The cost per outlet will be affected by the lengths of runs between outlets, kind of installation, and size of wires required. Job specifications may require rigid or flexible conduit, armored cable, or nonmetallic cable. The costs of these materials and the labor required to install them may vary considerably. Unless an estimator is able to apply a dependable cost factor, based on job conditions, to a particular job, the outlet method may result in a substantial variation from the true cost.

The best method of preparing a dependable estimate is to prepare an accurate material takeoff that lists the items separately by type, size, quantity, quality, and cost. When you are working from plans, it is good practice to indicate on the plans, using a colored pencil, each item as it is transferred to the material list. This procedure will reduce the danger of omitting items or counting them more than once. The application of appropriate costs for materials and labor to these items will give a dependable estimate.

ITEMS INCLUDED IN THE COST OF WIRING

When you are preparing a detailed estimate covering the cost of wiring a project, it is good practice to start with the first item required to bring electric service to the project. This may be the service wires from the transmission lines to the building, unless they are furnished by the utility. The following list may be used as a guide:

1. Service wires, if required
2. Meter box
3. Entrance switch box and fuses to circuit breakers
4. Wire circuits
   a. Conduit
   b. Wires
5. Junction boxes
6. Outlets for
   a. Switches, single-pole, double-pole, three-way
   b. Plugs, wall, and floor
   c. Fixtures
7. Bell systems, with transformers
8. Sundry supplies, such as solder, tape, etc.

ROUGHING IN ELECTRICAL WORK

TYPES OF WIRING

The types of wiring used in buildings may be governed by the National Electrical Code (NEC), published by the National Board of Fire Underwriters, or in some locations by state or municipal codes, which specify minimum requirements for materials and workmanship. Codes require that electrical wiring be protected against physical damage and possible short circuits by one or more methods such as installing them in conduit or by the use of wires that are furnished with factory-wrapper coverings.

RIGID CONDUIT

Rigid conduit is made of light galvanized pipe. Heavy conduit, which can be used for outside or inside installations, is threaded at each end. Since the walls of light conduit are too thin for threads, it is necessary to use special clamping fittings for making joints or entering boxes. The conduit extends into outlets or other boxes, where it is securely fastened with locknuts and bushings or with clamps. The conduit must be adequately supported and fastened in position along its run. The electric wires are pulled through the conduit at any desirable time after the conduit is installed.

Table 18.1 gives the sizes of conduit required for the indicated sizes and numbers of wires, as specified by the NEC.

TABLE 18.1 | Size conduit required for rubber-covered electric wires.

Size of wire, B & S gauge	Maximum number of wires in one conduit								
	1	2	3	4	5	6	7	8	9
	Size conduit required, in.								
14	$\frac{1}{2}$	$\frac{1}{2}$	$\frac{1}{2}$	$\frac{1}{2}$	$\frac{3}{4}$	$\frac{3}{4}$	$\frac{3}{4}$	1	1
12	$\frac{1}{2}$	$\frac{1}{2}$	$\frac{1}{2}$	$\frac{3}{4}$	$\frac{3}{4}$	1	1	1	$1\frac{1}{4}$
10	$\frac{1}{2}$	$\frac{1}{2}$	$\frac{3}{4}$	$\frac{3}{4}$	1	1	1	$1\frac{1}{4}$	$1\frac{1}{4}$
8	$\frac{1}{2}$	$\frac{3}{4}$	1	1	$1\frac{1}{4}$	$1\frac{1}{4}$	$1\frac{1}{4}$	$1\frac{1}{4}$	$1\frac{1}{2}$
6	$\frac{1}{2}$	1	$1\frac{1}{4}$	$1\frac{1}{4}$	$1\frac{1}{2}$	$1\frac{1}{2}$	2	2	2
5	$\frac{1}{2}$	$1\frac{1}{4}$	$1\frac{1}{4}$	$1\frac{1}{4}$	$1\frac{1}{2}$	2	2	2	2
4	$\frac{1}{2}$	$1\frac{1}{4}$	$1\frac{1}{4}$	$1\frac{1}{2}$	2	2	2	2	$2\frac{1}{2}$
3	$\frac{3}{4}$	$1\frac{1}{4}$	$1\frac{1}{4}$	$1\frac{1}{2}$	2	2	2	$2\frac{1}{2}$	$2\frac{1}{2}$
2	$\frac{3}{4}$	$1\frac{1}{4}$	$1\frac{1}{4}$	$1\frac{1}{2}$	2	2	$2\frac{1}{2}$	$2\frac{1}{2}$	$2\frac{1}{2}$
1	$\frac{3}{4}$	$1\frac{1}{2}$	$1\frac{1}{2}$	2	2	$2\frac{1}{2}$	$2\frac{1}{2}$	3	3
0	1	$1\frac{1}{2}$	2	2	$2\frac{1}{2}$	$2\frac{1}{2}$	3	3	3
00	1	2	2	$2\frac{1}{2}$	$2\frac{1}{2}$	3	3	3	$3\frac{1}{2}$
000	1	2	2	$2\frac{1}{2}$	3	3	3	$3\frac{1}{2}$	$3\frac{1}{2}$
0000	$1\frac{1}{4}$	2	2	$2\frac{1}{2}$	3	3	$3\frac{1}{2}$	$3\frac{1}{2}$	4

FLEXIBLE METAL CONDUIT

This conduit consists of an interlocking spiral steel armor, which is constructed in such a manner that it permits considerable flexibility. It can be used where frequent changes in direction are necessary. End connections must be made with special fittings.

ARMORED CABLE

Armored cables or conductors are made with or without a lead sheath under the armor. Special fittings are required to connect this cable into boxes, outlets, etc.

NONMETALLIC CABLE

The wires of this cable are covered with a flexible insulated braiding impregnated with a moisture-resisting compound. It is furnished in wire sizes 14 to 4 gauge. It is very popular for use in residences.

ELECTRIC WIRE

Electric wires are usually made of copper, either solid or standard. The wires may be plastic-covered for use indoors or weatherproof for use outdoors. The size of a wire is designated by specifying the Brown and Sharpe (B & S) gauge number, or for sizes larger than B & S gauge no. 0000, by the number of circular mils.

Table 18.2 gives the resistance in ohms per 1,000 feet and the electric current capacity in amperes, as specified by the NEC, for copper wire.

TABLE 18.2 | Resistance and current capacity of copper wire.

B & S gauge	Resistance, ohms/1,000 ft	Capacity, A	
		Plastic-covered	Weather-proof
18	6.374	3	5
16	4.009	6	10
14	2.527	15	20
12	1.586	20	25
10	0.997	25	30
8	0.627	35	50
6	0.394	50	70
5	0.313	55	80
4	0.248	70	90
3	0.197	80	100
2	0.156	90	125
1	0.124	100	150
0	0.098	125	200
00	0.078	150	225
000	0.062	175	275
0000	0.049	225	325

ACCESSORIES

In addition to the conduit and wire, numerous accessories will be needed to install wiring. Among the items needed will be:

1. Meter box
2. Entrance cap and conduit
3. Entrance box with switch and fuses or circuit breaker
4. Couplings, elbows, locknuts, bushings, and straps for conduit
5. Special fittings and clamps for flexible conduit and cable

6. Outlet boxes
7. Junction boxes
8. Light sockets
9. Rosettes
10. Solder, tape, straps, etc.

COST OF MATERIALS

The cost of materials for electric wiring will vary with the location, quality, quantities purchased, and time. Representative costs are given in Tables 18.3 to 18.8. The costs given in the tables should be used as a guide only. When estimating the cost of a project for contract purposes, the estimator should obtain current prices applicable to the particular project.

TABLE 18.3 | Representative costs of rigid conduit.

| Size, in. | Cost per 10 linear feet | | |
	Heavy rigid (galvanized)	Light thin wall—EMT (electrical metallic tubing)	Rigid PVC (schedule 40)
$\frac{1}{2}$	$26.10	$4.78	$1.79
$\frac{3}{4}$	33.65	9.42	2.79
1	44.30	19.08	3.15
$1\frac{1}{4}$	62.70	31.75	4.29
$1\frac{1}{2}$	87.60	38.00	6.99
2	123.50	49.20	7.54
$2\frac{1}{2}$	165.14	86.50	10.99
3	210.30	116.50	14.49

TABLE 18.4 | Representative costs per fitting for rigid conduit.

| Fitting | Size, in. | | | | | | | |
	$\frac{1}{2}$	$\frac{3}{4}$	1	$1\frac{1}{4}$	$1\frac{1}{2}$	2	$2\frac{1}{2}$	3
Heavy conduit								
Couplings	$4.60	$6.77	$11.41	$3.05	$3.50	$4.15	$7.12	$11.75
Elbows	2.89	3.79	5.29	6.99	10.09	14.20	19.30	25.00
Locknuts	0.12	0.29	0.39	0.49	0.59	0.79	1.29	1.59
Bushings	2.07	3.39	4.84	8.72	12.54	18.96	22.36	29.50
Straps, 2 holes	0.26	0.37	0.58	1.10	1.55	1.99	2.59	4.15
Thin-wall conduit								
Elbows	2.89	3.49	3.79	4.59	5.79	7.99	12.79	19.59
Setscrew connectors	0.26	0.39	0.69	1.15	1.69	2.29	5.99	7.99
Setscrew couplings	0.27	0.50	0.72	1.09	1.59	1.99	3.90	6.50
PVC schedule 40 conduit								
Couplings	0.52	0.65	1.00	1.33	1.85	2.41	4.03	7.03
Conduit bodies	1.94	2.48	2.73	4.64	5.52	9.56	16.53	21.10
Elbows	0.93	1.12	1.53	1.70	2.13	2.73	3.89	4.73
Caps	0.18	0.50	0.32	0.45	0.50	0.59	1.89	2.07

TABLE 18.5 |
Representative costs of
flexible metal conduit.

Size, in.	Cost per 100 lin ft
$\frac{1}{2}$	$67.00
$\frac{3}{4}$	95.00
1	148.00
$1\frac{1}{4}$	177.00
$1\frac{1}{2}$	254.00
2	323.00

TABLE 18.6 | Representative costs of
nonmetallic cable with ground (Romex).

B & S gauge	Number of wires	Cost per 1,000 lin ft
14	2	$280.00
12	2	390.00
10	2	660.00
8	2	1,050.00
14	3	480.00
12	3	590.00
10	3	930.00
8	3	1,270.00
6	3	1,800.00

TABLE 18.7 | Representative costs
of plastic-covered copper wire.

B & S gauge	Cost per 1,000 lin ft
14 THHN solid	$105.00
12 THHN solid	140.00
10 THHN solid	280.00
8 THW stranded	310.00
6 THW stranded	480.00
4 THW stranded	740.00
2 THW stranded	1,140.00
1 THW stranded	1,560.00
0 THW stranded	1,780.00
00 THW stranded	2,970.00
000 THW stranded	3,370.00
0000 THW stranded	3,730.00

TABLE 18.8 | Representative costs
of electrical accessories.

Item	Cost
Main breaker load center	
100 amps	$77.00
150 amps	138.00
200 amps	167.00
Outlet boxes, 4 in. each	7.00
Switch boxes, each	6.00
Wall plates, each	2.00
Switches:	
Toggle, single-pole, each	5.00
Toggle, three-way, each	7.00
Receptacles	11.00

LABOR REQUIRED TO INSTALL ELECTRIC WIRING

The labor required to install electric wiring can be estimated by one of three methods: by assuming a certain cost per outlet, by assuming that labor will cost a certain percentage of the cost of materials, and by assuming the time required to perform each operation of the work.

The first two methods may produce a sufficiently accurate estimate if dependable cost data are available from similar work previously done. If an estimate is prepared for a project that is unlike work previously done, it may be difficult to assume costs accurately.

The third method requires the preparation of a list of all materials needed, by quantity and quality. If the time required to install each item is known, the total time for the job can be estimated. Electric wiring is usually installed by two workers, an electrician and a helper, who work together as a team. The time per operation may be based on labor-hours or team-hours.

If the estimate is started with the service wires that enter a building, the estimate should include any work required to bring the wires to the building. Frequently the utility company will bring the service wires to a building, and the contractor will install an entrance cap, a conduit, and wires to the meter box, then install conduit and wires from the meter to a service fuse panel, with the main switch located within the building. From this panel, as many separate circuits as are necessary are installed throughout the building. Frequently a single conduit may be used for several circuits for at least a portion of the building, with individual circuits coming out of junction boxes installed in the main conduit line.

Wires must be installed from a circuit to each outlet, switch, and plug.

Heavy rigid conduit is installed with threaded couplings, elbows, and locknuts and bushings where it enters boxes. It should be supported with pipe straps or in some other approved manner, depending on the type of construction used for the building. This conduit is furnished in 10-ft lengths.

Lightweight rigid conduit is installed with special couplings and connectors.

Flexible conduit, armored, and nonmetallic cable can be bent easily as it is installed. Usually only end connectors are required where it enters boxes. It can be installed more rapidly than rigid conduit.

Where conduit, boxes, and accessories are installed in a building under construction, especially a concrete building, it may be necessary to keep one or more electricians on the job most of the time, even though they are unable to work continuously. This possibility should be considered by an estimator.

When 100 ft of straight rigid conduit is installed, the operations will include connecting 10 joints into couplings, plus the installation of any outlet boxes required. The installation of an outlet box requires the setting of a locknut and a bushing for each conduit connected to the box. More time will be required for large than for small conduit. Table 18.9 gives the labor-hours required for installing conduit.

The installation of a switch, fixture outlet, or plug requires a two-wire circuit from the main circuit to the outlet. The time required will vary with the length of the run and the type of building. For a frame building, it may be necessary to drill one or more holes through lumber, whereas for a masonry building it may be necessary to install the conduit through cells in concrete blocks, then chip through the wall of a block to install the outlet box.

The wires usually are pulled through the conduit just before the building is completed.

FINISH ELECTRICAL WORK

Under roughing in electrical work, the conduit, wires, and outlet boxes are installed. After the building is completed and the surfaces are painted, the fixtures are installed. The costs of fixtures vary so much that a list of prices in this book

TABLE 18.9 | Labor-hours required to install electric wiring per 100 lin ft.

Class of work	Electrician	Helper
Install service entrance cap and conduit*	0.5–1.0	0.5–1.0
Install conduit and fuse panel*	0.5–1.0	0.5–1.0
Install heavy rigid conduit with outlet boxes:		
$\frac{1}{2}$ and $\frac{3}{4}$ in.	5.0–10.0	5.0–10.0
1 and $1\frac{1}{4}$ in.	7.0–11.0	7.0–11.0
$1\frac{1}{2}$ in.	9.0–13.0	9.0–13.0
2 in.	12.0–17.0	12.0–17.0
$2\frac{1}{2}$ in.	15.0–21.0	15.0–21.0
3 in.	20.0–28.0	20.0–28.0
4 in.	25.0–35.0	25.0–35.0
Install thin-wall conduit with outlet boxes:		
$\frac{1}{2}$ and $\frac{3}{4}$ in.	4.0–6.0	4.0–6.0
1 in.	4.3–7.0	4.3–7.0
$1\frac{1}{4}$ in.	4.5–7.5	4.5–7.5
$1\frac{1}{2}$ in.	5.5–9.0	5.5–9.0
Install flexible conduit with outlet boxes:		
$\frac{1}{2}$ and $\frac{3}{4}$ in.	3.0–5.0	3.0–5.0
1 and $1\frac{1}{4}$ in.	4.0–6.0	4.0–6.0
Install nonmetallic cable with outlet boxes:		
14/2, 12/2, 10/2, and 8/2†	3.0–6.0	3.0–6.0
14/3, 12/3, 10/3, and 8/3†	3.5–6.5	3.5–6.5
6/3 and 4/3	4.0–7.0	4.0–7.0
Pull wire through conduit and make end connections, per circuit:		
14, 12, and 10 gauge	0.5–1.0	0.5–1.0
8 and 6 gauge	1.0–1.5	1.0–1.5
4 gauge	1.5–2.0	1.5–2.0
2 gauge	2.2–3.2	2.2–3.2
1 gauge	2.7–3.7	2.7–3.7

*The unit in this case is "each."

†The numbers 8/2 designate two wires, each no. 8 gauge in a cable.

would be of little value to an estimator. The costs should be obtained from a current catalog or jobber.

LABOR REQUIRED TO INSTALL ELECTRIC FIXTURES

Installing a fixture such as a switch or a plug involves connecting it to the wires which are already in the outlet box and attaching a cover plate. Installing a ceiling or wall fixture such as a light involves securing the fixture to a bracket and connecting it to the wires. Table 18.10 gives representative labor-hours required to install electrical fixtures.

TABLE 18.10 | Labor-hours per fixture required to install electric fixtures.

Class of work	Electrician	Helper
Install ceiling fixture	0.2–0.4	0.2–0.4
Install wall light	0.2–0.4	0.2–0.4
Install base or floor plug	0.1–0.2	0.1–0.2
Install wall switch	0.1–0.2	0.1–0.2
Install three-way switch	0.15–0.25	0.15–0.2

PROBLEMS

For these problems use the representative wage rates listed in Table 5.1 of Chapter 5 and the costs of materials listed in Chapter 18 of this book. Use an electrician and a helper to do the work.

18.1 Estimate the cost of furnishing and installing 420 lin ft of $1\frac{1}{2}$-in. heavy-duty conduit, including 36 couplings and 12 elbows under average conditions.

18.2 Estimate the cost of furnishing and installing 680 lin ft of 1-in. heavy rigid conduit, including 38 couplings and 20 elbows, when the installation is made in a concrete building under construction. The conditions are such that the electrician and the helper can work at only about 75 percent of normal efficiency.

18.3 Estimate the cost of furnishing and installing the following items in a frame residence:

1 service 100 amp breaker box with switch for six branches
180 lin ft of $\frac{3}{4}$-in. light conduit
130 lin ft of $\frac{1}{2}$-in. light conduit
23 outlet boxes
28 switch boxes
12 single-pole toggle switches
16 flush receptacles
8 $\frac{3}{4}$-in. elbows
12 $\frac{3}{4}$-in. couplings
24 $\frac{3}{4}$-in. connectors
18 $\frac{1}{2}$-in. couplings
12 $\frac{1}{2}$-in. elbows
18 $\frac{1}{2}$-in. couplings
48 $\frac{1}{2}$-in. connectors
28 wall plates
320 lin ft of no. 10 gauge plastic-covered copper wire, two wires per circuit
280 lin ft of no. 12 gauge plastic-covered copper wire, two wires per circuit

18.4 Estimate the cost of furnishing and installing the wiring for a residence, including the following items:

20 lin ft of $1\frac{1}{4}$-in. heavy rigid conduit

4 $1\frac{1}{4}$-in. elbows

2 $1\frac{1}{4}$-in. couplings

4 $1\frac{1}{4}$-in. locknuts and bushings

1 main breaker load center, 200 amps

980 lin ft of no. 12 gauge two-wire nonmetallic cable with ground wire

320 lin ft of no. 14 gauge two-wire nonmetallic cable with ground wire

26 junction boxes, 4 in.

36 switch boxes

36 wall plates

20 flush receptacles, wall plugs

16 single-pole switches

19

Sewerage Systems

ITEMS INCLUDED IN SEWERAGE SYSTEMS

Sewerage systems include one or more types of pipes and other appurtenances necessary to permit the system to collect waste water and sewage and transmit them to a suitable location where they can be disposed of in a satisfactory manner.

A sewerage system can be classified as one of two or more types depending on its primary function:

1. Sanitary sewer
2. Storm sewer
3. Other

This chapter is devoted to the construction of sanitary sewerage systems. Among the items included in this system are the following:

1. Sewer pipe and fittings
2. Manholes
3. Service connections
4. Cleanout services

SEWER PIPES

Sewer pipes can be made of one of the following materials:

1. Vitrified clay
2. Ductile iron
3. PVC (polyvinyl chloride)

The properties and other qualities of these pipes may be designated by such agencies as the American Society for Testing Materials (ASTM), the American Water Works Association (AWWA), or another appropriate agency.

Vitrified clay pipe. This pipe is available in variable diameters, lengths, and wall thicknesses to enable it to be used under different load conditions. The pipe has a rubberlike gasket installed in the bell end, which engages the spigot of the inserted pipe to produce a flexible watertight joint.

Ductile iron pipe. This pipe is available in various diameters, lengths, and wall thicknesses. Joints are made by inserting the spigot end of one pipe into the bell end of another pipe. An O-ring or a rubberlike gasket is used to produce a watertight flexible joint.

PVC pipe. This pipe is available in various sizes, lengths, and wall thicknesses. Joints are made by inserting the spigot end of one pipe into the bell end of the connecting pipe, to produce a flexible watertight joint.

Fittings for Sewer Pipe

Fittings for sewer pipe include single and double wyes, tees, curves, increasers, decreasers, and stoppers or plugs.

CONSTRUCTION OPERATIONS

When a sewerage system is constructed, the operations will include, but may not be limited to, these:

1. Clearing the right-of-way, if necessary
2. Excavating the trench for the pipe
3. Laying the pipe
4. Excavating the hole for the manhole
5. Placing the manhole in the hole
6. Installing service lines
7. Installing cleanout boots, if required
8. Backfilling with earth over the pipe and manhole
9. Removing surplus earth, if necessary

Equipment Required

The equipment required to install a sewerage system will include, but is not limited to, these items:

1. Trenching machines
   a. Wheel-type trencher
   b. Ladder-type trencher
   c. Backhoe
   d. Dragline

   2. Combination front-end loader, dozer, pipe handler
   3. Truck to dispose of surplus earth
   4. Power-type earth compactor
   5. Trench braces or shoring, if necessary
   6. Dewatering pumps, if necessary
   7. Power-operated pipe saw
   8. General utility truck
   9. Laser beam generator

CONSTRUCTING A SEWERAGE SYSTEM

Typically, sewer lines follow existing grades to ensure a gravity flow system, otherwise a lift station is necessary. In constructing a sewerage system it is common practice to begin at the outlet end of the pipeline and proceed to the other end. As the excavation progresses, shores or bracings are installed, if necessary, to protect against earth cave-ins.

   The depth of the trench as excavation progresses is determined by using a laser beam, which produces a light spot shining on a target mounted on a depth pole set in the trench from time to time. This beam is set at a slope parallel to the slope of the bottom of the trench, and it can be used from one setup for several hundred feet of trench.

   Depending on their weights, the pipe and fittings may be lowered into the trench manually or by a machine. The spigot end of a pipe should be lubricated with a specified compound, to enable easy insertion into the bell of the connecting pipe and to provide a watertight joint.

Manholes

Manholes are installed along sewer lines, usually 300 to 400 ft apart, to enable access to the lines for inspection and cleaning. Also, manholes should be installed at intersections with laterals, at changes in the grade or directions, and for changes in the size of the pipe.

   At one time manholes were constructed with bricks laid in place. But brick manholes have largely been replaced with precast (Fig. 19.1) or cast-in-place concrete manholes. When precast manholes are used, the base may be cast at the bottom of the hole first; then the remainder of the precast manhole will be set on the base to complete the installation. Another method of constructing a manhole is to cast the concrete base monolithically with a lower section of the manhole. This section of the manhole is set in place; then one or more sections are added to produce the desired depth. At the time the manholes are cast, holes are cast through the walls to permit pipes of the sewer line to enter or leave the manhole. Cast-in-place rubberized gaskets are installed to produce watertight seals for the pipes. The manhole may be cast with an invert at the bottom to permit the sewage to flow more effectively (see Fig. 19.2).

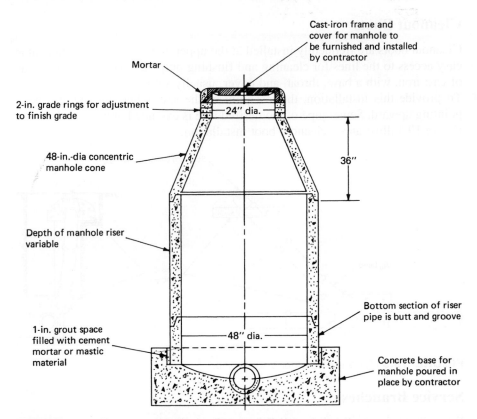

FIGURE 19.1 | A representative precast manhole.

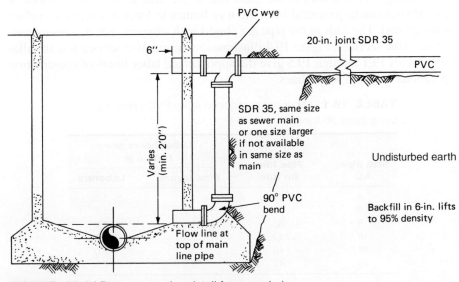

FIGURE 19.2 | Drop connection detail for a manhole.

Cleanout Boots

Cleanout boots are frequently installed at the upper end of sewer lines to enable easy access to the lines for cleaning and flushing purposes. The boots are made of cast iron, with a base, throat, and cover, usually set in concrete for stability. To provide this installation, the last joint of the sewer line ends with a wye, pointing upward. Sewer pipe of the desired size is extended upward to the boot. Figure 19.3 illustrates a cleanout boot installation.

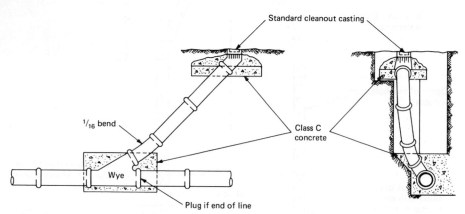

FIGURE 19.3 | Cleanout detail for a sewer line.

Service Branches

When the main pipeline is installed, it may be desirable to provide access to the pipe for future connections from users of the facility. This provision can be made by installing wye branches in the main line at the time it is installed. Pipes of desired sizes can be extended from the wye branch to locations near the surface of the ground. The ends of the pipe are closed with stoppers, which permits easy access to the services. Figure 19.4 illustrates representative service line installation. Tables 19.1 through 19.5 give the approximate labor-hours of construction operations to install sewer lines.

TABLE 19.1 | Labor-hours required to lay PVC sewer pipe, using joints 20 ft long.

Size pipe, in.	Pipe laid, lin ft/hr	Labor-hours per 100 lin ft	
		Pipe layers	Laborers
4	70	1.5	1.5
6	60	1.7	1.7
8	60	1.7	1.7
10	40	2.5	2.5
12	40	2.5	5.0
15	40	2.5	7.5

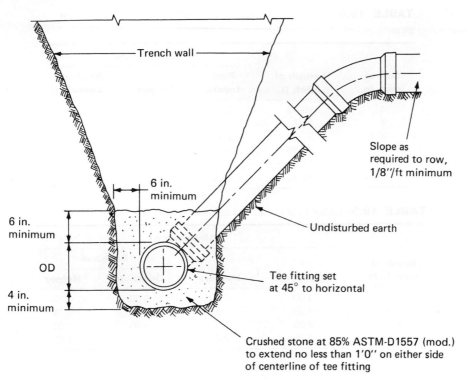

Trench wall

6 in. minimum

6 in. minimum

OD

4 in. minimum

Slope as required to row, 1/8″/ft minimum

Undisturbed earth

Tee fitting set at 45° to horizontal

Crushed stone at 85% ASTM-D1557 (mod.) to extend no less than 1′0″ on either side of centerline of tee fitting

FIGURE 19.4 | Sewer connection detail for a sewer line.

TABLE 19.2 | Labor-hours required to backfill and tamp earth by hand to 12 in. above the top of the sewer pipe.

Size of pipe, in.	Labor-hours per 100 lin ft*
4	7
6	9
8	10
10	11
12	12
15	14

*If the backfill is placed by machine but tamped by hand, the number of labor-hours can be reduced by 50 percent.

TABLE 19.3 | Machine-hours and labor-hours required to excavate the earth for manholes, using a backhoe.

Depth of manhole, ft	Hours per manhole		
	Machine	Operator	Labor
Up to 6	0.75	0.75	1.5
6–9	1.00	1.00	2.0
9–12	1.25	1.25	3.0
12–15	1.50	1.50	4.0

TABLE 19.4 | Labor-hours required to install service inlets to sewer pipe.

Size of inlet, in.	Length of inlet, ft	Labor-hours		
		Pipe layers	Helper	Backhoe operator
4	10–15	1.00	1.00	0.50
4	15–20	1.50	1.50	0.75
4	20–30	2.00	2.00	1.00
6	10–15	2.25	2.00	0.50
6	15–20	2.75	2.75	0.75
6	20–30	3.25	3.25	1.00

TABLE 19.5 | Labor-hours required to install a cleanout boot.

Depth of trench, ft	Labor-hours		Backhoe	
	Pipe layers	Helpers	Operator	Helper
4–6	1.00	1.00	0.50	0.50
6–8	1.50	1.50	0.75	0.75
8–12	2.00	2.00	1.00	1.00
12–15	2.75	2.75	1.25	1.25

EXAMPLE 19.1

Estimate the direct cost of installing a sewerage system requiring the installation of PVC pipe, fittings, manholes, service connections, and cleanout boots for the given conditions:

Length of project: 4,800 lin ft

Size of pipe: 6 in.

Average depth of trench: 7 ft

Number of manholes: 14

Average depth of manholes: 7 ft

The project will be constructed in an open and clear area with no obstructions, land clearing, utility line, or other items to delay progress. The soil will be hard clay with no groundwater. The trenches will be excavated by a wheel-type backhoe, which will enable the trench walls to assume normal slopes with no shoring required.

After the pipe is laid, the backfill around the pipe will be placed by a dozer and hand-tamped to a depth of 12 in. above the pipe by laborers. The balance of the backfill will be placed by the dozer. The time for dozer backfilling will be approximately $\frac{1}{4}$ of the time to excavate the trench.

The manhole will be precast concrete of the type illustrated in Figure 19.1. Service inlets will be spaced at 75-ft intervals along the pipeline. Each inlet will require a wye fitting. Service lines of 4-in. pipe will be 14 ft in length. Four cleanout boots will be required.

Cost of pipe:

Total length of pipe = 4,800 ft

Material costs:

Sewer pipe, 6 in. dia.: 4,800 ft @ $5.77/ft	=	$27,696.00
Wyes: (4,800 ft)/(75 ft/wye) = 64 wyes @ $23.50	=	1,504.00
Pipe stoppers: 64 × $4.86	=	311.04

Labor and equipment hours, reference Tables 19.1 and 19.2

Backhoe: 4,800 ft/60ft/hr = 80 hr @ $65.00/hr	=	5,200.00
Backhoe operator: 80 hr @ $30.00/hr	=	2,400.00
Pipe layers: (4,800 ft) × (1.7 hr/(100 ft) @ $24.00/hr	=	1,958.40
Laborers: (4,800 ft) × (1.7 hr/100 ft) @ $22.00/hr	=	1,795.20
Laser: (4,800 ft) × (1.7 hr/100 ft) @ $2.50/hr	=	204.00
Dozer backfilling trench: ($\frac{1}{4}$ × 80 hr) @ $55.00/hr	=	1,100.00
Dozer operator: ($\frac{1}{4}$ × 80 hr) @ $29.00/hr	=	580.00
Laborers tamping: (4,800 ft) × (4.5 hr/100 ft) @ $22.00/hr	=	4,752.00
Power tamper: (4,800 ft) × (4.5 hr/100 ft) @ $1.75/hr	=	378.00
Total cost of installed pipe	=	$47,878.64

Cost per ft of installed pipe: $47,878.64/4,800 ft = $9.97/ft

Cost of manholes:

Total number of manholes = 14

Material costs:

Manhole concrete base: 14 @ $310.00	=	$4,340.00
Manhole barrel: 14 @ $1,175.00	=	16,450.00
Cast-iron ring and cover: 14 @ $275.00	=	3,850.00

Labor and equipment hours, reference Table 19.3

Backhoe excavating hole: 14 × 1 hr @ $65.00/hr	=	910.00
Backhoe operator: 14 hr @ $30.00/hr	=	420.00
Backhoe setting manholes: 14 × 1 hr @ $65.00/hr	=	910.00
Backhoe operator: 14 hr @ $30/hr	=	420.00
Laborers installing manholes: 14 × 2 hr @ $22.00/hr	=	616.00
Dozer backfilling: 14 × 1 hr @ $55.00/hr	=	770.00
Dozer operator: 14 hr @ $29.00/hr	=	406.00
Laborers tamping earth: 14 × 2 hr × $22.00/hr	=	616.00
Power tamper: 14 × 2 hr × $1.75/hr	=	49.00
Total cost of manholes	=	$29,757.00

Cost per manhole: $29,757.00/14 = $2,126/manhole

Cost of service lines:

No. of 4-in. dia. 12-ft long service lines: (4,800 ft)/(75 ft/line) = 64 lines

Average cost for one service line:

Material costs:

Service pipe, 4-in. diameter: 12 ft @ $2.24/ft	=	$26.88
Bend, 4-in. dia. $\times \frac{1}{8}$: @ $6.45	=	6.45
Stopper, 4-in.: @ $7.56	=	7.56

Labor and equipment hours, reference Table 19.4

Backhoe trenching: 0.5 hr @ $65.00/hr	=	32.50
Backhoe operator: 0.5 hr @ $30.00/hr	=	15.00
Pipe layers: 1.0 hr @ $24.00/hr	=	24.00
Laborer helper: 1.0 hr @ $22.00/hr	=	22.00
Dozer backfilling: 0.5 hr @ $55.00/hr	=	27.50
Dozer operator: 0.5 hr @ $29.00/hr	=	14.50
Laborers tamping: (12 ft $\times$ 4.5 hr/100 ft) @ $22.00/hr	=	11.88
Power tamper: (12 ft $\times$ 4.5 hr/100 ft) @ $1.75/hr	=	0.95
Total cost per service line	=	$189.22

Total cost for 64 service lines: 64 lines $\times$ $189.22/line = $12,110.08

Cleanout boots:

Total number of boots = 4

Average cost for one boot:

Material costs:

Pipe, 6-in. dia.: 7 ft @ $5.77/ft	=	$40.39
Bend, 6.in. dia. $\times \frac{1}{16}$: @ $12.41	=	12.41
Cleanout casting	=	11.16
Plug, 6-in.: @ $3.40	=	3.40

Labor and equipment hours, reference Table 19.5

Pipe layer: 1.5 hr @ $24.00/hr	=	36.00
Laborer helper: 1.5 hr @ $22.00/hr	=	33.00
Backhoe: 0.75 hr @ $65.00/hr	=	48.75
Operator: 0.75 hr @ $30.00/hr	=	22.50
Helper: 0.75 hr @ $22.00/hr	=	16.50
Total cost of boot	=	$224.11

Total cost of boots: 4 boots $\times$ $224.11/boot = $896.44

Total cost of project:

Pipeline	=	$47,878.64
Manholes	=	29,757.00
Service lines	=	12,110.08
Cleanout boots	=	896.44
Total cost	=	$90,642.16

The $90,642.16 is the direct cost and does not include overhead and profit.

TRENCHLESS TECHNOLOGY

When it is necessary to install new underground utility systems that are located in congested areas near existing utility lines, it may be desirable to use one of the modern methods of trenchless technology, rather than the open cut trench method of utility construction. Trenchless technology includes all the methods of underground pipeline and conduit installation with minimum or no surface or subsurface disruption. It includes new installations and pipeline renewal.

Trenchless technology has been developed to address social costs, new safety regulations, and difficult underground job conditions such as deep ditches, high water tables, and the existence of natural or artificial obstructions. Social costs are the costs incurred to the general public: safety hazards and damage to the environment and public assets such as road pavements. Examples of social costs are loss of business activity due to road closings, reduced life of pavements due to trenching, cutting trees and damage to green areas, increased possibility of accidents due to trenching and trench cave ins, and noise and air pollution. Safety is also achieved because the workers remain above ground during the construction process, rather than entering an open trench or a pipe below grade.

There are several methods of trenchless technology, including horizontal directional drilling and microtunneling, which are presented in this book. Horizontal directional drilling is used primarily for installation of underground cables, gas and water lines, and sewer laterals (see Chapter 20). Microtunneling is ideally suited for installation of gravity sewer lines where a high degree of accuracy is required.

MICROTUNNELING

Microtunneling machines are laser-guided, remotely controlled, and have the capability to install pipeline on precise lines and grades. Microtunneling can be used to install pipes into almost any type of ground condition up to 150 ft below the ground surface and up to 1,000 ft in length from the entry drive shaft to the reception shaft. By definition, microtunneling is a remote-controlled pipe-jacking operation, and there is no maximum size diameter specified for microtunneling machines, although most jobs apply to pipe diameters from 24 to 60 in.

Microtunneling is a method of installing pipe below ground, by jacking the pipe behind a remotely controlled and steerable articulated microtunnel boring machine (MTBM). The MTBM, which is connected to and followed by the pipe that is being installed, bores through the ground as the excavated soil is transported out at the same rate as the boring operation. The minimum depth of cover above the pipe is generally greater than 6 ft, or 1.5 times the outer diameter of the pipe being installed. The overcut of the MTBM is usually 1 in. beyond the outside radius of the pipe. Based on the mode of operation, microtunneling methods can be subdivided into two major groups: slurry method and auger method. Both microtunneling methods consist of these major components:

1. Microtunnel boring machine
2. Remote control system
3. Active direction control
4. Automated spoil transportation and rate of excavation controls
5. Pipe-jacking equipment suitable for direct installation of the pipe
6. Jacking pipe

MICROTUNNEL BORING MACHINE

Figure 19.5 shows a typical microtunneling operation. An entry shaft, or sending shaft, is constructed where the boring operation starts and a receiving shaft is constructed at the end of the boring operation. Figure 19.6 is a view looking down an entry shaft that shows positioning of the MTBM for the boring operation. A view of the MTBM boring the tunnel is shown in Fig. 19.7. A concrete thrust block on the back side of the sending shaft provides resistance to the thrust during the pipe-jacking operation. A hydraulic jacking frame at the bottom of the sending shaft provides the mechanism to push the pipe. As the MTBM bores through the soil, additional pipe sections are added. Lowering of an additional pipe section into the entry shaft is shown in Fig. 19.8.

A laser system provides accurate guidance for the line and grade of the MTBM. In the slurry microtunneling, a hydraulic pumping system provides removal of the cuttings of the soil material. The solid soil material is separated

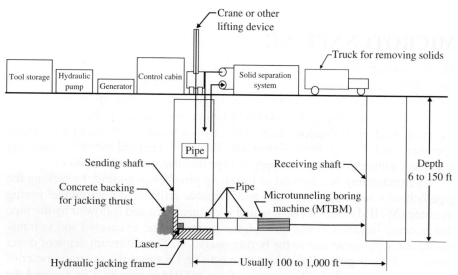

FIGURE 19.5 | Microtunneling operation.

FIGURE 19.6 | Lowering MTBM into entry shaft.

Courtesy: Luckinbill, Inc.

FIGURE 19.7 | MTBM boring tunnel for sewer pipe.

Courtesy: Luckinbill, Inc.

FIGURE 19.8 | Lowering an additional section of sewer pipe.

Courtesy: Luckinbill, Inc.

and removed by trucks. In auger microtunneling, the spoil is transported by rotating augers to the sending shaft.

The MTBM is capable of controlling rotation and roll by means of either a bidirectional drive on the cutter head or the use of fins or grippers. Electric or hydraulic motors power the MTBM cutter head. The MTBM is articulated to enable remote steering of the system.

A display showing the position of the shield in relation to a target is available to the operator on an operation console, together with other information such as face pressure, roll, pitch, steering attitude, and valve positions. The MTBM has a closed face system capable of supporting the full excavated area at all times.

REMOTE CONTROL SYSTEM

A remote control system enables operation of the microtunneling system without the need for personnel to enter the microtunnel. The control equipment simultaneously integrates the processes of excavation, soil removal, and replacement of pipe. As each pipe section is jacked forward, the control system synchronizes all of the operational functions of the system. The system provides complete and adequate ground support at all times.

ACTIVE DIRECTION CONTROL

Line and grade is controlled by a guidance system that relates the actual position of the MTBM to a design reference, by a laser beam transmitted from the jacking shaft along the centerline of the pipe to a target mounted in the shield. The MTBM is capable of maintaining grade to within ± 1 in. and line to within ±1.5 in.

The active steering information is monitored and transmitted to the operation console. The minimum steering information available to the operator on the control console usually includes the position relative to the reference, roll, inclination, attitude, rate of advance, installed length, thrust force, and cutter head torque.

AUTOMATED SPOIL TRANSPORTATION

The automated spoil transportation system should match the excavation rate to the rate of spoil removal, thereby maintaining settlement or heave within the tolerance specified in the contract documents. Balancing of groundwater pressures is achieved by using either a slurry pressure or compressed air for the auger MTBM system. The system continuously balances the groundwater pressure to prevent the loss of slurry and/or groundwater.

In a slurry spoil transportation system, the groundwater pressure is controlled by the use of variable-speed slurry pumps, pressure control valves, and a flow meter. A slurry bypass unit is included in the system to allow the direction of flow to be changed and isolated as necessary.

A separation process is provided when using the slurry transportation system. The process is designed to provide adequate separation of the spoil from the slurry so that the clean slurry can be returned to the cutting face for reuse. The type of process used is dependent on the size of the tunnel being constructed, the type of soil being excavated, and the space available for erecting the plant.

If an auger spoil transportation system is used, the groundwater pressures are managed by controlling the volume of spoil removed with respect to the advance rate (earth pressure balance methods) and the application of compressed air to counterbalance earth pressure and underground water.

JACKING PIPE

In general, pipe used for jacking is round, with a smooth uniform outer surface and with watertight joints that also enable easy connections between joints. Pipe lengths must be within specified tolerances and pipe ends must be square and smooth so the jacking loads will be evenly distributed around the entire pipe joint such that point loads will not occur when the pipe is jacked in a reasonably straight alignment.

Pipe used for pipe jacking must be capable of withstanding all forces that will be imposed by the installation process, as well as the loading conditions that will be applied to the pipe after final installation. The pipe manufacturer specifies the required protection against damage to the ends of the pipe and intermediate joints during the jacking process. The detailed method proposed to cushion and distribute the jacking forces is specified for each particular pipe material.

If a pipe shows signs of failure, it may be necessary to jack it through to the reception shaft and remove it. The pipe manufacturer's design jacking loads should not be exceeded during the installation process. In general, the ultimate axial compressive strength of the pipe must be a minimum of 2.5 times the design jacking loads of the pipe. The pipe materials specially manufactured for microtunneling operations include vitrified clay pipe, glassfiber reinforced polyester (GRP) pipe, reinforced concrete pipe, steel pipe, resin concrete pipe, and ductile iron pipe.

Pipe-Jacking Equipment

The main jacks are mounted in a jacking frame that is located in the drive (starting) shaft. The jacking frame successively pushes the MTBM, along with a string of connected pipes, toward a receiving shaft. The jacking capacity of the system must be sufficient to push the MTBM and the string of pipes through the ground. Calculations are made in advance of the boring operation to determine the excavation forces, frictional forces, and the weight of the MTBM and pipes.

For larger pipe, it may be necessary to use intermediate jacking stations. Spreader rings and packing are used to provide a uniform distribution of the jacking forces on the end of the pipe. A lubricant may be injected at the rear of the MTBM and through the pipe walls to lower the friction developed on the surface of the pipe during the jacking operation.

ADVANTAGES AND DISADVANTAGES OF MICROTUNNELING METHODS

Microtunnel methods are capable of installing pipes to accurate line and grade tolerances into all types of soil without extensive dewatering systems that are sometimes required of open trench construction. The need to load earth is minimized, compared to deep burial open trench construction. Utility lines can be installed without a drastic effect on the cost. The depth factor becomes increasingly important as underground congestion is increased, or when a high water table is encountered. Safety is enhanced because workers are not required to enter trenches or tunnels, only the entry and receiving shafts. The pipe with sufficient axial load capacity can be jacked directly in place without the need of a separate casing pipe.

One of the major disadvantages of microtunneling methods has been the inability to utilize flexible or low-strength pipes, such as PVC pipe. Other disadvantages include high capital cost necessary to purchase the microtunneling equipment and the problems caused by obstructions, such as large boulders, roots, or old manmade structures.

MICROTUNNELING PROCESS

A breakdown of the procedures for microtunneling include:

1. Mobilization, soil investigation, and setup of equipment
2. Design and construction of entrance and exit shafts
3. Equipment setup
4. Delivery and preparation of pipe section
5. Tunneling operation
6. Retrieving the MTBM
7. Pipe inspection, testing, and demobilization

Prior to any tunneling project, a soil investigation is conducted to determine the type of soil and the extent of water. The amount and mix of clays, sands, silts, or clayey gravel must be known to determine the average time for pushing the pipe. The presence of water determines the amount of water control that will be necessary for preparation of the entrance and exit shafts. Mobilization includes transportation of equipment to the jobsite, site preparation, temporary facilities, and setup of the equipment.

The contractor is responsible for the design and construction of the entrance and exit shafts. Where water is present, it may be necessary to install sheet piling, cofferdams, pumping facilities, or other methods of dewatering the shaft for the microtunneling operation. A shaft-sinking method may be desirable in areas with high water table and to eliminate the dewatering process. A hollow, bottomless structure (reinforced concrete) is built on top of the ground and is lowered into position by excavating material from within and below to enable the structure to settle by force of gravity. Once the desired depth is reached, a tremie seal is placed, the shaft is pumped free of water, and a structural bottom is constructed.

After mobilization and construction of the entrance and exit shafts, the micotunneling equipment is set up, pipe sections are prepared, and the tunneling operation can be started. A typical crew consists of the microtunneling machine operator, crane operator, technician, two laborers, and supervisor. Table 19.6 gives the equipment required for a microtunneling project.

TABLE 19.6 | Equipment for a microtunneling project.

Tunnel-boring machine
Control room with operator station and control cables
Laser guidance equipment
Pipe-jacking frame and hydraulic pump equipment
Water and slurry tanks for solids separation
Pumps, hoses, and flow meters for slurry and lubricants
Portable compressor, generator, and welding machine
Backhoe for excavating entrance and exit shafts and spoil removal
Crane for hoisting jacking frame and pipe sections into entrance shaft

TABLE 19.7 | Illustration of typical times and production rates for microtunneling.

Type of work	Time
Mobilization and installation of microtunneling equipment	5 days
Production rate for pushing a 24-in.-dia. 8-ft pipe section	
Clay	1.5 in./min
Silt	4.0 in./min
Sand	5.5 in./min
Clayey gravel	2.0 in./min
Time for installation of a 24-in.-dia 8-ft pipe section	
Clay	3.1 hr
Silt	1.7 hr
Sand	1.5 hr
Clayey gravel	2.2 hr
Demobilization and loading equipment	3 days

Since speed of installation in a microtunneling project is critical, everything required for the successful installation should be available when needed. Consideration must be given to alternative or corrective procedures that might be required if conditions at the site are different than anticipated. Table 19.7 provides an example of the time and production rates for a microtunneling operation. The actual production rates are dependent on the experience and skills of the contractor and the operator, and on the project-specific job conditions, such as surface and subsurface conditions, type of pipe and equipment used, and degree of planning and preplanning for the project.

CHAPTER 20

Water Distribution Systems

COST OF WATER DISTRIBUTION SYSTEMS

The cost of a water distribution system will include the materials, equipment, labor, and supervision to accomplish some or all of the following:

1. Clear the right-of-way for the trench
2. Removal and replacement of pavements
3. Excavate and backfill trenches
4. Relocate utility lines
5. Install pipe
6. Install fittings
7. Install valves and boxes
8. Install fire hydrants
9. Install service connections and meters
10. Bore holes under roadways and install pipeline
11. Test and disinfect water pipe

TYPES OF PIPE MATERIAL

The following types of pipes are used for water systems:

1. Polyvinyl chloride plastic pipe, commonly indentified as PVC Pipe
2. Polyethylene plastic pipe, commonly identified as PE Pipe
3. Ductile iron pipe, commonly identified as DIP Pipe
4. Reinforced concrete pipe, commonly identified as RCP Pipe

Polyvinyl Chloride (PVC) Plastic Pipe

Polyvinyl Chloride (PVC) plastic pipe is commonly used for transmission and distribution water lines. The pipe is manufactured in accordance with the AWWA C900 standard for pressure pipe, which applies to diameters from 4 in. through 12 in. for distribution lines. AWWA C905 applies to sizes from 14 in. to 48 in. for transmission lines.

The sizes commonly used for water pressure pipe range in diameters from 4 in. to 24 in. Sizes less than 4 in. are used for service lines that connect the distribution line to the customer's property. Table 20.1 gives representative dimensions of PVC pressure water pipe. PVC pipes are sealed with rubberized O-ring gaskets for water-tight connections. Fittings for PVC pipe include tees, bends, elbows, reducers, and pipe ends. Table 20.2 gives approximate labor-hours to install PVC pipe and fittings.

TABLE 20.1 | Dimensions of PVC Schedule 40 water distribution pipe.

Pipe size, in.	Outside diameter, in.	Minimum thickness, in.	Weight per foot, lb/ft
4	4.500	0.237	2.1
6	6.625	0.280	3.7
8	8.625	0.322	5.6
10	10.75	0.365	7.9
12	12.75	0.406	10.5
14	14.00	0.437	12.4
16	16.00	0.500	16.4
18	18.00	0.562	20.5
20	20.00	0.593	24.1
24	24.00	0.687	33.6

TABLE 20.2 | Labor-hours required to lay pipe and install fittings of PVC plastic water pipe.*

Pipe size, in.	Pipe laid, ft/hr	Labor-hours				
		Laying 100 lin ft pipe		Installing elbows, tees, plugs		
		Skilled laborer	Common laborers	Fittings per hr	Skilled laborer	Common laborers
4	46	2.1	4.2	11.3	0.08	0.08
6	39	2.5	5.0	10.0	0.09	0.09
8	34	2.9	5.8	8.8	0.12	0.12
10	31	3.2	6.4	6.2	0.17	0.17
12	28	3.6	7.2	3.7	0.27	0.27
14	25	4.0	8.0	2.8	0.35	0.35
16	22	4.5	9.0	1.9	0.52	0.52
18	19	5.2	10.4	1.8	0.55	0.55
20	16	6.3	12.6	1.2	0.83	0.83
24	13	7.6	15.2	1.1	0.91	0.91

* Labor-hours for sizes less than 4-in. are about the same as for 4-in. sizes

Polyethylene (PE) Plastic Pipe

Polyethylene (PE) pipe is more flexible than PVC, ductile iron, or concrete pipe. Pipe sizes are available in diameters from 4 in. to 63 in. Pipe less than 6 in. in size may be delivered to the jobsite coiled up in a reel, up to 1,500 ft in length. The pipe is unreeled and placed in the trench. On a typical day, a crew of three workers can install 300 to 500 ft of new PE pipe.

Pipe sizes are available in nominal sizes from 4 in. to 63 in. in diameter. PE pipe is given a dimension ratio (DR) rating, calculated as the outside diameter divided by the minimum wall thickness. A PE pipe with a low DR has a thicker wall than a pipe with a high DR. Dimension ratios for PE pipe range from 7.3 to 32.5. Pipes with a DR of 7.3 have working pressures of 198 to 254 psi, whereas pipe with a DR of 32.5 have working pressures of 40 to 51 psi.

All PE pressure pipe is given a DR rating, but there are different wall thicknesses for a given outside diameter. The sizing systems used for PE pipe are IPS (iron pipe size), DIPS (ductile iron pipe size), and CTS (copper tube size). For example, a pipe may be designated as an 8-in. PE IPS pipe. All DR ratings for 8-in. IPS pipe will have the same outside diameter, but the wall thickness will vary depending on the DR rating. For example, an 8-in. DR 32.5 has a wall thickness of only 0.265 in., whereas an 8-in. DR 7.3 pipe has a wall thickness of 1.182 in. Table 20.3 gives representative dimensions of HDPE (IPS) pipe.

PE pipe can be melted and resolidified, which makes it easy to join two sections of pipe. Two sections of pipe can be melted by heat and then pushed together and held, forming a single pipe. Thus, the two sections of HDPE pressure pipe are joined by a "butt weld." The flexibility of the pipe and the ability to fuse-weld sections make PE pipe ideal for horizontal directional drilling (HDD) pipe installation under roads, creeks, and rivers.

TABLE 20.3 | Dimensions of Polyethylene PE (IPS) water distribution pressure pipe, AWWA C906.

Pipe size, in.	Avg. OD, in.	Minimum wall thicknesses, in. (IPS sizing system)						
		DR 32.5 in.	DR 26.0 in.	DR 21.0 in.	DR 17.0 in.	DR 13.5 in.	DR 9.3 in.	DR 7.3 in.
4	4.500	0.138	0.173	0.214	0.265	0.333	0.482	0.661
6	6.625	0.204	0.255	0.316	0.390	0.491	0.710	0.908
8	8.625	0.265	0.332	0.411	0.507	0.639	0.927	1.182
10	10.75	0.331	0.413	0.512	0.632	0.796	1.156	1.473
12	12.75	0.392	0.490	0.607	0.750	0.944	1.371	1.747
14	14.00	0.431	0.538	0.667	0.824	1.037	1.505	1.918
16	16.00	0.492	0.615	0.762	0.941	1.185	1.720	2.192
18	18.00	0.554	0.692	0.857	1.059	1.333	1.935	2.466
20	20.00	0.615	0.769	0.952	1.176	1.481	2.151	2.740
24	24.00	0.738	0.923	1.143	1.412	1.778	2.581	3.288

Ductile Iron Pipe (DIP)

Ductile iron pipe (DIP) is manufactured as a centrifugal casting, where ductile iron is poured into a rapidly spinning water-cooled mold. The centrifugal force produces an even spread of iron around the circumference. Standard sizes are manufactured from 3 in. to 64 in. in diameter, and laying lengths of 18 ft and 20 ft. Table 20.4 gives dimensions and weights of ductile iron pipe.

A variety of linings are available to reduce internal corrosion of ductile iron pipe. For water pipe the most common lining is cement mortar, which is a mixture of Portland Cement and sand that is centrifugally applied during manufacture of the pipe.

Buried ductile iron pipe is subject to external corrosions and chemical attacks that may be present in some soils. It is common to protect the pipe with one or more external coatings. A loose polyethylene sleeve can be attached to the pipe for corrosion protection.

There are two methods of connecting ductile iron pipe, push-on-joint connection and mechanical-joint connection. The push-on connection applies to

TABLE 20.4 | Dimensions and weights of (DIP) ductile iron water distribution pipe, AWWA C-151.

Pipe size, in.	Pressure class, psi	Outside diameter, in.	Push-on-joint DIP 18-ft laying length average weight, lb/ft	Mechanical-joint DIP 18-ft laying length average weight, lb/ft
3	350	3.96	9.3	9.4
4	350	4.80	11.4	11.6
6	350	6.90	16.6	17.0
8	350	9.05	22.0	22.4
10	350	11.10	28.4	28.8
12	350	13.20	36.4	36.9
14	250	15.30	42.9	43.8
14	300	15.30	45.8	46.7
14	350	15.30	47.5	48.1
16	250	17.40	52.3	53.4
16	300	17.40	55.5	56.6
16	350	17.40	58.8	59.9
18	250	19.50	60.5	61.9
18	300	19.50	65.9	67.3
18	300	19.50	69.5	70.9
20	250	21.60	71.6	73.0
20	300	21.60	77.6	78.9
20	350	21.60	81.6	83.0
24	200	25.80	86.1	87.6
24	250	25.80	95.8	97.3
24	300	25.80	103.0	104.5
24	350	25.80	110.2	111.7

Note: Average lb/ft based on calculated weight of pipe before rounding.

bell-and-spigot ductile iron pipe. The connection of two pipes is accomplished by inserting (pushing) the spigot end of one pipe into the bell end of an adjacent pipe. A rubberized O-ring gasket in the spigot provides a water-tight seal between the connected pipes. This method of connection is commonly used for water service lines.

The mechanical joint connection is made by inserting the plain end of one pipe into the socket of an adjoining pipe, then forcing a gasket ring into the socket by means of a gland that is drawn to the socket by tightening bolts through the gland and socket. A mechanical-joint connection consists of a flange cast with a bell, a rubber gasket that fits in the bell socket, a gland or following ring to compress the gasket, and tee head bolts and nuts for tightening the joint. The joint assembly is more labor-intensive than push-on-joint connections. Table 20.5 gives labor-hours required to lay 100 lin ft of pipe.

TABLE 20.5 | Labor-hours required to lay 100 lin ft of ductile iron pipe (DIP).

| | | Labor-hours | | | | | | |
| | | Push-on connection | | | | Mechanical connection | | |
Pipe size, in.	Pipe laid, ft/hr	Skilled laborer	Common laborers	Crane operator	Pipe laid, ft/hr	Skilled laborer	Common laborers	Crane operator
4	50	2.0	4.0	3.0	25	4.0	8.0	6.0
6	41	2.4	4.8	3.6	20	5.0	10.0	7.5
8	25	4.0	8.0	6.0	16	6.3	12.6	9.4
10	22	4.5	9.0	6.7	14	7.1	14.2	10.6
12	20	5.0	10.0	7.5	12	8.3	16.6	12.4
14	16	6.2	12.4	9.3	10	10.0	20.0	15.0
16	14	7.1	14.2	10.6	9	11.1	22.2	16.7
18	12	8.3	16.6	12.4	8	12.5	25.0	18.8
20	11	9.0	18.0	13.5	7	14.3	28.6	21.4
24	10	10.0	20.0	15.0	6	16.7	33.4	25.1

Reinforced Concrete Pipe (RCP)

Reinforced concrete pipe (RCP) for large water transmission lines is available in diameter sizes from 30 in. to 144 in. Several methods are used to manufacture the pipe, which may be prestressed or non-prestressed.

One type is made with a cage of reinforcing steel, consisting of both circumferential and longitudinal steel bars, embedded in a concrete wall. Another type is made with a steel cylinder, lined in the middle with concrete, reinforced on the outside with steel bars or wire, which may or may not be prestressed with the cylinder, and reinforcing encasement in a concrete jacket.

Joints are made watertight with one or two rubber gaskets that completely encircle the spigot where it enters the bell of the adjacent pipe. Fittings, which permit changes in direction and sizes, are available or may be manufactured upon order.

TABLE 20.6 | Dimensions of reinforced concrete pipe (RCP), with concrete lining, AWWA C300.

Inside dia., in.	Minimum thickness	
	Total pipe wall in.	Concrete lining in.
30	3.5	1.0
36	4.0	1.0
42	4.5	1.0
48	5.0	1.25
54	5.5	1.25
60	6.0	1.25
66	6.5	1.5
72	7.0	1.5
78	7.5	1.5
84	8.0	1.5
90	8.0	1.5
96	8.5	1.75
102	8.5	1.75
108	9.0	1.75
114	9.5	1.75
120	10.0	1.75
132	11.0	1.75
144	12.0	1.75

AWWA C300 is the standard for reinforced concrete, steel-cylinder type, pressure pipe. RCP is manufactured to resist internal pressure, external loads, and bedding conditions. The pipe has a welded steel cylinder with steel joint rings welded to its ends and a reinforcing cage of steel bars, wire, or welded wire reinforcement. A layer of concrete encases the steel cylinder and reinforcing cage outside the steel cylinder. A joint with a preformed gasket of rubber is designed for a watertight connection. Table 20.6 gives dimensions of reinforced concrete pipe.

VALVES

Valves for cast-iron water pipes are usually cast-iron body, bronze-mounted, bell or hub type. Gate valves should be used. A cast-iron adjustable-length valve box should be installed over each wrench-operated valve to permit easy access when it is necessary to operate the valve.

SERVICE LINES

Service lines are installed from the water pipes to furnish water to the customers. These lines are tapped into the water pipe and extended from the water line to a meter adjacent to the customer's property. Customers install the service pipe from the meter to their house or business.

FIRE HYDRANTS

Fire hydrants are specified by the type of construction, size of valve, sizes and number of hose connections, size of hub for connection to the water pipe, and depth of bury. It is good practice to install a gate valve between each hydrant and the main water pipe, so that the water can be shut off in the event repairs to the hydrant are necessary.

TESTS OF WATER PIPES

Specifications usually require the contractor to subject the water pipe to a hydrostatic test after it has been laid, prior to backfilling the trenches. If any joints show leakage, they must be approprieately sealed. It is common practice to lay several blocks of pipe, install a valve temporarily, and subject the section to a test. If a test satisfies the specifications, the valve is removed and the trench is immediately backfilled. Additional lengths are laid and tested. This procedure is repeated until the system is completed.

STERILIZATION OF WATER PIPES

Prior to placing a water distribution system in service, it should be thoroughly sterilized. Chlorine is most frequently used to sterilize water pipes. It should be fed continuously into the water that is used to flush the pipe lines. After the pipes are filled with chlorinated water, the water is permitted to remain in the pipes for the specified time; then it is drained, and the pipes are flushed and placed in service.

LABOR REQUIRED TO LAY WATER PIPE

Installation of bell-and-spigot water pipe will include some or all of the following operations:

1. Excavating the trench
2. Cutting the pipe, if necessary
3. Lowering the pipe into the trench
4. Inserting the spigot into the bell
5. Installing fittings to connect service lines, if required
6. Pressure testing and sterilizing the pipe
7. Backfilling and compacting the trench

Each joint of pipe is lowered into the trench by hand for plastic pipe, or a tractor-mounted side boom for ductile-iron or reinforced concrete pipe. The pipe is aligned and the spigot end is forced into the bell to full depth. A rubberized O-ring gasket seals the pipe for a watertight connection. The size of crew required to lay the pipe and the rate of laying will vary considerably, with the following factors:

1. Class of soil
2. Extent of groundwater present
3. Depth of trench
4. Extent of trench shoring required
5. Extent of obstruction, such as utilities, sidewalks, pavements
6. Type and size of pipe
7. Type of pipe connection, push-on or mechanical-joint
8. Method of lowering pipe into the trench
9. Extent of cutting required for fittings and valves

A crew should install about four to six joints per hour, either pipe or fitting joints. The length of pipe laid will vary depending on the size and type of pipe. For plastic pipe the rate of laying pipe is about 15 to 45 ft/hr, depending on size of pipe. The length of pipe laid for ductile-iron pipe will vary from 10 to 50 ft/hr for push-on connections and 6 to 25 ft/hr for mechanical connections.

COST OF A WATER DISTRIBUTION SYSTEM

In estimating the cost of installing a water distribution system, the estimator must consider the many variables that will influence the cost of the project. No two projects are alike. For one project, there may be very favorable conditions, such as a relatively level terrain, free of trees and vegetation, out in the open with no obstructions, no rocks, no groundwater, no utility pipes, and no pavement, and little rain to delay the project. The equipment may be in good physical condition. The construction gang may be well organized and experienced. The specifications may not require rigid exactness in construction methods, tests, and cleanup.

For another project the conditions may be entirely different, with rough terrain, restricted working room (as in alleys), considerable rock, pavement, and unmarked utility pipes to contend with, as well as groundwater and rain. The equipment may be in poor physical condition, and the construction gang may be poorly organized, with inexperienced workers. The specifications may be very rigid regarding construction methods, tests, and cleanup. As a result of the effect of these variable factors, an estimator should be very careful about using cost data from one project as the basis of estimating the probable cost of another project, especially if the conditions are appreciably different.

The following example is intended to illustrate a method of estimating the cost of a water distribution system, but it should not be used as the basis of preparing an estimate without appropriate modifications to fit the particular project.

EXAMPLE 20.1

Estimate the direct cost of installing a 3,870-ft-long water distribution system. The pipe will be a 12-in. PVC water pressure pipe with push-on connections. Two 12-in. mechanical joint gate valves will be required.

The trench for the pipe will be dug in hard clay at an average depth of 42 in. and width of 30 in. using a ladder type trenching machine. The cost of the trencher, with operator, is $85.00/hr and the average digging speed of the trenching machine is 32 lin ft per hour.

After the pipe is laid, the backfill over the pipe will be placed by a dozer. The cost of the dozer, with operator, is $75.00/hr. The time for dozer backfilling will be approximately $\frac{1}{4}$ of the time to excavate the trench.

There will be 65 connections for domestic water service lines that will be attached to the 12-in. PVC pipe. The trenches for the water line will be dug by a small trenching machine that will operate at an average speed of 90 ft/hr. The trencher, with operator, will cost $55.00/hr. A blade on the front of the trencher will backfill each service line in about 0.5 hr. The pipe and fittings for each service line will be:

   30 ft of 1-in. PVC pipe

   1 mechanical joint tee

   1 meter box with lid

   1 water meter

Cost of water distribution line:

Pipe material and valves:
   Pipe delivered to the job: 3,870 ft @ $11.25/ft   = $43,537.50
   Mechanical joint gate valves: 2 valves @ $1,237.00   =   2,474.00

Excavating trench:
   Time for trenching: 3,870 ft/32 ft/hr = 121 hr
   Machine excavating trench: 121 hr @ $85.00/hr   =   10,285.00

Installing pipe, reference Table 20.2:
   Skilled laborer: 3,870 ft × (3.6 hr/100 ft) @ $29.00/hr   =   4,040.28
   Common laborers: 3,870 ft × (7.2 hr/100 ft) @ $25.00/hr =   6,966.00

Installing valves, reference Table 20.2:
   Skilled laborers: 2 valves × 0.27 hr @ $29.00/hr   =   15.66
   Common laborers: 2 valves × 0.27 hr @ $25.00/hr   =   13.50

Testing pipe and backfilling over pipe:
   Testing and sterilizing pipe: $1,400   =   1,400.00
   Dozer backfilling trench: ($\frac{1}{4}$ × 121 hr) @ $75.00/hr   =   2,268.75
                              Total direct cost = $71,000.69

   Direct cost per ft of water distribution line: $71,000.69/3,870 ft = $18.35/ft

Cost of 65 domestic service lines:

Pipe material, fittings, and meters:

Pipe for service lines: 65 × 30 ft = 1,950 ft @ $1.19/ft	=	$2,320.50
Mechanical joint tees: 65 @ $18.50	=	1,202.50
Meter box with lid: 65 boxes @ $35.86	=	2,330.90
Water meter (automatic meter read): 65 × $114.78	=	7,460.70

Excavating service lines:

Total length of service lines: 65 lines × 30 ft = 1,950 ft

Trenching machine: (1,950 ft)/(90 ft/hr) @ $55.00/hr = 1,191.67

Installing 1-in. dia. service line PVC pipe:

Skilled labor: 1,950 ft × (2.1 hr/100 ft) @ $29.00/hr = 1,187.55

Common laborers: 1,950 ft × (4.2 hr/100 ft) @ $25.00/hr = 2,047.50

Installing tees for service connection:

Skilled labor: 65 × 0.27 hr @ $29.00/hr = 508.95

Common laborers: 65 hr × 0.27 hr @ $25.00/hr = 438.75

Installing meter boxes with lid:

Skilled labor: 65 × 0.08 hr @ $29.00/hr = 150.08

Common laborers: 65 × 0.08 hr @ $25.00/hr = 130.00

Installing water meters:

Skilled labor: 65 × 0.08 hr @ $29.00/hr = 150.08

Common labor: 65 × 0.08 hr @ $25.00/hr = 130.00

Backfilling service lines:

Small trencher with blade: 65 × 0.5 hr @ $55.00/hr = 1,787.50

Total direct cost of service lines = $21,036.68

Direct cost per ft of service line: $21,036.68/1,950 ft = $10.79/ft

Summary of direct costs:

Distribution pipe = $71,000.69

Service lines = 21,036.68

Total direct cost = $92,037.37

The cost does not include job overhead and profit.

HORIZONTAL DIRECTIONAL DRILLING

Horizontal directional drilling (HDD) is frequently used for installation of water lines. It is also used for installation of gas lines and underground cables for electricity, phone, and fiber optic cables. In recent years, HDD has been used for gravity pipelines and, with increased accuracy of locating and tracking equipment and improvement in methods, this trend is growing. The term *directional drilling* is used to describe the unique ability to track the location of the drill bit and steer it during the drilling process. The accuracy of installation depends on the survey system that is used and the skill of the operator. Generally the accuracy is within 1 percent of the length.

Directional or horizontal directional drilling methods involve steerable tunneling systems for both small- and large-diameter lines. HDD usually is a two-stage process. The first stage consists of drilling a small-diameter pilot hole along the desired centerline of a proposed line. The second stage consists of enlarging the pilot hole to the desired diameter to accommodate the utility line and pulling the utility line through the enlarged hole.

All directional methods consist of a drilling unit to form the borehole and a survey system to locate the drill head. The drilling process is accomplished either by mechanical cutting using a drill bit or by fluid cutting with high-pressure jets, or a combination of both. There are a variety of survey systems that have been patented by different manufacturers. The choice of a particular system will largely depend on the type of job, the site conditions and accessibility, operator skill, finances available, etc.

Unlike microtunneling, presented in Chapter 19 for sewer construction, horizontal directional drilling usually does not require entrance shafts or receiving shafts. However, it is advisable to dig a small pit at the entrance and receiving locations to collect drilling fluids. The working area should be reasonably level, firm, and suitable for movement of the HDD machine. At the receiving end, sufficient room is required so the complete string of pipeline can be fitted together and aligned with the proposed borehole. Figure 20.1 shows a horizontal direction drilling machine.

Classifications of HDD

The HDD methods can be divided into four categories: micro-HDD, mini-HDD, medium- or midi-HDD, and large- or maxi-HDD. For comparison, micro-HDD applies to drive lengths less than 350 ft and pipe diameters up to 4 in. Mini-HDD is typically restricted to drive lengths less than 1,000 ft and pipe diameters less than 12 in. Micro- and mini-HDD are used primarily in the utility industry, for shallow-depth installation of underground cables, gas and water lines, and sewer laterals. Midi- and maxi-HDD are used for pressure pipelines.

Due to limited tracking capabilities of the smaller HDD systems, depths of installations are limited to less than 100 ft. In contrast, the maxi-HDD systems have elaborate guidance systems that can install pipes to almost any depth. The mini-HDD systems can be launched from a curb, whereas the maxi-HDD systems require substantial set-up space, 150 by 250 ft. Table 20.7 provides a comparison of the different HDD systems for common applications.

FIGURE 20.1 | Horizontal directional drilling machine.

Courtesy: The Charles Machine Works, manufacturer of DitchWitch equipment.

TABLE 20.7 | Comparison of horizontal directional drilling systems.

System	Product pipe diameter, in.	Utility depth, ft	Bore length, ft	Typical applications
Micro-HDD	2–4	2–20	200	Cables, gas, water, sewer
Mini-HDD	2–12	3–50	700	Cables, pressure pipelines
Midi-HDD	6–24	3–100	1,500	Pressure pipelines
Maxi-HDD	24–63	6–200	5,000	Pressure pipelines

PROCEDURE FOR HORIZONTAL DIRECTIONAL DRILLING

The directional drilling operation involves several operations: pilot hole, pre-ream, and pullback (see Fig. 20.2). The first stage involves drilling a pilot hole, from one side of the obstacle to the other, along the design centerline of the proposed pipeline. The second stage involves enlarging the pilot hole to the desired diameter to accommodate the product pipe. Usually, enlarging the borehole and pulling back the product pipe can be accomplished in one step. The pilot hole is drilled with a specially built rig with a setup angle for pushing the drill rods into the ground. The setup angle is typically adjusted between 5 and 20 degrees with respect to the ground surface.

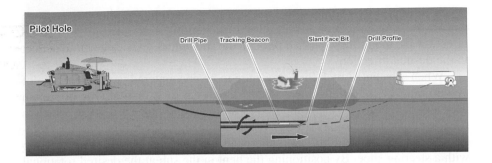

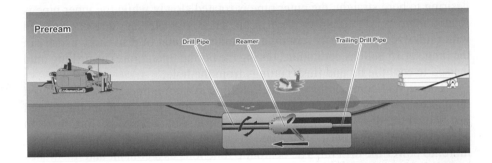

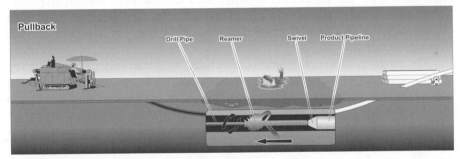

FIGURE 20.2 | Techniques of horizontal directional drilling.

Courtesy: The Charles Machine Works, manufacturer of DitchWitch equipment.

The optimum setup angle of entry of the pilot drill pipe is determined by the desired depth of the installed product pipe and the maximum bend radius of the drill pipe. The pilot hole continues at this setup angle until it passes through a sag-bend to level the drill rods horizontally under the obstacle at a desired depth. Once the hole has passed the obstacle, it rises through another sag-bend to exit on the far side at an angle of approximately 8 to 12 degrees with respect to the ground surface.

As the pilot hole is being drilled, bentonite drilling mud is pumped down the center of the drill rods. For soils, gravel, and light rock the drill head usually consists of a slanted type drill bit with nozzles for passage of the drilling mud.

When desiring to go straight, the soil is cut by the rotating pilot string. Then when a steering correction is needed, the slanted bit is oriented to the desired direction and then simply pushed through the soil, which causes the bit to deflect in the desired direction.

When substantial rock formations are to be encountered, a different type of system must be used to drill and steer through the rock. A first type of system uses a two pipe drill string composed of an inner and outer drill pipe, which is rotationally powered by a drill rig. When drilling, the inner pipe provides rotational power to the drill bit, while the outer pipe is used to position a bent sub with a steering shoe. By positioning the bent in the sub at the desired rotational position, the HDD operator can change the direction of the borehole. When desiring to bore straight, the outer pipe is rotated slowly while the inner pipe is rotated more rapidly to power the drill bit. These two pipe systems can be used with midi size rigs and only require a minimum amount of drill fluid, usually in the range of 15 gallons per minute, during the boring operation. The drilling mud is used to cool the bit and flush cuttings from the hole.

The second common type of rock drilling system is more commonly used on maxi-HDD rigs, where the bit is driven by a down-hole mud motor located just behind the drill bit from energy derived from the pumped drilling fluid. In these rigs, mud flows must be supplied in the range of 200 gallons per minute or more, to power the drill bit. The bentonite functions as a coolant and facilitates spoil removal. In addition to the drill head, for large-diameter crossings such as maxi-HDD rigs, the drill string consists of a slightly bent section (typically between 0.5 and 1.5 degrees) of a drill rod, called a bent housing, positioned close behind the drill head. Rotating and positioning the bent housing affect the steering operation.

The progress of the pilot hole is monitored by a specially designed survey system. One part of the system (usually for maxi-HDD operations it is located behind the bent housing) records the exact position, inclination, and orientation of the drill head. This information is transmitted by magnetic fields, or for larger HDD operations by a wireline system, to the other part of the system, which is located at the ground surface where a computer is used to interpret and plot this data. The actual position of the drill head is then compared to the required position on the design path and any deviations are corrected by moving the bent housing and steering the drill head to the desired location.

For mini-HDD and most midi-HDD systems, a walkover system is most commonly used, where the drill head is located through radio signals by a transmitter that is located behind the drill head. These signals are identified and interpreted on the surface by a receiving instrument, which is actually "walked over" the drill head location.

For larger diameters, the operator may make several passes to achieve the final diameter of the borehole. For each pass, a larger diameter reamer tool, called a "step reamer," is used. Alternatively, reaming devices can be directly attached to the drill pipe and pulled back through the pilot hole, enlarging it to the desired diameter suitable to accept the designed product pipe.

For diameters less than 20 in., the pipeline can sometimes be attached directly behind the reamer using a swivel device so the total assembly can be pulled back in one pass. However, even for diameters less than 20 in. many contractors choose to pream the borehole. Prereaming widens the pilot hole to a diameter slightly greater than the utility diameter. A circular cutting tool, attached to the drill pipe end, is then rotated by the drilling rig, simultaneously pulling it back along the drilled pilot hole. Bentonite is pumped down the drill pipe to carry the soil cuttings to the surface.

Prior to the pipeline pullback operation, the pipeline has to be made up into one full length. For steel pipe, which is usually used for larger diameters, it is welded, X-rayed, coated, and tested before installation. It is positioned on rollers in line with the drilled hole to minimize any axial loads that will be imposed on the line as it is being pulled, and to protect it from excessive stresses. The pipeline is then attached to the drill pipe with adapters. A fly cutter, or other type reamer, and swivel are installed between the drill pipe and the pipeline to increase the borehole to the desired diameter, to smooth the hole, and to ensure that the rotating action of the drill pipe is not transmitted to the pipeline. The pipeline is then pulled along the drilled path and installed in position. When polyethylene (PE) pipe is used, it is butt-fused using fusion devices.

Usually steel pipe is used for large diameters, whereas for small diameters PVC and more frequently polyethylene (PE) pipe is used for horizontal directional drilling. It is possible to install multiple lines (called a bundle) of small-diameter pipe in a single pull, such as a line crossing 2,800 ft with five separate lines with diameters ranging from 6 to 16 in. The installation procedure for a bundle is the same as a single unit along the prereamed profile.

Clay is an ideal material for directional drilling. However, cohesionless sand and silt generally behave in a fluid manner and typically require specialized additives in the drill mud to keep the sand or silt in suspension in the drilling fluid for a sufficient period for time to be washed out of the borehole. As the grain size increases into gravel, it is harder to maintain the fluid structure and installation may be more difficult.

PRODUCTION RATES

Most mini-HDD jobs are completed in a day or even a few hours. It is common to complete a 200-ft job in a few hours. For midi-HDD projects, and for lengths of 1,000 ft or longer, the job may take 2 to 5 days, including mobilization and setup of the rig and preparation of the product pipe. For many river crossings using maxi-HDD, the project may take up to several months to complete.

It should be noted that the duration of an HDD project is site-specific. The duration is dependent on many factors, such as experience of the operator and the contractor, project surface and subsurface conditions, equipment used, and the degree of planning and supervision of the work.

CHAPTER 21

Total Cost of Engineering Projects

The previous chapters of this book have discussed methods of estimating the cost of construction engineering projects. However, the cost of construction is not the only cost that the owner of the project must pay. The total cost to the owner may include, but is not necessarily limited to, these items:

1. Land, right-of-way, easements
2. Legal expense
3. Bond expense, or cost of obtaining money to finance the project
4. Permit expense
5. Bringing off-site utilities to project
6. Engineering and/or architects' expense
7. Cost of construction
8. Interest on money during construction
9. Contingencies

COST OF LAND, RIGHT-OF-WAY, AND EASEMENTS

If it is necessary to purchase land or obtain rights to use land in constructing an engineering project, the owner of the project must provide the money to finance these acquisitions. The land on which a project is to be constructed may be purchased by the project owner. The cost is negotiated between the land owner and the project owner.

If the project includes the construction of highways, pipelines, electrical power lines, underground cables, or other items that extend over a long distance, the owner of the project generally obtains a continuing right to construct and

446

maintain a facility through the property without actually purchasing it. This right is defined as an easement, for which the owner of the project must pay the owner of the land.

The purchase of land and/or easements can involve a considerable amount of time and legal expense, particularly if the land owner does not want the project constructed on their property. It is sometimes necessary to go through a condemnation process, where the project owner must go through hearings and the legal court system in order to gain access to the land.

The cost of acquiring land and/or easements should be included in the total cost of the project.

LEGAL EXPENSES

Owners in both the public and private sectors are involved in many business and legal issues related to engineering and construction projects. The construction of a project frequently involves actions and services that require the employment or use of an attorney. The actions requiring an attorney may be the acquisition of land and easements, establishing procedures for holding a bond election for a government agency, or representing the owner in the process of securing private money to finance the project.

Engineering projects often require the services of an attorney to ensure compliance with regulations related to health, safety, and environmental issues. An environmental assessment of a project can require extensive involvement of legal services, to review documents and make presentations to agencies that are responsible for enforcing governmental regulations.

The legal fees paid for these services should be included in the owner's total cost of the project.

BOND EXPENSE

Before a project can be constructed by a government agency, it is often necessary to hold a bond election for qualified voters to approve or reject the project. In the event voters approve the project, it is necessary to print, register, and sell the bonds, usually through a qualified underwriting broker, who charges a fee for these services. Private corporations frequently sell bonds to finance new construction. In any event, it is necessary to charge the cost of these services to the project.

PERMIT EXPENSES

City governments usually require building permits before construction can be started. The issue of a building permit starts the process of inspection of the project by city inspectors during various stages of construction.

The cost of building permits is established by the city in which the work is to be performed. Most cities base the cost of the permit on the total cost of the work.

As a general rule a building permit will cost from 0.10 to 0.25 of one percent of the total cost of the building. It is common practice for the construction contractor to secure the permit for a project and charge the cost to the owner.

BRINGING OFF-SITE UTILITIES TO A PROJECT

Some engineering projects are located in isolated areas where there are no utility services. For these types of projects it is necessary to bring off-site utilities to the project site. The utilities may involve water, sewer, gas, or electrical services. The responsibility for bringing utilities to the project is performed by the utility provider, who charges the cost for their work to the project owner.

The cost of bringing utilities to the project site should be included in the total cost to the project owner.

ENGINEERING EXPENSE

Prior to construction the owner engages an engineer or architect to make necessary surveys and studies, prepare the plans and specifications, and assemble the contract documents for bidding the work. Sometimes the engineer will assist in securing bids for construction. It is also common for the engineer to act as the owner's representative during construction, including inspection of the contractor's work, evaluation of pay requests from contractors, and evaluating change orders for approval by the owner. There are various methods of paying engineering services, including:

1. Lump sum
2. Salary times a multiplier
3. Cost plus a fixed fee
4. Percentage of construction

For simple or small projects, the cost of engineering design may be based on a single lump-sum payment. This method applies only when the work to be performed is well defined and the project has no unusual requirements.

The salary times a multiplier method is used for complex projects or for projects that do not have a well-defined scope of work. The designer provides a fee schedule to the owner that lists the classifications and salary costs of all personnel, and a rate schedule for all other costs that will be directly charged for engineering services. A multiplier, usually within a range from 2.0 to 3.0, is applied to direct salary costs to compensate the design organization for overhead, plus a reasonable margin for contingencies and profit.

The cost plus a fixed fee payment method is often used when an engineering design is required to start before the cost and scope of the project can be accurately determined. Such indeterminate projects generally occur when the owner wants to accelerate the design process, analyze special problems that require studies, or prepare estimates for alternate types of construction. The fixed fee usually varies from 10 percent for large projects to 25 percent for small projects that are short in duration.

The percentage of construction cost method may be used for certain types of engineering design, where design procedures, methods of construction, and types of material are well known and relatively standard. This methods applies only where engineering experience has established a correlation between engineering costs and construction costs. Generally the percentage is a sliding scale that decreases as the construction cost increases. The percentage also varies depending on the level of design services that are provided, such as design only; design and preparation of drawings; or full design services that include design, preparation of drawings, and observation during construction. The percentage generally will range from 5 to 12 percent of the anticipated construction costs.

COST OF CONSTRUCTION

The cost of construction is the largest cost item in the total cost of an engineering project. The cost of constructing a project is usually an estimate only, made in advance of receiving bids from construction contractors. Typically the estimate is prepared by the design professional, either an engineer for heavy/industrial projects or an architect for commercial building projects. This cost estimate is the amount of money the design professional believes the owner will have to pay for construction of the project. For building construction projects the estimate is usually a lump-sum cost.

For heavy/highway projects a unit-price estimate is prepared because the quantities of the various work items are not exactly known in advance of construction. For example, the contract for constructing a highway may specify a given payment to the contractor for each ton of asphaltic material placed in the pavement. The bid form provides a format for bidders to state the amounts that they will charge for constructing each specified unit of work.

INTEREST DURING CONSTRUCTION

Most construction contracts have a provision for the owner to make a payment to the contractor on a monthly basis. At the end of each month the contractor will submit a pay request to the owner for work completed during the month. Also, the contract typically has a provision that the owner will withhold a portion of the monthly pay request, usually 10 to 15 percent. This withheld money is called retainage. The cumulative monthly retainage of money is paid to the contractor after the contractor completes all of the work. In addition to the amounts of money the owner must pay to the contractor, the owner will have to pay other costs prior to completion of the project. Thus, the owner will have considerable money invested in the project while it is under construction. Since the owner must pay interest on money required to finance these costs, there will be an interest cost during construction, which should be included in the total cost of the project.

The amount of interest chargeable to the project during construction is usually estimated when the total cost of the project is estimated. A method sometimes used is to assume that one-half of the cost of construction will require payment of interest during the entire time of construction. However, if the full cost of the project is secured in advance of beginning construction, and if interest is paid on the full amount during this period, then the total cost of interest should be included in the cost of the project.

CONTINGENCY

The owner's total cost of a project includes expenses for conducting feasibility and economical studies, acquiring land and easements, surveying and engineering, construction contractors, procuring owner supplied permanent material and equipment, administration of the contract during construction, and final inspection and approval of the project. Thus, at the very beginning of any project there are many unknowns for the owner.

Although each project is unique, the cost of the construction contractor is generally the largest expense, compared to engineering, legal, and other expenses. For most projects construction represents about 80 percent of the total cost. Construction contractors apply a contingency to the cost of the work they quote to the owner because they want to cover risks that are involved in doing their work. Owners also need to apply a contingency to their budgets because they are exposed to risks other than just construction. Determining the amount of contingency is more difficult for owners than contractors because owners are involved in more phases of the project, such as feasibility studies, design alternatives during engineering, and the final phases of project closeout.

In the early stage of any project the exact cost of the project is not known in advance of raising funds to finance the project. Thus, the estimator must rely on personal experience and judgment to determine the appropriate amount of contingency.

EXAMPLE ESTIMATE FOR TOTAL COST OF AN ENGINEERING PROJECT

Table 21.1 is an example illustrating an estimate for the total cost of a project involving drilling three water wells and furnishing all materials, labor, and equipment required to provide additional sources of water for a city. The project will require acquiring land and easements, drilling wells, and installing pumps and water pipes, fittings, and valves to bring the water to the city.

A bond election will be held to provide the money to finance the total cost of the project. The estimated total cost is determined as shown in the table. The $1,128,585 in Table 21.1 is the minimum amount of money that should be included in a bond election for the project.

TABLE 21.1 | Example of total project cost.

Item		Estimated cost
1. Land and easements	=	$86,000
2. Legal expense	=	12,000
3. Bond expense	=	8,000
	Subtotal =	$106,000
4. Cost of construction		
a. Water wells with pumps: 3 @ $146,000	=	$438,000
b. Pump houses: 3 @ $7,500	=	22,500
c. Bring electrical power to wells: 9,420 ft @ $14.50	=	136,590
d. Ductile iron pipe, 8,780 lin ft @ $28.50	=	250,230
e. Gate valves: 4 @ $2,100	=	8,400
	Total cost of construction =	$855,720
5. Engineering expense: 7% × $855,720	=	59,900
6. Interest expense during construction, estimated to be		
10 months @ 6%/year: $855,720/2 × 0.06 × 10/12	=	21,393
7. Contingency: 10% × $855,720	=	85,572
8. Estimated total cost	=	$1,128,585

Computer Estimating

INTRODUCTION

The preceding chapters of this book presented the principles and concepts of estimating construction costs. Emphasis was placed on the thought process that is required of the estimator to analyze job conditions and access the required labor, equipment, and method of construction that will be necessary to perform the work. These are functions that can only be performed by the estimator who must exercise good judgment to pick from the myriad of options available.

To assemble a complete estimate for bid purposes, the estimator must combine knowledge of construction methods and techniques into an orderly process of calculating and summarizing the cost of a project. This process requires the assembly of large amounts of information in an organized manner and numerous calculations. The computer is an effective tool to facilitate this process by decreasing the time and increasing the accuracy of preparing cost estimates. Computer estimating enables the estimator to give more attention to alternative construction methods, assess labor and equipment productivity, obtain prices from subcontractors and material suppliers, and focus on bidding strategies.

This chapter introduces computer methods for estimating construction costs. Many commercial software programs are available. In this book, reference is made to the student version of the computer estimating software program, HeavyBid, published by Heavy Construction Systems Specialists, Inc. (HCSS) in 2012. The HeavyBid/Student version of HCSS software can be downloaded from the McGraw-Hill website (www.mhhe.com/peurifoy_oberlender6e) and used with this chapter.

IMPORTANCE OF THE ESTIMATOR

Although the computer is an effective tool for estimating, the estimator must still control the estimating process. The computer should work for the estimator—the estimator should not work for the computer. Estimates of any complexity tend to have unique problems that must be solved that the software may not have taken into account, in which case the estimator has to determine what to do rather than blindly enter the data into the software. It is important for all estimators and the estimator in charge to know the software being used, both its capabilities and limitations.

The estimator must use common sense and judgment and use the computer as a tool to assist the estimating process. Most estimates have enough complexity that many judgment calls will be required, which normally the estimator must make, not the computer. Once the estimator is relieved of the mundane calculations, the estimator can focus on "what if" analysis, such as the impact on the total cost if the production rate of a particular piece of equipment is changed or if different equipment is used. The estimator can also perform sensitivity analysis to determine which aspects of the bid contain the greatest risk, and thus more intelligently determine the best markup for contingency.

The estimator also plays a vital role as the estimate nears completion. In the final hours of competitive bidding, subcontractors and material suppliers frequently make cuts (reductions) to their previously quoted prices. They intentionally make these last minute cuts so there will be no time for the estimator to search for lower prices from other companies before finalizing the estimate into a bid. The estimator must effectively manage this exchange of information to track dozens or hundreds of quotes from subcontractors and suppliers, determine what is missing or unclear, enter the information into the computer, and then compare prices and qualifications to make the best possible choices to reduce risk and be competitive on the bid. Catastrophic errors can be made during this process without organized techniques and experienced estimators.

USE OF COMPUTERS IN ESTIMATING

The computer has many uses for construction estimating. The amount of usage depends on a company's philosophy with respect to computers and automation, the type and size of project for which the estimate is to be prepared, the complexity and level of detail anticipated in the estimate, and the knowledge and computer skills of the estimator. Computers can be used for the following tasks.

1. Maintaining master checklists
2. Maintaining an inventory of subcontractors, vendors, and suppliers
3. Maintaining bidding records of competitors
4. Performing material quantity takeoffs
5. Storing and retrieving historical cost data

6. Storing and retrieving labor and equipment productivity
7. Establishing consistent coding for resources and construction activities
8. Building the cost estimate
9. Summarizing costs at various levels
10. Distributing overhead and indirect costs
11. Analyzing subcontractor and suppliers quotes
12. Analyzing risk and assessing contingency for markup
13. Preparing and delivering the bid or proposal in electronic form
14. Sharing of cost data and information within an office or company

These tasks can be accomplished by a variety of software. Depending on the task, the estimator may use word processors, spreadsheets, or commercial construction software packages. It is common to use all of these types of software to prepare an estimate.

One of the benefits of computer estimating is consistency in the methods of preparing estimates and formatting of the final results. Also, the computer can be used as a master checklist to reduce the potential of omitting a cost item from the estimate, or including it twice.

Computer programs are available to assist in the quantity takeoff by helping the estimator to measure, count, compute, and tabulate quantities of lengths, areas, and volumes. Drawings are typically created in electronic media using computer-aided-design (CAD) software, which enhances the opportunity for the estimator to transfer quantities from drawings.

Since most construction projects are competitively bid, it is an advantage to contractors to have information and knowledge about their subcontractors, vendors, and suppliers as well as other contractors against whom they are bidding. For example, competitor records can be maintained on past jobs showing the amount of the bid, difference between low bidder and next higher bidder, etc. This type of information assists the estimator in determining an appropriate markup while still being competitive. It is also useful as a comparison tool for catching errors in an estimate. An example of this in 2012 is www.bidhistory.com which is a website of historical bid results maintained by HCSS and which contains many years of free bidding history.

One of the greatest benefits of computer estimating is the storage and rapid retrieval of historical cost data and productivity rates of labor and equipment. As illustrated in preceding chapters of this book, there are many ways that construction can be performed and there are many scenarios that can be analyzed in estimating construction costs. The computer's ability to store vast amounts of information and to retrieve the information almost instantaneously allows for examining alternatives that would otherwise not be possible.

The computer is an ideal tool for distribution of non-pay cost items. For unit-price projects, some, but not all, of the quantities that the estimator needs to know will be specified in the contract as the basis for payment. The cost of those items not specified as pay quantities must be prorated into some of the pay items, or prorated over all of the pay items. The computer should give the estimator many choices in distribution of non-pay items. For fixed-price projects, bid pay quantities are not specified, but the distribution of overhead and indirect costs are important in the measure for progress payments.

ELECTRONIC MEDIA

Today, many companies attempt to operate in a paperless environment by using electronic media and the Internet. Engineers produce drawings and specifications using CAD and word processing software. Thus, full sets of bid documents are sometimes supplied entirely in electronic media.

E-mail is often used for communications both inside and outside a company's office. It enables users to send written messages with optional attachments to anyone and everywhere at no cost. Attachments can include drawing files, word processing documents, spreadsheets, photos, multimedia clips, and web pages. Faxes can also be sent over the Internet instead of over regular voice lines. A request for information (RFI), request for quote (RFQ), request for proposal (RFP), letter of intent, transmittal, and similar type correspondence are often sent via electronic media. Many state Departments of Transportation (DOTs) provides biditems and quantities on their websites for access to prospective bidders, and some provide historical bid results. This information can be downloaded to the estimator's computer.

Security of information is always a concern of businesses. Security issues must be addressed with Intranets and Extranets. An Intranet is an Internet site set up for private use of a company that might include certain corporate financial data or other confidential information. The Intranet controls access of company sensitive information. An Extranet is an Internet site set up by a company for shared use by others. It might contain a variety of information with limited or full access to its employees, subcontractors, suppliers, or the general public. This information may be shared using FTP (File Transfer Protocol) or cloud-based apps accessed from desktops or mobile devices.

USING SPREADSHEETS FOR ESTIMATING

Spreadsheet programs are particularly effective in reducing repetitive arithmetic. Estimating was performed primarily by hand until the personal computer and spreadsheet programs were invented in the early 1980s. Enterprising estimators learned how to program spreadsheets to reduce their arithmetic and reduce the drudgery of estimating. However, for most companies spreadsheets are only an interim step between paper and a computer estimating system. Spreadsheets are essentially a duplication of the paper and calculator process that estimators have used for years to prepare estimates. Most estimators simply write down the items in a spreadsheet exactly as they did on paper before the advent of computers. The following example illustrates a traditional paper and calculator method to estimate the cost of excavating 1,500 cubic yards of earth.

Excavate 1,500 cy:

Backhoe, 24 hr @ $75/hr	=	$1,800
Operator, 24 hr @ $30/hr	=	$720
2 Laborers, 24 hr @ $25/hr	=	$1,200
Foreman, 24 hr @ $35	=	$840

Crew costs = $4,560/1,500 cy = $3.04/cy

Figure 22.1 is a spreadsheet of these hand calculations. The estimator only needs to type in the hours and wage rates of each item. Formulas in the spreadsheet extend the prices in the rows and sum the column total to immediately obtain the cost. For example, after the estimator enters the 24 hours in cell B7 for the backhoe and $75 in cell C7 for the backhoe's rate, the spreadsheet formula automatically calculates the $1,800 cost and places it in cell D7. As entries are made for the remaining rows, the computer performs all calculations automatically. This is a huge improvement over the paper and calculator method and only requires minimal typing skills. Not only does it save time in the initial entry and calculation, making changes is very quick.

	A	B	C	D	E	F
1						
2	Excavate					
3	1500 cy			Price		
4				Extensions		
5						
6	**Item**	**Hours**	**Rate**	**Cost**		
7	Backhoe	24	$75	$1,800		
8	Operator	24	$30	$720		
9	2 Laborers	48	$25	$1,200		
10	Foreman	24	$35	$840		
11			Total	**$4,560 / 1,500 cy = $ 3.04/cy**		
12						

FIGURE 22.1 | Spreadsheets are a duplication of the paper and calculator process.

Spreadsheets also enable the estimator to mimic the format and appearance of manual reports that were used before computers. This is an advantage to most construction companies because often managers and owners are more comfortable with computer printouts that look exactly like the handwritten reports they have used for many years.

There will always be a need for spreadsheet programs. Spreadsheets are tools for estimators to develop custom routines that otherwise could not get programmed into their estimating system. Like any tool, it is good to know which tool is appropriate for which job.

DISADVANTAGES OF SPREADSHEETS

Not only can spreadsheets be created to eliminate repetitive arithmetic, but programs like Excel™ are powerful enough to make elaborate estimating systems with many relationships that function as an entire estimating software system. It is not uncommon for companies, particularly smaller ones, to attempt developing spreadsheet systems rather than purchasing commercially developed estimating software. It is extremely time-consuming for an estimator to develop an entire estimating system, yet there are a significant number of companies doing estimates only by spreadsheets. Often it occurs because what starts out as a simple spreadsheet gets added to over time and becomes a complete system. Spreadsheet estimating systems have most of the following disadvantages:

1. The spreadsheet is probably not documented.
2. It is almost certainly not tested like a professional software product.
3. Changes in the spreadsheet over time likely result in no system testing.
4. It is likely the original author is the only person who knows the entire system.
5. The original author may become indispensable and strive to keep that status.
6. If the original author leaves, changes are likely difficult.
7. The spreadsheet may not work well in multi-user environments where several estimators work on the same estimate at the same time.
8. The author of an elaborate spreadsheet system is usually resistant to changing to more efficient processes in the future.

For these reasons, it is usually better for an estimator to devote time to estimating, rather than developing elaborate systems of spreadsheets.

COMMERCIALLY AVAILABLE ESTIMATING SOFTWARE

In the United States, computer estimating software is very specialized. The process of preparing a fixed-price estimate for a high-rise building is quite different than preparing a unit-price bid for a highway project. Thus, specialized estimating software is written for different sectors of the construction industry to reflect those differences. The companies tend to specialize into these areas:

1. Residential construction
2. Building construction
3. Infrastructure (heavy/highway) construction
4. Industrial process plant construction
5. Specific trades, such as electrical and mechanical
6. Takeoff systems

Commercial estimating software in the building sector is designed more for a quantity takeoff and pricing approach to estimating. Building type projects are generally fixed-price lump-sum bids. Much of the estimating effort involves pricing labor and material independent of the crew size, crew mix, and equipment spreads. Also, much of the work is subcontracted to contractors who specialize in a particular type of work. Popular estimating software packages in the general building sector include Sage Timberline Estimating and Interactive Cost Estimating® from Management Computer Controls (MC²).

Infrastructure projects include highways, streets, airports, bridges, tunnels, pipelines, waterways, underground utilities, and similar type projects. Underground utilities include water, sewer, gas, electrical, or communication cables. In general terms, infrastructure projects are built to move cars, planes, people, water, oil, gas, or other products. For telephones or cables, the product moved is audio or video signals.

Typically, infrastructure projects involve large amounts of construction equipment and the work is performed in an outside environment where weather often affects the work. The owner of many infrastructure projects is an agency of the government, including local, county, state, and federal agencies. However, many infrastructure projects are also in the private sector where the owner could be a residential developer needing land clearing, streets and utilities, an oil company needing cross country pipelines, or any business that has a parking lot.

For infrastructure (heavy/highway) construction projects, estimates are developed based on methods of construction and the resources needed to build the project. Analysis and selection of crew sizes and equipment spreads are important in developing the cost estimate for a particular job.

Commercially available estimating programs for infrastructure type projects include HCSS HeavyBid, Construction Link, Hard Dollar, and Bid2Win. Earthwork takeoff often required in infrastructure work is commonly supplied by separate software vendors, which include Agtek, Trimble Pay Dirt®, and Insite SiteWork. The estimating software furnished with this book is HeavyBid/Student whose opening screen is shown in Figure 22.2. This is actually a restricted student version of HCSS's estimating software for medium-size companies.

Usually infrastructure projects are unit-price bids. Each biditem (payitem) in the estimate is defined in the bid documents supplied by the owner. In a unit-price bid, the estimator must take the indirect costs, markup, and other costs not included in the biditems and spread those costs to derive the unit prices of the biditems. Unit-price, bid-oriented estimating software enables the contractor to unbalance, or shift costs, from items that are likely to underrun in quantity to those that may overrun, either to shelter risks or to exploit errors in the quantities. However, conspicuous unbalancing can get the bid disqualified on many jobs. Most programs also provide numerous analysis reports for labor, material, equipment, and subcontract costs, and summary reports to provide a better overall understanding and confidence in the accuracy of the numbers in the bid.

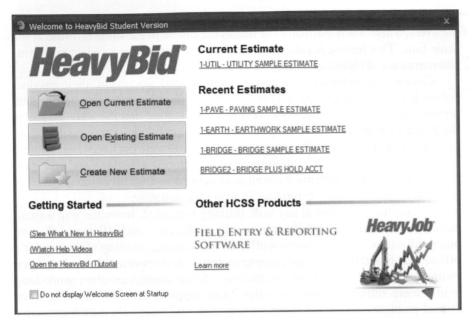

FIGURE 22.2 I Opening screen of the student version of HCSS HeavyBid.

Industrial process plant construction includes oil refining, gas plants, chemical manufacturing, pulp and paper, pharmaceutical manufacturing, and similar type process industry projects. Generally these types of projects involve large capital investments and engineering and construction are more tightly integrated than building or infrastructure projects. Cost estimating often overlaps the estimating design phase. As these projects can be highly specialized per industry, there are many different software packages available.

ADVANTAGES OF COMMERCIAL ESTIMATING SOFTWARE

Writing good-quality software is extremely time consuming and requires a staff of full-time professional software developers who are knowledgeable about computers as well as the tasks the software is to perform. These systems usually contain hundreds of thousands of lines of computer code and thus contain many options to improve estimating productivity.

Historical databases are vital to any computer estimating software. As well as databases of standardized codes and resources, these systems often have databases for past estimates, bidding history of competitors, job cost histories, databases of vendors, multiple labor rate tables, etc. These databases can contain thousands, or hundreds of thousands, of items. Usually a business will need to fill these databases over time with their own data, but some history can be purchased, such as databases from R. S. Means.

Unlike spreadsheets, commercial estimating software is designed for multiple users, where many estimators or managers can be in the same estimate at the same time. This feature is valuable for estimating large projects when numerous estimators are involved in putting a bid together.

Commercial software companies validate and document their product before it is released for sale. Writing software requires extensive testing and retesting to verify it produces correct results. Validating that the software meets the business need and improves productivity is a continuous process based on feedback from customers. Documentation of the software packages is extremely important because inevitably questions will arise during actual use and operation of the software. The systems are well documented and usually contain instruction and tutorials to assist estimators in using the software.

To become proficient at any task, training is needed. Investment in a small amount of training will greatly enhance the proficiency and efficiency of computer utilization. Many companies offer on-site training, training at the vendor's office, web-based training, and training materials and opportunities to help keep estimators up-to-date with the technology. Since estimators often work long hours, some software companies offer 24-hr support, which is essential when critical problems arise while using the software, and especially if problems arise at night before bid day.

MANAGEMENT OF DATA

Vast amounts of data may be used in computer estimating. The data must be created, stored, retrieved, and continually updated. Most data is stored on spinning disks, called hard disks, although solid state storage devices are becoming more popular. This data remains intact even when the power is turned off.

Data that is being temporarily worked on is stored in the computer's memory. These are electrical circuits that maintain their information only while power is being supplied to them. This data is processed at electrical speeds several orders of magnitude faster than data from the hard disks. Typically data is read from hard disks into memory, processed, and then written back to the hard disks.

Spreadsheet programs load an entire spreadsheet, such as an entire estimate, into memory where it can then be manipulated. Any new information added to the spreadsheet is kept in the temporary memory and therefore can be lost until the spreadsheet is saved again. If power is lost to the computer, any unsaved data will be lost. Therefore, the estimator should periodically save the spreadsheet to the hard drive, or have a default set in the spreadsheet program to automatically save the data every few minutes.

Database systems read only small amounts of data from the hard disk into memory and then write back to the hard disk when it is saved. For example, if the estimator calls up a material item, changes the price, and presses a Save or OK key, the data would be written back and safely saved to the hard disk.

Although hard disks are usually safe from power failure, they do occasionally wear out or *crash* for various reasons, including malicious pranks such as computer viruses. There is a substantial chance that eventually after several years of use, a hard drive will fail and the data on it will not be recoverable. Estimators should always plan for this possibility. Therefore, the active estimate on a hard drive must be copied to a separate physical device as a backup copy.

Many companies have a nightly backup performed by their computer department. Although this is a good practice, it is usually not sufficient to ensure against the loss of an estimate. Many estimators work late at night after normal work hours. Therefore, substantial work may be performed on an estimate after the nightly backup, so the work they have done may not get backed up. Also, a team of estimators can do an extensive amount of work on an estimate during the day prior to the nightly backup. Thus, active estimates should be backed up several times during the day, especially the final day or days before an important bid must be submitted. A good process is to have a routine that runs periodically to back up the estimate to the Internet.

If the estimator's personal computer is connected to a network that contains the estimate, the estimate can be backed up to the hard disk of the personal computer. Since the hard drive on a personal computer is a separate physical device from the drive on the main network computer, it constitutes a valid backup. If there is a copy of the estimating software on the stand-alone computer, it would be ready to use if the network failed for any reason. For stand-alone personal computers where the estimate resides on that computer, a backup to a USB device or the Internet is common.

Estimates should be backed up independently. If a company backs up a database while five estimates are in-progress and an estimator needs to restore one of the estimates, it is common in the midst of a crisis to not be able to find someone who knows how to restore one estimate without restoring the other four.

For an important bid, the estimator should keep at least three or four separate backup copies over time. The estimator should not keep backing up, or overwriting the last backup, because a serious error may not be discovered until *after* the latest backup. It is not uncommon for an estimator prior to an important bid to work 20 or 30 hours straight with no sleep. Tired estimators often make mistakes, one of which might be deleting the wrong estimate believing it was an old copy, when it fact it was the latest copy. The estimator in charge should always be prepared for such possibilities. Depending on how a particular backup routine works, it may be necessary to ensure that all estimators are out of the estimate before backing it up.

Regardless of where the backup resides, there must also be a copy of the estimating software that is ready to use. If the network goes down and there is a good backup on the Internet, it is of no value unless it can be restored somewhere and used. Therefore, there should be an executable program on a stand-alone PC, an alternate network, or the Internet that is ready to use. This whole procedure must be tested periodically. Unfortunately many companies have found that their backup procedure never worked when they attempted to restore what they thought was backed up.

TYPICAL STEPS IN COMPUTER ESTIMATING

The HeavyBid/Student software contains a tutorial to guide the estimator through the keystrokes of preparing an estimate. The best way to learn any software package is to explore features and try various scenarios to become familiar with its capabilities. This chapter is more concerned with presenting the concepts of computer estimating, rather than attempting to teach the reader how to use a specific software package, although the illustrations in this chapter and the problems at the end require learning a certain amount about the student version of HeavyBid.

The steps in making an infrastructure estimate might be more straightforward than other estimates because the owner typically provides biditems to bid on and has thus structured the bid for the contractor. Here are typical steps for an infrastructure computer estimate.

1. Create an estimate
2. Setup any parameters that apply to the estimate
3. Decide on the labor rate tables that apply to this job
4. Make modifications to crews that are applicable to this job
5. Type in or import the biditems
6. Perform a takeoff to determine quantities
7. Build the estimate, including documentation of anything that is unusual
8. Review the estimate, looking for better techniques, or to catch errors
9. Make changes as a result of one or more reviews
10. Study equipment requirements and make changes, if necessary
11. Spread indirect costs, add-ons, markups, etc., to the bid items
12. Present to management to get comments on techniques, alternatives, desired markup, how much risk they will undertake, and how badly they want the job
13. Take quotes for material and subcontract items
14. Analyze minority requirements, if there are any
15. Finalize the final biditem unit prices, and submit the bid

When learning how a system works for the first time and to check the computations, it can be helpful to make a fictitious estimate with only a few biditems, labor crafts, pieces of equipment, and materials and see how they work through the system without getting side-tracked in thinking about a real construction job. It is also helpful to use rates that are multiples of 10 and quantities that are multiples of 10 so the results can be easily checked without having to use a calculator to see what is happening. This is always a good practice, even for an experienced estimator when trying a new feature of the software for the first time.

STARTING AN ESTIMATE

A new estimate is started in HeavyBid by selecting "Create New Estimate" that is shown in the screen copy of Figure 22.2. The program walks the estimator through the steps, beginning with specifying the estimate code, name, and source of data for creating the estimate, which is shown in Figure 22.3. The normal source for the estimate in HeavyBid is the "Master," which is the directory with \ESTMAST in Figure 22.3, but an estimate can also be created from any other master template or prior estimate.

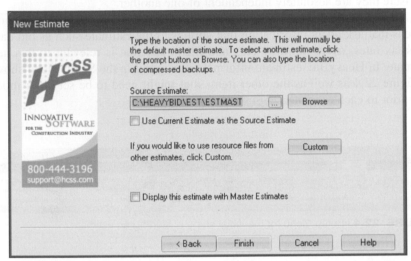

FIGURE 22.3 | Screen showing source options when creating new estimates.

A "Master" is a template estimate and contains data that is used in every estimate. Most software vendors use this technique and most people set up their spreadsheet programs the same way. Very few estimators start an estimate from scratch. Typically the Master contains a company's resources such as labor, equipment, materials, subcontractors, and crews. It may also contain standard overhead items with or without costs, and may even contain standard markups. These are then copied as the starting point for a new estimate. The Master does not contain anything specific to a particular estimate such as biditems, activities, or cost detail except possibly standard indirect items such as the project manager, office, phones, etc., that might be expected in every estimate.

The concept of copying data to start a completely new estimate makes each estimate totally independent of all others. Thus, any changes made to one estimate will have no effect on any other estimate. For example, changing a labor rate in one estimate would not change other estimates that are in progress. If it is desirable to change a labor rate throughout the system, the new rate would be made in the Master and in every estimate that had already been started.

Future estimates that started from the Master would then have the new rate. It is usually not possible to use the same set of rates for all estimates because any particular estimate can have rates imposed by owners, government agencies, unions, or they may be negotiated for any other reason.

Alternately, instead of having a separate rate table for each estimate, it is possible that there are a few tables shared by many estimates. For example, there could be five rate tables that are shared by 50 estimates. This may also work fine as long as a sixth rate table could be created for any estimate that needed different rates. For companies of any size, it could be hard to keep track of shared rate tables. Therefore, it is much easier to have different rate tables for each estimate to ensure they are absolutely independent of one another.

After creating a new estimate, the estimator should look for one or more screens that ask about information that applies to the estimate such as bid date, sales tax rates, overtime rules, etc., and fill in any that is relevant to this new estimate. In HeavyBid, estimate information is found in the Setup Menu shown in Figure 22.4 as well as the other items that might need to be set up or modified prior to each estimate.

FIGURE 22.4 | Setup Menu for the initial steps in starting an estimate.

After setting estimate parameters, it may or may not be necessary to make any adjustment to the resources for the conditions of this estimate. Then it is common to create the biditems provided by the owner, or create them if it is a private job in which the owner is depending on the contractor to lay out the job, or a design-build job where the items are continuously created and modified over time as the project works towards its final design.

BIDITEMS

To prepare a bid in HeavyBid, the structure of the estimate must be set up. Most infrastructure bids are unit-price bids in which the owner gives the contractor a list of biditems with quantities and the estimator furnishes the unit price for each item. Since the owner provides the biditems and quantities, it is logical to enter them into the system all at once. The estimator should find a screen similar to that shown in Figure 22.5 and then type or import the biditems into the estimate.

The import option button has been pressed in Figure 22.5, showing the import options. Note that one of the import options is D.O.T. import. A commercial software package specializing in infrastructure should have the

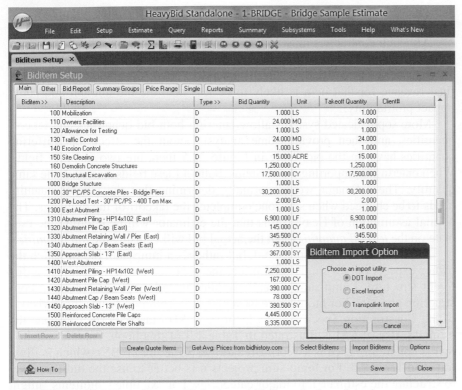

FIGURE 22.5 | Biditem Setup with the Import Biditems button clicked.

ability to import biditems and quantities from major organizations, such as state D.O.T.'s, which usually provide their information on the Internet. In this software, the D.O.T. code would be the client number in the illustration.

Virtually any owner who provides biditems in a bid document will have biditems in an electronic format. If the electronic document can be obtained and put into an Excel format, it should be possible to import it into the estimating software.

At the end of the estimating and bidding process, the amount of the bid is simply the summation of all items, where each item is the product of the owner's quantity multiplied by the estimator's unit price.

Sometimes the contractor may not agree with the quantity that is specified in the owner's bid documents. For example, the bid document may show 10,500 cubic yards of excavation, but the contractor's takeoff may show only 10,100 cubic yards. Thus, even though the contractor has to bid on 10,500 cubic yards, payment would only be expected on 10,100 cubic yards. The 10,100 quantity is referred to as the *"takeoff quantity,"* whereas the 10,500 quantity is referred to as the "bid *quantity*." The bid quantity is what the contractor must bid and what the owner expects to pay. The takeoff quantity is the quantity the contractor expects to be measured during construction and on which payment will be made upon completion.

Although contractors must bid the quantity that is given to them, they must take into account the financial consequences if the takeoff quantity turns out to be the actual quantity. If the takeoff quantity is less than the bid quantity, the final revenue for that item is likely to be less than the bid amount and could impact profitability.

More sophisticated systems allow entering both bid quantities and takeoff quantities, and then provide some analysis about the effect that differing quantities have on the bid. Note in Figure 22.5 there are two quantity fields. When entering or importing the biditems, both of these fields start with the same quantity. The estimator would normally determine that the takeoff quantity needs to be changed when someone completes the takeoff of the drawings. The normal assumption of HeavyBid is that the estimator will be building the estimate on the *takeoff quantities*, but the actual bid will be using the *bid quantities*.

If the bid is not a unit-price bid, but rather is a lump-sum bid in which the owner expects only a single total price, the contractor will need to use a different structure. Lump-sum typically applies to building and industrial construction projects. For building construction it is common to use the divisions of the Construction Specifications Institute (CSI) Master Format to code the estimate. Use of CSI is illustrated in Table 1.3 of this book. Another method that is commonly used is a Work Breakdown Structure (WBS), which is illustrated in Table 1.5. For lump-sum projects it is better to use the biditem level to structure the estimate into major categories. For example, "biditems" might be a division of CSI, such as Earthwork, Foundations, Structural, etc., as shown in Table 1.3, or they might be Building 1, Building 2, etc. Using the WBS, the "biditems" might be Transmission Line A, Transmission Line B, Distribution Line A, Distribution Line B, etc., as illustrated in Table 1.5. Ultimately an estimator working at a company that performs a wide variety of work must structure estimates in all of these formats. Therefore, the estimators must determine a method to structure estimates using their company's estimating software.

Many systems will allow the estimator to structure the bid in multiple levels and still provide the information required on the bid form. For example, HeavyBid allows splitting biditems down into up to eight levels of sub-biditems, none of which appear at the bid-pricing level. Figure 22.6 shows such a structure. A blue colored scroll indicates a bottom level biditem. Any other color means that biditem is broken down further. Before starting a bid with hundreds of biditems with the possibility of thousands of activities or tasks, it can be helpful to experiment with a few items in a test estimate for structure before doing a lot of actual estimating.

The biditem concept is the major difference between infrastructure estimating and other kinds of estimating. HeavyBid structures the entire estimate around biditems. Because biditems are rarely used in other types of estimating, they are not normally central to the design of such software.

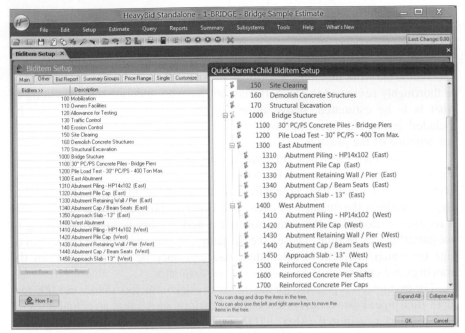

FIGURE 22.6 | Structuring biditems into various levels.

Because many infrastructure bids come from state D.O.T.s, their codes are likely to be used over and over. For example, the D.O.T. code for "clearing" could potentially be in every project the state puts out for bid. This suggests the possibility of building libraries of commonly used biditems with all of the associated resources and productions to accomplish that item. HeavyBid has a feature for building libraries of fully worked-up biditems that can be pulled automatically into future estimates by recognizing when that D.O.T. code is being used. The estimator should always be looking for shortcuts to reduce repetitive work. Some chief estimators, however, purposely don't want to use automated procedures on risky or unusual work because they want their estimators to spend more time on such items and think about what they are doing.

QUANTITY TAKEOFF

The term *quantity takeoff* is sometimes called *material quantity takeoff*, or simply *takeoff*. Takeoff is the process of reading the plans and determining the quantities of work required to build the project. It is the estimator's interpretation of the designer's intent. Takeoff can be performed by hand calculations or by the computer. Area and volume calculations, such as earthwork and paving, are usually more efficiently done with the computer. Structural takeoffs, such as concrete and reinforcing steel, are usually performed by hand calculations or with spreadsheets because the details are often spread over multiple drawings. Although some vendors have software for both takeoff and estimating/bidding, most construction software vendors specialize in one or the other.

The process of performing a takeoff is the estimator's opportunity to fully understand the work that must be estimated. An estimator who has not performed or thoroughly reviewed the takeoff may not have an accurate understanding of what is to be estimated. Items missed in the takeoff will obviously not be included in the bid price, but will still be costs that will be incurred during construction of the projects.

For unit-price estimates, the takeoff quantity may verify the bid quantity, but not include all of the work for which a takeoff is required. For example, an item to install 18-inch PVC pipe will state the length of pipe to be installed, but does not address the quantity of excavated earth or the gravel pipe bedding that is required to complete this item, nor will it address any waste from having to use partial pieces of pipe. These takeoff quantities are ancillary to the length of pipe on which payment is calculated, and can only be determined from the drawings and construction knowledge of the estimator.

For example, it is possible for the owner to ask contractors to bid on 1,000 feet of 18-inch pipe; however, a review of the drawings shows that an actual distance of 1,030 feet is required. Also, 1,100 feet of pipe may be needed because of frequent turns in the route, which required partial pieces of pipe. For this situation the bid quantity is 1,000 feet, and the takeoff quantity is 1,030 feet. The estimate would be prepared based on purchasing 1,100 feet of pipe but excavating and backfilling for 1,030 feet.

RESOURCE TYPES

Resources for construction projects include labor, material, equipment, and subcontractors. For infrastructure type projects, contractors typically categorize costs by the following cost categories:

1. Labor (or labor and labor burden)
2. Material (or permanent material and construction material)
3. Equipment (or company rent, outside rent, and operating expenses)
4. Subcontractors

These cost categories will be columns on reports as shown in Figure 22.7. The computer must use some technique to categorize the costs so it knows which column to use.

HeavyBid uses the following technique to code resources.

    Letters = Labor
        2 = Permanent material
        3 = Construction material
        4 = Subcontractors
        8 = Equipment

Cost for Item 150

Activity Resource	Desc	Pcs	Quantity Unit		Unit Cost	Labor	Perm Material	Constr Matl/Exp	Equip Ment	Sub- Contract	Total
BID ITEM = 150											
Description =	WATER LINE - 4" PVC SCH.40			Unit =	LF Takeoff Quan:		2,450.000			Engr Quan:	2,450.000
15	**4" PVC Pipe & Fittings - Sch.40**			Quan:	2,450.00 LF	Hrs/Shft:	10.00	Cal: 510	WC: TX6217		
2UWP04	4" PVC Pipe - Sch.40		2,450.00 LF		4.500		11,025				11,025
2UWPE04	4" PVC Ell - Sch.40		12.00 EA		18.000		216				216
2UWPT04	4" PVC Tee - Sch.40		22.00 EA		12.500		275				275
							11,516				11,516
							4.70				4.70
557104000	**Install Water Line - 4"**			Quan:	2,450.00 LF	Hrs/Shft:	10.00	Cal: 510	WC: TX6217		
UWM	Water Main Crew		98.00 CH		Prod:	250.0000 US	Lab Pcs:	7.00	Eqp Pcs:		5.00
8EX235	CAT 235 Excavator	1.00	98.00 HR		65.900				6,458		6,458
8L926	CAT 926E Wheel loader	1.00	98.00 HR		22.150				2,171		2,171
8RMX	Ramex	2.00	196.00 HR		10.490				2,056		2,056
8TPU	Pickup truck	1.00	98.00 HR		5.660				555		555
LG	General Laborer	2.00	196.00 MH		11.500	4,045					4,045
OP1	Operator - Crane, Exc	1.00	98.00 MH		20.000	3,229					3,229
OP4	Operator - Dozer, Ldr	1.00	98.00 MH		16.000	2,661					2,661
PFM	Pipe Foreman	1.00	98.00 MH		20.500	3,047					3,047
PL	Pipelayer	2.00	196.00 MH		15.750	5,251					5,251
	0.2800 MH/LF		686.00 MH			18,233			11,240		29,472
9.8000 Shifts	250.0000 Un/Shift	*	3.5714 Unit/MH			7.44			4.59		12.03
====> **Item Totals:**	150	- WATER LINE - 4" PVC SCH.40									
	0.2800 MH/LF		686.00 MH			18,233	11,516		11,240		40,988
	2450 LF					7.44	4.70		4.59		16.73

FIGURE 22.7 | Sample Cost Report showing cost types in columns.

For example, in Figure 22.7, LG (general laborer) is a labor code because it starts with a letter and 2UWP04 is a permanent material because it starts with the number 2. HeavyBid also allows using 1, 5, 6, 7, and 9 for miscellaneous cost types, for example, many companies use 5 for hauling. HeavyBid then provides reports that includes those cost types.

The designation used for resources will vary depending on the software. For example, some software packages use the following designations:

L = Labor

M = Material

E = Equipment

S = Subcontractors

Some software does not require a cost category as part of the code, or may not even use codes at all in an attempt to make the software easier to use. What ultimately matters is whether the estimator can find the necessary information for any resource or any groupings of that information by resource type. Because an estimate is simply made up of many resources entered over and over into an estimate, they are the fundamental building blocks of the estimate.

RESOURCES

Identifying and selecting resources is crucial in estimating infrastructure projects. For example, knowing the quantity of earth to be excavated is not enough information to estimate the cost of excavation. The cost will depend on the size of the crew and the type of equipment that will be used for excavation. Therefore, the estimator must select the crew mix and type of equipment to be used before the cost can be estimated.

Resources, particularly types of equipment, are much more important when estimating infrastructure projects compared to building type projects. Much of the work involved in building construction is independent of the number of workers or type of equipment. For example, knowing the square feet of painting is normally sufficient information to determine the cost of painting. Suppose a building requires 8,000 sf of painting. If the production rate of a painter is 1.5 hr per 100 sf and the hourly rate of a painter is $30/hr, then the estimator can simply multiply 8,000 sf × 1.5 hr per 100 ft × $30/hr = $3,600 to determine the total labor cost of painting. This total labor cost is independent of crew size. The total man-hours for painting is 8,000 sf × 1.5 hr per 100 sf = 120 hr. Thus, if only one painter is used the time for painting is 120 hours. If two painters are used the time to complete the painting would be 60 hours. Thus, the total estimated labor cost is independent of the crew size. The crew size only impacts the time for doing the work, not the total cost. If a gallon of paint will cover 400 sf and the cost of paint is $35/gal, then the estimator can simply multiply the 8,000 sf × 1 gal/400 sf × $35/gal = $700 to determine the total cost of paint material. Therefore, the total cost of painting 8,000 sf can be calculated as $3,600 labor plus $700 materials = $4,300.

A major difference in computer estimating software for infrastructure projects is the need for setting up resources in more detail than building construction software. Not only are there typically more resources such as labor, equipment, material, subcontractors, and crews, but also the treatment of each of these resources is more elaborate. Labor calculations take into account such details as taxes, union fringes, workers compensation rates, and overtime. Equipment can include such details as rental costs and components of operating costs; such as fuel, parts, and repairs.

Most contractors pick from the same pool of resources for every job. Thus, there should be a place in the software to set up company resources that can be used for every job. In HeavyBid it is common to open the "Master" estimate and set up the most common resources under the Setup Menu. This will then be copied to start each new job. However, a company might have more than one master as a time-saving feature to create a job with different characteristics. This master estimate is found by selecting "Open Existing Estimate," selecting the master estimates tab, and selecting the estimate with the code of ESTMAST or any other master as shown in Figure 22.8. The "Master" concept is simply a short-cut to quickly identify the estimates that are typically used for creating new estimates.

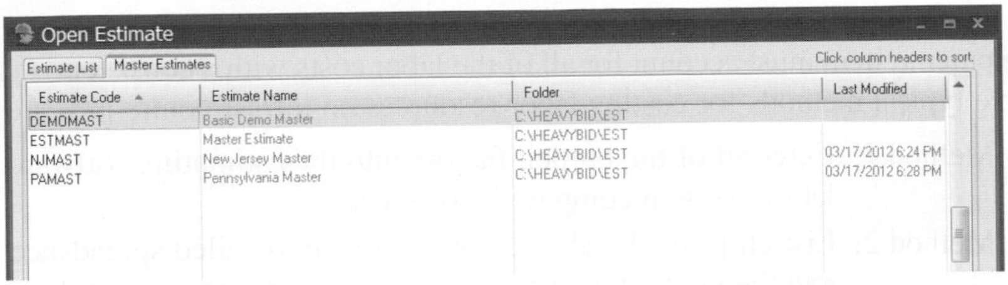

FIGURE 22.8 | Opening "master" estimates to set up reusable resources.

LABOR RESOURCES

Labor is typically estimated by craft, rather than by individual people, because most contractors bid much more work than they expect to win, and therefore do not know how they will be staffing their jobs until after winning a bid.

In computer estimating, labor crafts are usually selected from pop-up lists. These crafts should appear in a manner that enables the estimator to quickly find them. For example, if all operator codes start with the letter "O," then the estimator is able to pick and select all of the operators without scanning through possibly dozens of unrelated labor codes.

Estimates of labor cost can vary from relatively simple for non-union 40 hour/week jobs, to extremely complex union jobs with various combinations of union wage scales, overtime rules, taxes, fringe benefit rules, per diem and shift differentials. Figure 22.9 shows a labor screen from HeavyBid for a job with a medium level of complexity. The drop-down on the fringe field is pressed to show the tax and overtime issues related to labor fringes.

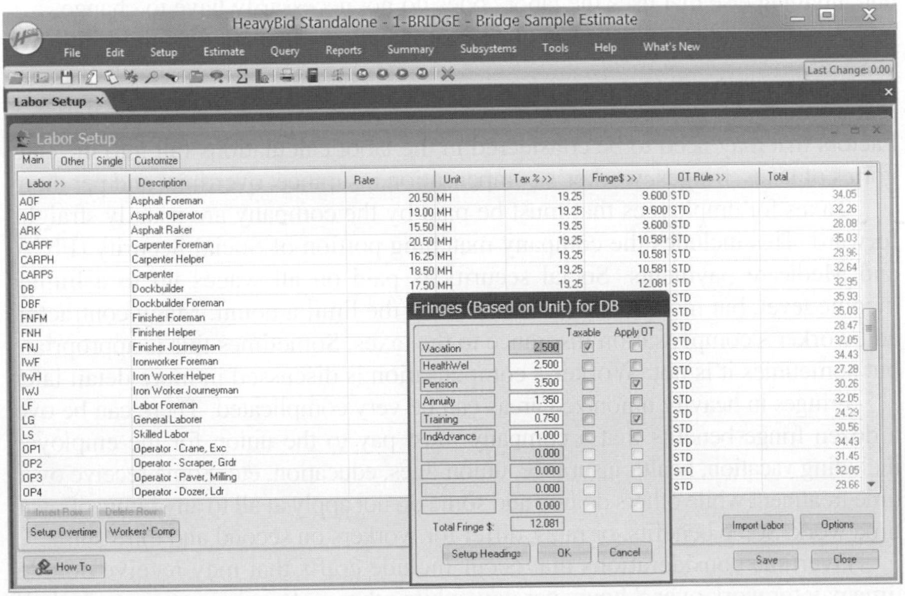

FIGURE 22.9 | Labor setup with fringe drop-down of possible itemizations.

Regardless of the simplicity or complexity of the actual labor calculation, the computer system must account for all of the labor costs with a consistent method. Three typical methods for costing labor resources on infrastructure jobs are:

Method 1: Enter all of the burden factors into the estimating software and let the system compute labor rates

Method 2: List all possible labor combinations in detailed spreadsheets and then enter the average rates that apply to a particular estimate as a simple rate

Method 3: Simply add a historical percentage to the labor base rate to account for taxes, overtime, and other labor burdens

The methods used will depend on the complexity of the labor conditions, the capabilities of the estimating software, the level of accuracy desired by the estimator, and the flexibility desired while estimating.

For example, the screen shown in Figure 22.9 shows a base rate, a total or itemized tax rate, a total or itemized fringe rate, and an overtime rule. For most contractors with straightforward labor computations, this information is sufficient to be able to use Method 1, which is preferable when possible. Method 1 is particularly desirable on a large, complicated union job with many unusual conditions such as evening and weekend work while traffic is shut down. However, a contractor with complex labor conditions who wishes to use Method 1 must have a very sophisticated software package, or else will have to use Method 2.

Regardless of the method used, if possible, the labor codes in the system should remain consistent across estimates. It is much easier to establish estimating standards when the codes remain the same and only the rates change when bidding in different union areas or different states. If the codes do not change, then the crews and anything else that uses the labor codes do not necessarily have to change.

A more detailed explanation of each of the labor calculation methods follows.

Method 1: All Labor Factors

Factors that may need to be considered in the labor calculations fall into the categories of taxes, fringes, worker's compensation insurance, overtime, and per diem.

Taxes for employees that must be paid by the company are usually straightforward. This includes the company matching portion of Social Security (FICA) and Medicare payments. Social security is paid on all wages up to a limited income level, but most workers do not reach the limit amount. Many contractors add worker's compensation insurance to the taxes. Sometimes this is appropriate, but sometimes it is not. Worker's compensation is discussed in more detail later.

Fringes in heavily unionized areas can be very complicated. There can be over a dozen fringe benefits that a company must pay to the union for the employee, including vacation, health insurance, union dues, education, etc. Some receive overtime treatment while others do not, and some do not apply at all to any hours of overtime work. Some benefits, or rates, differ for workers on second and third shifts.

Overtime considerations that occur include crafts that may receive double-time pay for work over 8 hours per day, while other crafts receive time and a half.

There can also be shift differential pay and premium pay for Saturday and Sunday. Some contractors also pay overtime over 8 hours in a day, while others pay overtime over 40 hours per week. These overtime rates may or may not apply to fringes. There may also be bonus time on second and third shifts, which complicates productions because the worker must be paid 8 hours, but may actually only work 7 or 7.5 hours.

Worker's compensation premiums are not calculated the same in all states. Some states use gross wages, others use total wages at the straight time rate (no premium), and others use labor-hours.

Worker's compensation insurance is often a function of the type of work being performed, not the labor craft, because as insurance it is associated with risk. Sometimes the entire job is classified by the major type of work, such as "earthwork" job, and sometimes it is necessary to charge the appropriate insurance rate for each type of work. In the latter, a laborer doing concrete work has a different rate than a laborer helping in steel erection. For simplicity, many companies add the insurance rate to the labor tax, but in the labor example above that would not be accurate unless two different labor codes were set up, one for laborers doing concrete work, and another for laborers doing steel erection. Rates are typically higher for activities performed over water compared to the same activities performed over land.

Method 2: Prepare All Labor Combinations

Since most estimating packages do not handle all of the factors that have been described, some companies with complex union requirements prepare spreadsheets of all possible combinations from which they can select the variations they need for each particular estimate. They are able to program into the spreadsheet all of the unusual requirements of their particular union, which might otherwise be difficult to accomplish by a software company that develops estimating software for a nationwide clientele.

The following is an example of what might be in such a spreadsheet, which would typically include all of the calculations and a final hourly rate:

Carpenter journeyman: 8-hr day
Carpenter journeyman: 9-hr day
Carpenter journeyman: 10-hr day
Carpenter journeyman: Saturday
Carpenter journeyman: 8-hr day, second shift
Carpenter journeyman: 8-hr day, third shift

If an estimator is preparing an estimate for a project that works one shift of 10 hours/day, Monday through Saturday, then the estimator has two choices:

1. Setup two carpenter journeymen and use both the 10-hr day and the Saturday entries where appropriate, or

2. Average those two rates to get a third rate by using 5 times the 10-hr rate plus the Saturday rate, divided by 6 (days) and enter that into the estimate.

Alternatively, the estimator could include more spreadsheet entries, such as:

Carpenter journeyman: 10-hr day + Saturday

This method is very tedious and could become a very elaborate and minimally tested spreadsheet, but if the estimating software does not handle the actual situation, this method may have to be used.

Method 3: Simple Percentage Based on Historical Rate

Some contractors are able to use one all-inclusive burden rate provided by the accounting department as a percentage of the base rate and enter it in the labor tax field. For example, if the labor burden of all jobs for last year was 26 percent, then 26 percent would be used again this year as the tax on all labor crafts and no attempt would be made to fine-tune an estimate unless it was exceptional.

This method is valid only if all jobs are similar. For example a company that always works 10-hr days, 5 days/week, might find this method reasonable. However, if a job is being bid that required 7 days/week, these rates would no longer be valid, and it might be difficult to determine what the rates should be without doing it the detailed way. If the company wins the bid, the accounting department should consider excluding the actual labor figures for this job from their annual computations. This is because it would increase their burden rate and thus penalize all of their normal jobs with too high a labor cost, possibly causing future estimates to be noncompetitive.

PRECISION IN LABOR COSTING

When choosing a labor-costing strategy, the estimator should recognize that labor-hours in the estimate are truly an estimate and thus too much precision in the rate makes little practical sense. For example, an item may be estimated for 7.2 hours which is less than a full 8-hr day. The question arises: will the workers go home when they finish or will they start something else? The difference between 8 and 7.2 represents over a 10 percent margin of error. The amount of potential error may exceed any attempt to develop a precise labor cost figure.

This last example brings up the issue of rounding. For example, if in a job with 8-hr workdays, a task computes to 31 crew-hours of work, most estimators have a tendency to change the crew-hours to 32 hours to make it an even 4 days. This roundup represents a 3 percentage margin of error compared to the computed production rate, and it would be a true error if this activity started at noon one day and actually finished in 31 hours.

The estimator should always consider reasonableness when assessing labor-hours. For example, if an estimator is using 8-hr days and the estimated time for an activity is computed as 76 hours (9.5 days), rounding may be introducing an error. Often it is not possible to estimate with sufficient accuracy to know whether it would take 9 days, 9.5 days, or 10 days to do the work. Therefore, 9.5 days may be the best estimate. Even if the hours were 78 and rounded to 80, an error of $2/78 = 2.6\%$ is introduced unless the estimator is certain this task will take exactly 10 days.

Thus, fine tuning the labor burden rates to take into account situations that have 0.1% impact on the total estimate, and then regularly making rounding guesses of over 2.0% of labor may not make a lot of sense. However, the more accurate the costs, the better—just don't negate the effort by making guesses that may have a much greater impact on costs.

EQUIPMENT RESOURCES

Setting up equipment has many of the same considerations as labor. Equipment is usually generic as opposed to any specific piece of equipment because the estimator typically does not know exactly which machine will be used, only the required size and capacity. Equipment should be coded in like groups. For example, all dozers should appear together in selection lists. If specific pieces of equipment are used in estimating jobs, they should be coded such that they appear near generic equipment of the same type. The equipment codes kept by the accounting department are often inappropriate to use in an estimating system because they were often coded in historical sequence, rather than with quick lookups in mind. Similar to labor, equipment can be treated as a simple rate or be broken down into complex detail.

Fine-tuning an estimate based on job conditions can be accomplished more accurately when the equipment rates can be broken down into two components: ownership costs and operating expenses. For example, suppose a piece of equipment costs $25 per hour and the operating cost is $20 per hour for a total of $45 per hour. Suppose also that this piece of equipment is going to be on-site for 100 hours, but only used for 50 of them.

100 hours on-site times $45/hour = $4,500, which is too high

50 hours actually used times $45/hour = $2,250, which is too low

Actual cost is (100 hr $\times$ $25/hr) + (50 hr $\times$ $20/hr) = $3,500

If the computer has this information, some software can make use of it to help prepare a more accurate cost estimate. HeavyBid has a simple screen for entering one ownership cost (rent) and one operating cost (EOE), which added together gives the total equipment cost. This is sufficient for many contractors. A contractor only wanting to use one combined cost would simply put the total cost in the rent field. However, Figures 22.10 and 22.11 are screens for a more detailed treatment of the rent and operating expenses respectively, to be used for discussing some of the complicated issues related to estimating equipment costs.

Equipment Ownership Costs

Equipment purchased by a contractor requires cost recovery calculations that include purchase price, interest, insurance, etc., which are independent of the utilization of the machine. These costs must be recovered through jobs in which the equipment is used.

The contractor may be making installment payments on some equipment, so it is obvious that the equipment is not free; however, other purchased equipment

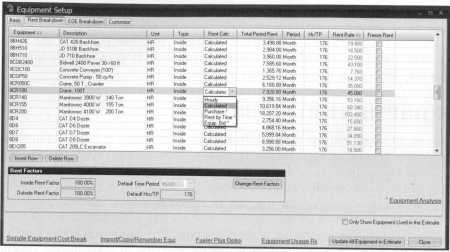

FIGURE 22.10 | Equipment setup showing possible rent complexity.

may be owned with no debts, and thus the lack of payments may make the equipment appear to be free. However, all equipment must be charged a rental cost during its use to cover the cost of ownership or replacement, and ideally should bear some relationship to cost of ownership and the competitive rates other contractors are using. From a cash flow perspective however, equipment totally paid for may require much less cash outlay on the particular job, and if the ownership costs are broken down in an estimate, the contractor can determine how much of the equipment ownership cost may not require immediate cash. Because cash flow is critical on some jobs, this information may be valuable to know.

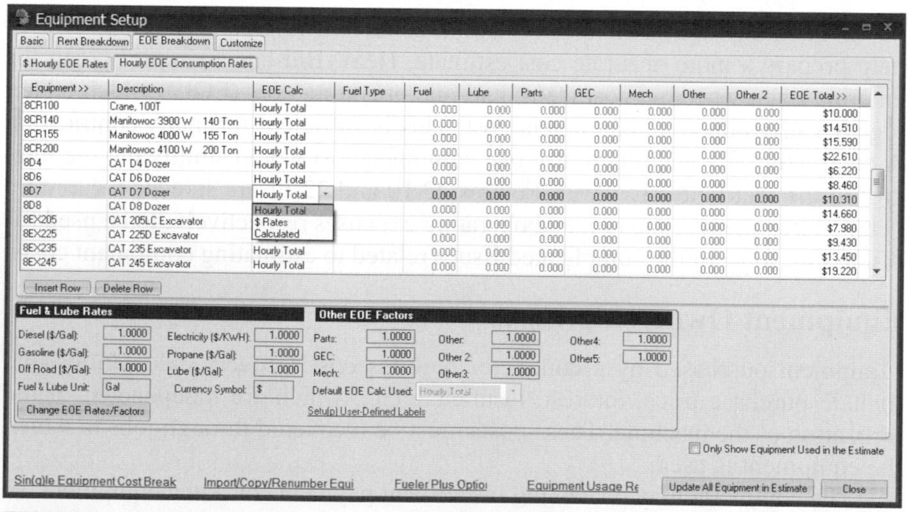

FIGURE 22.11 | Setting up equipment to use consumption rates and factors.

Another issue related to construction equipment is whether to charge for days the equipment is on the job, but not used. If operating expenses are broken out separately from ownership costs, the estimator has the ability to recover equipment ownership costs independently of how much the equipment is actually operated. The impact of separately charging equipment ownership and operating costs based on usage can be illustrated by two scenarios.

For scenario 1, suppose a piece of equipment is being charged at $4,000 per month and it will be on a job all month, the operating expense (excluding the operator) is $20 per hour, it is being used for 15 days, and there are 8 hours per day and 176 work-hours per month. For these conditions the equipment cost will be:

Scenario #1 costs:

For this scenario the operating hours = 15 days × 8 hr/day = 120 hr
Operating cost = 120 hr × $20/hr = $2,400
Ownership cost could be calculated as one of the following:
1) $4,000, by charging the equipment for a full month
2) $4,000/176 hr = $22.73/hr × 120 hr = $2,727 by charging for a partial month, consisting of 15 days

The total equipment cost will range from $5,127 to $6,400 per month, calculated by combining the operating and ownership costs as follows ($2,400 + $2,727 = $5,127) and ($2,400 + $4,000 = $6,400).

Scenario #2 costs:

For Scenario #2, all of the conditions are the same as Scenario #1 except the equipment will be used 10 hr/day, instead of 8 hr/day.
Operating hours = 15 days × 10 hr/day = 150 hr
Operating cost = 150 hr × $20/hr = $3,000
Ownership cost could be calculated as one of the following:
1) $4,000, by charging the equipment for a full month
2) $4,000/176 hr = $22.73/hr × 150 hr = $3,410
3) $4,000 / [(10/8) × 176 hr] = $18.18/hr × 150 hr = $2,727

The total equipment cost will range from $5,727 to $7,000, calculated as ($3,000 + $2,727 = $5,727) and ($3,000 + $4,000 = $7,000).

Notice Scenario #2 had three unique costs for the equipment ranging from $5,727 to $7,000, which represents a variation of 22% in the equipment costs. Because every one of these techniques is used regularly by contractors, it has a major impact on the bid results. Ideally, the method chosen should be based on the decision of the estimator and/or company management and not because of limitations of the software. It also may vary for different kinds of equipment.

This is further complicated by issues such as constantly making changes that will change the percentage of time actually used, by changing the length of the day, or running equipment for multiple shifts in a day. The estimator needs to be able to make changes for any number of reasons including "what if" scenarios and immediately see the effect on the estimate.

Some of the ways HeavyBid attempts to handle these issues is shown in Figure 22.10. In the dropdown of "Rent Calc," the hourly option means an hourly rate was entered. The "Calculated" option means to calculate the hourly rate by dividing a period rent (usually monthly) by the typical number of hours in that time period. The remaining options have an asterisk (*) beside them because they are computed at points in time with the Equipment Analysis option at the bottom of the screen. The purchase option means to compute the rent as purchase less salvage and adjust the costs everywhere it is used to recover the net cost. The "Rent by Time" means to determine how long the equipment will be on the site, compute its cost and adjust the costs everywhere it is used to recover those costs. While many companies do not need this capability, large companies, or companies frequently involved in joint ventures, will likely encounter all of these situations.

Another helpful feature in equipment setup is the ability to specify the operator who typically runs each piece of equipment. Then whenever a piece of equipment is entered into an estimate, the corresponding operator is automatically brought into the estimate.

Equipment Operating Costs

Some software systems allow entering many equipment operating components, such as fuel consumption, lube consumption, GEC (ground engaging components, such as tires and treads), mechanic time, minor repairs, etc. In HeavyBid, each of these factors can be individually adjusted for the particular job, including fuel rates and mechanic rates. Figure 22.11 is a screen where consumption rates could be entered into HeavyBid (if the calculated drop-down option were used). These cost components can be obtained from actual job records, or from equipment manufacturer's literature such as the *Caterpillar Performance Handbook*, or from rental rate sources such as the *Dataquest Blue Book*.

The fuel, for example, might be entered as gallons per hour in the grid and down at the bottom is the cost per gallon of the various fuel types which multiplies times the consumption to calculate the hourly cost. The other operating expense components work the same way. The total is on this screen, but a separate tab shows the actual cost of each column. The advantage of this technique, particularly for highly volatile prices such as fuel, is that changing the fuel price here will update the operating cost for all equipment in the job. Other reasons for adjusting the factors are rough working conditions, such as in rock or in the desert, where equipment is going to need more repair.

MATERIAL

Some companies lump all material together, while others break material down into permanent materials that are incorporated into the work, and other materials, often called construction materials.

A significant problem with automating materials is the large number of materials that may be used. It is not uncommon for a company to have from 1,000 to 5,000 materials that are consistently used. One technique for entering material is simply to create a code and description only for the material that is used in the estimate. However, it is more common to have a database setup from which to choose material. It is also common to have "plug" prices for many of those items. For example, "Class A Concrete" could be set up with a unit of cy and a price of $105.00.

An example of how a material database might be coded and structured is:

Code	Description
2	Material
2C	Concrete
2CR	Reinforcement
2CRS	Reinforcing Steel
2CRS60	Grade 60 Rebar

It is important for the estimator to be able to find the appropriate material code quickly. A good structure for material codes improves estimating speed and accuracy, and the structure is a good one if that objective is achieved. Establishing material codes must also work in conjunction with the software. Different software may handle material codebooks differently, so the structure may need to depend some on the retrieval techniques the software uses. Before setting up thousands of material codes, the estimator should test one branch of the proposed database structure, such as the rebar example above to see that it is easy to find and enter into the estimate. If the accounting department already has material codes, they should only be used if they are effective for estimating, assuming the estimating software has a way to map the estimated codes to the accounting codes.

When setting up material codes for the first time, codes for the most tedious entry operations should be set up and tested by pretending to enter them into the estimate. A poor coding structure could cause estimators to waste countless hours while estimating in the future; whereas, a good coding structure can make estimators much more productive. If there are codes already in use, but they are causing substantial wasted estimator time, consideration should be given to changing the code structure. To change an existing structure may require mapping the old codes to new codes and possibly converting all past estimates, or some important estimates, to the new codes.

There is another issue with material when estimating. If an estimator enters an item from a company-wide database containing a plug price into an estimate and changes the price, what will happen when the estimator does it a second time in the same estimate? Does the item get the plug rate or the new rate? HeavyBid addresses this issue by making a copy of the material in the estimate that will override the company price. If the rate is changed in the new estimate, that rate will be used; otherwise the company rate will continue to be used.

CREWS

Most infrastructure systems use the concept of a crew to link together labor, equipment, and optionally, material. Figure 22.12 shows the crew setup in HeavyBid. Using crews is a very fast way to enter labor and equipment into an estimate. When the estimator selects one of these crews and enters a production rate, the computer can bring in many laborers and pieces of equipment and compute their hours and total costs. The estimator can do in 5 seconds what takes 2 or 3 minutes on handwritten estimates, and there are no arithmetic errors.

Most companies should set up anywhere from a few to 20 or 30 standard crews that can then be used over and over in future estimates. However, some software does not allow modification of a crew where it is used, which might require the estimator to have hundreds of crews to account for every possible variation in the crew.

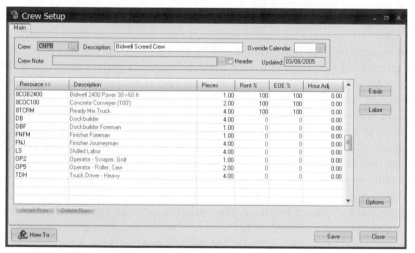

FIGURE 22.12 | Using crews to group resources for quickly building an estimate.

The "Rent%" fields and the "EOE%" fields shown in Figure 22.12 are used for adjusting rental and operating costs for the equipment in this crew. The "Pieces" field simply tells how many members are in the crew and thus the number of hours per crew-hours. For example, if there are 3 carpenters, then if the crew is entered for 40 hours, there will be 120 hours of carpenters. The "Pieces" (or number, or members) field should not require just whole numbers. For example, a foreman running two crews could be put into each as 0.5 pieces. If one operator takes turns driving a backhoe and dozer, the backhoe and dozer might be entered as 0.5 each, depending on how equipment rent is being charged, or they could be entered as one each with 50% in the "EOE%" field. Also, the equipment may not have the same number of hours as the operators. For example if there are 4 dozers and 1 is standby, there could be 4 dozers, but only 3 operators.

STRUCTURING THE ESTIMATE

One characteristic of infrastructure software is that it is built around the concept of a "biditem." For unit-price projects, the owner usually provides the biditems which, for most contractors, determines the first level of structure in the estimate. If the job is lump sum, then the contractor can use the first level, referred to as a "biditem" in HeavyBid, as the highest level of work, such as a parking lot, a building, a type of work, etc. HeavyBid uses three levels of structure for an estimate:

    Level 1 – Biditem

    Level 2 – Activity

    Level 3 – Resource

As discussed in the section on biditems, HeavyBid allows breaking the biditem down into as many as 9 levels of sub-biditems.

The second level is referred to by HeavyBid as an "activity," which is not a universal construction term. Other commonly used terms are tasks, phases, operations, etc. An activity is something being performed during construction and is usually represented by a verb, such as "Buy Pipe," "Place Pipe," "Excavate Trench," etc. In Figure 22.13, "Install Silt Fence" is an activity under biditem 140. The third level is the resources of labor, equipment, material, and subcontractors. In Figure 22.13, 2EC10 is the material resource for silt fence, 2EC14 is netting, and the items below it are equipment and labor resources.

Some software packages have more levels in their structure, which is useful when biditems regularly contain several activities, and need to be grouped to make them more understandable. For example, Figure 22.13 shows biditem 1000

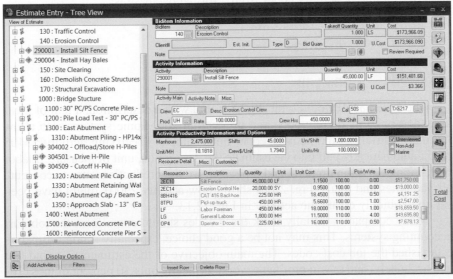

FIGURE 22.13 | Entry screen showing a possible structure of a cost estimate.

(Bridge Structure) broken down into many "sub-biditems" and item 1300 is broken down still further. Item 1310 has three activities in it. All levels of the structure tree have subtotals and unit costs that are often meaningful to the people creating or reviewing the estimate.

Some contractors rarely have more than one activity per biditem and thus could use very simple software that omits the activity level. However, such software would not be appropriate for a company that intends to expand into more complex work.

After the biditems have been entered, the estimator must then itemize the activities that must be performed to complete each biditem. Typically they will either type in the activities or select them from a database. While it might seem obvious that activities should be selected from a database, many contractors do not have one because it takes a lot of work to set up hundreds or thousands of activities that a company typically uses in its construction operations. When a database does not exist, the activities can be created as illustrated below, where A, B, and C represent the activities:

10 Furnish and Install 18-in. CMP 650 LF

A	Excavate	2342 cy
B	Lay pipe	650 lf
C	Backfill	2020 cy

Rather than A, B, and C, the activity codes could have been 1, 2, and 3, but those may be confused with biditems 1, 2, and 3. In either case, there is no historical data associated with these items because they do not represent any kind of unique code. There likely would be an activity "A" in every single biditem each having a different description.

An example of activity codes from a formal database is:

10 Furnish and Install 18-in. CMP

55150	Excavate
55405	Lay 18-in. pipe
55720	Backfill

The code 55150 would typically be stored with a description, a unit of measure, and possibly a crew code and production rate. Some systems also allow linking materials to the activity code. For example, specifying activity 55405 might automatically pull into the estimate the description, unit of measure, 5 labor crafts, 2 pieces of equipment, and pipe, and use the production rates to compute all quantities and totals. Fig. 22.14 shows how activity 29001 might have been set up to quickly create the activity in Fig. 22.13.

Many companies initially think these codes should come from the job cost accounting systems. This may not be a good choice, especially if non-estimators established the job cost codes. The main objective of the estimator is to bid fast and accurately, and to win profitable work. Typically only 10 percent of jobs estimated are going to be won. Therefore, the structure of an estimate or database of activity codes should be developed to help the estimator, not necessarily the

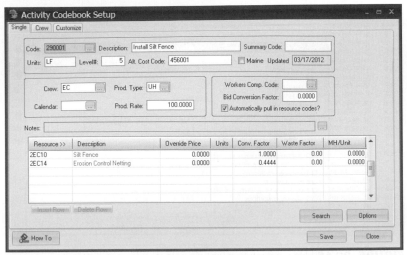

FIGURE 22.14 | How activity 29001 might be stored in a database for repetitive use.

project manager or accountant. When an estimator has 5,000 codes to choose from, it is essential to be able to find one quickly.

For the jobs that are won, many companies do not necessarily have a one-to-one correspondence between estimated activities and cost codes on the job. For example an estimate with 200 activities might result in a job with only 100 cost codes. This is due to the impracticality of getting information correctly recorded on vague or minor items compared to the ease with which an estimator can determine a cost for anything.

Nevertheless, coordination between estimate and job cost codes is desirable so estimators can benefit from feedback on the job to improve their estimating accuracy. It is also expected that the estimate can be translated into the starting budget if the bid is accepted and that process should be automated between the estimating software and the job-costing software. HeavyBid allows a second code which can be used as an accounting code to be setup for each activity. For example, in Figure 22.14, the alternate cost code of 456001 could be set up as the accounting code. Thus it is possible to set up a coding structure for estimating and then send a different code to accounting for job costing if the bid is accepted by the owner and a contract is awarded.

ENTERING THE ESTIMATE

Entering Activities

The previous section showed biditems containing several activities to illustrate the structure of the estimate. Some contractors enter all of the activities without resources as an outline, and then work on activities later, while others complete each activity as they build the estimate. The best technique is the one that keeps the estimator's thought processes organized so that nothing is omitted.

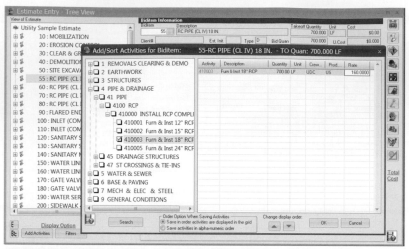

FIGURE 22.15 | How activities might be selected from a database.

Figure 22.15 illustrates a screen in HeavyBid for entering multiple activities at once. There is no particular reason to assume other software would work exactly the same way, but the end objective is to get activities into the estimate as quickly as possible. This screen can be obtained by clicking the "Add Activities" button at the bottom left. The checkbox field allows selecting activities from a database to move to the right of the screen to work on. If the database contains a crew and perhaps a production, those will be automatically filled in on the right, otherwise the estimator can fill them in. Note that a well-structured database can make it very easy for an estimator to check off all of the items needed for a biditem, whereas a poor one can result in the estimator having to search for the right code.

If the estimator is not using a database, or wishes to ignore it, the activities can be typed in directly on the right of the screen. After entering activities here, the estimator would examine them one at a time on the main entry screen such as that of 22.16.

Using Crews and Productions

Using crews is an essential time-saver in estimating for any kind of construction that is performed by common crews. Over time, experienced estimators learn what kind of crew is required to do a particular type of work and the productivity that can be expected. While the crew for a particular type of work may be fairly constant, and thus could be set up in the company-wide activity database, the productivity likely will vary for any number of reasons, including the part of the country, the weather, union vs. nonunion workers, and ground conditions. Therefore it may not be as useful to store the production rate in the database, although a useful average might at least allow an inexperienced estimator to build an estimate that has plausible costs.

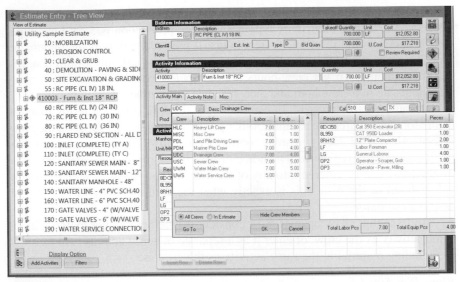

FIGURE 22.16 | Efficient screen for finding and selecting crews.

Figure 22.16 shows the crew dropdown of HeavyBid when the button beside the crew field is pressed. This is the same dropdown that would appear if entering the crew in Figure 22.15. This screen is designed to show the estimator the composition of each crew while scanning the crew choices to see which one to use. As the Up and Down arrows are used to move through the crews, the members are shown to the right. If a company has many crews, it might be difficult to remember what is in each crew and the estimator should not be wasting time looking them up. This is one feature that most estimators use constantly, therefore it should be efficient.

After selecting the crew, the production type and production rate must be entered just below the crew. The HeavyBid software allows the following production types:

UH and HU	− Units per Hour and Hours per Unit
MU and UM	− Man-hours per Unit and Units per Man-hour
US and SU	− Units per Shift and Shifts per Unit
S	− Shifts
CH	− Crew- hours
$U	− $ per unit

There are many production types because different estimators think in terms of different productivities and some types of work are more appropriately thought of with different productivities. Since the objective of a good estimating system is to help the estimator estimate quickly, it should not be necessary to reach for a calculator when the computer can do the work.

The crew uses the production type and production rate to calculate crew hours. For example, if the productivity type is shifts, the production is 5, and hours per shift is 8, this would result in 40 crew-hours. If the productivity is man-hours per unit, the productivity is 12 and the activity quantity is 100, this would result in 1,200 man-hours. To get crew-hours, the computer would calculate how many man-hours are in a crew-hour and then calculate the number of crew-hours.

Because the reviewer of an estimate may think in different terms than the estimator who entered the estimate, other productivities should show on the entry screen. HeavyBid has a display option that allows specifying up to five additional productivities from a choice of nine in addition to the main one. The productivity can be entered into any of these fields and the computer will back-calculate all of the rest. This is exactly what the computer should be doing to help the estimator.

An issue that all companies must consider is whether to enter material in the same activity as the crew, or enter it as a separate material activity. An example of including material in the same activity as the crew is:

Place Pipe 500 LF

Foreman	40 mh
Pipe layer	80 mh
Laborer	40 mh
Operator	40 mh
Backhoe	40 mh
Pipe	500 LF

Some companies separate material activities from those containing labor and equipment, such as:

Buy Pipe   *500 LF*

Pipe	500 LF

Place Pipe   *500 LF*

Foreman	40 mh
Pipe layer	80 mh
Laborer	40 mh
Operator	40 mh
Backhoe	40 mh

The latter technique may be used because there are years of historical data of unit costs that do not include material, or simply because the company wishes to isolate high-risk labor items from low-risk material items. There is also an advantage to separate the "buy" activities from the "place" activities to reduce problems in costing and forecasting the progress of work during construction. For example, when 100% of the pipe is received, but only half of the pipe is placed, forecasting that item at the cost code level with only one quantity is difficult and requires a more tedious method of forecasting. This last reason, though, is not necessarily sufficient to establish an estimating technique because even if the material is in the crew, it can be separated out should the bid be won and the project moves to the construction phase.

Overtime

HeavyBid uses "calendars" and work rules to compute overtime. It is likely that every estimating software package will have a different way of handling overtime. The calendar field, really a weekly calendar, is on the same line as the crew line (shown in the middle of the right side of Figure 22.16) with a field name of CAL and a value of 510, which designates "five 10-hour days." Contractors often like meaningful codes so it is not necessary to see the description to know what it is. This calendar code, in conjunction with the labor rules for each craft, determine the overtime for each labor craft.

For example, if the work rule is time-and-a-half over 8 hours for carpenters, then the 510 calendar is going to result in two hours of overtime each day, which means carpenters will work 10 hours but be paid 11 hours (two ½ hours of overtime.) HeavyBid uses this 11/10 = 110% in the resource calculation to increase the labor cost to account for overtime. If the work rule is double-time over 8 hours for operators, which happens with some unions, 10 hours of work would have 2 extra hours of pay and thus result in 12/10 = 120%. Thus, in the same crew with a 510 calendar, the carpenter cost is being increased by 110% and the operator cost is being increased by 120%. This combination of weekly calendars and work rules provides an almost infinite capability to calculate overtime for any task.

If overtime is used, the calculations should be tested and understood by everyone using them. If a new set of circumstances occur, such as weekend shift work or a union with more complex rules, the new variations should be tried before relying on them in an important estimate.

Notes

It is very important to enter notes explaining anything that people reviewing the estimate might not understand. There are several reasons why leaving notes may prove useful:

1. Someone, other than the estimator who put the notes in, may be reviewing it.
2. The job could be postponed for months or years and even the estimator who placed the notes cannot remember why something is done the way it is.
3. The job may end up being used in estimate history for future estimates and the estimator is trying to determine if the situation of the historical estimate applies to the new estimate.
4. If a bid is successful, the estimate may be given to the project manager or superintendent who may need to know what the estimator intended when the estimate was prepared.

Figure 22.16 shows several places where notes may be entered to support the estimate. In the Activity section there is a note line with a button following it and a paper clip. This allows entering notes for the activity or attaching other files such as word processing documents, spreadsheets, or photographs. The first line of the note shows on the screen so the estimator can quickly determine if there is a note, in which case the drop-down button could be pressed to see the entire note. If there is an attachment, the paperclip will be blue and clicking on it will cause the appropriate program to open the document.

There is also a note for each biditem with dropdown button and paper clip that works just like the activity. In addition, there are the same two options on the estimate information screen where some contractors sometimes put their job documents.

Calculation Routines

Calculations are inevitable and crucial in preparing estimates. Because the computer can do calculations well, it should be helping the estimator as much as possible. Estimators can always create spreadsheets outside of the estimating software, but it is desirable for the calculations to be a part of the estimate, get backed up with the estimate, and stored historically with the estimate.

Figure 22.17 is an example of the integrated spreadsheet in HeavyBid. Integrating spreadsheets into estimating software is difficult, so it is likely that no two software vendors will handle spreadsheets the same way, if they do it at all. Figure 22.17 is an example of a template calculation routine that can be used over

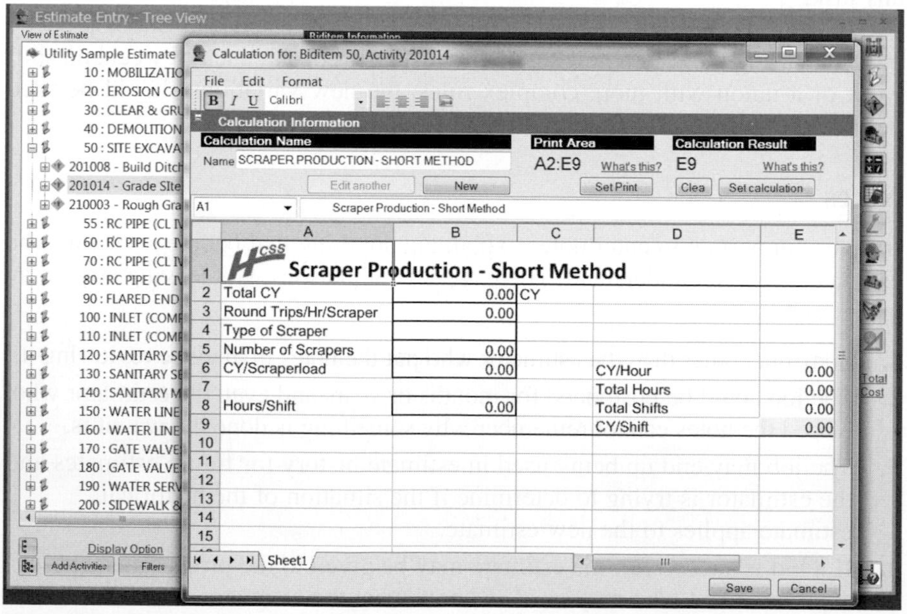

FIGURE 22.17 | Using an integrated spreadsheet for various calculation routines.

and over in an estimate and a copy automatically saved with the appropriate activity. Another nice feature is that a copy of the spreadsheet will be saved as text in the notes so that it can be reviewed and printed without opening the spreadsheet. In this case, the result of the calculation in cell E9 will be copied into the estimate at the point of the cursor when the routine was executed.

There are a number of calculation routines furnished with the software, and because it is a spreadsheet, estimators can create their own templates for use in the future or create one for a one-time use within their estimate.

Selecting from a list of calculation routines, filling in the numbers, having the calculation drive a quantity or production in the estimate, and having the assumptions copied into the notes is a very powerful tool for rapidly building well-documented estimates.

Some estimating software attempts to link cells in spreadsheets to fields in the estimate. It makes the most sense for a spreadsheet of takeoff quantities when the values in the takeoff are linked to drive quantities in the estimate. If someone changes the takeoff values in the spreadsheet, the changes would automatically update the estimate. The estimator should be aware that linking spreadsheets in commercial software raises the following issues:

1. If someone changes the spreadsheet in a program such as Excel™ unrelated to the estimating software, does the estimating software know that this has occurred and when does it know it?

2. To avoid the problem above, does the software only allow changes to the spreadsheet while in the estimate? Can this be enforced? Does everyone doing takeoff have access to the estimating software?

3. Are there limits to where the spreadsheets can reside? Is it possible that an estimator can copy the estimate to a laptop or send it to a joint venture partner and not include the spreadsheet? If so, what happens when running the estimate or trying to update from the links?

4. How does an estimator in the estimate know that the spreadsheet has been changed and that all of the links need to be checked to see if they are current?

5. If a row is inserted into a spreadsheet, does that affect existing links?

6. If a spreadsheet with 200 links has been updated, how long will it take to recheck and update the estimate?

7. How easy is it for someone not knowledgeable about the linking to accidently remove some links to make 95% accurate what the chief estimator thought was 100% accurate?

8. What happens if a quantity with a spreadsheet link is changed in the estimate? Is that allowed or does it break the link?

9. What happens if an item within the estimate linked to a spreadsheet is copied? Does it carry the link with it or start with no link?

10. If the spreadsheet quantity is linked to the estimate quantity, how does the estimator handle waste?

These are just some of the issues that a software developer has to address with a spreadsheet linking program and it is very difficult to make it fool-proof; therefore, the estimator would want to thoroughly test the spreadsheet linking using some of the preceding points to see what happens. HeavyBid contains a linking ability to any Excel™ spreadsheet and checks the links and updates them at points in time when the estimator requests it.

COPYING FROM PAST ESTIMATES

The estimator will want to use copying features because of their time-saving power. However, the estimator should know as much as possible about how they work because much, if not all, of the estimate may be built using copying features. The more features an estimating system has, the more questions the computer has to address during the copy process, and the options chosen by the software may not work the way the estimator thinks they should. A good way to learn copying features is to create two small estimates with manageable numbers and copy items with several scenarios to see what happens. Another issue with copying is to know what has been copied and reviewed, and what has not. For this reason HeavyBid sets any activity to "Unreviewed" when it is copied and requires that it be changed when someone reviews it.

Another real time-saver in computer estimating is to set up a library that contains fully worked-up biditems that an estimator typically uses. Some of the biditems in the library could have many activities. Copying those biditems into an estimate could save a lot of the estimator's time, compared to looking for them one-at-a-time from a database while estimating. A library can be built gradually over time. For example, an estimator may wish to save 11 out of 60 biditems from the first computer estimate and copy them into a library estimate. A "library" estimate is an estimate just like any other, except it is used to collect parts of previous estimates. Any software program that allows copying from other estimates can use this concept. If 8 biditems were saved after the second computer estimate, and 13 biditems from the third computer estimate, there would then be 32 items from which an estimator could select in future estimates. In this manner, a contractor can easily acquire an extensive library of biditems for building new estimates.

Another way a library of biditems can be used is to develop a pick-list of all possible activities within a biditem, even those that may be mutually exclusive. This requires software that allows selecting from the activities without necessarily selecting the entire biditem, although software without this feature can be used in this manner by copying all activities and then deleting or zeroing the quantities of those that do not apply. The pick-list concept serves as a reminder of choices so something is not overlooked.

When anticipating a library that may have hundreds of items, there should be a coding structure so estimators can quickly find the desired item. Also, significant figures must be considered because huge rounding errors can occur. For example, an item with a unit of 1 cubic yard might have a mathematical computation of 0.000449 units/hour and be rounded by the computer to 0.0004 if only 4 decimal places are used in productivity calculations. This innocent-looking rounding represents an 11 percent error, which results in a huge error when copied into an estimate with 500,000 cubic yards. The estimator would probably never know this has happened because no one knew about the initial 0.000049 that the computer discarded. Normally the quantity in the library should be 100 or greater except items that are typically bid in low quantities.

Figure 22.18 shows a split-screen feature to copy from one estimate to another. Because copying estimates is something the estimator does constantly, it should be as efficient as possible. At the top left of this screen are shortcuts to quickly find an estimate to copy from. One button is for a library estimate containing parts of prior estimates that have been collected specifically to be copied. One button shows the most recent estimate used, assuming it might be needed again immediately. The "Other Recent" button shows several recent ones copied from, and "Favorites" shows ones that have been listed as likely to be copied from regularly. The whole idea is that if there are thousands of past estimates in the system, the most likely ones to be needed can be found instantly.

To actually copy, individual items can be dragged from the left side of the screen and dropped to the current estimate on the right side, or multiple items can be checked and then dragged as a group. An entire biditem can be copied or it can be broken down as shown in the example, where only parts of the item are copied. This capability can be used to create pick-lists. For example, the office staff for a job could have dozens of people to choose from, but the particular project only needs a project manager, superintendent, and surveyor, therefore only those three would be checked to be copied.

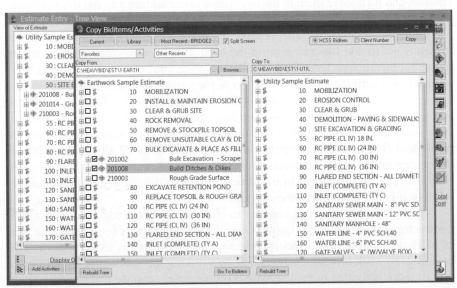

FIGURE 22.18 | Copying from other estimates helps build new estimates quickly.

There are many other issues related to copying that make it more complicated than it might at first seem. Some of these are:

1. Should the labor and equipment rates used in the "from" estimate be replaced with the ones in the "to" estimate?
2. What material rates should be used, the rates from the "from" estimate or the plug rates from the company-wide database?
3. Should quantities be adjusted as they are copied? What about lump sum items?
4. If worker's compensation codes are being used, should they be copied from the "from" estimate or should the default value for the current estimate be used?
5. If the "from" estimate has a parent-child structure, will it copy the entire structure into the new estimate?
6. Is it possible to determine what was copied and does it matter?

This is a feature that should be thoroughly tested to see what happens in various scenarios before using it to quickly build an important estimate.

ALTERNATES

Estimators have many opportunities, particularly within infrastructure estimates, to compare alternate ways of building a job. In fact, comparing "what if" scenarios should be encouraged as the results of this type of activity are often what will give a contractor a competitive advantage. There needs to be a way in the estimating software to work up alternate methods to get their costs for comparison and discussion, and then leave them in the estimate without adding into the estimate total.

HeavyBid allows making both biditems and activities "non-additive" to accomplish that objective. This is probably not a universal term in software, but it means not to add these items into the total. It is used to work up the costs of items but then not include them into the estimate total. An example might be to work up two activities, one with the costs for a truck haul and the second using a scraper haul. If it were determined that the truck haul was more economical, the scraper haul could then be marked non-additive but it is retained in the estimate so that anyone viewing it can tell why the decision was made to use the truck haul. The non-additive field can be seen in Figure 22.13 in the middle of the screen on the right-hand side. If it were checked, there would also be a red indicator marking that item in the structure tree so the estimator could quickly tell that it is not included in the estimate.

HeavyBid also uses this concept to compare a contractor's work to a subcontractor's work. Often the quotes from the subcontractors for a particular type of work do not appear to be acceptable and the estimator has to work up the cost of doing the work instead. HeavyBid allows linking the subcontractor activity to one or more activities in the estimate. For example, suppose there is an activity for subcontract work, but it would take 3 self-performed activities to do the work instead. The one subcontract activity can be linked to the 3 contractor activities such that one group is additive into the estimate and one group is not. Thus if the subcontractor was chosen to do the work, the other 3 activities would be marked non-additive.

An issue to remember when using alternates is to make sure by some method that only one of two mutually exclusive choices is added into the estimate to prevent double-adding costs. In the link of the subcontractor vs. self-performed described in the preceding paragraph, HCSS recommends initially entering the subcontractor with zero costs so that if someone forgets to link them, the cost is not roughly double-added into the estimate.

Another situation that occurs is very large sets of alternate items in the estimate, for example, a steel alternate vs. a concrete alternate. Each alternate may consist of as many as 100 biditems. It is helpful if there is a way to compare each alternate and have one set additive and one set non-additive and to be able to quickly change from one set to the other without having to change every single biditem.

SHOWING HOW COSTS WERE CALCULATED

It is important to tediously check the calculations in an estimate, especially the first time an estimator uses software, or for experienced estimators the first time they use a new feature. Every estimator should learn the calculations in the software because knowing exactly how something works makes it easier to creatively use that knowledge and to overcome problems that may arise in the future. Figure 22.19 shows the calculations for a labor foreman. This screen is obtained by right clicking the mouse when the cursor is on a resource code (LF in this illustration), and selecting "Show Calculations."

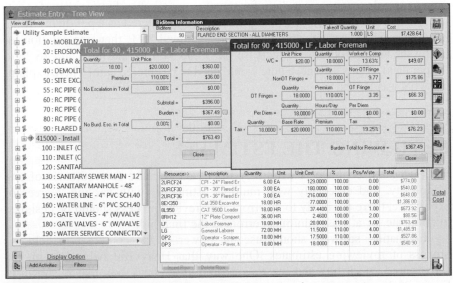

FIGURE 22.19 | Checking the calculations of the labor foreman cost.

Labor cost can be a very complicated calculation. The left side of the screen for this labor resource shows the calculation with labor burden entered as one number. To the right of the burden is a button which, if pressed, shows the calculation of the burden. Burden in HeavyBid includes taxes, fringes, worker's compensation, and per diem.

The example illustrated in Figure 22.19 shows the labor foreman has labor premium (overtime), no escalation, there is worker's comp, there are fringes that consider overtime and those that do not consider overtime, there is no per diem, and there is tax. The calculations and results of these labor burdens are shown on the left-hand side of the screen. It should be possible from this information to determine if the computer calculated the cost the way the estimator expects, given all of the information that was entered into the estimate pertaining to the labor foreman that is entered into this specific activity.

It should be noted that regarding labor calculations, there is not a clear right or wrong calculation that is agreed on by all estimators in all states and unions. It is the estimator's responsibility to understand how the computer calculates costs, and then use those calculations as necessary to reduce estimating time and improve estimating accuracy.

If the company decided to use per diem for the first time, some examples should be entered as simple test estimates, and then the calculations should be checked to see exactly what kind of results occurred. Just by looking at the calculations above, it would appear that per diem is not taxed. Is this actually what happens and is this the desired result? If it is not the desired result, what will the estimator do to get the desired result? It is quite common for calculations to not work exactly as the estimator wants, or to fit an unusual situation that requires some additional thinking about how to take the unusual condition into account. As an example, suppose per diem for travel is not taxable, but per diem for eating lunch in the tunnel is (assuming the lunch was entered as a per diem). If the lunch were $5.00 per day and the tax rate averaged 25%, then one solution for taxing the lunch would be to increase the $5.00 by 25% and then add $6.25 to the per diem. If this is done, a note would need to be entered somewhere explaining what was done.

REVIEWING THE ESTIMATE

Reviewing the estimate is extremely important because for estimates of any complexity, and especially with multiple estimators with varying degrees of experience, there are likely to be errors. The review process should be designed to eliminate, or at least minimize, errors in the final bid.

Estimating software should use the power of the computer to help the process of reviewing estimates. One way that HeavyBid helps is with the *Unreviewed* check box and the *Review Required* check box that is shown in Figure 22.20.

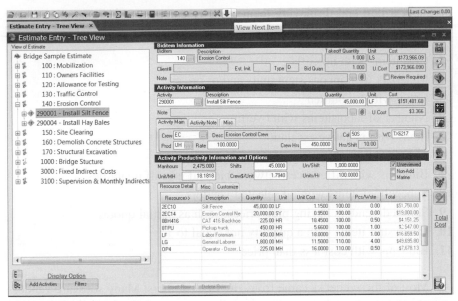

FIGURE 22.20 | Using the unreviewed feature to find activities needing to be reviewed.

The review feature in HeavyBid gives the chief estimator several options to systematically review an estimate. One way is to set an option to display all "unreviewed" items one at a time. Another way is, after all estimators have reviewed their own work, the chief estimator or reviewer can reset all activities back to "unreviewed" and then go through all of the items one at a time, resetting them again as each item is reviewed.

The "unreviewed" checkbox in HeavyBid is more than a simple data field. It can be set in several different ways depending on the objectives of the chief estimator. The normal method set by HCSS is that items brought in from codebooks, or copied in, are set to "unreviewed" by putting a check in that box. This is because those routines can be automated to such an extent that dozens or hundreds of costs can come into the estimate without anyone actually looking at them. The idea is that the estimator would study the item and then uncheck the box. Should the estimator forget, a senior estimator or chief estimator reviewing the estimate would see that the item had not even had a preliminary review. This feature was added to the software because people in charge of estimating often objected to too much automation for fear that no one is actually studying the job and that they are blindly relying on computer reports.

CHECKING THE ESTIMATE FOR REASONABLENESS

In addition to the typical cost reports one would expect from estimating software, the power of the computer should be used to look for possible errors. In HeavyBid, there is a display and report that runs through the entire system looking for things that appear odd and displaying them to be researched. Some examples of the nearly 100 checks are:

1. Cost items with no costs
2. Materials with inconsistent units and/or prices
3. Labor with no burden
4. Number of labor items with overtime
5. Items with negative escalation
6. Activities with no costs
7. Biditems with differences between bid and takeoff quantities
8. Subcontract items with no selected vendors
9. Selected vendors having plug prices, such as partial quotes
10. Lump sum items with a quantity not equal to 1
11. Biditems with cost less than price

Most of these checks came from HCSS customers. Many software vendors appreciate getting such requests, which they can use to make their customers' estimates more reliable. No vendor likes to see their customers make errors, even if it is clearly the customer's fault.

The objective is to identify something that is not right, or does not appear to be correct. For example, if the estimate was prepared with no overtime and item 4 above showed 21 labor items with overtime, this would be an item that should be researched. If a lump sum item had a quantity of 4.77, this most likely is an error because it was probably copied from somewhere where all quantities were multiplied by 4.77, including the lump sum item.

In addition to features provided specifically to highlight potential mistakes, the estimator should cross-reference the totals in various places. For example, the cost total on a cost report should equal the cost total on summary screens and the bid total should be significantly more than the cost total. Some checks may appear trivial, but the better an estimator is at correlating numbers, the more likely an error will be uncovered. Although computers may seem infallible, the human factor must always be considered and a careful check of the numbers cannot be overemphasized.

TURNING A COST ESTIMATE INTO A BID

After an estimate has been prepared that contains all of the costs for the proposed project, the company costs not related to the project and profit sufficient to make the project undertaking worthwhile must also be added. All costs and profit must then be converted into a bid.

In HeavyBid, the process of converting from a cost estimate to a bid is referred to as the "Bid Summary." This term is not universal, but the processes that have to occur would be common to all but the simplest of software. The Bid Summary of HeavyBid has two parts:

1. Summarize all costs to the biditem level, and
2. Compute indirect costs, bonds, add-ons, markup, and profit, and then spread them back to the various biditems to arrive at bid unit prices.

It is difficult to keep all totals current in the software, particularly as more features become available in a software package and especially if some items become related to and/or dependent on other items. Estimates can become very large. A typical estimate might have 1,000 to 2,000 uses of resources, a very large one could have 10,000, and mega projects can have over 20,000. HCSS has elected to not keep totals current in order to provide maximum speed while working on the estimate and to avoid the programming difficulties associated with trying to keep totals in synchronization with all of the raw data.

The second part of the Bid Summary, spreading costs and markup back to the biditems, is crucial because the way it is done affects the final bid prices which can have an effect on profitability and cash flow. Many estimators are still so worried about controlling this process that they may go to a manual procedure the last hour or so before a bid is submitted to the owner.

Figure 22.21 shows the Bid Summary screen from HeavyBid after the first part of the summary has computed totals. This screen allows the estimator to enter further costs and markup and specify how to spread everything back to the biditems. This illustrates some of the complexity of what contractors do and why infrastructure software is specialized.

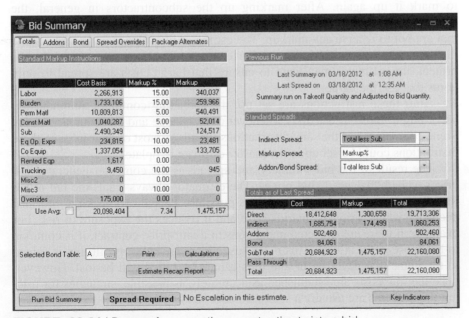

FIGURE 22.21 | Process for converting a cost estimate into a bid.

The most straightforward use of this screen is to add markup by cost type, set up a bond table to compute bond, and then spread all costs and markup using the standard spreads on the right of the screen. Many estimates are that simple. However, many projects will not be that simple and thus there are many other options on this screen.

One item that contractors worry about is risk. To mitigate risk, it is common to markup labor items more than other cost items because labor is the biggest variable in estimating how a job is to be built. Thus, a company with 5% on other items might markup labor 15%, or even more if attempting risky work they have never done before.

The "Standard Spreads" section has several options for prorating various groups of costs or markup across the biditems. These are simple prorations. In this illustration, the indirect cost will be prorated to all biditems based on the biditem total cost, minus any subcontract cost that is in the biditem. If a biditem is 100% subcontract, it will receive no proration of indirect costs. The markup% option on this screen means to spread the markup the way it was computed. In this example, a biditem that is 100% subcontract would be marked up 5% because that is the markup amount specified on the left of the screen for subs.

Regardless of how simple an estimator might think the spreading should be, jobs always seem to demand exceptions. The "spread overrides" tab allows the estimator to go to any biditem and override the spread—by stopping the spread completely or overriding the amount of the spread. For example, part of the work may be "sub-contracted" to another division of the estimator's company that may have already marked up the job, and thus the estimator does not want to mark it up again. After marking up the subcontractors in general, the estimator would go to the override screen and specify zero markup for the items that are not to receive any markup. The estimator may also want to target some indirect costs to specific biditems. For example, if a barge is required for the project, the estimator may want to ensure that barge costs are spread to only items that use the barge.

After these instructions are set up, the spreading continues automatically according to those instructions whenever costs change and the summary is run. Some companies use the same instructions for all, or most of their jobs, so these instructions could be set up in the master estimate template.

The result of the bid summary and spreads is the computer-generated bid prices, called "balanced" bid prices. The estimator always wants the bid prices to reflect the true cost as accurately as possible. This allows making informed decisions when various scenarios arise and it reduces the exposure to risk. For example, if there are expected under-runs in a unit-price contract, it is common to bid those items at cost. This is because any overhead or profit that is spread to the items that never occur and thus cannot be billed will be lost. However, once the true costs are established as accurately as possible, strategies for pricing the biditems are usually a senior management function.

BID PRICING

The end result of the estimate is a bid. Figure 22.22 shows a sample of a pricing screen used to arrive at the final bid. Most unit-price estimating software will have a screen very similar to Figure 22.22 but will differ in the number of features and amount of information available.

The "Balanced Price" column is the computer-generated unit-price, obtained by adding all of the biditem costs and spreading any costs or markup to it and then dividing by the biditem *takeoff* quantity. HeavyBid normally assumes that the estimate was prepared on the takeoff quantity and thus the costs in the estimate are the costs to perform the takeoff quantity. Normally the takeoff and the bid quantity are the same so there is no ambiguity. However, suppose one biditem in a unit-price bid is to move 500,000 cubic yards of dirt, but the contractor's takeoff showed only 480,000 cubic yards of dirt. The assumption is made that the estimator put costs in to move 480,000 cubic yards of dirt. Dividing the cost plus markup for 480,000 cubic yards by 480,000 gives the price per cubic yard. This unit price will then be extended by the 500,000 cubic yards to get the bid total for this biditem.

The "Bid Price" column is the final price that will be used in the bid proposal. Normally the bid price and the balanced price start the same and remain the same unless the estimator enters a bid price to override the balanced price. The "Biditem Total" column is the bid price extended by the *bid* quantity.

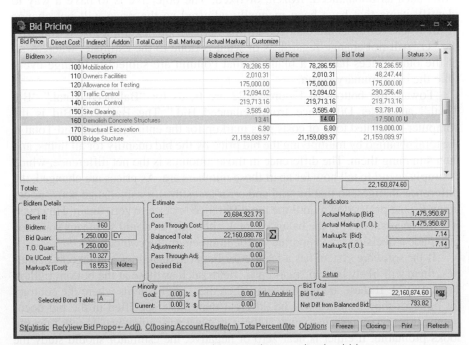

FIGURE 22.22 | Pricing screen used to complete a unit price bid.

There are at least five reasons contractors override the price:

1. The item finishes early and the contractor increases the price to obtain early cash flow of money.
2. The item quantity is expected to overrun or underrun and the contractor will gain or lose indirect costs, add-ons, and markup depending on how the item is priced.
3. The contractor wants to round prices to even numbers.
4. It is one of the few items remaining open in the last few minutes of the bid and the item must absorb changes to items that are already closed.
5. The owner mandates a price regardless of what the computer generates.

It should be noted that many government entities forbid unbalancing prices. However, because of the many ways that estimators cost, markup, spread, and round, it is difficult to determine if unbalancing has occurred unless it is blatant. There is also an element of gambling in unbalancing biditems. The contractor who unbalances most effectively has a competitive advantage, and if their bid is not thrown out, there is a better chance of winning the bid. A contractor who plays by the rules can only win if the rules are enforced, which is not consistently the case.

In Figure 22.22 the last column, labeled "Status," indicates whether an item has been unbalanced, or frozen. Some software will use various combinations of terms such as unbalanced, frozen, or fixed. The objective is to have a way to identify items that have been manually adjusted so the price will not change again whenever costs are changed in this item. In this example, Item 160 with the "U" will not change regardless of what is done to the estimate, whereas all of the other prices having a blank in that column are subject to change every time the bid summary is run.

Toward the bottom of Figure 22.22 are fields for "Balanced Total," "Bid Total," and "Net Difference from Balanced Bid." The Balanced Bid is the computer-generated total of costs plus markup for the entire job. The Bid Total is the total obtained by multiplying the bid price by the bid quantity for all items, and then summing them. The difference in those amounts is the net difference, which is simply an aid to tell how far apart those numbers are. As prices are changed, the difference changes, so that by trial and error an estimator can change some prices up and some prices down until the net difference is close to zero again. However, it is not necessary to bid the balanced total, although not doing so means someone has made a decision not supported by the numbers in the computer.

If the software system has a feature for both bid quantities and takeoff quantities, there will likely be two "markup%" fields such as those at the bottom right in Figure 22.22. If these are not the same, then there are biditems where the contractor believes the quantities will be different than those specified in the bid documents. It is very important to look at these as it is a very quick way to tell if the bid looks profitable. If either is near zero or negative, the estimator should study the estimate carefully as this may represent an element of risk that must be discussed.

The terms *markup* and *profit* are often used interchangeably in estimating and their meanings are not always clear. One marks up an estimate by adding to it. The markups may or may not be all profit. For example, 2% of the markup may be company overhead. HeavyBid does not actually use the word "profit" anywhere, so it is the responsibility of the contractor to understand what is included in the final figure referred to as *markup*.

Most unit-price estimating systems have a "cut and add" sheet for accumulating last-minute changes that are not to be entered directly into the cost estimate, but rather to be added at the end as an adjustment to the total bid. This was more valuable in the past when it might take hours to re-compute the estimate from new cost information. Since computers now allow estimators to re-compute bids for even very large estimates in less than a minute, it is more common to make the changes in the estimate and rerun the bid summary. However, there are still contractors whose policy is to "freeze" the estimate several hours before the bid and make all last-minute changes as cuts and adds. They do this to control the changes that are occurring in an attempt to reduce mistakes in the confusion of bid day. Estimators who have lost a bid in the past because they lost control can become skeptical of automated computer procedures where they cannot tell what is happening.

For the estimator in charge of the bidding, it is extremely important to experiment with the bid summary and pricing features of the computer software. A simple test estimate can be set up with two or three biditems with only a few costs of different types, such as one laborer, one piece of equipment, one material, one subcontractor, etc. Simple numbers should be used that are easy to work with mentally, such as 1, 10, 50, 100, etc. These numbers should be followed all the way through the summary, making several rounds of changes to see how the numbers flow through the system. This should give the chief estimator confidence in understanding what is happening without being confused by too many numbers.

TAKING QUOTES

A major part of most bids is quote taking. It is difficult to have a profitable bid that is capable of winning without the co-operation of reliable subs and suppliers that provide quality work at competitive prices. It is important to determine in advance the companies that are expected to provide quotes as bid day approaches. If the estimator is not sure that several quotes will be received for every type of work, then additional companies must be solicited for bids.

Finding contractors to quote work can be particularly important if the work has minority requirements. It is not uncommon for the contractor to be required to use minorities or other "disadvantaged" types of contractors for as much as 20 to 25 percent of the job value. Obtaining qualified minorities to quote this much of the job is often challenging and cannot be left to chance. Therefore it is important to determine how many of the subcontractors that have agreed to quote are classified as minority contractors.

Traditionally, subcontractors and suppliers wait until the last few minutes to supply their best prices to the contractors to make it difficult for them to have enough time to call other subcontractors for better prices. To process many quotes in a short period of time, a good procedure and computer system for handling the quotes is essential to ensure the best prices make it into the system and are reflected in the bid. Figure 22.23 shows a quote sheet for entering quotes for a type of work, RCP in this example. The items to be quoted are down the left side and the quotes are on the right side. As suppliers have indicated their desire to quote, they can be entered on the quote sheet and thus the system can be ready for quote entry.

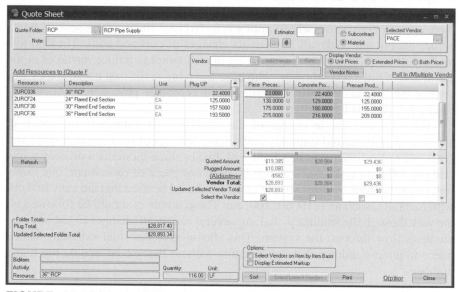

FIGURE 22.23 | Quote sheet for entering and comparing quotes.

Taking quotes from subcontractors and suppliers is complicated by these factors:

1. Quotes are coming in via phone, fax, and e-mail, and possibly for many jobs at once.
2. The fax machines may not be able to handle the load and suppliers may keep getting the busy signal.
3. The e-mail coming in may get delayed because of traffic on the Internet.
4. Some suppliers do not quote all of the items and the missing items must be obtained from someone else in order to be able to make useful comparisons.
5. Some suppliers quote several classes of items, such as electrical, mechanical, guardrails, and signage. Often they are low on one category and not low on

the total, but will not do the one category of work without the others (like the prime contractors, they have spread their overhead and unbalanced their bids). This complicates determining what combination is both low and agreeable to the suppliers.

6. The contractor may be forced to select suppliers that are not low in order to meet minority requirements.

7. Sometimes all of the quotes seem too high and the estimator has to make an estimate of what it would cost to self-perform the work, and then decide at the last minute whether to use the sub's price or their own.

8. There are also inclusions and exclusions on quotes such that often the estimator is not comparing exactly the same thing with the quotes and may have to add or deduct money from each in order to put them all on the same basis for comparison.

Figure 22.24 shows an overview of the quote status in several quote folders at once. An overview such as this is helpful to quickly see if a new quote that is coming in should be considered and thus entered into the software, or quickly set it aside and to go onto the next quote. If the quote is one that is worth taking the time to enter, the quote sheet option would be selected that would bring up the quote screen shown in Figure 22.23.

As the estimator decides on suppliers to use, there should be a mechanism to indicate to the system that they have been selected, and if minority goals are applicable, there should be a way to instantly see the status of the bid with respect to that goal. For suppliers with incomplete quotes, there must be a way to supply a "plug" price for missing items so that their quotes can be compared with those of the other suppliers. There should also be a place to add adjustments to any quote to make it comparable to the other quotes.

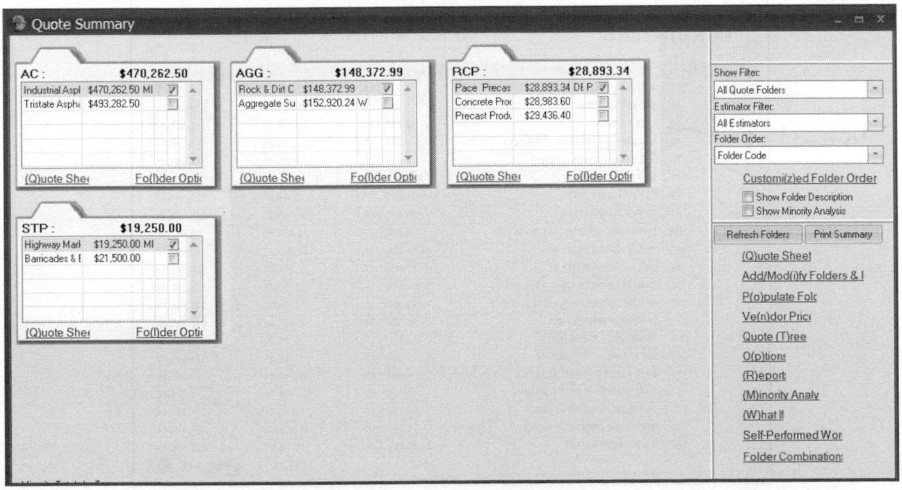

FIGURE 22.24 | Quote summary to quickly analyze quotes.

TURNING IN THE BID

After days or weeks of estimating, and hours of quote taking, analysis, and completing the bid, the estimator cannot relax until the bid has been successfully submitted to the owner. Most contractors have at one time or another failed to complete this last step, even before the advent of the computer. Computers have simply increased the number of things that can go wrong. Figure 22.25 shows the final bid from HeavyBid for a hypothetical bid.

Regardless of the method used for submitting bids, the estimator should test the bidding procedure *for each particular bid* at least a day prior to bid day. The estimator should print out the bid, even if the prices are still zero, and then compare the computer printout to the official bid document to verify that the bid to be submitted looks exactly like the bid in the owner's specifications. If there is anything unusual, it is possible to investigate as long as there are several hours or a day left prior to the bid. If the second-place bidder wants the job after seeing everyone else's bid, they often study the winning bid for irregularities in hope of getting the winning bid disqualified.

When relying on a computer-generated bid, it is advisable to have a nearly completed bid ready a couple of hours ahead of bid time in case anything goes wrong with the technology. If paper bids are to be submitted, the estimator should

PROPOSAL

HCSS
Building Efficiency Together

HCSS
13151 West Airport Blvd.
Sugar Land, TX 77478
713-270-4000

03/18/2012

QUOTE   Christopher Davis
TO   Project Owner, Inc.
        1734 Main St.
        Houston, TX 77001
        (713) 111-2222

JOB	LOCATION	START DATE
1-UTIL - Utility Sample Estimate	Harris County, TX	04/23/2012

ITEM	DESCRIPTION	QUANTITY	UNIT	UNIT PRICE	EXT PRICE
10	MOBILIZATION	1.000	LS	$21,202.70	$21,202.70
20	EROSION CONTROL	1.000	LS	$20,785.08	$20,785.08
30	CLEAR & GRUB	12.000	AC	$1,035.77	$12,429.24
40	DEMOLITION - PAVING & SIDEWALKS	1.000	LS	$11,364.25	$11,364.25
50	SITE EXCAVATION & GRADING	1.000	LS	$84,923.49	$84,923.49
60	RC PIPE (CL IV) (24 IN)	840.000	LF	$ 45.28	$38,035.20
70	RC PIPE (CL IV) (30 IN)	436.000	LF	$ 54.23	$23,644.28
80	RC PIPE (CL IV) (36 IN)	260.000	LF	$ 64.15	$16,679.00
90	FLARED END SECTION - ALL DIAMETERS	1.000	LS	$8,697.99	$8,697.99
100	INLET (COMPLETE) (TY A)	16.000	EA	$2,003.17	$32,050.72
110	INLET (COMPLETE) (TY C)	12.000	EA	$2,278.77	$27,345.24
120	SANITARY SEWER MAIN - 8" PVC SDR 35	1,124.000	LF	$ 23.31	$26,200.44
130	SANITARY SEWER MAIN - 12" PVC SDR 35	1,890.000	LF	$ 26.03	$49,196.70
140	SANITARY MANHOLE - 48"	96.000	VLF	$ 486.60	$46,713.60
150	WATER LINE - 4" PVC SCH.40	2,450.000	LF	$ 20.30	$49,735.00
160	WATER LINE - 6" PVC SCH.40	1,135.000	LF	$ 25.21	$28,613.35
170	GATE VALVES - 4" (W/VALVE BOX)	4.000	EA	$1,260.24	$5,040.96
180	GATE VALVES - 6" (W/VALVE BOX)	8.000	EA	$1,488.51	$11,908.08
190	WATER SERVICE CONNECTIONS	64.000	EA	$ 691.01	$44,224.64
200	SIDEWALK & CURB RESTORATION	1.000	LS	$6,420.04	$6,420.04
210	PAVING RESTORATION	2,240.000	SY	$ 10.61	$23,766.40
				TOTAL:	**$588,976.40**

FIGURE 22.25 | Example of a simple bid document.

have a completed paper bid with only a few items left open and no bid total. It is important to ask appropriate technology people what can go wrong and have backup plans to ensure the bids can be submitted. It is not acceptable to spend man-weeks or more working on an estimate and then fail to turn it in. If an electronic bid is to be turned in over the Internet, if the bid can be submitted and then revised later, a bid should be submitted as early as possible to constitute a test of the procedure.

LOADING HEAVYBID FROM THE INTERNET

HeavyBid/Student is a freeware estimating program for use by students with this book to work the problems that follow and for semester projects that involve computer estimating. This software is the student version of HeavyBid/ Advanced, the mid-level estimating product that is offered by HCSS in 2012. The student version has most of the functionality of the full version, but a few menu items are commented out and it has limitations on the size of the estimate. The software is located at the McGraw-Hill website www.mhhe.com/peurifoy_oberlender6e.

PROBLEMS

Before attempting the homework problems in this chapter, the reader should become familiar with HeavyBid/Student by going through the tutorial that is in the software. This will assist the student in understanding the features of the software and assist in preparing estimates for the problems.

The master estimate template, ESTMAST, has cost rates for resources, including labor, equipment, and materials. It also has standard crews for various types of work. The drop-down for a crew shows the composition of the crew, including labor and equipment. When a crew is selected, all of the resources in the crew are automatically brought into the estimate.

A thorough review of the contents of ESTMAST is important before using the software. In particular the crews should be printed and studied, to see the code, name, and the types of labor and equipment that are in each standard crew. Also, the material resources in ESTMAST should be printed and reviewed, particularly to see the subcontractors that are available. Knowing and understanding the data in ESTMAST will save substantial time in using the software.

Other information required for preparing an estimate includes production rates, company bidding rules, markup instructions, and bond rates. Tables 22.1 through 22.4 give this information and should be used in preparing estimates for the homework problems in this chapter.

Table 22.1 provides historical production rates for various types of work for use in the homework problems. The historical production rates that are given as quantities per hour refer to the crew-hours of the standard crews used in HeavyBid/Student. Table 22.2 gives company bidding rules that apply to all bids. Markup instructions are given in Table 22.3, and the payment and performance bond rates are given in Table 22.4.

Crews and historical production rates are essential to estimating. The process of developing a computer estimate involves selecting a crew from ESTMAST for each type of work and then selecting a production rate for the crew from Table 22.1. To build an estimate, this process is repeated for each type of work in the project.

TABLE 22.1 | Historical production rates.

Work item	Production rate
Concrete—Structural	
Formwork—walls	4.50 sf/mh
Formwork—slabs	8.25 sf/mh
Formwork—footings	12.00 sf/mh
Formwork—chamfer strip	50.00 lin ft/mh
Formwork lumber—job built	2.5 bf/SFCA
Install embedded misc. metal	1.00 hr/piece
Reinforcing steel—walls	125.00 lb/mh
Reinforcing steel—slabs	190.00 lb/mh
Reinforcing steel—footings	175.00 lb/mh
Place concrete—walls	30 cy/hr
Place concrete—slabs	45 cy/hr
Place concrete—footings	15 cy/hr
Place concrete—bridge deck slabs	66 cy/sh
Setting precast bridge girders	2 units/hr
Setting precast bridge deck planks	4 units/hr
Rub and finish walls	0.02 mh/sf
Concrete paving	
Set and strip 6-in. forms	120 lin ft/hr
Set and strip 9-in. forms	110 lin ft/hr
Reinforcing mesh	50 sy/mh
Slip formed curb—9-in. × 18-in.	300 lin ft/sh
Place stone base	80 ton/hr
Place and finish 6-in. concrete paving	130 sy/hr
Place and finish 9-in. concrete paving	120 sy/hr
Clean construction joint (labor only)	10.00 sf/mh
Spray cure surface	100.00 sf/mh
Earthwork	
Site clearing	1.5 ac/sh
Bulk excavation—over 100,000 cy	175 cy/hr
Bulk excavation—under 100,000 cy	140 cy/hr
Structural excavation–backhoe/FEL	80 cy/hr
Place and compact fill (bulk with equip.)	125 cy/hr
Place and compact fill (structures)	50 cy/hr
Hand place and compact fill	10 cy/hr
Fine grade for paving	300 sy/hr
Rip rap—hand placed	1.25 mh/ton
Drainage (includes average trenching and backfill)	
Install 12-in. RCP or CMP	200 lin ft/sh
Install 15-in. RCP or CMP	180 lin ft/sh
Install 18-in. RCP or CMP	170 lin ft/sh
Install 24-in. RCP or CMP	150 lin ft/sh
Install 36-in. RCP or CMP	120 lin ft/sh
Install 48-in. RCP or CMP	100 lin ft/sh
Install precast inlet	2 hr/piece
Perforated drains (install pipe only)	50 lin ft/hr

TABLE 22.2 | Company bidding rules.

1. All bids will include a superintendent with pickup truck for the duration of the project.
2. All bids will include one surveyor with a pickup truck for 25% of the project duration.
3. All bids will include project overhead costs for a portable toilet, office costs, mobile phone charges, and office/engineering expenses.
4. Mobilization will include one load of equipment haul each way by subcontract for each off-road piece of equipment required on the jobsite.
5. All bids containing imported borrow material will include 15% additional volume purchased to allow for compaction.
6. Purchased stone products will be calculated at 1.5 ton/cy in-place volume.
7. All bids will be estimated at straight-time rate (40hr/week) unless schedule concerns dictate overtime to be required.
8. Any indirects, markups, and bond should be spread to all the biditems based on the total cost of each biditem.
9. Any variation from these guidelines must be approved in advance by the vice president of the Estimating Department.

TABLE 22.3 | Markup instructions.

Item	Markup
Labor and burden	25%
Materials	5%
Equipment	10%
Subcontracts	5%

TABLE 22.4 | Payment and performance bond rates.

Amount	Rate
First 100,000	$7.25 per thousand
$100,000 to $500,000	$6.50 per thousand
$500,000 to $2,000,000	$5.50 per thousand
$2,000,000 to $5,000,000	$4.70 per thousand
$5,000,000 and higher	$4.20 per thousand

22.1 This problem is designed to illustrate the use of the HeavyBid/Student software. It is representative of what an estimator would do when going from a manual estimate to a computer estimate. This problem provides an already completed estimate that needs to be entered into the computer. The activities required to perform the work in each biditem are given in this problem.

Create the estimate from the master estimate entitled "Student Problem No. 1 Master Estimate" which contains the necessary resources and crews. Follow the company bidding rules of Tables 22.2 through 22.4 for preparing this estimate. Table 22.1 is not needed because the production rates for this project are given in this problem. In the bid summary spread the indirects, markups, and bond on the total direct cost less subcontract cost.

Project Information:

Estimate No.	E-001
Estimate Name:	Flood Plain Diversion Channel
Owner:	Adventure Development Company
Engineer:	Site-Work Engineers, Inc.
Project Duration:	6 months
Bid Date:	3 weeks from today

Biditems:

Code	Description	Quantity
1	Clear site and strip topsoil	4.6 acre
2	Rough site grading	37,155 cy
3	Excavate—cut channel to grade	27,782 cy
4	Concrete channel lining	3,334 cy
5	Furnish and install motorized sluice gates	1 set
6	Electrical work	ls

Activities for each Biditem:

Biditem 1
  A. Remove and stack brush—4.6 acres; clearing crew; 62 mh/acre
  B. Strip and stockpile topsoil—3,710 cy; grading/clearing crew; 1,100 cy/shift

Biditem 2
  A. Level site—excavation crew; 0.075 mh/cy
  B. Haul material offsite—subcontractor using 18-cy trucks

Biditem 3
  A. Cut channel—24,520 cy; excavation crew; 0.09 mh/cy
  B. Cut slopes and final grade—3,262 cy: excavation crew; 0.115 mh/cy
  C. Haul material offsite—subcontractor using 18-cy trucks

Biditem 4
  A. Buy concrete—3% waste; include curing compound @1 gal per 400 sf
  B. Place concrete in walls—1,445 cy; concrete placing crew; 0.14 mh/cy
  C. Place concrete in slab—51,300 sf; concrete paving crew; 5,600 sf/shift
  D. Paving forms, 12-in. high—650 linear feet for 3 months
  E. Formwork, steel panel—5,800 sf for 4 months
  F. Set and strip forms from walls—31,450 sf; formwork crew; 0.3 mh/sf
  G. Reinforcing steel—416,800 lb; buy steel and install by subcontract

Biditem 5
  A. Mechanical subcontractor—$290,000

Biditem 6
  A. Electrical subcontractor—$84,000

22.2 Adventure Development Company is adding an additional access road between sections of their Pleasant Valley Project. The road will join existing pavements at both ends. Excavation and preparation of the subgrade have been performed by others. The project owner is a private client and does not require a payment and performance bond.

Prepare a computer estimate for the project using labor and equipment resources, crews, material and subcontract prices from ESTMAST. Use the historical production rates, company bidding rules, and markups in Tables 22.1 through 22.3. Company rules in Table 22.2, Items 1 through 3 (superintendent, surveyor, and overhead cost), will be included in the Mobilization Biditem. Project information for preparing the estimate is detailed next.

Project Information:

Estimate No.:	E-002
Estimate Name:	New Road and Drainage
Owner:	Adventure Development Company
Engineer:	Road Engineers, Inc.
Project Duration:	7 weeks
Bid Date:	2 weeks from today

Scope of Work:

Paving dimensions:	2,560-ft long × 27-ft wide (exclusive of curb) × 9-in. thick
Road details:	Reinforced mesh (6 × 6 × W2.9) and 3,500 psi concrete, surface to be sprayed with concrete curing compound
Contraction joints:	Saw cut across width at 15-ft on centers along the entire length
Curb:	9-in. wide × 18-in. high along both edges, non-reinforced, installed by slip-form equipment
Base course:	6-in. of crushed stone
Drainage under road:	Corrugated metal pipe; 2 @ 18-in. dia. and 3 @ 24-in. dia. All culverts are 32 ft long with flared end sections at each end. A small size drainage crew will be used.
Landscaping:	Grassing to be included along 30 ft on both sides for the entire length of the pavement. Grass to be watered for 2 months after installation.

Biditems:

Code	Description	Quantity	Unit
10	Mobilization	1	ls
20	Furnish and place aggregate base course	???	cy
30	Furnish and place 9-in. reinforced concrete paving	???	sy
40	Saw cut contraction joints	???	lin ft
50	Concrete curb	???	lin ft
60	Drainage culverts under roadway	1	ls
70	Landscaping (grassing)	???	sy

Note that ??? denotes quantities are to be determined from your takeoff.

22.3 Radio station WBBC is adding a new transmitter tower and is asking for a bid to build the concrete base and anchor blocks. The project owner is a private client and does not require a payment and performance bond.

Prepare a computer estimate for the project using labor and equipment resources, crews, material, and subcontract prices from ESTMAST. Use the historical production rates, company bidding rules, and markups in Tables 22.1 through 22.3. Company rules in Table 22.2, Items 1 through 3 (superintendent, surveyor, and overhead cost), will be included in the Mobilization Biditem. Project information for preparing the estimate is detailed next.

Project Information:

Estimate No.:	E-003
Estimate Name:	New radio tower anchor blocks—WBBC
Owner:	Multi-City Communications, Inc.
Engineer:	Transmission Engineers, Inc.
Project Duration:	3 weeks
Bid Date:	2 weeks from today

Scope of Work:

All footings are concrete cast directly on the excavated subgrade. To allow for formwork, the excavation needs to be 5 ft larger than the dimensions of the concrete blocks. The top of the concrete is even with the surrounding ground level. Each footing is cast monolithically with 3,500 psi concrete and receives a spray-on curing compound on the upper surface and on all four sides. Granular backfill is placed around the footings.

Tower footing	1 each at 24-ft wide × 24-ft long × 6-ft deep with 16 steel embeds at 125 lb each with an average of 100 lb/cy of grade 60 reinforcing steel.
Anchor blocks	4 each at 18-ft wide × 18-ft long × 5-ft deep with 4 steel embeds at 175 lb each with an average of 95 lb/cy of grade 60 reinforcing steel.

Biditems:

Code	Description	Quantity	Unit
10	Mobilization	1	ls
20	Site clearing	3	ac
30	Excavate for anchor blocks and footing	???	cy
40	Concrete anchor blocks and footing	???	cy
50	Furnish and install steel embeds	???	lb
60	Hand place granular fill and compact	???	cy

Note that ??? denotes quantities are to be determined from your takeoff.

22.4 A rural county road project requires demolition of an existing 30-ft wide and 60-ft long concrete bridge over a small dry stream and replacing it with new abutments, six precast girders, precast deck planks, and reinforced concrete deck slab.

    Prepare a computer estimate for the project using labor and equipment resources, crews, material, and subcontract prices from ESTMAST. Use the historical production rates, company bidding rules, markups, and payment/performance bond rates in Tables 22.1 through 22.4. Project information for preparing the estimate is detailed next.

Project Information:

Estimate No.:	E-004
Estimate Name:	Bridge Modernization
Owner:	Rural Garfield County
Engineer:	Bridge Engineers, Inc.
Project Duration:	8 weeks
Bid Date:	2 weeks from today

Scope of Work:

Demolition: Based on a visit to the jobsite it will require 10 days to demolish the existing bridge and haul the aggregate to a yard where it will be recycled. The existing foundation piles are in good condition and will be reused, but the abutment walls to support new girders and decking will be replaced.

Earthwork: A backhoe will excavate 200 cy of earth for installation of the new abutment. After the abutment is installed, the 200 cy of excavated earth will be backfilled and compacted by hand. Excavation around the abutment will be slow; therefore, use 10 cy/hr for excavation and use 5 cy/hr for compaction, rather than the values shown in Table 22.1.

Abutments: The new abutments will be cast on-grade where the existing piles are protruding out of the ground. The new abutments will be 30 ft long, 6 ft wide, and 4 ft high. A 12-in.-thick back wall will rest on the abutment that will rise 4 ft 8 in. to the level of the deck slab. The back wall will extend 8 ft straight back at the ends of the abutment to contain the fill. Abutments and walls will require 200 lb of rebar per cy of concrete. Job built forms will be constructed of lumber and $\frac{3}{4}$-in. plywood. Form lumber is 2.5 bf per sq. ft of forms and plywood is 1.15 sf per sq. ft of forms based on 1 use. For this job the forms will be used 4 times. For the abutment the rebar is grade 60 and the concrete is 5,500 psi.

Precast girders: The six precast girders are 2-ft wide, 4-ft high, and 60 ft long.

Precast deck planks: Precast deck planks 4-ft long, 3-ft wide, and 6-in. thick will be placed on the girders. The 4-ft dimension of the planks will be placed in the transverse direction of the roadway. A plank will rest 2.4 in. on adjacent girders.

Deck slab and curbs: The deck slab will be 8-in.-thick cast-in-place reinforced concrete with a broom finish. There will be a cast-in-place concrete curb on each of the 60-ft dimensions of the bridge. The curbs will be 18-in. wide and 12-in. high. The deck slab and curbs will require 200 lb of rebar per cy of concrete. Rebar is grade 60 steel and concrete is 5,500 psi. Use 3% waste for concrete. Form lumber from the abutments and walls will be used for the deck slab and curb.

Cross-section of bridge for calculating reinforcing steel and concrete in the bridge:

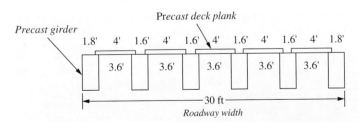

Precast girders are 2-ft wide and 4-ft high × 60-ft long
Precast deck planks are 4-ft long × 3-ft wide × 6-in. thick
Cast-in-place concrete is 8-in. thick over precast deck planks

Biditems:

Code	Description	Quantity	Unit
1	Demolition/removal of existing bridge	1	ls
2	Excavation and backfill of abutments	200	cy
3	Abutments with walls	70	cy
4	Precast girders	6	each
5	Precast deck planks	100	each
6	Cast-in-place concrete deck and curb	63	cy

Note 1 – There is no mobilization biditem, therefore mobilization costs will need to go into an indirect item that ultimately gets spread into other biditems.

Instructions to Estimator:

For this project a special crane will be used, a 3900 Crane. Because of the size of the crane it will require 7 loads to transport the crane to and from the job. It will require 8 hours to set up the crane and 8 hours to take down the crane.

Some of the equipment that will be used for this project is different than the equipment shown in the standard crews in the computer software. One example is the special 3900 Crane. For this project five of the crews will use equipment different than the equipment in the standard crews. Therefore, the estimator must modify the following standard crews. It is helpful to have a print-out of the crews before modifying them to be able to compare the crews before and after the changes.

CPP Crew – Remove the truck and driver. Add a 3900 Crane and operator and a 2-cy bucket.

CSF Crew – Remove the truck and driver. Add a 185 cfm air compressor.

CSP Crew – Replace the hydraulic crane with 3900 Crane and add a 185 cfm air compressor and a 2-cy bucket. Also add 2 more concrete placing laborers (CPL).

EXS Crew – Remove the excavator and roller, and replace it with a backhoe.

FHC Crew – Replace the 8B416 backhoe with a 8B426 backhoe.

Example Bid Documents

This appendix contains the plans (drawings) and Division 0 of the written specifications for a project. A complete set of the technical specifications (Divisions 0 through 16) can be obtained from the McGraw-Hill website at www.mhhe.com/ peurifoy_oberlender6e.

MAINTENANCE FACILITY PROJECT
PROJECT OWNER'S NAME
CITY, STATE

INDEX of SHEETS

	COVER	
CIVIL	C1	SITE PLAN
ARCHITECTURAL	A1	INTERIOR ELEVATION
	A2	INTERIOR ELEVATION
	A3	EXTERIOR ELEVATION
	A4	SECTIONS
	A5	SECTIONS
	A6	FOUNDATION PLAN
MECHANICAL	M1	HVAC FLOOR PLAN
PLUMBING	P1	PLUMBING PLAN
ELECTRICAL	E1	ELECTRICAL POWER
	E2	ELECTRICAL LIGHTING

ARCHITECT - ENGINEER

ABC DESIGN FIRM
ARCHITECTURE - ENGINEERING SERVICES

3456 STREET ZIP
CITY, STATE
PHONE NUMBER

CONTACT: Architect / Engineer Name

BUILDING CODES

This project has been designed in
accordance with:
1996 BOCA National Building Code.
1995 International Plumbing Code
    with 1996 Supplement.
1996 International Mechanical Code.
1996 National Electrical Code.
1994 Life Safety Code.

All work shall be performed in
accordance with these codes.

Project
Owner's
Name

Street Address

City, State

John Smith, P.E.
Owners Rep.
Street Address
City, State
Phone Number

SEAL

DRAWN BY JDC M-D-YR
ISSUE DATE M-D-YR
REVISION NUMBER 0

LOCATION PLAN

SHEET NAME:

MAINTENANCE FACILITY PROJECT
STREET ADDRESS
CITY, STATE

PROJECT NAME:

BUILDING NO. 637

PROJECT NO.
WORK ORDER

DRAWING NO.   OF 12

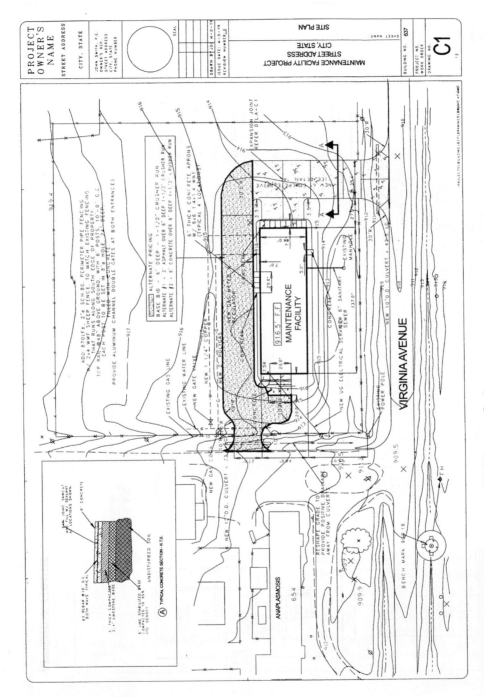

APPENDIX FIGURE 1 | Site Plan (Civil Drawing C–1)

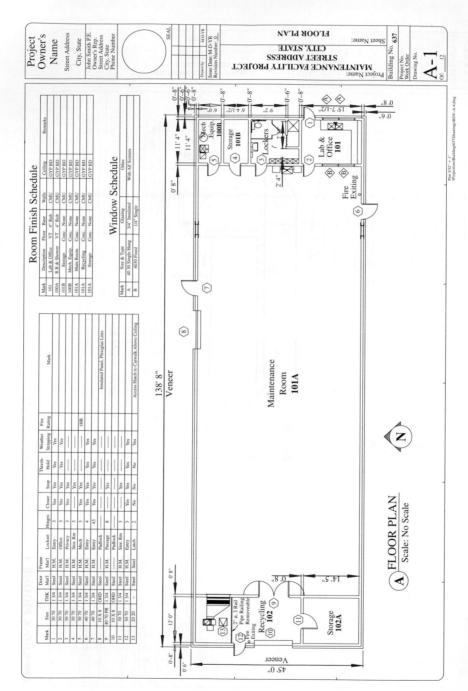

APPENDIX FIGURE 2 | Floor Plan (Architectural Drawing A–1)

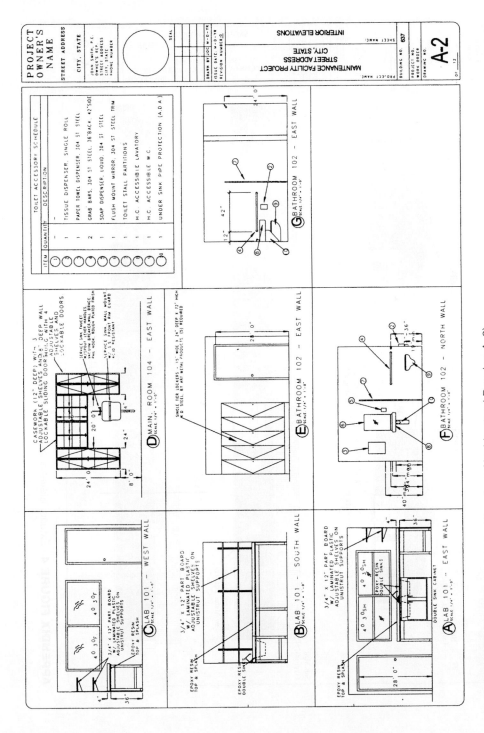

APPENDIX FIGURE 3 | Interior Elevations (Architectural Drawing A-2)

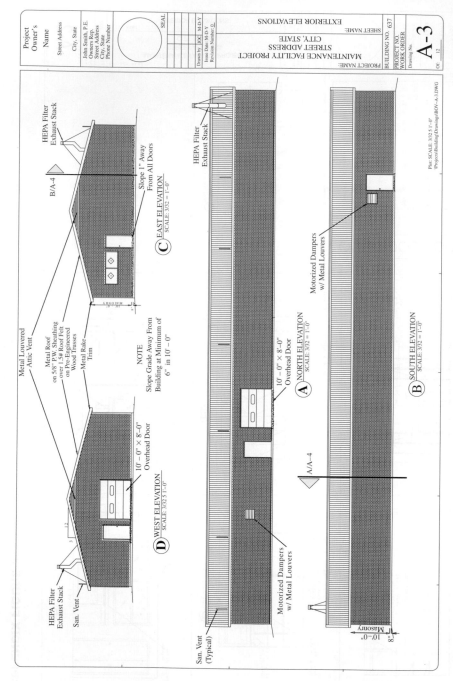

APPENDIX FIGURE 4 | Exterior Elevations (Architectural Drawing A–3)

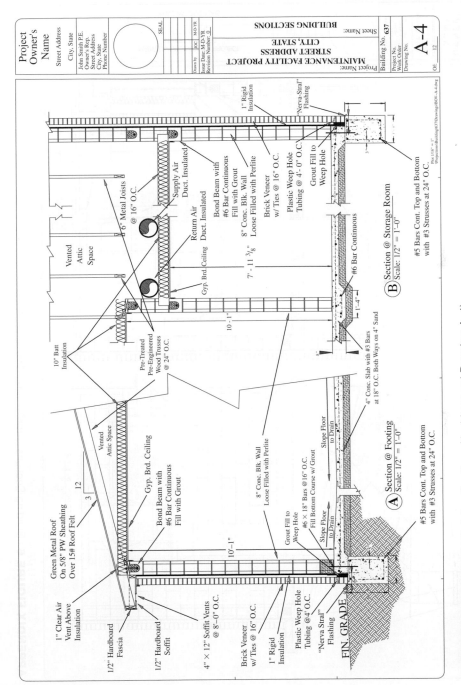

APPENDIX FIGURE 5 | Building Sections (Architectural Drawing A–4)

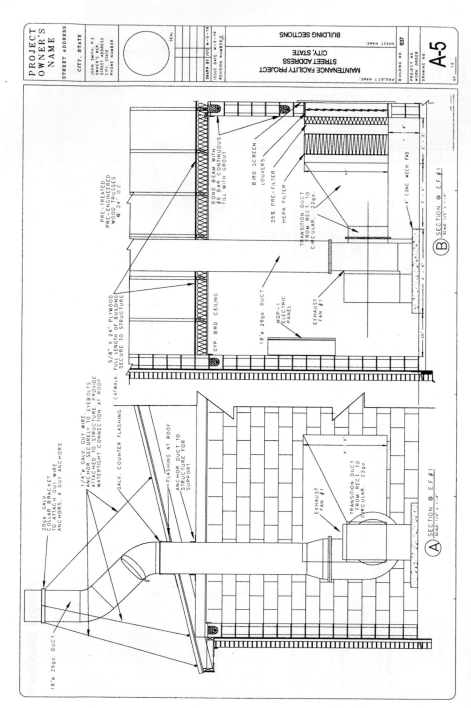

APPENDIX FIGURE 6 | Building Sections (Architectural Drawing A–5)

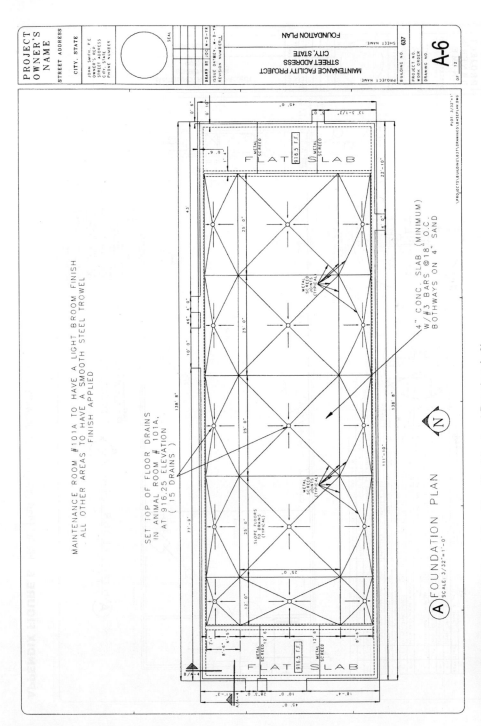

APPENDIX FIGURE 7 | Foundation Plan (Architectural Drawing A–6)

521

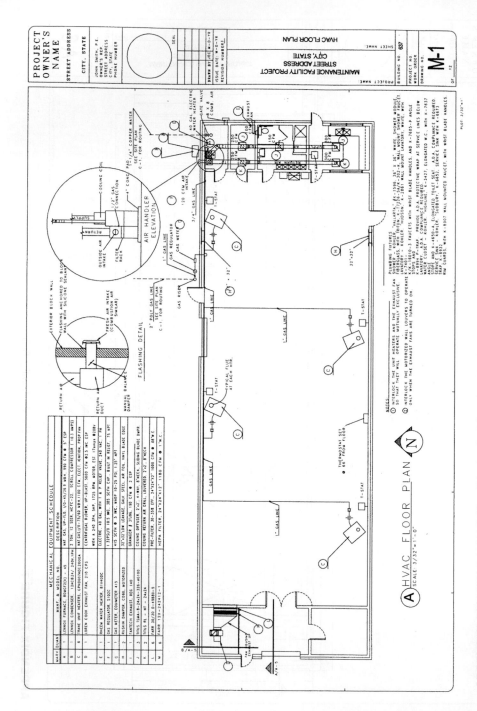

APPENDIX FIGURE 8 | HVAC Plan (Mechanical Drawing M–1)

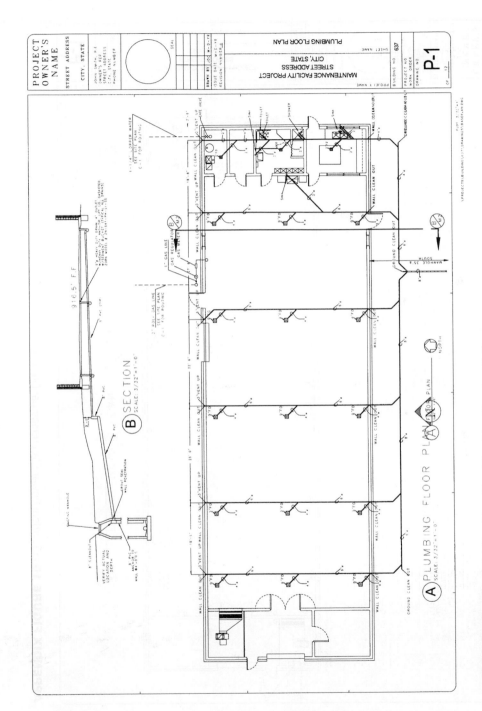

APPENDIX FIGURE 9 | Plumbing Plan (Plumbing Drawing P-1)

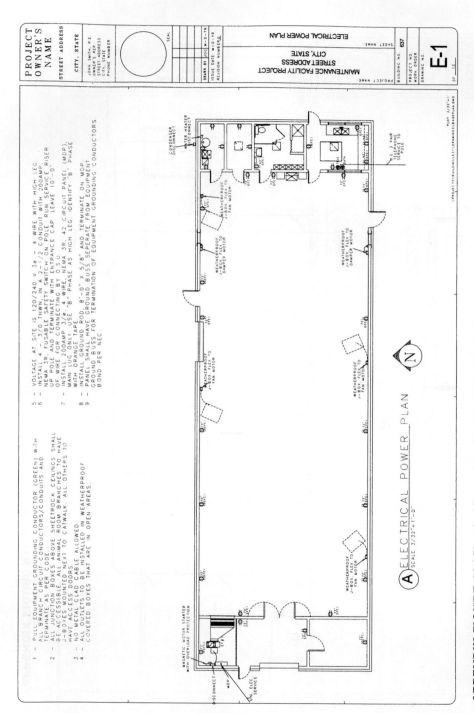

APPENDIX FIGURE 10 | Electrical Power Plan (Electrical Drawing E–1)

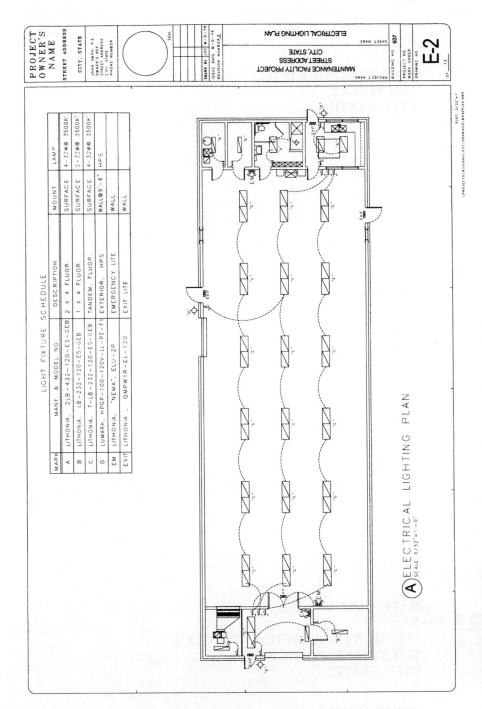

APPENDIX FIGURE 11 I Electrical Lighting (Electrical Drawing E–2)

BID DOCUMENTS

FOR:

MAINTENANCE FACILITY PROJECT
1234 STREET
CITY, STATE

DATE: JANUARY 20xx

MAINTENANCE FACILITY PROJECT
1234 STREET, ROOM 200
CITY, STATE

SET NUMBER _____

	INDEX
DIVISION 0—BIDDING AND CONTRACT DOCUMENTS	*Pages*
SECTION 00001: ADVERTISEMENT FOR BIDS	1
SECTION 00002: INSTRUCTIONS TO BIDDERS	1–5
SECTION 00003: BID FORM	1–4
SECTION 00004: AFFIDAVIT OF NONDISCRIMINATION, NONSEGREGATED FACILITIES, NONCOLLUSION, AND BUSINESS RELATIONSHIPS	1–2
SECTION 00005: BID BOND	1–2
SECTION 00006: CONTRACT	1–2
SECTION 00007: AFFIDAVIT	1
SECTION 00008: PERFORMANCE PAYMENT BOND	1–2
SECTION 00009: STATUTORY BOND	1–2
SECTION 00010: WARRANTY BOND	1–2
SECTION 00011: AFFIDAVIT (RELEASE OF LIENS, ETC.)	1
SECTION 00012: CERTIFICATE OF APPROVAL (SURETY'S CONSENT OF RELEASE OF LIENS)	1
SECTION 00013: GENERAL CONDITIONS	1
SECTION 00014: SUPPLEMENTARY CONDITIONS	1–9
DIVISION 1—GENERAL REQUIREMENTS	1–11
DIVISION 2—SITEWORK	
SECTION 02200—EARTHWORK	1–4
SECTION 02282—TERMITE CONTROL	1–2
SECTION 02511—HOT-MIXED ASPHALT PAVING	1–4
DIVISION 3—CONCRETE	
SECTION 03300—CAST-IN-PLACE CONCRETE	1–6
DIVISION 4—MASONRY	
SECTION 04200—UNIT MASONRY	1–9
DIVISION 5—METALS (NONE)	
DIVISION 6—WOOD AND PLASTICS	
SECTION 06105—MISCELLANEOUS CARPENTRY	1–2
SECTION 06192—WOOD TRUSSES	1–2
DIVISION 7—THERMAL AND MOISTURE PROTECTION	
SECTION 07210—BUILDING INSULATION	1–2

SECTION 07411—MANUFACTURED ROOF PANELS   1–2
SECTION 07901—JOINT SEALANTS   1–2

DIVISION 8—DOORS AND WINDOWS

SECTION 08111—STEEL DOORS AND FRAMES   1–2
SECTION 08360—SECTIONAL OVERHEAD DOORS   1–2
SECTION 08520—ALUMINUM WINDOWS   1–2
SECTION 08710—DOOR HARDWARE   1–3

DIVISION 9—FINISHES

SECTION 09250—GYPSUM BOARD ASSEMBLIES   1–2
SECTION 09650—RESILIENT FLOORING   1–2
SECTION 09678—RESILIENT WALL BASE   1
SECTION 09900—PAINTING   1–2

DIVISION 10—SPECIALTIES

SECTION 10155—TOILET COMPARTMENTS   1
SECTION 10500—METAL LOCKERS   1
SECTION 10522—FIRE EXTINGUISHERS AND
                       WALL BRACKETS   1
SECTION 10800—TOILET AND BATH ACCESSORIES   1–2

DIVISION 11—EQUIPMENT (NONE)

DIVISION 12—FURNISHINGS

SECTION 12346—WOOD LABORATORY CASEWORK
AND FIXTURES   1–10

DIVISION 13—SPECIAL CONSTRUCTION (NONE)

DIVISION 14—CONVEYING SYSTEMS (NONE)

DIVISION 15—MECHANICAL

SECTION 15100—PLUMBING   1–3
SECTION 15500—HEATING, VENTILATING, AND
AIR CONDITIONING   1–2

DIVISION 16—ELECTRICAL

SECTION 16100—ELECTRICAL   1–4

****** END OF INDEX******

DIVISION 0—**BIDDING REQUIREMENTS, CONTRACT FORMS, AND CONDITIONS OF THE CONTRACT**

<u>SECTION 00001</u>—**ADVERTISEMENT FOR BIDS:**

ADVERTISEMENT FOR BIDS:

BIDS: 3:00 P.M., FEBRUARY 19, 20XX

The OWNER, will receive bids on:

MAINTENANCE FACILITY PROJECT
1234 STREET
CITY, STATE

at the office of **OWNER, located in the Room 200 at 1234 Street in City, State until 3:00 P.M., FEBRUARY 19, 20XX,** and then publicly opened and read aloud at a designated place. Bids received after this time or more than ninety-six (96) hours excluding Saturdays, Sundays, and holidays before the time set for the opening of bids will not be accepted. Bids must be turned in at the above office during the time period set forth. All interested parties are invited to attend.

The contract documents may be examined in the office of ABC DESIGN FIRM, 3456 Street, City, State, Zip Code, and copies may be obtained there upon receipt of a letter of intent to pay **$100.00** for two (2) sets of drawings and specifications in the event that they are not returned in good condition within ten (10) days of the date of bid opening. Additional sets may be obtained by paying the cost of printing the drawings and assembly of specifications.

The OWNER reserves the right to reject any or all bids or to waive any minor informalities or irregularities in the bidding.

Each bidder must deposit, with his bid, security in the amount and form set out in the contract documents. Security shall be subject to the conditions provided in the Instruction to Bidders.

No bidder may withdraw his bid within sixty (60) days after the date of opening thereof. Attention of bidders is particularly directed to the statutory requirements and affidavit concerning nondiscrimination, nonsegregated facilities, noncollusion, and business relationship provisions.

OWNER

By _____ Chairman's Name _____
                   Chairman

1st Publication: January 29, 20XX

2nd Publication: February 5, 20XX

***** END OF SECTION 00001—ADVERTISEMENTS FOR BIDS *****

SECTION 00002—INSTRUCTIONS TO BIDDERS:
RECEIPT AND OPENING OF BIDS:

The OWNER invites bids on the referenced project on the Bid Form included in the Contract Documents, all blanks of which must be appropriately filled in. Bids will be received by the Owner at the time and place set forth in the Advertisement for Bids and then at said place be publicly opened and read aloud.

The Owner may reject any bid not prepared and submitted in accordance with the provisions hereof and may reject any and all bids or may waive any minor informalities or irregularities in the bidding.

Bids received prior to the time of opening will be kept unopened in a secure place. The officer whose duty it is to open them will decide when the specified time has arrived, and no bid received thereafter will be considered. No responsibility will attach to an officer for the premature opening of a bid not properly addressed and identified. Unless specifically authorized, *telegraphic, oral, facsimile or telephonic bids will not be considered,* but modifications by telegraph or facsimile of bids already submitted will be considered if received prior to the hour set for opening.

BID SUBMISSION:

Each bid must be submitted in a sealed envelope bearing on the outside the name of the bidder, his address, and the name of the project for which the bid is submitted, and addressed as specified in the Bid Form.

If forwarded by mail, the sealed envelope containing the bid must be enclosed in another envelope addressed as specified in the Bid Form.

The Owner shall not be responsible for failure by the postal service to timely deliver any bid.

SPECIAL BIDDING DOCUMENTS:

Attention is particularly called to those parts of the contract documents which must be attached to the Bid Form:
a. **Affidavit of Nondiscrimination, Nonsegregated Facilities, Noncollusion, and Business Relationships (pgs. 1 and 2, Section 00004).**
b. **Bid Bond (pgs. 1 and 2, Section 00005).**
c. **Power of Attorney.**

SUBCONTRACT:

Any person, firm, or other party to whom it is proposed to award a subcontract under this contract must be acceptable to the Owner.

TELEGRAPHIC MODIFICATION:

Any bidder may modify his bid by telegraphic or facsimile communication at any time prior to the scheduled closing time for receipt of Bids, provided such telegraphic or facsimile communication is received by the Owner prior to the closing time, and provided further, the Owner is satisfied that a written confirmation of the telegraphic or facsimile modification over the signature of the bidder was mailed prior to the closing time. The telegraphic or facsimile communication should not reveal bid price, but should provide the addition or subtraction to the Owner until the sealed bid is opened. If written confirmation is not received within two days from the closing time, no consideration will be given to the telegraphic or facsimile modification.

WITHDRAWAL OF BIDS:

Bids may be withdrawn on written or telegraphic or facsimile request received from bidders prior to the time fixed for opening. Negligence on the part of the bidder in preparing the bid confers no right for the withdrawal of the bid after it has been opened.

BASIS OF BID:

The bidder must include all unit cost items and all alternatives shown on the Bid Forms; failure to comply may be cause for rejection. No segregated bids or assignments will be considered.

QUALIFICATION OF BIDDER:

The Owner may make such investigation as he deems necessary to determine the ability of the bidder to perform the work, and the bidder shall furnish to the Owner all such information and data for this purpose as the Owner may request. The Owner reserves the right to reject any bid if the evidence submitted by, or investigation of, such bidder fails to satisfy the Owner that such bidder is properly qualified to carry out the obligations of the contract and to complete the work contemplated therein. Conditional bids will not be accepted.

BID SECURITY:

Each bid must be accompanied by a certified check made payable to the Owner in the sum of not less than five percent (5%) of the base bid, plus add alternates, if any, or a bond with sufficient sureties, to be approved by Owner, in a penal sum equal to five percent (5%) of the base bid, plus add alternates, if any, and naming the Owner as obligee, or for any bid of $2,000,000 or less an irrevocable letter of credit containing such terms as may be prescribed by the Owner and issued by a financial institution insured by the Federal Deposit Insurance Corporation or the Federal Savings and Loan Insurance Corporation for the benefit of the state, on behalf of the Owner in an amount equal to five percent (5%) of the bid. Such security shall be returned to all except the three (3) lowest responsible bidders within three (3) days after the opening of the bids, and the remaining security within three (3) days after the Owner and the bidder to whom the contract is awarded have executed the contract.

BID DEFAULT:

If the successful bidder fails or refuses to enter into a contract as required by the Owner or fails to provide the required bonds and insurance to the Owner, within the time limited, said bidder shall forfeit to the Owner the difference between the low bid of said defaulting bidder and the amount of the bid of the bidder to whom the contract is subsequently awarded and the cost, if any, of republication of Notice to Bidders and all actual expenses incurred by reason of the bidder's default. The amount of said forfeiture shall not exceed the total amount deposited as security and shall be forfeited to the Owner as liquidated damages and not as a penalty.

Negligence on the part of the bidder in preparing or submitting the bid confers no right for the withdrawal of the bid after it has been opened and shall not constitute a defense to or excuse from the requirements of this provision.

TIME OF COMPLETION:

Bidder must agree to commence work on or before a date to be specified in a written "Notice to Proceed" of the Owner and to substantially complete the project **within 180 consecutive calendar days thereafter.**

CONDITIONS OF THE WORK:

Each bidder must inform himself fully of the conditions relating to the construction of the project and employment of labor hereon. Failure to do so will not relieve a successful bidder of his obligation to furnish all material and labor necessary to carry out the provisions of his contract. Insofar as possible, the contractor, in carrying out his work, must employ such methods necessary to prevent any interruption of or interference with the work of any other contractor.

ADDENDA AND INTERPRETATIONS:

No interpretation of the meaning of the plans, specifications or other contract documents will be made to any bidder orally. Every request for such interpretation should be in writing addressed to:

ABC Design Firm
3456 Street
City, State Zip
CONTACT PERSON: Designer's name and phone number

and to be given consideration must be received at least ten days prior to the date fixed for the opening of bids. Any and all interpretations and any supplemental instructions will be in the form of written addenda to the specifications, which if issued, will be mailed to all prospective bidders (at the respective addresses furnished for such purpose), not later than five days prior to the date fixed for the opening of bids. Failure of any such bidder to receive any such addenda or interpretation shall not relieve such bidder from any obligation under his bid as submitted. All addenda so issued shall become a part of the contract documents.

POWER-OF-ATTORNEY:

The Attorney-in-Fact who signs bid bonds or contract bonds must file with each bond a certified and effectively dated copy of the Power-of-Attorney.

LAWS AND REGULATIONS:

The bidder's attention is directed to the fact that all applicable State Laws, Municipal Ordinances, and the rules and regulations of all authorities having jurisdiction over construction of the project shall apply to the contract throughout, and they will be deemed to be included in the contract the same as though herein written out in full.

METHOD OF AWARD:

If the order of acceptance of alternates is not a factor in determining the low bidder, the Owner may accept any or all alternates in any order of preference. If the acceptance or rejection of alternates or the order of acceptance of alternates is a factor the low bidder will be determined by the application of alternates in the order listed on the proposal form, at the time bids are opened, to the extent that the amount of budgeted funds will permit. After the low bidder has been so established, the Owner may elect to revise the extent or order of acceptance of alternates or may adjust the amount of budgeted funds so long as the selection of the low bidder is not affected.

The Owner reserves the right to reject any and all bids.

OBLIGATION OF BIDDER:

At the time of the opening of bids, each bidder will be presumed to have inspected the site and to have read and to be thoroughly familiar with the plans and contract documents (including all addenda). The failure or omission of any bidder to examine any form instrument or document shall in no way relieve any bidder from any obligation in respect of his bid.

Bidders shall carefully examine the documents and the construction site to obtain first-hand knowledge of the existing conditions, make actual examination of site, existing construction, and construction conditions prior to submission of bids. Examination of conditions includes sizes, quantities, functions, etc. Examination prior to bidding is the Contractor's responsibility.

Coordinate requests for examination through: Tim Mitchel, Ph. AC & Number

NOTICE OF REQUIREMENTS FOR AFFIDAVIT OF NONDISCRIMINATION, NONSEGREGATED FACILITIES, NONCOLLUSION, AND BUSINESS RELATIONSHIPS:

Bidders must execute and submit with the bid the affidavit pertaining to Nondiscrimination, Nonsegregated Facilities, Noncollusion, and Business

Relationships. The Certification of Nonsegregated Facilities must be executed in contracts exceeding $10,000 which are not exempt from the Provision of the Equal Opportunity Clause. Failure to execute the required form properly shall result in rendering the bid proposal non-responsive to the solicitation requirements.

As used in this certification, the term "segregated facilities" means in waiting rooms, work areas, restrooms and wash rooms, restaurants and other eating places, time clocks, locker rooms and other storage or dressing areas, parking lots, drinking fountains, recreation or entertainment areas, transportation, and housing facilities provided for employees which are segregated by explicit directive or are in fact segregated on the basis of race, creed, color or national origin, because of habit, local custom, or otherwise.

NOTICE OF REQUIREMENTS FOR AFFIDAVIT OF NONCOLLUSION:

The successful bidder must execute and submit with the contract the affidavit pertaining to noncollusion. Failure to execute the required form properly shall result in delay of execution of the contract.

SUBSTITUTIONS AND PRODUCT OPTIONS:

Where material is mentioned in the Specifications by trade name or manufacturer's name, the same is not a preference for said material, but the intention of using said name is to establish a type or quality of material. Material of other trade names or of other manufacturer's which is equivalent or better in type or quality **shall be submitted for approval 10 days prior to bidding** and shall be accompanied by actual full sized samples and such technical data as the Architect may need in order to compare the proposed materials with the materials which were specified. Approved substitutions shall be made by the Architect through the issuance of an Addendum. Any substitutions without prior approval are submitted at the Contractor's risk and may be rejected.

Before submitting materials or equipment for approval, the Contractor shall ascertain that it can be installed in the manner indicated and in the space available, and that it complies with the requirement of the Contract. Failure of the Contractor to do so shall not relieve him of the responsibility for furnishing suitable materials and equipment.

The Owner reserves the right to reject any proposed substitution which is submitted after bidding.

IRREVOCABLE LETTERS OF CREDIT:

The successful bidder for contracts in amounts of $100,000 or less may provide an irrevocable letter of credit for *each* bond in lieu of the performance payment bond, statutory bond, and the warranty bond as herein specified. *Each* letter of credit shall contain such terms as may be prescribed by the Owner and issued by a financial institution insured by the Federal Deposit Insurance Corporation or the Federal Savings and Loan Insurance Corporation for the benefit of the state, on behalf of the Owner in an amount equal to the full contract amount. Letters of credit may not be substituted for bonds required by contracts in excess of $100,000.

SURETY:

The Surety Company required to execute all bonds shall be authorized to transact business in the State in accordance with Title XX, State Statutes, Sections 481 et. seq.

MINORITY STATUS:

The contractor shall indicate on the bid form if a claim is being made for minority bid preference in accordance with the State Minority Business Enterprise Assistance Act (Title XX, Section YY of the State Statutes). Such a claim, if desired, must be indicated on the bid form at the time of receipt of bids. Claims for preference after receipt of bids will not be allowed.

****** SECTION 00002—INSTRUCTIONS TO BIDDERS *****

<u>SECTION 00003</u>—**BID FORM:**

Bidders shall note that bid must be made on this form. Amounts of the bid shall be completely filled in, both in figures and in writing. Enclose the bid form in a sealed envelope plainly marked and addressed as follows:

1. The name and address of the bidder shall appear in the upper left hand corner of the envelope.

2. The lower left hand corner of the envelope shall be marked as follows:

BID FOR: **MAINTENANCE FACILITY PROJECT**
                        1234 STREET
                        CITY, STATE

3. The envelope shall be addressed to:

                        OWNER
                        1234 Street, Suite 200
                        City, State Zip Code

4. Failure of the bidder to execute and submit the following forms with his proposal will result in disqualification of his bid:

a. Affidavit of Nondiscrimination, Nonsegregated Facilities, Noncollusion, and Business Relationships.

b. Bid Security.

c. Power of Attorney (with Surety Bonds).

SECTION 00003—**BID FORM:**

Place:

Date:

PROPOSAL OF

(hereinafter called "Bidder"), a

(corporation/partnership/an individual) doing business as

TO: The Owner.

The bidder, in compliance with your invitation for Bids for the:

MAINTENANCE FACILITY PROJECT
1234 STREET
CITY, STATE

having examined the plans and specifications with related documents and the site of the proposed work, and being familiar with all of the conditions surrounding the work, including the availability of materials and labor, hereby agree to furnish all labor, materials, equipment and supplies, and to perform the work required by the project in accordance with the contract documents, within the time set forth in Instructions to Bidders, and at the prices stated below. These prices are to cover all expenses incurred in performing the work required by the contract documents, of which this bid is a part.

Bidder acknowledges receipt of the following addenda: _____.

All Bid amounts shall be shown in both WORDS and FIGURES. In case of discrepancy, the amount shown in words will govern.

BASE BID:

Bidder agrees to perform all of the work described in the plans and specifications as being in the Base Bid for the sum of: _____ Dollars.

$_____

The Bidder agrees to perform all of the work described in the Drawings and Specifications, which has been designated as Alternate Bids. Alternate Bids shall be add amounts to the Base Bid. Changes shall include any modifications of the work or additional work that the Bidder may require to perform by reason of Owner's acceptance of any or all Alternate Bids.

ALTERNATE NO. 1 :

Add 3″ thick Type "C" asphalt paving to compacted crushed limestone as shown on Sheet C-1.

ADD: $_____.

_____ Dollars.

ALTERNATE NO. 2:

Added cost difference to provide 6″ thick reinforced concrete paving in lieu of asphalt paving as shown on Sheet C-1.

ADD: $_____.

_____ Dollars.

The Bidder hereby agrees to commence work under this contract on or before a date to be specified in a written "Notice to Proceed" by the Owner and to substantially complete the project **within 180 consecutive calendar days thereafter.**

The Owner reserves the right to reject any and all bids.

Bidder agrees that this bid shall be good and shall not be withdrawn for a period of sixty (60) calendar days after the opening thereof.

In the event a contract is awarded by the Owner to the Successful Bidder, it shall be executed within thirty (30) days. The Bidder shall return with his executed contract the Performance Payment Bond, Statutory Bond and Warranty Bond as required by the Supplementary Conditions.

If the successful bidder fails or refuses to enter into a contract as required by the Owner or fails to provide the required bonds and insurance to the Owner, within the time limited, said bidder shall forfeit to the Owner the difference between the low bid of said defaulting bidder and the amount of the bid of the bidder to whom the contract is subsequently awarded and, the cost, if any, of republication of notice to bidders and all actual expenses incurred by reason of the bidder's default. The amount of said forfeiture shall not exceed the total amount deposited as security and shall be forfeited to the Owner as liquidated damages and not as a penalty. Negligence on the part of bidder in preparing or submitting the bid confers no right for the withdrawal of the bid after it has been opened and shall not constitute a defense to or excuse from the requirements of this provision.

(SEAL) if bid
is by Corporation

Respectfully submitted,

By: _____

Title

_____

FEI/SS Number_____

****

Is a claim for minority preference being made?

Yes _____ No _____

If so, give certificate number _____.

****** END OF SECTION 00003—BID FORM *****

SECTION 00004—AFFIDAVIT OF NONDISCRIMINATION, NON-SEGREGATED FACILITIES, NONCOLLUSION, AND BUSINESS RELATIONSHIPS

STATE OF   _____)

COUNTY OF   _____)

The undersigned, of lawful age, being first duly sworn upon oath, deposes and states that I am the duly authorized agent of the bidder submitting the attached bid and am authorized by said bidder to execute the within affidavit.

I further swear that if said bidder is successful on this project, it will not discriminate against anyone in employment or employment practice because of race, color, religion, sex, or national origin. The undersigned further states that said bidder will comply with all federal and state laws and executive orders concerning the subject of nondiscrimination.

The undersigned further states that said bidder does not and will not maintain or provide for its employees any segregated facilities as defined in the instructions to bidders for this project at any of its establishments, nor permit its employees to perform their services at any location under its control, where segregated facilities are maintained. The bidder further agrees that a violation of this certification is a breach of the equal opportunity clause of this bid and any contract awarded pursuant thereto. Said bidder further agrees that (except where it has obtained identical certification from proposed subcontractors for specific time periods), it will obtain identical certifications from proposed subcontractors prior to the award of subcontracts exceeding Ten Thousand Dollars ($10,000) which are not exempt from the provisions of the equal opportunity laws, and that said bidder will retain such certifications in its files.

The undersigned further states that, for the purpose of certifying the facts pertaining to the existence of collusion among bidders and between bidders and state officials or employees, as well as facts pertaining to the giving or offering of things of value to government personnel in return for special consideration in the letting of any contract pursuant to the bid to which this statement is attached; that I am fully aware of the facts and circumstances surrounding the making of the bid to which this statement is attached and have been personally and directly involved in the proceedings leading to the submission of such bid; and that neither the bidder nor anyone subject to the bidder's direction or control has been a party: (a) to any collusion among bidders in restraint of freedom of competition by agreement to bid at a fixed price or to refrain from bidding; nor (b) to any collusion with any state official or employee as to quantity, quality or price in the prospective contract, or as to any other terms of such prospective contractor; (c) in any discussions between bidders and any state official concerning exchange of money or other thing of value for special consideration in the letting of a contract.

The undersigned further states that any partnerships, joint ventures, or other business relationships that are not in effect, or existed within one (1) year prior to this statement, with the architect, engineer or other party to this project; or any such business relationships between any officer or director of the bidder and any officer or director of the architectural or engineering firm or other party to the project are described as follows:

_____

NAME OF BIDDER: _____

BY: _____

Affiant

Subscribed and sworn to before me this _____ day of _____, 20____.

My commission expires: _____

Notary Public

****** END OF SECTION 00004—NONCOLLUSION AFFIDAVIT *****

SECTION 00005—**BID BOND**

KNOW ALL PERSONS BY THESE PRESENTS, that we, the undersigned _____

_____

as Principal, and _____

as Surety, authorized to transact business in the State, are hereby held and firmly bound unto Owner, in the penal sum of _____ for the payment of which, well and truly to be made, we hereby jointly and sever-ally bind ourselves, our heirs, executors, administrators, successors and assigns.

Signed this _____ day of _____, 20_____.

The condition of the above obligation is such that whereas the Principal has sub-mitted to **OWNER** a certain bid, attached hereto and hereby made a part hereof to enter into a contract in writing, for the

MAINTENANCE FACILITY PROJECT
1234 STREET
CITY, STATE

NOW THEREFORE,

(a) If said Bid shall be rejected, or in the alternate,
(b) If said Bid shall be accepted and the Principal shall execute and deliver a Contract in the Form of Contract attached hereto (properly completed and in accordance with said Bid) and shall furnish all bonds required by the specifications and shall in all other respects perform the agreement created by the acceptance of said Bid,

then this obligation shall be void, otherwise the same shall remain in force and effect; it being expressly understood and agreed that the liability of the Surety for any and all claims hereunder shall, in no event, exceed the penal amount of this obligation as herein stated.

The Surety, for value received, hereby stipulates and agrees that the obliga-tions of said Surety and its bond shall be in no way impaired or affected by any extension of the time within which the Owner may accept such Bid; and said Surety does hereby waive notice of any such extension.

IN WITNESS WHEREOF, the Principal and the Surety have hereunto set their hands and seals, and such of them as are corporations have caused their cor-porate seals to be hereto affixed and these presents to be signed by their proper officers, the day and year first set forth above.

_____
Principal

_____
Surety

SEAL   BY _____
Attorney-in-Fact

****** END OF SECTION 00005—BID BOND *****

SECTION 00006—**CONTRACT**

THIS AGREEMENT made this _____ day of _____, 20_____, by and between the OWNER and

_____

an individual, doing business as a partnership, a corporation, of the

City of _____,   County of _____ and,

State of _____,   hereinafter called "Contractor."

WITNESSETH: That for and in consideration of the payments and agreements hereinafter mentioned, to be made and performed by the Owner, the Contractor hereby agrees with the Owner to commence and complete the construction described as follows:

MAINTENANCE FACILITY PROJECT
1234 STREET
CITY, STATE

hereinafter called the Project, for the sum of _____

_____ Dollars (_____)

as stated in the Contractor's base proposal as attached hereto and made a part of these contract documents and all extra work in connection therewith, under the terms as stated in the General and Supplementary Conditions of the Contract, and at his (its or their) own proper cost and expense to furnish all the materials, machinery, tools, superintendence, labor, insurance, and other accessories and services necessary to complete the said project in accordance with the conditions and prices stated in the Bid Form, the General Conditions, and Supplementary Conditions of the Contract, the plans which include all maps, plats, blueprints, and other drawings and printed or written explanatory matter thereof, the specifications and contract documents for:

MAINTENANCE FACILITY PROJECT
1234 STREET
CITY, STATE

as prepared by **ABC Design Firm of 3456 Street, City, State, Zip,** herein entitled the Architect, and as enumerated in Para. 4.1 of the Supplementary Conditions, all of which are made a part hereof and collectively evidence and constitute the contract.

TIME OF COMPLETION
The Contractor hereby agrees to commence work under this Contract on or before a date to be specified in a written "Notice to Proceed" of the Owner, and to substantially complete the project **within 180 consecutive calendar days thereafter.**

PAYMENTS

The Owner agrees to pay the Contractor from current funds for the performance of the Contract, subject to the additions and deductions, as provided in the General Conditions of the Contract and to make payments on account thereof as provided in the General Conditions.

IN WITNESS WHEREOF, the parties to these presents have executed this Contract in eight (8) counterparts, each of which shall be deemed as original, in the day and year first above mentioned.

(SEAL) if bid
is by Corporation

THE OWNER
1234 Street, Suite 200
City, State, Zip Code

ATTEST:

By: _____   By: _____

SEAL: _____
Contractor

ATTEST:

By: _____   By: _____
Corporate Secretary

FEI/SS Number _____

****** END OF SECTION 00006—CONTRACT *****

SECTION 00007—AFFIDAVIT

(The successful bidder must submit this affidavit with the contract)

STATE OF _____)

COUNTY OF _____)

The undersigned, of lawful age, being first duly sworn upon oath, deposes and states that I am the duly authorized agent of the contractor under the contract which is attached to this statement, for the purpose of certifying the facts pertaining to the giving of things of value to government personnel in order to procure said contract; that I am fully aware of the facts and circumstances surrounding the making of the contract to which this statement is attached and have been personally and directly involved in the proceedings leading to the procurement of said contract; and that neither the contractor nor anyone subject to the contractor's direction or control has paid, given or donated or agreed to pay, give or donate to any officer or employee of this State any money or other thing of value, either directly or indirectly, in the procuring of the contract to which this statement is attached.

NAME OF CONTRACTOR: _____

_____

Affiant and Agent of Contractor

Title: _____

Subscribed and sworn to before me this _____ day of _____ 20_____ .

My commission expires: _____

Notary Public

****** END OF SECTION 00007—AFFIDAVIT *****

<u>SECTION 00008</u>—**PERFORMANCE PAYMENT BOND**

KNOW ALL PERSONS BY THESE PRESENTS:

That we (1) _____

a (2) _____ hereinafter called "Principal"

and (3) _____ of _____

State of _____, and authorized to transact business in this State hereinafter called the "Surety" are held firmly bound unto The Board of Regents for the Agricultural and Mechanical Colleges Acting for and on Behalf of **OWNER,** in the penal sum of _____ Dollars ($_____) in lawful money of the United States, for the payment of which sum well and truly to be made, we bind ourselves, our heirs, executors, administrators and successors, jointly and severally, firmly by these presents.

THE CONDITION OF THIS OBLIGATION is such that: WHEREAS, the Principal entered into a certain contract with the Owner, dated the _____ day of _____, 20_____ a copy of which is hereto attached and made a part hereof, for the construction of:

MAINTENANCE FACILITY PROJECT
1234 STREET
CITY, STATE

NOW THEREFORE, if the Contractor shall promptly and faithfully perform said contract in accordance with the plans, specifications, and other construction documents furnished by the Owner, then this obligation shall be null and void; otherwise, it shall remain in full force and effect.

Surety hereby waives notice of any alteration or extension of time made by Owner.

Whenever Contractor shall be, and declared by Owner to be in default under the contract, Owner having performed Owner's obligations thereunder, Surety may promptly remedy the default or shall promptly complete the contract in accordance with its terms and conditions or, at the option of the owner, obtain a bid or bids for submission to Owner for completing the Contract in accordance with its terms and conditions, and upon determination by Owner and Surety of the Lowest responsible bidder, arrange for a contract between such bidder and Owner, and make available as work progresses (even though there should be a default or a succession of defaults under the contract or contracts of completion arranged under this paragraph) sufficient funds to pay the cost of completion less the balance of the contract price; but not exceeding, including other costs and damages for which Surety may be liable hereunder, the amount set forth in first paragraph hereof. The term "Balance of the contract price", as used in this paragraph shall mean the total amount payable by Owner to Contractor under the

Contract and any amendments thereto, less the amount properly paid by Owner to Contractor.

Any suit under this Bond must be instituted before the expiration of three (3) years from the date on which final payment under the contract falls due.

No right of action shall accrue on this Bond to or for the use of any person or corporation other than the Owner named therein or the heirs, executors, administrators or successors of Owner.

IN WITNESS WHEREOF, this instrument is executed in 5 counterparts, each of which shall be deemed an original, this the _____ day of _____, 20_____.

{SEAL:}   _____
                          Name of Principal

ATTEST:

By: _____   By: _____

Title: _____   Title: _____

****

{SEAL:}   _____
                          Name of Surety

ATTEST:

By: _____   By: _____
                                          Attorney-in-Fact

Title: _____   Title: _____
                                          Address

SURETY CLAIMS REPRESENTATIVE

(1) Contractor
(2) Corporation, Partnership   Name: _____
    or Individual
(3) Name of Surety   Address: _____

                                      Telephone: _____

****** END OF SECTION 00008—PERFORMANCE BOND ******

SECTION 00009—**STATUTORY BOND**

(To be used in this State as required by State Statute Title 61, Sections 1 and 2 as amended.)

KNOW ALL PERSONS BY THESE PRESENTS:

(1) _____

(2) _____, as principal, and

(3) _____, a Corporation,

organized under the laws of the State of _____and authorized to transact business in this State, as Surety, are held firmly bound unto the **OWNER** in the penal sum of _____ Dollars ($_____) lawful money of the United States, for the payment of which sum well and truly be made, said Principals and Surety bind themselves, their heirs, administrators, executors, successors and assigns, jointly and severally, by these presents.

Signed, sealed and delivered this _____ day of _____, 20_____.

THE CONDITION OF THE FOREGOING OBLIGATION IS SUCH THAT,

WHEREAS, said Principal has entered into a written contract with

OWNER dated _____, 20_____, for the construction of or making of the following described improvements:

MAINTENANCE FACILITY PROJECT
1234 STREET
CITY, STATE

In accordance with the detailed plans and specifications and other contract Documents on file in the office of the ABC Design Firm.

NOW, THEREFORE, if the principal or the subcontractor or subcontractors of said Principal, shall pay all indebtedness incurred by such Principal of the subcontractors who perform work in the performance of such contract for labor and materials, and repairs to and parts for equipment used and consumed in the performance of said contract, this obligation shall be void; otherwise it shall remain in full force and effect.

The said surety, for value received, hereby stipulates and agrees that no change, extension of time, alterations, or additions to the terms of the contract or to the work to be performed, thereunder or the specifications accompanying the same, shall in anywise affect its obligation on this Bond, and it does hereby waive notice of any such change, extension of time, alteration or addition to the terms of the contract or to the specifications.

This Bond is executed in 5 counterparts.

{SEAL:}

_____
Name of Principal

ATTEST:

By: _____   By: _____

Title: _____   Title: _____

****

{SEAL:}

_____
Name of Surety

ATTEST:

By: _____   By: _____

                                Attorney-in-Fact

Title: _____   _____

                                Address

SURETY CLAIMS REPRESENTATIVE

Name: _____

Address: _____

Telephone: _____

NOTE: Date of Bond must *not* be prior to the date of Contract.
NOTE: If Contractor is partnership, all partners should execute the Bond.

(1) Name of Contractor.
(2) Corporation, Partnership, or Individual.
(3) Name of Surety.
(4) Full amount of Contract.

****** END OF SECTION 00009—STATUTORY BOND *****

SECTION 00010—WARRANTY BOND

KNOW ALL PERSONS BY THESE PRESENTS, that on this _____ day of _____, 20_____,

we as principal, and _____, authorized to transact business in this State, as Surety, are held and firmly bound unto the **OWNER** in the full and just sum of the full contract price of

_____ Dollars ($_____)

for payment of which, well and truly to be made, we and each of us, bind ourselves, our heirs, executors and assigns, themselves, and its successors and assigns, jointly and severally, firmly by these presents, as follows:

WHEREAS, the conditions of this obligation are such that said principal has, by a certain contract between _____ and the **OWNER,** dated the _____ day of _____, A.D., 20_____ agreed to construct **MAINTENANCE FACILITY PROJECT,** located at 1234 Street, City, State.

AND WHEREAS, Principal and Surety agree to correct any defects in workmanship or materials appearing during a period of one (1) year from the date of acceptance of the project as complete by the **OWNER.**

NOW THEREFORE, said _____, Principal, agrees for a period of one (1) calendar year from and after completion of said project and acceptance by **OWNER** to perform all acts necessary to correct any and all defects appearing in workmanship and materials and to bear the costs of all labor and materials, including the prime contractor and all subcontractors, which may develop during the term of this bond.

IT IS FURTHER AGREED that if said Principal and/or Surety shall fail to correct defects which may develop due to faulty workmanship and/or materials, then the cost of said corrective work shall be determined by **OWNER,** and if, upon thirty (30) days notice to Principal and Surety of said amount so ascertained, the necessary corrections are not made, the said amount shall become due and payable to the **OWNER** and suits may be maintained in any Court of competent jurisdiction to recover the amount so determined, together with court costs and reasonable attorney fees.

The amount so determined shall be conclusive upon the parties as to the amount due on this bond for the corrections included therein. The cost of all corrections shall be so determined from time to time during the life of this bond, as the condition of the improvements may require.

This obligation is for a period of one (1) year from the date of acceptance of improvements.

SIGNED, SEALED AND DELIVERED the day and year first above written.

{SEAL:} _____
Name of Principal

ATTEST:

By: _____ By: _____

Title: _____ Title: _____

****

{SEAL:} _____
Name of Surety

ATTEST:

By: _____ By: _____
Attorney-in-Fact

Title: _____ _____
Address

SURETY CLAIMS REPRESENTATIVE

Name: _____

Address: _____

Telephone: _____

****** END OF SECTION 00010—Warranty Bond *****

SECTION 00011—RELEASE OF LIENS AFFIDAVIT

(Attach to FINAL Application and Certificate for Payment)

STATE OF _____)

COUNTY OF _____)

_____ of lawful age, being first duly sworn upon oath, deposes and says:

That he or she is the _____ of the _____, a corporation organized and existing under the laws of the State of _____; that he or she makes this Affidavit for and on behalf of said corporation; and, that he or she has authority from the corporation to make this Affidavit.

That the corporation named herein is the same corporation that entered into an Agreement with OWNER on the _____ day of _____, 20_____, for:

MAINTENANCE FACILITY PROJECT
1234 STREET
CITY, STATE

That the said corporation has completed the work set forth in said agreement; and that in accordance with said agreement, affiant further says under oath that there are no existing claims, judgements or liens, outstanding for labor and/or materials furnished under said agreement and that all persons, firms or corporations who have performed work or furnished materials under this agreement have been fully paid. Affiant says nothing further.

_____

Affiant

Subscribed and sworn to before me this _____ day of _____ 20_____.

My commission expires: _____

Notary Public

****** END OF SECTION 00011—RELEASE OF LIENS AFFIDAVIT ***

SECTION 00012—**CERTIFICATE OF APPROVAL**

(Attach to FINAL Application and Certificate for Payment)

The _____ Surety Company, hereby certifies through its constituted Attorney-in-Fact, _____, that it has seen the attached Affidavit of _____, a corporation, sworn to on the _____ day of _____, 20_____, stating that there were no existing claims, judgements or liens outstanding against said corporation for labor and/or materials furnished under its agreement with the OWNER and that all persons, firms or corporations who have performed work or furnished materials under said agreement have been fully paid; and furthermore, approved and becomes legally bound to said Board of Regents under the terms of its surety bond agreement with the Board, by virtue of the execution of said Affidavit without qualification, condition or exception.

Further, it consents that the said OWNER shall make its final payment under the above mentioned agreement to said corporation upon the showing of such affidavit.

_____

Surety Company

_____

Attorney-in-Fact

Dated this _____ day of _____, 20 _____.

(SEAL)

****** END OF SECTION 00012—CERTIFICATE OF APPROVAL *****

SECTION 00013—**GENERAL CONDITIONS**

The provisions of AIA Document A201, GENERAL CONDITIONS OF THE CONTRACT FOR CONSTRUCTION shall apply to this project, except as hereinafter amended or altered. Copies of this document are available for inspection in the ABC Design Firm office.

****** END OF SECTION 00013—GENERAL CONDITIONS *****

SECTION 00014 – **SUPPLEMENTARY CONDITIONS**

The following supplements modify the "General Conditions of the Contract for Construction," AIA Document A201, where a portion of the General Conditions is modified or deleted by these Supplementary Conditions, the unaltered portions of the General Conditions shall remain in effect.

REFER ARTICLE 1:

Add the following to subparagraph 1.1.1:

ENUMERATION OF DRAWINGS, SPECIFICATIONS AND ADDENDA

Following are the Drawings and Specifications, which form a part of this contract:

Drawings:

Cover
C1 Site Plan
A1 Floor Plan
A2 Interior Elevations
A3 Exterior Elevations
A4 Building Sections
A5 Building Sections
A6 Foundation Plan
P1 Plumbing Plan
M1 HVAC Plan
E1 Electrical Power Plan
E2 Electrical Lighting Plan

Contract Documents:

Index to Specifications
Advertisement for Bids
Instructions to Bidders
Bid Form
Affidavit of Nondiscrimination, Nonsegregated Facilities, Noncollusion, and
    Business Relationships
Bid Bond
Contract
Affidavit
Performance Payment Bond
Statutory Bond
Warranty Bond
Release of Liens Affidavit
Certificate of Approval
AIA General Conditions
Supplementary Conditions (where applicable)

Specifications:

Divisions 1–16, Sections as indicated on the Index.

Addenda:

All Addenda issued before Bid Opening also form a part of the contract.

Add the following sub-subparagraph 1.2.3.1:

In the event of conflicts or discrepancies among the Contract Documents, interpretations will be based on the following priorities:

1. The Agreement.
2. Addenda, with those of later date having precedence over those of earlier date.
3. The Supplementary Conditions.
4. The General Conditions of the Contract for Construction.
5. Drawings and Specifications.

In the case of an inconsistency between Drawings and Specifications or within either Document not clarified by addendum, the better quality or greater quantity of Work shall be provided in accordance with the Architect's interpretation.

REFER ARTICLE 2:

Add the following to subparagraph 2.1.1:

The term Owner as used in the following specifications and the contract documents shall be construed to mean OWNER.

Omit subparagraphs 2.1.2 and 2.2.1 in their entireties.

Subparagraph 2.2.5 is hereby modified as follows:

The Contractor will be furnished, free of charge, **10** copies of the drawings and specifications for the execution of the work.

REFER ARTICLE 3:

Add the following to subparagraph 3.3.1:

No work shall be done for any trade during "off hours," weekends, or any other time aside from the Owner's normal working hours, unless the Contractor's Superintendent is present, and unless express permission is obtained from the Owner for accomplishing such work.

Add the following to subparagraph 3.6.1:

TAXES:

All sales taxes and any other Municipal, State and Federal taxes applicable to this work shall be paid by the Contractor. An exemption may be obtained from state and municipal sales tax on purchases of tangible personal property which is incorporated into and becomes a part of the project and where title thereto passes directly from the vendor to the Owner. If a Contractor desires to endeavor to take advantage of this sales tax relief, the Owner will designate the Contractor as an agent of the Owner for the purpose of purchasing tangible personal property. This agency will be created for the sole and exclusive purpose of avoiding sales tax and will contain a provision stating that it is so limited and that the Owner is not responsible to the vendor or any other person dealing with the designated agent for the payment of the purchase price. Furthermore, said agency shall not be deemed to apply to minor isolated sales, purchases of small or random items, or to property used only incidentally in connection with the project.

Add the following to subparagraph 3.7.1:

A building permit will not be required if building constructed on State property.

Subparagraph 3.17.1 is hereby modified as follows:

The Contractor shall acquire all necessary licenses and pay all necessary royalties and license fees and shall defend all suits or claims for infringement of any patent rights and shall indemnify and save the Owner harmless from all loss on account thereof whether or not such suits, claims, or losses arise from the specification by the Owner of a particular design, process, or the product of a particular manufacturer or manufacturers.

REFER ARTICLE 4:

Add the following to subparagraph 4.1.1:

The term Architect, as used on the contract documents, shall be construed to be ABC Design Firm, who have prepared the plans and specifications and who will review the work.

Omit subparagraph 4.1.4 in its entirety.

Subparagraphs 4.3.2, 4.3.4, 4.4.4 are hereby modified as follows:

Omit "arbitration" from these subparagraphs.

Subparagraph 4.3.7 is hereby modified as follows:

Omit "(2) an order by the Owner to stop the Work where the Contractor was not at fault," "(4) failure of payment by the Owner, (5) termination of the Contract by the Owner, (6) Owner's suspension," from this subparagraph and add "as determined by the Owner" after "reasonable grounds."

Subparagraph 4.3.8.1 is hereby modified as follows:

Omit "of cost and" from this subparagraph and add the following:

"Requests for time extensions shall be submitted with each request for progress payment and shall cover the same time period as the Application for Payment.

Failure to submit such request waives Contractor's right to claim time extensions for that designated period.

It is agreed that the Owner's liability for delay or any cost incurred therefrom is limited to granting a time extension to the Contractor, and there is no other obligation expressed or implied on the part of the Owner to the Contractor for the delay or any cost incurred therefrom.

Time extension requests may not be acted upon until substantial completion of the project has been reached as defined in Article 9.8 of the General Conditions.

The maximum extension of time granted will not exceed the total requested. The Owner reserves the right to approve or disapprove with cause such requested extensions."

Omit subparagraph 4.3.8.2 and replace with the following:

No extension of time for completion of the work will be granted on account of weather during the contract time.

Omit Paragraph 4.5 in its entirety.

REFER ARTICLE 7:

Delete subparagraph 7.2.2 and replace with the following:

7.2.2 When the need for a change order has been justified, the Architect will send the Contractor drawings, descriptions, and specifications which will define the proposed changes. The Contractor shall then establish a cost for accomplishing the changes and shall notify the Architect in writing of the amount on forms furnished by the Architect. The amount shall then be reviewed. If a decision is reached agreeing to pursue the changes for the established amount, then the Architect will request approval from the Owner. Upon approval of the changes by the Owner a Formal Change Order will be prepared and issued to the Contractor for the stated amount. Processing time for change orders normally requires several weeks from the time the Contractor's estimate is received by the Architect until issuance of formal change order. Processing change orders of a critical nature may be expedited with permission of the Owner. Section 121 of the STATE Public Competitive Bidding Act of 1974, as amended provides that all change orders shall contain a unit price and total for each of the following items:

a. All material with cost per item.
b. Itemization of all labor with number of hours per operation and cost per hour.
c. Itemization of all equipment with the type of equipment, number of each type, cost per hour for each type, and number of hours of actual operation for each type.
d. Itemization of insurance cost, bond cost, social security, taxes, workers compensation, and overhead cost.
e. Profit for the contractor.

Individual Change Orders less than $10,000 may be based on an acceptable unit price basis in lieu of the itemization as listed in items a through e in the preceding paragraph.

Accordingly, we must request that the foregoing information be provided as an attachment to the change order proposal request.

REFER ARTICLE 8:

Subparagraph 8.3.1 is hereby modified as follows:

Omit "or by delay authorized by the Owner pending arbitration," from this subparagraph.

Omit subparagraph 8.3.3 in its entirety.

REFER ARTICLE 9:

Subparagraph 9.2.1 is hereby modified as follows:

Delete "Before the first application for payment" and replace with "Immediately after formal award of Contract."

Omit subparagraph 9.3.1 and replace with the following:

9.3.1 Applications for monthly progress payments shall be executed on AIA forms G702 and G703. The cost breakdown shall be itemized in a schedule of values as approved by the Architect. Submit seven (7) copies of Payment Request including the signed and notarized original.

Applications shall be reviewed by the Architect. If approved, they will be forwarded to the appropriate office of the institution of record. If not approved, it will be returned to the Contractor for revisions as noted. Application will then be reviewed by the appropriate State agency for disbursement of funds. Approximately three weeks are required for processing of a properly prepared application. All stored materials must be on the project site to be certified for payment.

Omit subparagraph 9.6.1 and replace with the following:

9.6.1 After the Architect has issued a Certificate for Payment, the Owner shall make monthly progress payments amounting to ninety percent (90%) of the portion of the contract sum properly allocable to labor, materials and equipment incorporated into the work or suitably stored at the site or at some other location agreed upon in writing by the parties up to five days prior to the date on which the application for payment is submitted, less the aggregate of all payments in each case. Provided, however, that at any time the Contractor has completed in excess of fifty percent (50%) of the total contract amount, the retainage may be reduced to five percent (5%) of the amount earned to date. Provided further, however, that the Owner or the Owner's duly authorized representative has determined that satisfactory progress is being made and upon approval by the Surety.

The Contractor may, from time to time, withdraw any part, or the whole, of the amount which has been retained from partial payments upon depositing with or delivery to the Owner of (1) United States Treasury bonds, United States Treasury notes, United States Treasury bills, or (2) general obligation bonds of this State, or (3) certificates of deposit from a state or national bank having its principal office in this State. The withdrawal of such retainage shall be accomplished in accordance with procedures established by the business office of the Owner and Section 113.2 of this State's Competitive Bidding Act.

Add the following to subparagraph 9.9.1:

If such use prior to the Contract time for completion increases the cost of the work or delays its completion, the contractor shall be entitled to extra compensation or extension of time, or both provided that the contractor shall notify the Owner of the potential for extra cost and time extension upon giving written consent for such use.

    The Contractor's claim for such extra compensation shall be in writing, with vouchers and other supporting data attached.

    After the Contract time for completion has expired, the Contractor shall not be entitled to extra compensation or extension of time due to such use, neither shall the amount of the liquidated damages, if required by these documents, be reduced because of partial use or occupancy.

Add the following to subparagraph 9.10.1:

The following items must be received before final payment will be processed:

a.   Monthly Progress Payment Application,
b.   Affidavit (Release of Liens, Judgements, Etc.),
c.   Certificate of Approval (Surety),
d.   As-Builts,
e.   Operating, Maintenance, and Instruction Manuals,
f.   Warranties, Bond, and Guarantees.

REFER ARTICLE 10:

Subparagraph 10.1.2 is hereby modified as follows:

Omit the portion of the last sentence beginning with . . . "or in accordance with" . . . and ending with . . . "under Article 4."

Omit subparagraph 10.1.4 in its entirety.

REFER ARTICLE 11:

Add the following to Article 11:

"The Contractor shall not commence any work under this contract until he has obtained all the insurance required under this paragraph and such insurance has been approved by the Owner, nor shall the Contractor allow any subcontractor to commence work on his subcontract until the insurance required of the subcontractor has been obtained and approved."

Add the following to subparagraph 11.1.2:

"The Contractor shall procure and shall maintain during the life of this contract, Public Liability Insurance, Property Damage Insurance, and Vehicle Liability Insurance, in the amounts of not less than $100,000 for injuries, including accidental death, to any one person and in an amount of not less than $500,000 for each accident. Limits of Liability for Contractor's Property Damage Insurance shall not be for less than $100,000 each occurrence. Workers' Compensation and Employer's Liability Insurance limits of liability shall be as established by Oklahoma Statutes."

Modify paragraph 11.2 as follows:

"The Contractor shall name the Owner and the Architect as additional insured under said Public Liability Insurance as specified in subparagraph 11.1.2. above. This insurance shall protect the Owner and the Architect from claims arising from operations under this contract."

Modify paragraph 11.3 as follows:

"The Contractor shall purchase the property insurance in lieu of the Owner. At the option of the Contractor, this insurance may be All-Risk type of insurance policy." This insurance shall cover the entire work at the site equal to the full contract amount.

Modify subparagraph 11.3.9 as follows:

Delete the phrase beginning with "or in accordance with" and ending with "as provided in paragraph 4.5."

Modify subparagraph 11.3.10 as follows:

Delete the portion of the subparagraph beginning with "if such objection" and ending with "as provided in Paragraph 4.5."

Omit subparagraph 11.4 and replace with the following:

11.4.1 The Contractor to whom this work is awarded will be required to furnish a Performance Payment Bond and Statutory Bond each in the Principal sum of the contract. These bonds to be payable to OWNER and shall be executed on forms adopted by the Owner which are a part of the contract documents. The surety company executing these bonds shall be one that is authorized to do business in this State, and shall be subject to the approval of the Owner.

REFER ARTICLE 12:

Add the following to subparagraph 12.2.2:

The Contractor to whom this work is awarded will be required to furnish a Warranty Bond in the Principal sum of the contract. This bond to be payable to OWNER and shall be executed on a form adopted by the Owner which is a part of the contract documents. The surety company executing this bond shall be one that is authorized to do business in this State, and shall be subject to the approval of the Owner.

The Contractor does hereby warrant and/or guarantee against and shall remedy any defect due to faulty materials or workmanship and shall pay for any damages to other work resulting therefrom, which may appear within a period of one (1) year from the date of Substantial Completion of the project.

Supplemental, special and extended warranties or guarantees which are also required are indicated in the various Sections of the Specifications.

REFER ARTICLE 13:

Delete subparagraph 13.6.1 and replace with the following:

13.6.1 Interest shall only be paid according to Section 113.3 of this State's Public Competitive Bidding Act of 19XX, as amended.

Delete paragraph 13.7 in its entirety.

Add the following paragraph 13.8:

13.8 STATUTORY AND REGULATION COMPLIANCE

13.8.1 COMPLIANCE WITH TITLE 68, STATE STATUTES, Section 1701 et. seq.:

All contractors and subcontractors shall comply with all requirements of Title 68, State Statutes, Section 1701 et. seq. A copy of this document is on file in the Owner's office, for any contractor wishing to check same.

13.8.2 BUY IN THIS STATE AND THE U. S. A.:

It is the policy of this State, as expressed by the Legislature in Title YY, State Statutes, Sections 9 and 10, to "Buy in this State;" which means preference shall be given to this state's labor, equipment, materials and products produced and/or manufactured in this State, when quality and quantity are available and the price thereof is equal to or less than that of such labor, materials, and products available from other sources. Also, preference shall be given to materials and equipment produced and/or manufactured in the United States of America. Foreign made items shall not be incorporated in the finished work unless approval thereof is obtained from the Owner.

13.8.3 EMPLOYMENT LAWS COMPLIANCE:

The Contractor shall comply with all pertinent federal laws, regulations, and executive orders as well as state laws, pertaining to employment practices as delineated in his bid certificate on this subject.

13.8.4 OCCUPATIONAL SAFETY AND HEALTH ACT OF 1970 (OSHA):

The contractor shall comply with the latest edition and revisions of the Federal Occupational Safety and Health Act for construction.

13.8.5 BIDDING ACT:

The Contractor shall comply with all provisions of the "Public Competitive Bidding Act" of Title 61, State Statutes, Sections 101 et. seq. and the amendments thereto.

13.8.6 HAZCOM:

The contractor shall comply with all the requirements of the STATE Hazard Communications Standard and 29 CFR (1910.1200) and OWNER Policy and Procedures.

REFER ARTICLE 14:

Add the following to subparagraph 14.2.1:

14.2.1.5 is adjudged bankrupt, or if he makes a general assignment for the benefit of his creditors, or if a receiver is appointed on account of his insolvency.

***** END OF SECTION 00014—SUPPLEMENTARY CONDITIONS *****

INDEX

A

Accessories for brick veneer walls, 347
Accuracy of conceptual estimates, 64–65
Addenda, 14, 32
Adjustments for location, 71
Adjustments for size, 71
Adjustments for time, 70
Aesthetic exterior siding, 318
Aggregates for asphalt mixes, 195–196
Alternates, 32
Approximate estimates, 5–6
Area of roof, 321
Armored electrical cable, 406
Arrangement of bid documents, 22–23
Arrangement of contract documents, 22
Asphalt pavements, 195–205
    aggregates for, 195–196
    asphalt, 196
    asphalt plants, 196–198
    compacting asphalt mixes, 201
    cost of asphalt pavements, 202–205
    equipment for asphalt pavements, 202
    transporting and laying asphalt, 199–201
Asphalt plants, 197–198
    batch–type, 197
    drum–mix type, 197–198
Asphalts, 196
Asphaltic shingles, 330–335
    labor required, 331
Assessing estimate sensitivity, 61

B

Backhoe excavation, 149–152
Batching and hauling concrete, 182–183
Batch–type asphalt plants, 197
Bid and contract documents, 22–23
Bid bonds, 33–34
Bidding requirements, 25–32
    bid forms, 28
    bid solicitations, 26
    information for bidders, 26–32
    instruction to bidders, 26

Bid documents, 13–14, 22–23
    arrangement of, 23–24
    bid vs. contract documents, 22–23
    building specifications, 24–25
    example of, 519–563
    heavy/highway specifications, 24–25
Bid forms, 28–32
    for lump–sum contracts, 29–30
    for unit–price contracts, 29–33
Bid solicitations, 26
Bolting structural steel, 287–288
Bolts and screws, 300–303
Bond patterns, 343–345, 356–357
    for brick, 343–345
    for stone, 356–357
Bonds, 16, 33–35
    bid bond, 33–34
    material and labor payment
        bond, 34–35
    performance bond, 34
Brick veneer walls, 343–352
    accessories for, 247
    bond patterns for, 343–345
    cleaning brick masonry, 347–348
    estimating mortar for, 345–346
    joints for brick masonry, 345
    labor–hours laying bricks, 349–350
    sizes and quantities of bricks, 343
Bridging for steel joists, 361–362
Broad–scope conceptual
    estimates, 68–69
Builder's risk insurance, 35
Building construction drawings, 37–43
Building construction projects, 9–10
Building construction
    specifications, 24–25
Build–up roofing, 334–347
    felt underlayment for, 335
    gravel and slag, 335
    laying on concrete, 336–337
    laying on wood decking, 335–336
    pitch and asphalt, 335

C

Capacity ratio estimates, 75–76
Carpentry, 295–325
    classification of lumber, 295–298
        plywood, 298
        sizes of lumber, 295–296
        species and grades of, 296–298
    exterior finish carpentry, 317–319
        aesthetic exterior siding, 318–319
        fascia, frieze, and corner
            boards, 317–318
        soffits, 318
        wall sheathing, 318
    fabricating lumber, 304–305
    fasteners, 299–304
        bolts and screws, 300–303
        nails and spikes, 299–300
        timber connectors, 303–304
    floors, 306–308, 324–325
        finishing, 324–325
        framing, 307–308
        girders, 306–307
        subfloors, 316
    framing walls and floors, 305–306
    framing windows and doors, 311
    heavy timber structures, 319–320
    joists for floors and ceilings, 307
    plywood, 298
    roofs, 311–320
        decking, 315
        prefabricated trusses, 314–315
        rafters, 311–312
        wood shingles, 315–316
    walls, 309–311, 318–319, 322–323
        aesthetic exterior siding, 318–319
        gypsum wallboard, 322–323
        paneling, 323
        sheathing, 318
        sills, 306
        studs, 309–310
        windows and door openings, 311
Cast-in–place concrete piles, 220–221
    cost of, 220
Cellular–steel floor systems, 265
Change orders, 14, 32
Checklist for estimating, 13, 52–53
Clamshell excavator, 142–143
Classifications of lumber, 295–298
    plywood, 298

    sizes of, 295–296
    species and grades of, 296–298
Clay tile roofing, 334
Cleaning brick masonry, 347–348
Cleanout boots for sewers, 418
Clearing and grubbing land, 174–181
    estimating time to pile trees, 179–180
    disposal of brush, 180–181
    intermediate clearing, 175
    large clearing, 175–176
    light clearing, 174–175
    rates of clearing land, 176–180
Column capitals, 254
Column clamps, 248–249
Column materials, 248
Commercial estimating software, 457–459
    advantages of, 459–460
Commercial prefabricated forms, 345–346
Compacting asphalt mixes, 201
Comprehensive general liability, 36
Computer applications, 18–20
Computer estimating, 452–512
    alternates, 492–493
    biditems, 464–467
    bid pricing, 499–501
    checking for reasonableness, 494–496
    commercial software, 457–460
        advantages of, 459–460
    copying past estimates, 490–492
    electronic media, 455
    entering the estimate, 483–490
        calculation routines, 488–490
        entering activities, 483–484
        notes, 487
        overtime, 487
        using crews, 484–487
    importance of estimator, 453
    loading HeavyBid from
        internet, 505
    management of data, 460–461
    precision in labor costing, 474
    quantity takeoff, 467–468
    resources, 470–480
        crews, 480
        equipment, 475–478
        labor, 471–475
        material, 478–479
    reviewing the estimate, 494–495
    showing how costs calculated, 493–494

Computer estimating—*Cont.*
  spreadsheets in estimation, 455–456
    disadvantages of, 457
  starting an estimate, 463–464
  steps in computer estimating, 462
  structuring the estimate, 481–483
  taking quotes, 501–503
  turning an estimate into a bid, 496–498
  turning in the bid, 504–505
  using computers, 453–455
Conceptual estimating, 65–74
  accuracy of, 64–65
  broad–scope estimates, 68–69
  liability of, 65
  location adjustments, 71
  narrow–scope estimates, 71
  parametric estimating, 65–68
  preparation of, 65–66
  size adjustments, 71
  time adjustments, 70
  unit–cost adjustments, 72–74
Concrete beams, 256–259
  cost of concrete beams, 258–259
  material and labor–hours
      for, 256–258
Concrete columns, 246–254
  capitals, 254
  clamps, 248–249
  cost of concrete columns, 249–253
  economy of reuse, 252–253
  materials and labor–hours for, 250
Concrete floors, 259–265
  cellular–steel floor systems, 265
  corrugated–steel forms, 264–265
  cost of concrete floors, 261–262
  forms for flat–slabs, 259–262
  material and labor–hours, 260, 263
  patented forms floor slabs, 262–263
Concrete footings, 236–239
Concrete formwork materials, 232–235
  form liners, 234
  form oil, 233–234
  form ties, 234
  lumber, 233
  nails, 233
  plywood and Plyform, 232–233
Concrete masonry units, 352–356
  cost of, 353–356
  joints for brick masonry, 345

  labor–hours laying, 353–354
  quantities of, 352
Concrete materials, 272–277
  labor and equipment placing, 274–276
  lightweight concrete, 276
  perlite concrete aggregate, 276–277
  quantities for, 273–274
Concrete pavements, 181–194
  batching and hauling concrete, 182–183
  concrete pavement joints, 186–188
  construction methods used, 181–182
  curing concrete pavements, 189
  placing concrete pavements, 183–186
  preparing subgrade for, 161–166
Concrete piles, 218–221
  cast–in–place, 220–221
  prestressed, 218–220
Concrete reinforcing steel, 267–273
  cost of, 269–270
  estimating quantity of, 268–269
  labor placing, 270–272
  properties of, 268
  types of, 267–268
  welded–wire fabric, 272
Concrete slabs on grade, 236
Concrete stairs, 265–267
  cost of, 266–267
  labor required, 267
  lumber required, 265–266
Concrete structures, 231–277
Concrete tilt–up walls, 277
Concrete walls, 236–246
  commercial forms, 245–246
  cost of concrete walls, 240–244
  materials and labor–hours
      for, 239–240
  prefabricated forms, 244–245
Conduit for electrical wiring, 407–408
Connections for steel structures, 231–277
Construction equipment, 90–105
  equipment costs, 91–92
  depreciation costs, 92
  depreciation methods, 92–96
    declining balance, 94–95
    straight line, 93–94
    sum of the years, 95–96
  fuel consumption, 101–102
  investment costs, 95–97
  oil consumption, 102

operating costs, 100–103
  cost of rubber tires, 102–103
  fuel consumption, 101–102
  lubricating oil, 102
  maintenance and repairs, 100–101
ownership costs, 98–100
  depreciation and investment, 98–99
  time–value–of–money, 99–100
renting vs. owning, 90–91
sources of, 90
Construction labor, 82–89
  cost of labor, 83
  fringe benefits, 84–85
  production rates for, 87–89
  public liability insurance, 84
  social security tax, 83
  sources of labor rates, 82–83
  unemployment tax, 83
  workers' compensation tax, 84
Contingency, 57–62
  assessing estimate sensitivity, 61–62
  expected net risk, 59–60
  percentage of base estimate, 58–59
  simulation, 60–61
Contract documents, 22–23
Contractor's equipment floater, 36
Contractor's protective liability, 36
Contract requirements, 23
Copper pipe, 390–391
Corrugated–steel forms, 264–265
Cost capacity curve estimates, 75–76
Cost of
  asphalt pavements, 195–202
  carpentry, 295–325
  concrete pavements, 191–194
  concrete structures, 231–277
  electrical wiring, 404–412
  equipment, 90–105
  fabricating steel, 280–293
  floor finishes, 372–384
  floor systems, 361–371
  foundations, 208–229
  labor, 82–87
  lumber, 299
  masonry, 341–359
  painting, 380–385
  plumbing, 387–403
  reinforcing steel, 269–270
  roofing, 328–340

sewerage systems, 414–430
shop drawings for steel, 281–283
steel structures, 279–293
transporting steel, 283
water distribution systems, 431–451
Covering capacity of paints, 381–382
Curing concrete pavements, 189
Cycle times, 109
Cycle times and production rates, 127–128

D

Decision to bid, 44
Demolition, 181
Depreciation costs, 92–96
  declining balance, 94–95
  straight–line, 93–94
  sum–of–the–year's digits, 95–96
Detailed estimates, 7–8
Disposal of brush, 180
Documentation of estimates, 53–54
Dozers, 152–154
Dragline excavation, 139–141
Drawings, 36–43
  building construction projects, 37–40
  examples of, 514–525
  heavy/highway projects, 36
  line work, 41
  plans, elevations, sections, 38–40
  scales, 41
  schedules, 41–42
  symbols and abbreviations, 42–43
Drilled shaft foundations, 224–229
Drilling and blasting rock, 166–167
Drum–mix asphalt plants, 197

E

Earthwork, 126–171
  backhoes, 149–152
  clamshells, 142–143
  dozers, 152–154
  draglines, 139–141
  drilling and blasting rock, 166–168
  estimating production rates, 127
  front shovels, 143–147
  graders, 159–161
  hand excavation, 133–134
  hauling earth, 147–149
  methods of excavation, 128–129

Earthwork—*Cont.*
  physical properties of, 129–131
    shrinkage factors, 130–131
    swell factors, 130–131
  preparing subgrades, 165–166
  scrapers, 154–159
  shaping and compacting, 161–165
  trenching machines, 135–139
Easements and right–of–ways, 446–447
Electrical wiring, 408–412
  accessories for wiring, 407–408
  armored cable, 406
  cost of wiring materials, 408–409
  electric wire, 407
  factors affecting cost, 404–405
  finish electrical work, 410–412
  flexible metal conduit, 406
  items in wiring cost, 405
  labor installing wiring, 409–410
  nonmetallic cable, 407
  rigid conduit, 406
  types of wiring, 405
Equipment–factored estimates, 77–79
Equipment for erecting steel, 286–287
Equipment for hot–mix asphalt, 202
Equipment operating costs, 100–103
  fuel consumption, 101–102
  lubricating oil, 102
  maintenance and repairs, 100–101
  rubber tires, 102–103
Equipment ownership costs, 98–100
  depreciations costs, 92–96
  investment costs, 96–98
Erecting structural steel, 286–291
  bolting steel, 287–288
  equipment used, 286–287
  labor and equipment cost, 291–292
  welding steel, 288–290
Estimates, 4–8, 44–81
  approximate estimates, 5–6
  broad–scope estimates, 68–69
  checklists, 52–53
  conceptual estimates, 64–79
  detailed estimates, 7–8
  documentation of, 53–54
  forms for, 20–21
  narrow–scope estimates, 74
  organization of, 8–11
  overhead, 14–15

  parametric estimates, 66–68
  preparation of, 50–51
  representative estimates, 17
  reviews of, 54–56
  types of, 4
  work plan for, 47–49
Estimating, 2–3, 13, 44–63
  checklist of operations, 13, 52–53
  contingency, 57–62
    assessing estimate sensitivity, 61
    expected net risk, 59–60
    percentage of base estimate, 58
  documentation, 53–54
  estimating team, 46–47
  feedback for improvement, 62–63
  kick–off meeting of team, 47
  methods and techniques, 49–50
  procedures for preparing, 51–52
  process industry estimates, 74–79
    capacity ratios, 74, 76
    cost capacity curves, 74–71
    equipment–factored, 77–80
    plant cost per unit, 77
  process of preparing, 44–46
  purpose of, 3–4
  quantity takeoff, 11–12
  risk analysis, 57
  risk assessment, 57
Estimator, 3, 46–47
  importance of, 3
  teams of estimators, 46–47
Excavating, 133–171
  backhoes, 149–152
  dozers, 152–154
  draglines, 139–141
  front shovels, 143–147
  hand, 133–134
  scrapers, 154–159
  trenching machines, 135–139
Expected net risk, 59–60
Exterior finish carpentry, 317–319
  aesthetic exterior siding, 318–319
  fascia, frieze, and corner boards, 317–318
  soffits, 318
  wall sheathing, 318

F

Fabricating lumber, 304–305
Factors affecting cost records, 74

Fascia, frieze, and corner
    boards, 317–318
Fasteners for lumber, 299–304
    bolts and screws, 300–303
    nails and spikes, 299–300
    timber connectors, 303–304
Felt for built–up roofing, 335
Field painting structural steel, 293
Finish electrical work, 410–412
Finishing wood floors, 324–325
Fire hydrants, 347
Fittings for sewer pipes, 415
Fittings for water pipes, 394
Flashing, 338–340
    flashing roofs at walls, 338–340
    flashing valleys and hips, 340
    labor installing flashing, 340
    metal flashing, 338
Flexible metal conduit, 406
Floor and ceiling joints, 307–308
Floor finishes, 372–377
    concrete, 372–375
    labor finishing, 372–376
    monolithic topping, 372–373
    separate concrete toppings, 374–375
    terrazzo, 376–377
    vinyl tile, 378–379
    wood, 324–325
Floors, 306–308, 316
    finishing wood, 324–325
    girders and joists, 306–308
    subfloors, 316
Floor systems, 361–371
    bridging, 361–362
    corrugated–steel forms, 368–371
    joists end supports, 363
    metal decking, 362–363
    sizes of steel joists, 363
    steel–joist system, 361
Footings, 209, 236–237
Form liners, 234
Form oil, 233–234
Forms for, 239–265
    concrete beams, 256–259
    concrete columns, 246–253
    concrete walls, 239–244
    flat–slab floors, 259–262
    patented forms for floors, 262–265
    slabs on grade, 236

Forms for concrete structures, 231–277
    form liners, 234
    form oil, 233–234
    form ties, 234
    lumber, 233
    materials, 232–235
    nails, 233
    plywood, 232–233
Forms for preparing estimates, 20–21
Forms for slabs on grade, 236
Form ties, 234
Foundations, 208–229
    concrete piles, 219–221
    cast–in–place, 221–222
    prestressed, 219–220
    drilled shafts, 224–229
    pile–driving equipment, 211–214
    sheeting trenches, 209–211
    sheet piling, 214–217
    steel piles, 221–224
    wood piles, piles, 217–219
Framing lumber, 304–305
Framing windows and doors, 311
Fringe benefits, 84–85
Front shovels, 143–147
Fuel consumption, 101–102

G

General conditions, 33
Girders, 303–307
Graders, 159–160
Gravel and slag, 335
Gypsum wallboards, 322–323

H

Hand excavation, 133–134
Handling cast-iron pipe, 119–120
Handling materials, 108–123, 147–149
    bricks, 122–123
    cast iron pipe, 119–120
    excavated earth, 147–149
    lumber, 120–121
    sand and aggregate, 113–119
Haul distances for scrapers, 157–158
Heavy engineering projects, 10–11
Heavy/highways drawings, 24–25
Heavy/highway specifications, 24–25

Heavy timber structures, 319–320
Highway pavements, 174–206
    clearing and grubbing, 174–181
    hot–mix asphalt plants, 196–198
    portland cement pavements, 182–183
    subgrade preparation, 165–166
Horizontal directional drilling, 440–445
    classifications of, 441–442
    procedures for, 442–445
    production rates, 445
Hydraulic excavators, 143–152
    backhoes, 149–152
    front shovels, 143–147

I

Information available to bidders, 26
Installing corrugated sheets, 369
Instructions to bidders, 26
Instructions to readers, 17
Insurance, 16–17, 35, 36
    basic builder's risk, 35
    builders risk extended coverage, 35
    comprehensive general liability, 36
    contractor's equipment floater, 36
    contractor's protective liability, 36
    public liability/property damage, 35–36
    worker's compensation, 36
Interest during construction, 449–450
Interior finish carpentry, 320–325
Interior finishes, 321–325
    doors and windows, 321
    gypsum wallboards, 322–323
    trim molding, 324
    wall paneling, 323
    wood floors, 324–325
    wood furring strips, 322
Intermediate clearing of land, 175
Investment costs, 96–98

J

Jacking pipe, 427–428
Jetting piles into position, 224
Job factors, 126
Joists for floors and ceilings, 307–308

L

Labor, 16, 82–89
    burden, 16

cost of, 83, 85–87
    fringe benefits, 84–85
    labor/equipment crews, 12–13
    labor rates, 83
    production rates, 87–88
    public liability/property damage, 84
    social security tax, 83
    sources of labor rates, 83
    unemployment compensation tax, 83
    workers' compensation insurance, 84
Labor and equipment crews, 12–13
Labor and equipment placing concrete, 274–276
Labor burden, 16
Labor erecting steel joists, 364–365
Labor erecting structural steel, 291–293
Labor finishing concrete floors, 372–377
Labor–hours to set and trim doors, 291
Labor–hours to set and trim windows, 321
Labor installing metal decking, 365
Labor installing welded–wire fabric, 272
Labor laying bricks, 349–350
Labor laying built–up roofing, 335–337
Labor setting stone masonry, 358–359
Labor rates, 82–83
Labor taxes, 83
Ladder–type trenching machines, 138–139
Land–clearing operations, 174–181
    intermediate clearing, 175
    large clearing, 175–176
    light clearing, 174–175
Laying asphalt pavements, 199–201
Laying built–up roofing on concrete, 336–337
Laying built–up roofing on wood deck, 335–336
Laying vinyl tile, 378
Legal expenses, 447
Liability of conceptual estimates, 65
Lightweight concrete, 276–277
Line work of drawings, 41
Lubricating oil, 102
Lumber, 233, 295–299
    for concrete formwork, 233
    for wood framing, 295–299

M

Maintenance and repair costs, 100–101
Management factors, 126–127
Manholes, 416–417
Mortar for stone masonry, 356

Masonry, 341–359
  bond patterns, 343–345, 356–357
    for brick and CMU, 343–345
    for stone, 356–357
  brick veneer walls, 346–348
    accessories for, 347
    bond patterns, 343–345
    cleaning brick masonry, 347–348
    joints for brick masonry, 345
    labor–hours laying bricks, 349–350
    mortar for, 345–346
    sizes and quantities of bricks, 343
  concrete masonry units, 352–356
    labor–hours for, 353–354
    mortar for, 352
    weights and quantities of, 52
  solid brick walls, 348–350
Material and labor payment bonds, 34–35
Materials for
  concrete beams, 356–358
  concrete columns, 248–250
  concrete floors, 259–260
  concrete walls, 239–240
  footings, 236–239
  forms, 232–235
  painting, 380–381
  steel structures, 279
Material taxes, 15
Metal decking, 362–363
Metal flashing, 338
Microtunneling, 323–340
  active direction control, 427
  advantages and disadvantages, 428
  automated spoil transportation, 427
  jacking pipe, 427–428
  microtunnel boring machine, 424–426
  pipe–jacking equipment, 428
  process of, 429–430
  remote control system, 426
Monolithic topping, 372–375
Mortar, 342, 345–347, 352
  for masonry, 342
  quantity for bricks, 346–347
  quantity for CMU, 352
  for stone masonry, 356–357

N

Nails and spikes for carpentry, 299–300
Nails for concrete formwork, 233

Narrow scope cost estimates, 74
Negotiated work, 32
Nonmetallic electric cable, 407

O

Operating costs of equipment, 100–103
  fuel consumption, 101–102
  lubricating oil, 102
  maintenance and repairs, 100–101
  rubber tires, 102–103
Overhead, 14–15
Ownership costs of equipment, 98–100
  depreciation costs, 92–96
  investment costs, 96–98

P

Painting, 380–385
  covering capacity of, 381–382
  labor–hours applying, 383–384
  materials, 380
  preparing surface for, 382
Parametric estimating, 66–68
Patented forms for floor slabs, 262–263
  materials and labor for, 263
Pattern bonds, 343–345
Pavement joints, 186–188
Pavements
  hot–mix asphalt pavements, 195–205
  portland cement pavements, 181–194
Performance bond, 34
Physical properties of earth, 129–133
  shrinkage factors, 130–131
  swell factors, 130–131
Pile–driving equipment, 211–214
Pipe, 391–392, 431–436
  copper, 390–391
  ductile iron pipe (DIP), 434–435
  polyethylene (PE) plastic pipe, 433
  polyvinyl chloride (PVC), 391–392, 432
  reinforced concrete pipe (RCP), 435–436
  soil, waste, and vent, 393
  steel, 390
  types of, 431
Pitch and asphalt, 335
Placing concrete pavements, 183–186
Plans, elevations, and sections, 37–40
  example of drawings, 514–525
Plant cost per unit of production, 77

Plumbing, 397–403
  codes for, 388–389
  copper tubing, 390–391
  finish plumbing, 400–403
  fittings, 394
  house drain pipe, 393
  indoor CPVC plastic water pipe, 392
  labor installing fixtures, 402–403
  labor installing plastic pipe, 392
  labor installing rough in plumbing, 395–396
  piping used for, 389
  plastic drainage pipe and fittings, 395–396
  PVC water supply pipe, 391–392
  requirements for, 387
  rough–in plumbing, 394–400
  solid, waste, and vent pipe, 393
  steel pipe, 390
  traps, 394
  valves, 394
Plywood for carpentry, 98
Plywood for formwork, 216–217
Prefabricated form panels, 244–245
Prefabricated roof trusses, 314–315
Preparing a surface for paint, 382
Preparing conceptual estimates, 65–66
Preparing estimates, 50–51
Preparing subgrades, 165–166
Prestressed concrete piles, 219–220
Process industry estimates, 74–79
  capacity rations, 76–77
  cost capacity curves, 75–76
  equipment factored, 79
  plant cost per unit, 77
Production rates, 18
Production rates for labor, 87–88
Production rates for scrapers, 155–156
Properties of reinforcing steel, 268
Property damage insurance, 35–36, 84
Public liability insurance, 35–36, 84
Purpose of estimating, 3–4
PVC water supply pipe, 391–392

Q

Quantity takeoff, 11–12

R

Rafters, 311–314
Rates of clearing land, 176–181
  clearing light vegetation, 176
  estimating time to fell trees, 177
  estimating time to pile trees, 179
Rates of drilling holes, 169
Reinforcing steel
  cost of, 269–270
  estimating quantity of, 268–269
  labor placing, 270–272
  properties of, 268
  types and sources of, 267–268
  welded–wire fabric, 272
Renting vs. owning equipment, 90–91
Representative estimates, 17
Review of estimates, 54–56
Rigid electrical conduit, 406
Risk analysis, 57
Risk assessment, 57
Roofing and flashing, 328–340
  areas of roofs, 328
  asphalt shingles, 330
    cost of, 332–333
    labor–hours required, 331
  built–up roofing, 334–337
    on concrete, 335–336
    labor–hours, 337
    pitch and asphalt, 335
    on wood, 335–336
  clay tile, 334
  felt for roofing, 329
  metal flashing, 338–340
    labor–hours required, 340
    at valleys and hips, 340
    at walls, 338–340
  slate roofing, 333–334
  steepness of roof, 329
  wood shingles, 315–329
Roofs, 311–316
  decking, 315
  prefabricated trusses, 314–315
  rafters, 311–314
  wood shingles, 315–316
Rough carpentry, 307–316
  floor and ceiling joists, 307–308
  floor girders, 306–307
  framing door openings, 311
  framing window openings, 311
  rafters, 311–314
  roof decking, 315
  sills, 306
  studs, 309–310
  subfloors, 316

Roughing in electrical, 405–410
  accessories, 407–408
  armored cable, 406
  cost of materials, 408–409
  electric wire, 407
  flexible metal conduit, 406
  labor required, 409–410
  nonmetallic cable, 407
  rigid conduit, 406
  type of wiring, 405
Roughing in plumbing, 394–400
  cost of materials, 395
  estimating the cost of, 394–395
  labor required, 396–398
  lead, oakum, and solder, 395
  plastic drainage pipe, 395–396
Rubber tires, 102–103

S

Scales, 41
Schedules, 41–42
Scrapers, 154–159
  excavating and hauling, 154–159
    haul distances for, 157–158
    production rate for, 155–156
Separate concrete topping, 374–375
Service branches for sewers, 418–419
Service lines for water lines, 436
Sewage systems, 414–430
  cleanout boots, 418
  equipment required, 415–416
  items included, 414
  manholes, 416–417
  microtunnel boring machine, 424–426
  microtunneling, 423–430
    active direction control, 427
    advantages of, 428
    automated spoil transportation, 427
    boring machine for, 424–426
    jacking pipe, 427–428
    process of, 429–430
    remote control system, 428
  service branches, 418
  sewer pipes, 414–415
  trenchless methods, 423
Sewer pipes, 414–415
Shaping and compacting earth, 161–165
Sheeting trenches, 209
Sheet piling, 214–217

Shores and scaffolding, 254–256
Shop drawings for steel, 281–283
Sills, 306
Simulation, 60–61
Size adjustments for estimates, 71
Sizes and quantities of bricks, 343
Sizes of lumber, 295–296
Slabs on grade, 236
Slate roofing, 333–334
Social security tax, 83
Soffits, 318
Soil, waste, and vent pipes, 393
Solid brick walls, 348
Sources of equipment, 90
Sources of labor rates, 82–83
Species and grades of
    lumber, 296–299
Specifications, 24–25, 526–563
  for building projects, 24–25
  example of, 526–563
  for heavy/highway, 24
Spreadsheets, 455–456
Steel piles, 221–224
Steel pipe, 390
Steel structures, 279–293
  connections for, 280
  erecting steel, 286
    bolting steel, 287–288
    equipment used, 286–287
    labor and equipment
        cost, 291–292
    welding steel, 288–290
  estimating the cost of, 280–281
  fabricating steel, 283
  field painting, 293
  materials used, 279
  shop drawings, 281–283
  standard shapes, 281
  transporting steel, 283
  types of steel structures, 279
Steepness of roofs, 329
Sterilization of water pipes, 437
Stone masonry, 356–359
  bonds form, 356–357
  labor setting, 358
  mortar for, 356
  weights of, 356
Studs for wall framing, 309–310
Subfloors, 316
Symbols and abbreviations, 42–43

T

Taxes, 15, 83
  labor, 15
  material, 15
  social security, 83
  unemployment, 83
Terrazzo floors, 376–377
Test of water pipes, 437
Tilt–up concrete walls, 277
Timber connectors, 303–304
Time adjustments for estimates, 70
Total cost of engineering
    projects, 446–451
Track loaders, 113–117
Transporting and laying asphalt, 199–201
Transporting sand and
    gravel, 113–119
Traps, 394
Trenching machine, 135–139
Trenchless technology, 423–430,
    440–445
  horizontal directional
     drilling, 440–445
  microtunneling, 423–430
Types of
  estimates, 4–5
  foundations, 208
  joints for brick masonry, 345
  reinforcing steel, 267–268
  steel structures, 279
  wiring, 405

U

Unemployment compensation tax, 83
Use of computer in estimating, 453–455

V

Valves for sewer systems, 394
Valves for water systems, 436
Vinyl tile floors, 378–379

W

Walls, 305–311
  aesthetic exterior siding, 318–319
  gypsum and paneling, 322–323
  sheathing, 308
  sills, 306
  studs, 309–310
  window and door openings, 311
Warranties, 33
Water distribution systems, 430–445
  cost of water systems, 438–440
  fire hydrants, 437
  horizontal directional drilling, 440–445
  service lines, 436
  sterilization of water pipes, 437
  tests of water pipes, 437
  types of pipe, 431–436
  valves, 436
Welded–wire fabric, 272
Welding structural steel, 288–290
Wheel loaders, 113–117
Window and door openings, 311, 323
Wood floors, 324–325
Wood furring strips, 322
Wood piles, 217
  driving wood piles, 217–219
Wood shingles, 315–316
Worker's compensation insurance, 15–16, 36, 84
Work plan for estimates, 47–49

Units

acre = 43,560 square feet
cubic foot = 7.48 gallons
cubic yard = 27 cubic feet
foot = 12 inches
mile = 5,280 feet
square yard = 9 square feet
ton = 2,000 pounds
yard = 3 feet

A = ampere
bf = board foot, measure for lumber
c = 100
cf = cubic feet
cu. in. = cubic inches
cwt = hundredweight, 100 lb of steel
cy = cubic yard
D and M = dressed and matched, of lumber
d = pennyweight, for nails
dia. = diameter
deg. = degree
F = Fahrenheit, measure of temperature
f.o.b. = freight on board
ft = feet
ga. = gauge
gal = gallon
hp = horsepower
hr = hour
in. = inch

kW = kilowatt
kWh = kilowatt-hour
lb = pound
lin. ft = linear feet
ls = lump sum
M = 1,000
MM = million
mh = man-hour
mi. = mile
min. = minute
mph = miles per hour
No. = number
pc = pieces
psf = pounds per square foot
psi = pounds per square inch
pt = pint
sec = second
sf = square feet
SFCA = square feet contact area, of formwork
sq. in. = square inches
sy = square yard
S4S = surfaced four sides, of lumber
T = ton (2,000 pounds)
T&G = tongue and groove, of lumber
WH = work-hour
yd = yard

Acronyms

AASHTO—American Association of State Highway and Transportation Officials

ABC—Associated Builders and Contractors

ACEC—American Consulting Engineer's Council

ACI—American Concrete Institute

AGC—Associated General Contractors

AIA—American Institute of Architects

AISC—American Institute of Steel Construction

AITC—American Institute of Timber Construction

ANSI—American National Standards Institute

ARTBA—American Road and Transportation Builders Association

ASA—American Standards Association

ASHRAE—American Society of Heating, Refrigerating, and Air Conditioning Engineers

ASME—American Society of Mechanical Engineers

AWS—American Welding Society

AWWA—American Water Works Association

BIA—Brick Institute of America

BOCA—Building Officials and Code Administrators

CRSI—Concrete Reinforcing Steel Institute
CSI—Construction Specification Institute
DIP—Ductile Iron Pipe
DOT—Department of Transportation
EJCDC—Engineer's Joint Contract Documents Committee
EPA—Environmental Protection Agency
FERC—Federal Energy Regulatory Commission
FHWA—Federal Highway Administration
NCMA—National Concrete Masonry Association
NDS—National Design Specification
NEC—National Electrical Code
NEMA—National Electrical Manufacturers Association
NFPA—National Fire Protection Association
NSPE—National Society of Professional Engineers
OSHA—Occupational Safety and Health Administration
PCA—Portland Cement Association
PCI—Prestressed Concrete Institute
PE—Polyethylene Pipe
PVC—Polyvinyl Chloride Pipe
RCP—Reinforced Concrete Pipe
SAE—Society of Automotive Engineers
SDI—Steel Deck Institute
SFPE—Society of Fire Protection Engineers
SJI—Steel Joist Institute
SMACNA—Sheet Metal and Air Conditioning Contractors' National
　　　　　Association
TCA—Tile Council of America
UL—Underwriter's Laboratory